ADVANCES IN PROTEIN CHEMISTRY

Volume 66

Protein Simulations

ADVANCES IN PROTEIN CHEMISTRY

EDITED BY

FREDERIC M. RICHARDS
Department of Molecular Biophysics
and Biochemistry, Yale University
New Haven, Connecticut

DAVID S. EISENDBERG
Department of Chemistry and Biochemistry
University of California, Los Angeles
Los Angeles, California

JOHN KURIYAN
Department of Molecular Biophysics
Howard Hughes Medical Institute
Rockefeller University
New York, New York

VOLUME 66

Protein Simulations

EDITED BY

Valerie Daggett
Department of Medicinal Chemistry
University of Washington
Seattle, Washington

ELSEVIER
ACADEMIC
PRESS

AMSTERDAM • BOSTON • HEIDELBERG • LONDON
NEW YORK • OXFORD • PARIS • SAN DIEGO
SAN FRANCISCO • SINGAPORE • SYDNEY • TOKYO
Academic Press is an imprint of Elsevier

This book is printed on acid-free paper. ♾

Elsevier Academic Press
525 B Street, Suite 1900, San Diego, California 92101-4495, USA
84 Theobald's Road, London WC1X 8RR, UK
http://www.academicpress.com

International Standard Book Number: 0-12-034266-9

PRINTED IN THE UNITED STATES OF AMERICA
03 04 05 06 07 08 9 8 7 6 5 4 3 2 1

CONTENTS

PREFACE ix

Assessment of the Role of Computations in Structural Biology

Irwin D. Kuntz and David A. Agard

I. Introduction 1
II. Comparison of the Role of Computation in the Study of Small Molecules and Macromolecules 2
III. Comparison of Theory and Experiment 5
IV. Molecular Recognition and Docking 8
V. Error Estimation and Error Propagation in Experiment and Theory 10
VI. The Role of Models in Protein Simulations 13
VII. Parameterization 15
VIII. Testing of Theoretical Models for Structural Biology 16
IX. Conclusions 19
References 19

Force Fields for Protein Simulations

Jay W. Ponder and David A. Case

I. Introduction 27
II. Protein Force Fields, 1980 to the Present 30
III. Beyond Fixed Atomic Point-Charge Electrostatics 45
IV. Modeling the Solvent Environment 62
V. Conclusions 77
References 78

Protein Simulation and Drug Design

Chung F. Wong and J. Andrew McCammon

I. Introduction 87
II. Fixed-Conformation Models 88

III. Including Protein and Ligand Flexibility in Modeling Molecular Recognition 95
IV. Fast Approximate Simulation Methods for Calculating Binding Free Energies 103
V. Including Protein Flexibility in Virtual Screening 111
VI. Enhanced Sampling via the Replica-Exchange Method 115
VII. Conclusions 117
References 118

Free Energy Calculations and Ligand Binding

BJØRN O. BRANDSDAL, FREDRIK ÖSTERBERG, MARTIN ALMLÖF, ISABELLA FEIERBERG, VICTOR B. LUZHKOV, AND JOHAN ÅQVIST

I. Introduction 123
II. Free Energy Perturbation and Thermodynamic Integration 125
III. Extrapolation of Free Energies 135
IV. Linear Interaction Energy Approaches 137
V. MM-PBSA 147
VI. PROFEC 149
VII. λ-Dynamics and Chemical MC/MD 151
VIII. Conclusions 154
References 155

Membrane Protein Simulations: Ion Channels and Bacterial Outer Membrane Proteins

CARMEN DOMENE, PETER J. BOND, AND MARK S. P. SANSOM

I. Introduction 159
II. Simulation Methods 162
III. Ion Channels 166
IV. Outer Membrane Proteins 174
V. Future Prospects 183
References 185

Large Scale Simulation of Protein Mechanics and Function

EMAD TAJKHORSHID, ALEKSIJ AKSIMENTIEV, ILYA BALABIN, MU GAO, BARRY ISRALEWITZ, JAMES C. PHILLIPS, FANGQIANG ZHU, AND KLAUS SCHULTEN

I. Introduction 195
II. Technology for Simulation and Visualization of Large Biomolecular Systems 198
III. Aquaporins–Membrane Water Channels 201
IV. Energy Conversion in ATP Synthase 211
V. Mechanical Signaling in Fibronectin 232
VI. Outlook 237
References 239

Structure/Function Correlations of Proteins Using MM, QM/MM and Related Approaches: Methods, Concepts, Pitfalls, and Current Progress

A. SHURKI AND A. WARSHEL

I. Introduction 249
II. Classical Force Fields and Their Use 250
III. Evaluation of Electrostatic Free Energies of Macromolecules 258
VI. Methods for Simulation of Chemical Processes in Enzymes 263
V. Studying Enzyme Catalysis 274
IV. Concluding Remarks 301
References 302

Catalysis and Specificity in Enzymes: A Study of Triosephosphate Isomerase and Comparison with Methyl Glyoxal Synthase

QIANG CUI AND MARTIN KARPLUS

I. Introduction 315
II. Computational Methods 320
III. Preparation of the Active Site: The TIM "Lid" Transition 324
IV. Analysis of Mechanisms of the TIM Catalyzed Reactions: The Effect on the Activation Barrier 327
V. Dynamic Effects in the TIM Reaction 344

VI. Catalytic Specificity: Comparison of TIM and MGS 358
VII. Conclusions 366
References 368

All-Atom Simulations of Protein Folding and Unfolding

RYAN DAY AND VALERIE DAGGETT

I. Background 373
II. Overview of Results 377
III. Conclusions/Summary 395
References 399

AUTHOR INDEX 405
SUBJECT INDEX 429

PREFACE

This is the first volume on protein simulations for the *Advances in Protein Chemistry* series. This is noteworthy for a couple of reasons. Firstly, it reflects the maturity of computational techniques. Simulations must be continually validated, but force fields, upon which dynamics simulations are based, have improved and been applied successfully to thousands of systems. Secondly, simulations have reached the point that they can provide insight into interesting biological phenomena.

It is tempting to think of proteins as static due to the large number of beautiful crystal structures that are now available and the difficulty in comprehending such structures with thousands of atoms moving simultaneously. But, there is no doubt that proteins move and that they exhibit a variety of complex motions. This motion may just be a byproduct of thermal energy or it may be vital to function and "designed" into a protein. Proteins are, then, necessarily dynamic, able to change conformation in the presence of a substrate, other proteins, effectors or merely changing conditions. The process by which the proteins move from one conformation to another is much more difficult to ascertain experimentally than the static structures.

Detailed information regarding the dynamic behavior of the native state alone can aid in the determination of the forces important to function, such as binding, recognition, and allostery. The availability and use of dynamic views of protein structure should aid in drug design, since a current limitation is the focus on static crystal structures. Realistic molecular dynamics simulations and other computational techniques are attractive because of their ability to provide detailed information about protein dynamics, function and folding. The information from such studies is complementary and extends what can be gleaned from experiment. In fact, there is synergy between theory and experiment: theory provides detailed and testable molecular models and experiment-provides the means to test and validate the models, such that the combination can yield a much richer view of the system in question than either approach could alone. In fact, a well-known experimentalist told me recently that he thinks that it won't be long before experimental papers submitted for publication are required to include simulations. This view might be a bit extreme, but there is no doubt that protein simulations are becoming increasingly popular.

This volume of *Advances in Protein Chemistry* provides detailed and authoritative reviews by top practitioners in the field. The first chapter in this volume begins by outlining the many ways in which force fields and computation are used, and critical, in the area of structural biology. In fact, such approaches are pervasive in even ''experimental'' studies, such as structure determination using X-ray crystallography and NMR spectroscopy. Force fields, or potential energy functions, are fundamental to most protein calculations, and the second chapter provides detailed information about force fields and different factors that can affect their reliability and use. The next two chapters deal with different aspects, both historical and cutting-edge, of simulations and free energy perturbation calculations as applied to ligand binding. The next two chapters show simulations at their extreme with respect to system size and complexity. The first of these highlights studies of membrane proteins, providing an unprecedented view of the movement of molecules through channels and the role of protein motion in this process. The second examines similar issues and describes studies of proteins as machines, focusing on some of the largest systems conquered to date. The next two chapters illustrate how dynamic behavior is important to enzymatic catalysis both as applies to the actual chemical mechanism and more indirectly through the protein's ability to trap and harbor the substrate. Finally, one of the most extreme manifestations of protein dynamics is described, protein folding and unfolding.

These chapters show that simulation studies are really coming of age. They can provide much needed insight into a variety of biologically interesting phenomena. Computational approaches continue to improve and one cannot discount the importance of technology to this field. Increased computer power is now making it possible simulate dynamic behavior on longer and longer timescales for larger and larger systems. The future is bright.

ASSESSMENT OF THE ROLE OF COMPUTATIONS IN STRUCTURAL BIOLOGY

By IRWIN D. KUNTZ* AND DAVID A. AGARD†

*Department of Pharmaceutical Chemistry, University of California, San Francisco, California 94143-2440, and †Department of Biochemistry and Biophysics, University of California, San Francisco, California 94143-2240

I.	Introduction	1
II.	Comparison of the Role of Computation in the Study of Small Molecules and Macromolecules	2
III.	Comparison of Theory and Experiment	5
IV.	Molecular Recognition and Docking	8
V.	Error Estimation and Error Propagation in Experiment and Theory	10
VI.	The Role of Models in Protein Simulations	13
VII.	Parameterization	15
VIII.	Testing of Theoretical Models For Structural Biology	16
IX.	Conclusions	19
	References	19

Abstract

Computations are now an integrated part of structural biology and are used in data gathering, data processing, and data storage as well as in a full spectrum of theoretical pursuits. In this review, we focus on areas of great promise and call attention to important issues of internal consistency and error analysis.

I. Introduction

Computations are now intertwined with almost every aspect of experimental and theoretical structural biology. For example, laboratory instrumentation contains data collection and data analysis programming; every structure in the Protein Databank has required the use of data processing and refinement techniques; the Human Genome was assembled with computational algorithms; our understanding of protein folding kinetics, macromolecular structural analysis, and molecular recognition have been greatly advanced by theoretical treatments and computer graphic displays of data. In this review we consider the multifaceted roles and challenges for computation in protein simulations and the analysis of protein function. We cannot begin to construct a comprehensive survey. PubMed retrieved over 1000 titles under "Molecular Simulation" alone for 2002.

ADVANCES IN PROTEIN CHEMISTRY, Vol. 66

0065-3233/03 $35.00

Rather, we focus on particular themes that we believe have importance for the entire field over the next 5 years. Specific questions include the following: In what areas of structural biology can theory lead experiment? What factors affect quantitative comparisons between theory and experiment? How is the concept of error applied to molecular simulations? What are the roles of computational models? And, importantly, what are the most promising sources of data for testing current theories and for stimulating new developments? Many of these issues are treated in detail in other chapters in this volume.

II. Comparison of the Role of Computation in the Study of Small Molecules and Macromolecules

Computational studies on the gas-phase properties of molecules of a few dozen atoms can be expected to yield highly accurate descriptions of the geometries and energy levels for ground state and many excited-state species (Helms, 2002; Schleyer, 1998). In many cases, the accuracy of the calculations approaches or even exceeds the accuracy of measurement. Furthermore, the geometries and energies of closely related molecules such as positive and negative ions, protonated and deprotonated species, and tautomers are readily calculated. It is also possible to determine the distribution of molecules among the energy levels so that statistical and thermodynamic properties can be derived. Absolute entropies can be calculated from spectroscopic data. Clearly, the primary reason for these successes is the quantitative application of quantum mechanics to systems of increasing size and complexity. (Schleyer, 1998)

Condensed phases bring more complexity and less accuracy because of the many-body coupling that underlies most physical properties. Nonetheless, it is feasible to calculate the geometries and relative energies of conformers in pure liquid, solution, and crystalline states, the enthalpy of vaporization, and, in favorable cases, the crystal packing of small molecules through a combination of quantum mechanics and parameterized molecular mechanics force fields (Schleyer, 1998). Many spectroscopic features can be assigned and condensed-phase excited state features elucidated (Peyerimhoff, 1998).

These results are in strong contrast to the situation with macromolecules that we loosely define as containing >100 non-hydrogen atoms (e.g., larger than a decapeptide). First, except for mass spectrometry, there has been little interest in gas phase properties of macromolecules, and only isolated structural or spectroscopic data are available for peptides or proteins. This is unfortunate, because there is no doubt that isolated molecules in a vacuum are the easiest to treat theoretically. However, even if we knew experimental structures and energies of gas phase species, there is no guarantee that the current theoretical treatments could reproduce them accurately. The major roadblocks are well-known: the technical difficulties in extending quantum mechanical treatments to large heterogeneous molecules; uncertainties in empirical force fields, especially electrostatic treatments; lack of knowledge of "van der Waals" forces on the scale of 10 to 1000 Å; the need for extensive conformational sampling; and the difficulties in quantitative evaluation of conformational entropy. Kinetic properties of macromolecules in the gas phase are even more obscure.

For these reasons, along with the dearth of biological questions for gas phase environments, major computational efforts for structural biology have focused on aqueous solution or heavily hydrated crystals. Such systems present all the challenges of the gas phase plus additional concerns: water is a particularly complex solvent both in its self-interactions and its interactions with solutes. The well-studied maximum density, electrostriction, and hydrophobic effects are almost unique among molecular liquids (Eisenberg and Kauzmann, 1969). While there has been much progress in modeling the properties of water (Mahoney and Jorgensen, 2001; Silverstein, Haymet, and Dill, 1999), it seems that none of the current theoretical models captures the full range of water's properties. The computational problems are compounded by the limited experimental data detailing interactions between macromolecular solutes and the solvent. Complex solute size–dependent effects will surely have an important bearing on the next round of theory development (Silverstein, Haymet, and Dill, 1999). The near singularity, low symmetry, and complex energy landscape of the "native" state of proteins (Dill, Alonso, and Hutchinson, 1989) have also posed unusual difficulties for standard statistical mechanics approaches.

Given these special challenges, the successes of molecular calculations in structural biology are indeed impressive, but they have generally relied more heavily on ad hoc and empirical theory than have small molecule studies. For example, theory in the form of lattice models (Dill, 1985), analytical treatments (Dill and Chan, 1997; Hardin *et al.*, 2002) and large-scale simulations (Shea and Brooks, 2001; Snow *et al.*, 2002) has greatly expanded our understanding of protein folding kinetics (Fersht, 2002; Schuler, Lipman, and Eaton, 2002). Atomic force fields are directly embedded in the thousands of crystallographic (Brünger *et al.*, 1998) and NMR structural determinations (Linge and Nilges, 1999). Other achievements include free energy perturbation techniques to assess the strength of binding of ligands (Kollman, 1996; Straatsma and McCammon, 1992), the application of Ewald sums (Darden, York, and Pedersen, 1993) and continuum solvent models to improve quantitative treatment of electrostatics

(Bashford and Case, 2000; Honig and Nicholls, 1995), and the use of hybrid formulations of quantum mechanics and molecular mechanics to deduce enzymatic mechanisms (Tantillo, Chen, and Houk, 1998).

However, almost by historical accident, developments in macromolecular calculations have not followed the usual scientific paradigm of thorough study of simple cases before progressing to more complex ones. A striking example is the lack of a bridge between small peptides (e.g., containing 2 to 5 amino acids) that can often be studied with quantum mechanical approaches and small proteins. There is a very limited amount of work on polypeptides in the 10 to 30 amino acid size range. Of course, this regime is a difficult one for which to get definitive structural data because these systems typically explore a wide range of geometries rather than having a well-defined "folded" state. Only recently have detailed simulations been carried out to yield thermodynamic properties (Daura *et al.*, 2002) for peptides of this intermediate size. Without a comfortable "build up" principle to rely on, most large-scale protein studies must seek verification through perturbation strategies or by using surrogate markers of success.

If we examine plausible *a priori* goals for molecular computation in structural biology, we must admit that it is not currently possible to calculate, from first principles, the folded structure of an arbitrary protein (Bonneau and Baker, 2001; Duan and Kollman, 1998), to locate the accessory water molecules or counter ions, or to determine the energy of either the isolated or the condensed system with any accuracy. Nor can one enumerate the accessible conformers of a small protein, calculate the thermodynamic values of the phase-like transitions of folded to unfolded states, or predict the detailed spectroscopic or kinetic properties of proteins, although helical spectroscopic features have been qualitatively understood for some time (Cooper and Woody, 1990). Very recent progress has been reported in the literature in some of these areas (Gogonea *et al.*, 2001; Higo and Nakasako, 2002; Snow *et al.*, 2002), and the current status of many of these issues is developed further in the other chapters in this volume.

What can be done now, in relatively routine fashion, includes (1) simulations of nanosecond trajectories of small hydrated proteins, with heroic efforts yielding microseconds (Duan and Kollman, 1998); (2) rapid refinements of crystallographic and NMR data in an automated manner (Brünger *et al.*, 1998); (3) extensive modeling of ionic and nonionic aqueous solutions (Jorgensen, Maxwell, and Tirado-Rives, 1996); (4) low–resolution treatments of macromolecular vibrational and electronic states (Helms, 2002); (5) differential methods for assessing the thermodynamic and kinetic effects of amino acid substitutions on protein stability and folding dynamics and nucleic acid binding (Fersht and Daggett, 2002; Huo, Massova, and Kollman, 2002a; Reyes *et al.*, 2001; Wang and Kollman, 2002). At a similar level, there are semi-quantitative treatments of molecular recognition and ligand binding that work particularly well in the limit of small perturbation (Chipot, 2002; Huo, 2002b; Åqvist, this volume); combined quantum mechanical/molecular mechanics treatments of enzyme kinetics (Gogonea *et al.*, 2001); normal mode analyses of molecular vibrations (Cornell and Louise-May, 1998); qualitative prediction of small protein structures (Bonneau and Baker, 2001; Tramontano, Leplae, and Morea, 2001); and well-developed efforts in protein homology models (Baker and Sali, 2001) and secondary structure models for RNA (Zucker, 1989).

What are the prospects for the future? In most of the cases cited above, the accuracy and/or precision of the results are clearly limited by the available computer resources. Faster computers or better algorithms will, in almost every area, allow either better statistical treatments or the use of already existing more accurate models. Two important consequences of the limitations of computer speed/time are that frontier calculations can only be run once and methodology development is often restrained because rigorous testing regimens are not easily established. The future will bring significant relief, processor speeds have improved exponentially for several decades and are likely to continue to do so for the next decade. With the cost per cycle and cost per storage element still dropping rapidly, and with the great advances in parallel computing, there is reason to expect much more extensive calculations in both size and sophistication in the next few years. For example, we are, by some estimates, very close to the time scales needed for *ab initio* folding trajectories for small proteins (Snow *et al.*, 2002; Sullivan and Kuntz, 2002). Further, as such calculations are repeated, we will have a direct method of assessing systematic discrepancies in the underlying procedures (Burgi, Pitera, and van Gunsteren, 2001; Fersht and Daggett, 2002).

III. Comparison of Theory and Experiment

It might seem to be a scientific truism that any theory must be compared to experiment. However, in the structural biology field, careful consideration is required for useful comparisons to be made. To what extent are experimental structural determinations biased by the theoretical models/force fields used in the data processing? Do ensembles from NMR represent physically meaningful conformations or are they biased by the manner in which the geometric constraints are introduced?

Similarly, to what degree are the thermal factors (''B'' factors) in diffraction indicative of conformational distributions and to what extent are they simply operational parameters, subsuming errors in the models and inconsistencies in the force field constraints?

Consider, for example, a typical NMR structural experiment. A large number of reasonably pure molecules of a particular protein along with auxiliary materials to adjust pH and salt conditions are dissolved in water to provide a 1 to 10 millimolar solution and placed in a magnetic field typically near room temperature. Measurements that report population-average and time-average values of properties related to dozens of internal dihedral angles and hundreds to thousands of interatomic separations are obtained (Burgi, Pitera, and van Gunsteren, 2001). These data are used with mathematical procedures (Havel, 1998) and in conjunction with a molecular force field (Brünger *et al.*, 1998) to obtain a small set of molecular structures, each of which meets the experimental constraints within some error or uncertainty. This set of structures, or perhaps a single structure closest to the average structure is reported as ''the'' structure of the protein under the experimental conditions.

Now examine a typical calculation for the same protein: a single protein molecule is surrounded by several thousand water molecules, providing a molar concentration of 5 to 50 millimolar. The pH of the system is established in a static manner by fixing the protonation state of each ionizable group. Normally there is no consideration of the *in situ* pKs. Counter ions to provide overall electrical neutrality may be included. A simulation, ordinarily using molecular dynamics, is carried out at a reasonable temperature that is maintained constant by using coupling to a temperature bath through an empirical procedure that compensates for changes in energy over the course of the calculation (Leach, 2001). Periodic boundary conditions are used to avoid edge effects. The boundary conditions have no direct physical meaning and their impact, if any, on detailed properties of the system under study have not been determined. The simulation is started either in some random state or with an initial protein conformation established through geometric constraints (Havel, 1998). These constraints are monitored throughout the calculation, often using an ''error'' or ''objective'' function that compares the instantaneous values with the constraints. The functional form is typically chosen for mathematical convenience (e.g., harmonic constraints) and with an arbitrary weight compared with the other forces acting on the system. The simulation is run for a time determined by computational resources. The simulation may be repeated several times with varying starting conformations. The system is monitored for excessive fluctuations in temperature, energy, and geometry, but no strict criteria for equilibrium are invoked.

Normally, the root-mean-square differences in geometry are reported as a measure of internal consistency and to determine agreement with experiment. To the best of our knowledge, there are no studies that demonstrate true convergence or thermal equilibrium in such systems (Smith, Daura, and van Gunsteren, 2002). An interesting early calculation (Kitson and Hagler, 1988) suggested that the use of non-crystallographic symmetry relationships (crystal unit cells with molecular copies that are not related by crystal symmetries) would provide a powerful test for (geometric) convergence. Renewed interest in simulations with crystal boundary conditions (Jacobson *et al.*, 2002) is encouraging. Monitoring kinetic properties such as diffusion would also be revealing on selected systems. Clearly the time has come to use the relatively large expansion in computer resources to provide a more quantitative assessment of the extent of sampling in molecular simulation calculations and the convergence of the underlying thermodynamic properties.

The current state of these procedures has the potential for significant systematic differences between experiment and theory. Among them are the use of the trajectory of one molecule to represent the (experimental) equilibrium ensemble, the impact of the simulation boundary conditions, and the quality of the molecular force field and the convolution of errors in the force field with the geometry-based objective function that is also being minimized. Notice that there is no guarantee that the force field for the calculations is the one that has been used to refine the experimental data. In the NMR field, the primary experimental constraints are often given in semi-quantitative terms rather than derived self-consistently from full relaxation data (Oppenheimer and James, 1989). Another set of concerns could address the pressure, pH, temperature, and/or energy fluctuations in the simulations; most of these variations are not reported. Recent progress in dynamic, conformation-dependent pK determination is encouraging (Burgi, Kollman, and Van Gunsteren, 2002; Gorfe *et al.*, 2002). Although this example has dealt with the details of NMR, similar issues can be raised with the standard crystallographic procedures.

Given so many differences between experiment and theory, it is remarkable that the crystallographic and NMR experimental results generally agree well with each other, especially around the active sites of enzymes, and that simulations that start near the experimental conformations do not diverge significantly over many nanoseconds. The simplest interpretation of this correspondence is that the native structures of proteins and nucleic acids are stable, lying in moderately deep (>10 kT) minima under normal conditions. Further, these structures do not change continuously as a function of the normal thermodynamic variables. Rather the "native" state population increases or decreases with corresponding decreases or increases in a more amorphous ''unfolded'' or ''denatured'' state whose properties do shift progressively (Dill, Alonso, and Hutchinson, 1989). Thus, events conspire to make the determination of the *geometry* of the native state a relatively robust (but unsolved) problem. Evaluation of the thermodynamic properties of macromolecules is more difficult. Both native and denatured states must be evaluated. Further, the computational evaluation of enthalpy, entropy, and heat capacity lag well behind the experiments (Leavitt and Freire, 2001; Shafer *et al.*, 2001).

IV. Molecular Recognition and Docking

A simple question is whether current computational protocols are effective enough to be of practical use in the discovery/design of new ligands. There is no single ''gold standard'' for utility, but the dominant discovery technology in the pharmaceutical industry is high throughput screening (HTS). This methodology is well suited to processing very large numbers of compounds in standard format for enzyme-based assays. Virtual screening can parallel this process, either with or without structural information on the target. Using structure-based docking, various groups have reported successes (Charifson, 1997; Kick *et al.*, 1997) that yield compounds of appropriate activity (generally, inhibitory constants below micromolar) that can be carried forward in a discovery pipeline, often through a direct link to chemical strategies (Haque *et al.*, 1999; Kick *et al.*, 1997). Several groups have reported enrichment factors of detecting known inhibitors from databases using computational methods (Diller and Merz, 2001; Knegtel and Wagener, 1999). There are also reports of quite novel inhibitors of processes as diverse as viral fusion and kinesin binding to microtubules (Hoffman, Kuntz, and White, 1997; Hopkins, Vale, and Kuntz, 2000). Although these examples are only suggestive of the applications of virtual screening to drug discovery, they seem sufficient to support the view that computational efforts, especially in conjunction with coordinated chemical strategies, can add efficiency and quality to ligand discovery and design (Kick *et al.*, 1997).

On a more theoretical level, we can ask whether the approximate potential energy expressions are adequate for calculating the binding affinities of ligand-receptor complexes. In brief, current approaches predominantly use molecular mechanics force fields, usually parameterized against experimental or quantum mechanical data. These treatments certainly have limitations: they completely omit entropic considerations, are restricted to pair-wise potentials, and assume transferability at an atomic level. Even if the force fields provided exact energetic contributions to complex formation, they would not be expected to calculate accurate free energies or even to rank order diverse compounds effectively. Energies or enthalpies would, at best, be able to describe differences only among a narrow range of closely related ligands. The obvious solution is to extend the force field terms to include entropic terms. There is no exact way to "add entropy" to a molecular configuration (e.g., a snapshot of the geometry of a solvated macromolecular complex) because entropy is a distributive quantity containing components from the conformational choices of ligand and receptor and from the configurations of the solvent. The two basic strategies to solve this problem are (1) to emulate the physical situations by using molecular dynamics or Monte Carlo simulations to generate an ensemble of thermally equilibrated structures that include large numbers of solvent molecules and counter ions and (2) to use the continuum solvent models (Poisson-Boltzmann, Generalized Born) along with an empirical correction for hydrophobic interactions based on buried surface area (Bashford and Case, 2000; Tsui and Case, 2000; Zou, Sun, and Kuntz, 1999). The major concern with the first class of approaches is that it is difficult to assure sufficient sampling to obtain good averaging of thermodynamic properties. As noted above, the only criteria used routinely to track sampling are to monitor the geometric stability and monitor the energetic stability of a simulation. These are inadequate to demonstrate that the system is at equilibrium or that thermodynamic properties have converged. Running longer simulations is one solution, albeit an inefficient one. With parallel processor facilities, multiple trajectories are likely to be more productive for perturbation calculations (Kuhn and Kollman, 2000), as they have been for protein folding (Snow *et al.*, 2002) given a fixed amount of processor time.

What can be said about the accuracy of these calculations in reproducing experimental results? Full simulation calculations aimed directly at molecular interaction energies are sufficiently computer-intensive that they are relatively rare. (Head, Given, and Gilson, 1997; Luo and Sharp, 2002). Perturbation procedures, which yield differential values, are much more common. Free energy perturbation or thermodynamic integration methods are presently the most accurate calculations available to assess binding free energies (McCammon, this volume; Gouda, 2002). In favorable cases, average unsigned errors of less than 1 kcal/mole have been reported (Kollman, 1993; McCammon and Åqvist, this volume). These errors contain contributions from inaccuracies in the force field, from sampling, and—of course—any experimental uncertainties. Kuhn and Kollman examined the reproducibility of MMPBSA results from two different trajectories (Kuhn and Kollman, 2000). It seems unlikely that they can be reduced much further without careful attention to each of these areas. It should also be noted that we have focused on the "best cases." Agreement between perturbation simulations and experiments is not always as good, and "poor" results do not get published! A major challenge is to uncover the causes, be it insufficient conformational sampling, systematic force field errors, or other sources that can be revealed by particularly difficult applications. Perhaps lists of "challenging" problems can be maintained at a suitable web site.

If we use the perturbation methods as the "best available" standard, it is possible to assess the more approximate free energy calculations by using continuum solvent models (Bashford and Case, 2000; Zhou *et al.*, 2001) or the admixture of molecular solvent and continuum solvent in MM-PBSA (Srinivasan *et al.*, 1998). In several papers, results comparable to the perturbation methods are achieved, but the authors note anticipated limitations: failure to account for specific water interactions; sensitivity to trajectories; sensitivity to "induced-fit" (Gouda *et al.*, 2002; Huo *et al.*, 2002b; Kuhn and Kollman, 2000). We must infer that the errors in comparing experimental and calculated free energies of interaction for MM-PBSA, LIE, and other such approaches will normally be worse than the 1 kcal/mol level, but these methods should still be quite useful in screening mid-sized databases or sorting among proposed synthetic candidates.

The overall impression is that the better quality force field/simulation procedures are reasonably adequate for estimating ligand-binding free energies, but accuracy could benefit from increased computational power. They may be as good as we can get without expanding the number of "parameters" and may not be the limit in accurate experiment/theory comparisons. As a test of this idea, it will be interesting to see if the newer "polarizable" force fields provide significant improvements in the ligand binding estimations (Cieplak, Caldwell, and Kollman, 2001; Kaminski *et al.*, 2002).

V. Error Estimation and Error Propagation in Experiment and Theory

A critical component of scientific studies is the analysis of errors, both random and systematic. Error analysis for experimental structural biology has been largely treated in an ad hoc manner with the major exception of the classic work of Luzatti (Luzatti, 1952); Stroud (Stroud and Fauman, 1995); and recent insightful efforts by Brüger, who introduced statistical methods to the refinement of diffraction data (Brünger, 1992). An initial analysis of specific physical models that underlie NMR data interpretations has been reported (Chalaoux, O'Donoghue, and Nilges, 1999; Linge and Nilges, 1999). A major hurdle throughout structural biology is the amalgamation of many types of data, each with its own underlying physical assumptions and limitations. Consider, as an example, the use of geometric constraints (e.g., fixed bond lengths and bond angles) in lower-resolution macromolecular crystallography. Hendrickson suggested that such variables be weighted with respect to each other by normalization with the standard deviation of the variable, determined in some independent manner (Hendrickson and Konnert, 1981). This procedure works well for chemical bonds, which can be obtained to very high resolution from small molecule crystallography. However, identifying random or systematic errors associated with the use of a more complete force field—including non-bonded terms such as electrostatics, hydrogen bonds, and van der Waals terms—is much more difficult, and the concept of the standard deviation of a molecular energy is largely unexplored (see below). Moreover, the reliance on restraints should probably be altered dynamically throughout the refinement process in response to estimates of the errors in the structure (Stroud and Fauman, 1995).

Our major point is, for the most part, error analysis is given relatively little attention *by those who solve the structures* in spite of recent attention to error sources (Kleywegt, 1999; Kleywegt, 2000; Tickle, Laskowski, and Moss, 1998) and monitoring by the Protein Databank (Abola *et al.*, 1997). At a minimum, the diffraction experiments should routinely examine the range of structures (e.g., conformers) consistent with the experimental data, the influence of the force field on the outcome, and the internal consistency of the structure (Kleywegt, 2000; Network, 1998; Word *et al.*, 1999). These are potentially even more serious issues for NMR structures. The NMR community must look more closely at the conflicts between matching geometric constraints and the need to have a Boltzmann-distributed set of structures; constraints must be evaluated as ensemble averages rather than applied to individual conformations. It is now routine in the x-ray community to examine a number of secondary markers of structural quality (e.g., statistics for bond length, bond angle, and dihedral angle deviations, number and quality of hydrogen bonds, heavy atom clashes). A standard set should be supplied with all structures as they are reported or deposited. For complexes with ligands, it would be helpful to have separate statistics compiled for the ligand, since these structures are often more uncertain than the protein core atoms. It is also essential that the critical experimental data—nOe intensities, dihedral values, residual dipolar couplings, and diffraction structure factors—be available to the research community when the structures are deposited, since the essential canon of reproducibility of results cannot be assured without this information.

In a similar vein, there is currently no way that molecular simulations can be directly reproduced from laboratory to laboratory. The essential problem is the "chaotic" nature of molecular dynamics and Monte Carlo calculations. The large number of parameters, the use of random numbers at various phases of the computation, and the sensitivity to systems hardware (word length) and systems software (arithmetic routines) makes it very unlikely the same coordinates will be generated even if the same version of a standard program is used with all nominal settings preserved. Of course, for many purposes, the *conclusions* from the calculation may be unchanged even if the "details" differ. But the field should seek more robust and more transferable software with standard routines for evaluating the quality of the results. We can borrow standard methods from statistics and from computational efforts on simpler systems. For example, it is standard practice throughout science to carry out replica studies, averaging the results. In these days of large processor "farms," it seems straightforward to examine the stability of the results as the number of "identical" runs is increased (Snow *et al.*, 2002). The standard deviation among runs is likely to be more informative than the standard deviation from within a single simulation. It is also common, in many areas of physics and chemistry, to examine the sensitivity of results to the independent parameters—either from knowledge of the underlying functionality or from variation of the parameters. Even with a large number of parameters, advances in multivariate analysis make it feasible to think about perturbation strategies that could lead to a better understanding of the reliability of computational protocols and the sensitivity of the derived quantities to the parameterization (Martin *et al.*, 1995; Rabitz, 1989; Weiss and Ferretti, 1985). Finally, there are many secondary markers that can be used, as suggested above, to confirm the soundness of a molecular simulation. These include stability of temperature and/or energy, assessment of pressure fluctuations in constant volume simulations; and measurement of kinetic quantities such as thermal and mass diffusion, order parameters, and B-factors. Since it is such a likely source of error, it would be of great value to have some monitor of sampling efficacy. For Monte Carlo and some molecular dynamics simulations, it might be possible to track the heat capacity at constant volume or constant pressure. Experimental heat capacities of protein solutions are known to moderate accuracy. Calculated values in theoretical studies of simple liquids would be expected to lie below the experimental values, providing a direct indication of the completeness of the thermodynamic sampling

(Beveridge *et al.*, 1983). At a less rigorous level, one could monitor the conformational sampling through parallel simulations or by keeping track of the number of conformational transitions for each dihedral angle in the system. Some report of the average time of transition (for dynamics) and the number of occurrences of a transition for each degree of freedom would allow the sampling of different procedures to be compared in an objective way.

In brief, we note that there are serious obstacles to overcome if we are to make an accurate comparison of experiment and theory for macromolecular systems. More sophistication in error analysis would serve the field well.

VI. The Role of Models in Protein Simulations

Theoretical models of physical phenomena play a critical role in the development and application of scientific principles. A model is a set of assumptions about the relevant features of a well-defined system that can generally be reduced to equations that describe the behavior of the system as a function of its independent variables. All models are incomplete in the sense that they cannot include the full complexity of the real world. Many models recognize these limitations explicitly by identifying variables or relationships among variables that will not be treated (e.g., neglect of relativistic effects in classical mechanics). For complex systems, the full physical laws may not be known. In these cases, the strategy of comparing models to the experimental systems can often reveal some of the neglected terms. It is often desirable to identify the minimum number of assumptions required to reproduce complex behavior (Silverstein, Haymet, and Dill, 1999). Both theorists and experimentalists should seek to understand the limitations and applicabilities of models in structural biology and to select models that have the appropriate accuracy and granularity for the range of properties of interest (Leach, 2001).

"First principles" or "ab initio" models in structural biology that attempt a complete description based on quantum mechanics or thermodynamics/statistical mechanics are rare. Much more common are partial descriptions of complex systems (e.g., molecular mechanics) or full descriptions of highly simplified systems (e.g., lattice models of proteins). The range of validity of these approaches is relatively easy to determine and is often included in the initial assumptions.

Semi-empirical models make use of empirical parameters that are introduced to facilitate comparison in experiments. The whole field of molecular mechanics is a good example. It uses first principles formulations of Newton's laws coupled to a large set of empirical parameters (bond lengths, bond angles, van der Waals radii, Lennard-Jones potentials, "partial atomic charges," etc.). The values for the empirical parameters are established through the use of training sets of experimental or quantum mechanical data (Cornell *et al.*, 1995; Kaminski *et al.*, 2002; Onufriev, Case, and Bashford, 2002). It is more difficult to establish the range of validity of semi-empirical theories. A Lennard-Jones treatment of a hard sphere fluid could be exact for a toy system of atomic billiard balls, whereas it would be less successful for liquid helium or liquid xenon than for liquid argon because of quantum mechanical terms that are not readily included directly into the model. Molecular mechanics descriptions of proteins and nucleic acids raise additional issues. How can we tell if the fundamental formulas for electrostatics and van der Waals interactions work well on the scale of 10 to 100 Å? Are the simulations of these systems at least quasi-ergodic? What is the proper procedure to establish the hundreds of parameters needed to describe the mechanical behavior? What solvent model (generally a water model) should one use among the many water models to choose from (Mahoney and Jorgensen, 2001)? Each has been developed to match certain physical properties of water (density, diffusion constants, enthalpy of vaporization, temperature dependence of the density, etc.). Regardless of the accuracy of these formulations over the range of physical variables, additional testing is needed to be sure that the properties of the water in the *presence of the macromolecular solutes* are sensible. And, of course, the validation is made much more difficult by our lack of knowledge about the experimental values for water molecules in such solutions or in the heavily hydrated crystals used for diffraction studies.

Within the broad sweep of protein computations, there are a number of completely empirical models that are based on linear or non-linear regression techniques. These are most prominent in studies of molecular recognition and biological activities and are often called "quantitative structure-activity relationships (Hansch, 1969). These approaches are often used in ligand design and drug discovery protocols. On occasion, semi-empirical formulations are extended with the use of additional empirical terms. Examples include the addition of hydrophobic effects in continuum solvent models (Zou, 1999; Majeux, 2001; Kuhn, 2000; Case, this volume) and the Linear Interaction Energy model of Aqvist (Aqvist, Medina, and Samuelsson, 1994; Zhou *et al.*, 2001). Neural nets and other means of abstracting or clustering trial data are also useful in addition to, or in place of, regression techniques (Agrafiotis, Cedeno, and Lobanov, 2002; Agrafiotis and Xu, 2002). In a general way, one should be aware that it is all too easy to add large numbers of variables indiscriminately. Cross validation is critical. Another good strategy for empirical protocols is to carry out a principal components analysis (Stahle and Wold, 1988). The number of non-zero eigenvalues provides a good count of the minimum number of parameters required, and the magnitude of the eigen vectors indicates the potential contribution of each parameter to the variance. One can also examine the distribution of the data along each principal mode (Stahle and Wold, 1988; Wold 1991) to see, in advance, how robustly each mode is sampled. In general, empirical models are limited to a domain fixed by the training set. They can provide accurate interpolation within the domain but rarely can be generalized or extrapolated beyond this domain. Semi-empirical or ab initio models may provide superior extrapolation but often do not provide the same precision for interpolation.

VII. Parameterization

Many aspects of molecular calculations require the generation of large sets of (typically) atom-based parameters. Most of these parameters represent simplifications of complex electron distributions (e.g., point charges, van der Waals radii, generalized Born radii). There is often disagreement over the computational procedures or precise physical meaning of the resulting terms, even though they are in widespread use. Further, as noted earlier, many parameters have intrinsic coupling terms that are often neglected at a given level of treatment (bond stretching-bending coupling, atomic polarization effects). To add to the problem, some physical phenomena such as hydrogen bonding, the hydrophobic effect, and atomic polarization do not lend themselves to representation as simple two-body interactions. All of these difficulties are further confounded when one seeks a standard procedure for setting the optimum values of the parameters. Historically, those skilled in the art identified specific issues (dihedral barriers are too low, dipole moments are too high) and adjusted the first-order terms accordingly based on a small number of experiments or high-level quantum mechanical calculations. Too often, the search is for unique values of the parameters that fit the data without regard for other parameter choices that do equally well. Force fields have been repeatedly reparameterized (Cornell *et al.*, 1995; Lazaridis and Karplus, 1999; Mackerrell, Banavali, and Foloppe, 2001), but what is missing is a protocol that has statistical validity with some understanding of the variance in the parameter values (Weiss and Ferretti, 1985). It is frequently the case that parameterization problems are significantly under-determined, with more parameters than independent pieces of data. A step in the right direction is offered by the RESP procedure (Bayly and Kollman, 1993) for point charges and recent work on introducing atomic polarization parameters (Cieplak, Caldwell, and Kollman, 2001; Kaminski *et al.*, 2002). The growing number of very high-resolution protein structures provide another data set that could be used for setting geometric and energetic parameters in a constrained multivariate regression (see below).

A final comment about parameters is aimed at both the developers and the users of molecular simulation software. Both academic and commercial packages include, as a service to the research community, a very large number of options and parameter choices to allow the end user to customize the program to a particular task. It is almost impossible for the developers to test all possible combinations of choices. It is sometimes equally difficult for the user, even the expert user, to be sure that all program parameters and option flags have been set correctly. A good dose of user interface design principles would be very helpful to all concerned.

VIII. Testing of Theoretical Models For Structural Biology

We have argued that the extension of classical theory to macromolecular systems carries an extra burden because it is often difficult to provide high-quality explicit tests of the many parameters that such theories often entail. While experimental data on small molecules in solution, for example (Still *et al.*, 1990), and quantum mechanical studies of small molecules in the gas phase can play important roles (Kaminski *et al.*, 2002), critical tests remain difficult to devise. We have put together a short list of important experimental data sets that could have a significant impact on the testing of current theories and the development of new efforts.

1. Electrostatic properties of proteins can be probed in a direct way by studying the changes in electromotive force (EMF) in redox active systems. Of particular note are the cytochromes, many of which are available as naturally occurring or cloned variants. EMF measurements are straightforward and very sensitive—changes of 0.01 kcals/mol are easily discerned. While absolute calculations of EMF would be challenging, the relative values as single or multiple amino acids are changed might provide a very appropriate test bed for both atomic and continuum models of protein electrostatics (Blouin, Guillemette, and Wallace, 2002; Lo, Wong, and Hill, 1999; Martell *et al.*, 1999; Xue *et al.*, 1999). Fluorescent readouts of dielectric properties are also available (Cohen *et al.*, 2002).

2. Van der Waals parameters raise two fundamental issues. First, there are many atom types, especially as one moves into bioorganic or bioinorganic ligands. Second, the usual Lennard-Jones formulation is clearly not accurate for molecular scattering in the gas phase and remains a historical

artifact based on numerical ease in the days when "computers" referred to the people carrying out the computations. Additionally, there is direct experimental evidence for long-range interactions on the 1000 Å scale (Leckland and Israelachvili, 2001). It would be interesting to see how well current parameters can reproduce the energetics of small molecule crystals—especially the enthalpy and entropy of crystallization. Software to estimate these values is available for small molecule systems (DeTar, 2001).

3. The overall quality of force fields can be explored in a much more critical fashion using very-high-resolution (<1 Å) protein crystal structures, especially if a complete set of anisotropic thermal factors, occupancies, and structure factors is also available. The simplest tests would be to minimize the unit cell components *in the crystalline environment* with/without explicit water molecules. Examining the initial forces on the system and the ensuing deformations might directly point to deficiencies in the force field. Comparing molecules in the unit cell that are *not* related by formal symmetry (e.g., non-crystallographic symmetry) would provide a way to explore sampling issues (Kitson and Hagler, 1988). It would be important to ensure that the structures were generated directly from the experimental data and that no force field terms were used in the final refinement. One standard program, developed for small molecule crystallography, provides this capability (Sheldrick and Schneider, 1997). Simultaneous fitting to multiple conformers would be an important component (Gorler, Ulyanov, and James, 2000; Pellegrini *et al.*, 1997; Rader and Agard, 1997).

4. High-quality NMR data provide a number of opportunities for testing computational techniques. First, the ensemble of structures for a given experiment can each be used as a starting point for molecular dynamics or Monte Carlo calculations to see if they converge or diverge with a particular force field and simulation conditions. It should be straightforward to inspect the trajectories or MC runs for time-average or ensemble-average constraint violations. Second, the collection of trajectories can also be used to calculate dynamical properties, especially backbone and side chain order parameters that can be compared with increasingly available experimental data (Palmer, 2001). Third, when possible, the averaged structure, with a list of alternative conformations accessed under thermal equilibration, can be compared with high-resolution diffraction studies to identify similarities and differences that are due to the aqueous solution and crystal environments.

5. Thermodynamic assessments would greatly benefit from more experimental data. Of particular value would be sets of ligand-binding data for related ligands bound to families of proteins carried out, as nearly as possible, under the same experimental conditions. Crystal structures of several representative complexes would also be extremely helpful. Of some

importance is the need for the data to span as many orders of magnitude of binding as can be feasibly measured. Starts have been made in both academic and industrial laboratories (Holloway *et al.*, 1995; Kick *et al.*, 1997), but there are very few well-documented data sets available to the general community (Chen and Gilson, 2002; Roche, Kiyama, and Brooks, 2001; Wilson, Mace, and Agard, 1991). The related area of protein-protein complexes has been much illuminated by the exemplary efforts of Laskowski and co-workers, who have produced voluminous data on the binding of ovomucoid protease inhibitors to serine proteases (Lu *et al.*, 2001; Lu *et al.*, 1997). These experiments meet all of the standards set out above and are limited only by the number of crystal structures. A wish list should also include more calorimetric data on both kinds of complexes. An excellent beginning has been made by Freire and coworkers, whose data provide important insights into the enthalpy and entropy of ligand binding (Leavitt and Freire, 2001).

A closely related issue is the comparison of the physical conditions for obtaining the structural data with those for the binding data. Obviously, there will be shifts in the structure as the pH, salt, and solvent conditions are changed. Further, the diffraction structures are derived under very concentrated conditions with physical contact among the macromolecular constituents, whereas the binding experiments are in highly dilute solutions. Both the experimental and computational communities would be well served if there were direct measurements of the change in binding as a function of these variables—especially pH, macromolecular concentration, cosolvent, and ionic strength for some representative systems. Conversely, if it were possible to measure structural properties as a function of the same variables, perhaps by NMR, one could build up more powerful tests of theory and experiment.

6. Mass spectrometry of proteins now provides a variety of data sets for theorists. First, the use of dual spectrometers offers unambiguous fragmentation patterns of individual species for those interested in the computational challenge of bond cleavage. Second, the charged states of gas-phase proteins can be probed experimentally (Horn, Breuker, and McLafferty, 2001). Third, hydrogen-deuterium exchange phenomena can be determined to atomic resolution. Fourth, new strategies of chemical cross-linking invite the use of computational methods to explore protein fold family classification and rapid, low-resolution, structure determination (Young *et al.*, 2000).

7. Ultra-high-resolution crystallographic studies of small molecules can provide direct measurement of electron density that, in turn, can offer experimental parameterization of atomic charges, van der Waals radii, and atomic polarization (Lecomte, 1995).

IX. Conclusions

We close with a general comment about computational problems and what can be learned from the experience of physics, statistics, and computer science experiences in dealing with complex systems. First, many classical problems in theoretical and applied physics have aspects of structural biology problems: incomplete theories, coupled parameters, systematic differences between what theory can calculate and what experiment can measure. A useful paradigm is to take extra care with the underlying models to focus on the critical features and to organize the theory so additional terms can be added later without forcing a complete new formulation of the problem. From the statistics community, we can borrow strategies for multivariate analysis, cross validation, error propagation techniques and sensitivity analysis. Computer science provides guidance in algorithms but also introduces the fruitful concept of the complexity or difficulty of problems. Many of the problems of interest to the structural biologists (e.g., protein folding, molecular recognition, multisequence alignment) have been demonstrated to be of the most difficult nonpolynomial class (Salomaa, 1991; Wilson, 1972). However, this realization just recasts the challenge from finding a general "elegant" solution to finding appropriate approximations to solve the specific task at hand. Such solutions, coupled with improved computational resources, often provide practical answers even to the most difficult questions.

References

Abola, E. E., Sussman, J. L., Prilusky, J., and Manning, N. O. (1997). Protein data bank archives of three-dimensional macromolecular structures. *Methods Enzymol.* **277,** 556–571.

Agrafiotis, D. K., Cedeno, W., and Lobanov, V. S. (2002). On the use of neural network assemblies in QSAR and QSPR. *J. Chem. Inf. Comput. Sci.* **42,** 903–911.

Agrafiotis, D. K., and Xu, H. (2002). A self-organizing principle for learning nonlinear manifolds. *Proc. Natl. Acad. Sci. USA* In press.

Aqvist, J., Medina, C., and Samuelsson, J-E. (1994). A new method for predicting binding affinity in computer-aided drug design. *Protein Eng.* **7,** 385.

Baker, D., and Sali, A. (2001). Protein structure prediction and structural genomics. *Science* **294,** 93–96.

Bashford, D., and Case, D. A. (2000). Generalized Born models of macromolecular solvation effects. *Annu. Rev. Phys. Chem.* **51,** 129–152.

Bayly, C. I., and Kollman, P. A. (1993). A well-behaved electrostatic potential based method using charge restraints for deriving atomic charges: the RESP model. *J. Phys. Chem.* **97,** 10269–10280.

Beveridge, D. L., Mezei, M., Mehrotra, P. K., Marchese, F. T., Ravi-Shanker, G., Vasu, T., and Swaminathan, S. (1983). "Monte Carlo Computer Simulation Studies of the Equilibrium Properties and Structure of Liquid Water." American Chemical Society.

Blouin, C., Guillemette, J. G., and Wallace, C. J. (2002). Probing electrostatic interactions in cytochrome c using site-directed chemical modification. *Biochem. Cell Biol.* **80,** 197–203.

Bonneau, R., and Baker, D. (2001). Ab initio protein structure prediction: progress and prospects. *Annu. Rev. Biophys. Biomol. Struct.* **30,** 173–189.

Brünger, A. T. (1992). Free R value: a novel statistical quantity for assessing the accuracy of crystal structures. *Nature* **355,** 472–475.

Brünger, A. T., Adams, P. D., Clore, G. M., DeLano, W. L., Gros, P., Grosse-Kunstleve, R. W., Jiang, J.-S., Kuszewski, J., Nilges, M., Pannu, N. S., Read, R. J., Rice, L. M., Simonson, T., and Warren, G. L. (1998). Crystallography and NMR system: a new software suite for macromolecular structure determination. *Acta Cryst.* **D54,** 905–921.

Burgi, R., Kollman, P. A., and Van Gunsteren, W. F. (2002). Simulating proteins at constant pH: an approach combining molecular dynamics and Monte Carlo simulation. *Proteins* **47,** 469–480.

Burgi, R., Pitera, J., and van Gunsteren, W. F. (2001). Assessing the effect of conformational averaging on the measured values of observables. *J. Biomol. NMR.* **19,** 305–320.

Chalaoux, F. R., O'Donoghue, S. I., and Nilges, M. (1999). Molecular dynamics and accuracy of NMR structures. *Proteins* **34,** 453–463.

Charifson, P. S. (1997). "Practical Application of Computer-Aided Drug Design." Marcel Dekker, New York.

Chen, X., and Gilson, M. K. (2002). The binding database: overview and user's guide. *Biopolymers* **61,** 97–98.

Cieplak, P., Caldwell, J., and Kollman, P. A. (2001). Molecular mechanical models for organic and biological systems going beyond the atom-centered two-body additive approximation: aqueous free energies of methanol and N-methyl acetamide, nucleic acid base, and amide hydrogen bonding and chloroform/water partition coefficients of nucleic acid bases. *J. Comput. Chem.* **22,** 1048–1057.

Cohen, B. E., McAnaney, T. B., Park, E. S., Jan, Y. N., Boxer, S. G., and Jan, L. Y. (2002). Probing protein electrostatics with a synthetic fluorescent amino acid. *Science* **296,** 1700–1703.

Cooper, T. M., and Woody, R. W. (1990). The effect of conformation on the CD of interacting helices: a theoretical study of tropomyosin. *Biopolymers* **30,** 657–676.

Cornell, W., and Louise-May, S. (1998). Normal modes. "Encyclopedia of Computational Chemistry." John Wiley & Sons, Chichester.

Cornell, W. D., Cieplak, P., Bayly, C. I., Gould, I. R., Merz, K. M., Ferguson, J., Spellmeyer, D. M., Fox, T., Caldwell, J. W., and Kollman, P. A. (1995). A second generation force field for the simulation of proteins, nucleic acids and organic molecules. *J. Amer. Chem. Soc.* **117,** 5179–5197.

Darden, T., York, D., and Pedersen, L. (1993). Particle mesh Ewald: an N log (N) method for Ewald sums in large systems. *J. Chem. Phys.* **98,** 10089–10092.

Daura, X., Glattli, A., Gee, P., Peter, C., and van Gunsteren, W. E. (2002). Unfolded states of peptides. *Adv. Protein Chem.* **62,** 341–360.

DeTar, D. F. (2001). Thermochemical values of oxygen-containing compounds from ab initio calculations. *J. Phys. Chem. A* **105,** 2073–2084.

Dill, K. (1985). Protein folding on lattices. *Biochemistry* **24,** 1501.

Dill, K. A., Alonso, D. O. V., and Hutchinson, K. (1989). Thermal stabilities of globular proteins. *Biochemistry* **28,** 5439–5449.

Dill, K. A., and Chan, H. S. (1997). From Levinthal to pathways to funnels. *Nature Structural Biology* **4,** 10–19.

Diller, D. J., and Merz, K. M. Jr. (2001). High throughput docking for library design and library prioritization. *Proteins* **43,** 113.

Duan, Y., and Kollman, P. A. (1998). Pathways to a protein folding intermediate observed in a 1-microsecond simulation in aqueous solution. *Science* **282,** 740–744.

Eisenberg, D., and Kauzmann, W. (1969). "The Structure and Properties of Water" Oxford University Press, London.

Fersht, A. R., and Daggett, V. (2002). Protein folding and unfolding at atomic resolution. *Cell* **108,** 573–582.

Fersht, A. R. (2002). On the simulation of protein folding by short time scale molecular dynamics. *Proc. Natl. Acad. Sci. USA* **99,** 14122–14125.

Gogonea, V., Suarez, D., van der Vaart, A., and Merz, K. M. Jr. (2001). New developments in applying quantum mechanics to proteins. *Curr. Opin. Struct. Biol.* **11,** 217–223.

Gorfe, A. A., Ferrara, P., Caflisch, A., Marti, D. N., Bosshard, H. R., and Jelesarov, I. (2002). Calculation of protein ionization equilibria with conformational sampling. *Proteins* **46,** 41–60.

Gorler, A., Ulyanov, N., and James, T. L. (2000). Determination of the populations and structures of multiple conformers in an ensemble by NMR. *J. Biomol. NMR* **16,** 147–164.

Gouda, H. I., Kuntz, D., Case, D. A., and Kollman, P. A. (2002). Free energy calculations for theophylline binding to an RNA aptamer: comparison of MM-PBSA and thermodynamic integration methods. *Biopolymers* **68,** 16–34.

Hansch, C. (1969). A quantitative approach to biochemical structure-activity relationships. *Acc. Chem. Res.* **2,** 323.

Haque, T. S., Skillman, A. G., Lee, C. E., Habashita, H., Gluzman, I. Y., Ewing, T. J. A., Goldberg, D. E., Kuntz, I. D., and Ellman, J. A. (1999). Potent, low-molecular-weight non-peptide inhibitors of malarial aspartyl protease plasmepsin II. *J. Med. Chem.* **42,** 1428–1440.

Hardin, C., Eastwood, M. P., Prentiss, M., Luthey-Schulten, Z., and Wolynes, P. G. (2002). Folding funnels: the key to robust protein structure prediction. *J. Comput. Chem.* **23,** 138–146.

Havel, T. F. (1998). Distance geometry: theory, algorithms, and chemical applications. *In* "Encyclopedia of Chemical Computations." (P. V. R. Schyler, Ed.), J. Wiley & Sons.

Head, M. S., Given, J. A., and Gilson, M. K. (1997). "Mining Minima"; direct computation of conformational free energy. *J. Phys. Chem. A* **101,** 1609.

Helms, V. (2002). Electronic excitations of biomolecules studied by quantum chemistry. *Curr. Opin. Struct. Biol.* **12,** 169–175.

Hendrickson, W. A., and Konnert, J. H. (1981). Stereochemically restrained crystallographic least-squares refinement of macromolecule structures. *In* "Biomolecular Structure, Conformation, Function, and Evolution." (E. S. N. Y. R. Srinivasan, Ed.), pp. 43–57. Pergamon Press, Oxford.

Higo, J., and Nakasako, M. (2002). Hydration structure of human lysozyme investigated by molecular dynamics simulation and cryogenic x-ray crystal structure analyses. *J. Comput. Chem.* **23,** 1323–1336.

Hoffman, L. R., Kuntz, I. D., and White, J. M. (1997). Structure-based identification of an inducer of the low-pH conformational change in the influenza hemagglutinin: irreversible inhibition of infectivity. *J. Virol.* **71,** 8808–8820.

Holloway, M. K., Wai, J. M., Halgren, T. A., Fitzgerald, P. M., Vacca, J. P., Dorsey, B. D., Levin, R. B., Thompson, W. J., Chen, J. J., and deSolms, S. J. (1995). A priori prediction of activity for HIV-1 protease inhibitors employing energy minimization in the active site. *J. Med. Chem.* **38,** 305–317.

Honig, B., and Nicholls, A. (1995). Classical electrostatics in biology and chemistry. *Science* **268,** 1144–1149.

Hopkins, S. C., Vale, R. D., and Kuntz, I. D. (2000). Inhibitors of kinesin activity from structure-based computer screening. *Biochemistry* **39,** 2805–2814.

Horn, D. M., Breuker, K., and McLafferty, F. W. (2001). Kinetic intermediates in the folding of gaseous protein ions characterized by electron capture dissociation mass spectrometry. *J. Amer. Chem. Soc.* **123,** 9792–9799.

Huo, S., Massova, I., and Kollman, P. A. (2002a). Computational alanine scanning of the 1:1 human growth hormone-receptor complex. *J. Comput. Chem.* **23,** 15–27.

Huo, S., Wang, J., Cieplak, P., Kollman, P. A., and Kuntz, I. D. (2002b). Molecular dynamics and free energy analysis of cathepsin D-inhibitor interactions: insight into structure-based ligand design. *J. Med. Chem.* **45,** 1412–1419.

Jacobson, M. P., Friesner, R. A., Xiang, Z., and Honig, B. (2002). On the role of the crystal environment in determining protein side-chain conformations. *J. Mol. Biol.* **320,** 597.

Jorgensen, W. L., Maxwell, E. S., and Tirado-Rives, J. (1996). Development and testing of the OPLS all-atom force field on conformational energetics and properties of organic liquids. *J. Amer. Chem. Soc.* **118,** 11225–11236.

Kaminski, G. A., Stern, H. A., Berne, B. J., Friesner, R. A., Cao, Y. X., Murphy, R. B., Zhou, R., and Halgren, T. A. (2002). Development of a polarizable force field for proteins via ab initio quantum chemistry. *J. Comput. Chem.* **23,** 1515–1531.

Kick, E. K., Roe, D. C., Skillman, A. G., Liu, G., Ewing, T. J. A., Sun, Y., Kuntz, I. D., and Ellman, J. A. (1997). Structure-based design and combinatorial chemistry yield nanomolar inhibitors of cathepsin D. *Chem. Biol.* **4,** 297–307.

Kitson, D. H., and Hagler, A. T. (1988). Theoretical studies of the structure and molecular dynamics of a peptide crystal. *Biochemistry* **27,** 5246–5257.

Kleywegt, G. J. (1999). Experimental assessment of differences between related protein crystal structures. *Acta Cryst. D* **55,** 1878–1884.

Kleywegt, G. J. (2000). Validation of protein crystal structures. *Acta Cryst. D* **56,** 249–265.

Knegtel, R. M. A., and Wagener, M. (1999). Efficacy and selectivity in flexible database docking. *Proteins* **37,** 334.

Kollman, P. A. (1996). Advances and continuing challenges in achieving realistic and predictive simulations of the properties of organic and biological molecules. *Accounts of Chem. Res.* **29,** 461–469.

Kuhn, B., and Kollman, P. A. (2000). Binding of a diverse set of ligands to avidin and streptavidin: an accurate quantitative prediction of relative affinities by a combination of molecular mechanics and continuum solvent models. *J. Med. Chem.* **43,** 3786–3791.

Lazaridis, T., and Karplus, M. (1999). Effective energy function for proteins in solution. *Proteins* **35,** 133–152.

Leach, A. R. (2001). "Molecular Modeling: Principles and Applications." 2nd ed. Prentice Hall.

Leavitt, S., and Freire, E. (2001). Direct measurement of protein binding energetics by isothermal titration calorimetry. *Curr. Opin. Struct. Biol.* **11,** 560–566.

Leckland, D., and Israelachvili, J. (2001). Intermolecular forces in biology. *Quart. Rev. Biophysics* **34,** 105–267.

Lecomte, C. (1995). Experimental electron densities of molecular crystals and calculation of electrostatic properties from high resolution x-ray diffraction. *Adv. Mol. Struct. Res.* **1,** 261–302.

Linge, J. P., and Nilges, M. (1999). Influence of non-bonded parameters on the quality of NMR structures. *J. Biomol. NMR* **13,** 51–59.

Lo, K. K-W., Wong, L-L., and Hill, H. A. O. (1999). Surface-modified mutants of Cytochrome p450-cam. *FEBS Lett.* **451,** 342.

Lu, S. M., Lu, W., Qasim, M. A., Anderson, S., Apostol, I., Ardelt, W., Bigler, T., Chiang, Y. W., Cook, J., James, M. N. G., Kato, I., Kelly, C., Kohr, W., Komiyama, T., Lin, T.-Y., Ogawa, M., Otlewski, J., Park, S.-J., Qasim, S., Ranjbar, M., Tashiro, M., Warne, N., Whatley, H., Wieczorek, A., Wieczorek, M., Wilusz, T., Wynn, R., Zhang, W., and Laskowski, M. (2001). Predicting the reactivity of proteins from their sequence alone: Kazal family of protein inhibitors of serine proteinases. *PNAS* **98,** 1410–1415.

Lu, W., Apostol, I., Qasim, M. A., Warne, N., Wynn, R., Zhang, W. L., Anderson, S., Chiang, Y. W., Ogin, E., Rothberg, I., Ryan, K., and Laskowski, M. Jr. (1997). Binding of amino acid side-chains to S1 cavities of serine proteinases. *J. Mol. Biol.* **266,** 441–461.

Luo, H., and Sharp, K. A. (2002). On the calculation of absolute macromolecular binding free energies. *Proc. Natl. Acad. Sci. USA* **99,** 10399.

Luzatti, V. (1952). The statistical treatment of errors in crystal structures. *Acta Cryst.* **5,** 802–810.

Mackerrell, A. D. Jr., Banavali, N., and Foloppe, N. (2001). Development and current status of the CHARMM force field for nucleic acids. *Biopolymers* **56,** 257–265.

Mahoney, M. W., and Jorgensen, W. L. (2001). Quantum, intramolecular flexibility and polarizability effects on the reproduction of the density anomaly of liquid water by simple potential functions. *J. Chem. Phys.* **115,** 10758–10768.

Martell, P. J., Soares, C. M., Baptista, A. M., Fuxreiter, M., Naray-Szabo, G., Louro, R. O., and Carronde, M. A. (1999). Comparative redox and pK calculations in cytochrome c3. *J. Biol. Inorgan. Chem.* **4,** 73–86.

Martin, E. J., Blaney, J. M., Siani, M. A., Spellmeyer, D. C., Wong, A. K., and Moos, W. A. (1995). Measuring diversity: experimental design of combinatorial libraries for drug discovery. *J. Med. Chem.* **38,** 1431–1436.

Network, E. U. (1998). 3-D Validation. Who checks the checkers? *J. Mol. Biol.* **276,** 417–436.

Onufriev, A., Case, D. A., and Bashford, D. (2002). Effective Born radii in the generalized born approximation: the importance of being perfect. *J. Comput. Chem.* **23,** 1297–1304.

Oppenheimer, N. J., and James, T. L. (1989). Nuclear magnetic resonance. *In* "Methods in Enzymology." (J. N. Abelson and M. I. Simon, Eds.), Academic Press, San Diego.

Palmer, A. G., III. (2001). NMR probes of molecular dynamics: overview and comparison with other techniques. *Annu. Rev. Biophys. Biomol. Struct.* **30,** 129–155.

Pellegrini, M., Gronbeck-Jensen, N., Kelly, J. A., Pfluegl, G. M. V., and Yeates, T. O. (1997). Highly constrained multiple copy refinement of protein structures. *Proteins* **29,** 426–432.

Peyerimhoff, S. D. (1998). Spectroscopy: computational methods. *In* "Encylcopedia of Computational Chemistry" (P. V. R. Schleyer, Ed.), John Wiley & Sons, Chichester.
Rabitz, H. (1989). Systems analysis at the molecular scale. *Science* **246,** 221–226.
Rader, S. D., and Agard, D. A. (1997). Conformational substates in enzyme mechanism: the 120 K structure of alpha-lytic protease at 1.5 A resolution. *Protein Science* **36,** 1375–1386.
Reyes, C., Nifosi, R., Frankel, A., and Kollman, P. A. (2001). Molecular dynamics and binding specificity analysis of the Bit Tat-TAR complex. *Biophys. J.* **80,** 2833–2842.
Roche, O., Kiyama, R., and Brooks, C. L., III. (2001). Ligand-protein database: linking protein-ligand complex structure to binding data. *J. Med. Chem.* **44,** 3592–3598.
Salomaa, A. (1991). A determinate algorithm for modular knapsack problems. *Theor. Comput. Sci.* **88,** 127–138.
Schleyer, P. von Ragué (1998). *In* "Encyclopedia of Computational Chemistry." (P. V. R. Schleyer, Ed.), John Wiley & Sons, Chichester.
Schuler, B., Lipman, E. A., and Eaton, W. A. (2002). Probing the free-energy surface for protein folding with single-molecule fluorescence spectroscopy. *Nature* **419,** 743–747.
Shafer, H., Daura, X., Mark, A. E., and van Gunsteren, W. F. (2001). Entropy calculations on a reversibly folding peptide. *Proteins* **43,** 45–56.
Shea, J. E., and Brooks, C. L. (2001). From folding theories to folding proteins: a review and assessment of simulation studies of protein folding and unfolding. *Annu. Rev. Phys. Chem.* **52,** 499–535.
Sheldrick, G. M., and Schneider, T. R. (1997). High resolution refinement. *Methods Enzymol.* **277,** 319–343.
Silverstein, K. A. T., Haymet, A. D. J., and Dill, K. A. (1999). A simple model of water and the hydrophobic effect. *J. Amer. Chem. Soc.* **120,** 3166–3175.
Smith, L. J., Daura, X., and van Gunsteren, W. F. (2002). Assessing equilibration and convergence in biomolecular simulations. *Proteins* **15,** 487–496.
Snow, C. D., Nguyen, H., Pande, V. S., and Gruebele, M. (2002). Absolute comparison of simulated and experimental protein-folding dynamics. *Nature* **420,** 102–106.
Srinivasan, J., Cheatham, T. E., Kollman, P. A., and Case, D. A. (1998). Continuum solvent studies of the stability of DNA, RNA and phosphoramidate-DNA helices. *J. Am. Chem. Soc.* **120,** 9401–9409.
Stahle, L., and Wold, S. (1988). Multivariate data analysis and experimental design in biomedical research. *Prog. Med. Chem.* **25,** 291–338.
Still, W. C., Tempczyk, A., Hawley, R. C., and Hendrickson, T. (1990). Semianalytical treatment of solvation for molecular mechanics and dynamics. *J. Am. Chem. Soc.* **112,** 6127–6129.
Straatsma, T. P., and McCammon, J. A. (1992). Free energy perturbation. *Annu. Rev. Phys. Chem.* **43,** 407–435.
Stroud, R. M., and Fauman, E. B. (1995). Significance of structural changes in proteins: expected errors in refined protein structures. *Protein Science* **4,** 2392–2404.
Sullivan, D. C., and Kuntz, I. D. (2002). Protein folding as biased conformational diffusion. *J. Phys. Chem. B* **106,** 3255–3262.
Tantillo, D. J., Chen, J., and Houk, K. N. (1998). Theozymes and compuzymes: theoretical models for biological catalysis. *Curr. Opin. Chem. Biol.* **2,** 743–750.
Tickle, I. J., Laskowski, R. A., and Moss, D. S. (1998). Error estimates of protein structure coordinates. *Acta Cryst. D* **54,** 243–252.
Tramontano, A., Leplae, R., and Morea, V. (2001). Analysis and assessment of comparative modeling predictions in CASP4. *Proteins* **Suppl. 5,** 22–38.

Tsui, V., and Case, D. A. (2000). Molecular dynamics simulations of nucleic acids using a generalized Born solvation model. *J. Am. Chem. Soc.* **122,** 2489–2498.

Wang, W., and Kollman, P. A. (2002). Computational study of the molecular basis of the HIV-1 protease drug resistance. *Proc. Natl. Acad. Sci.* In press.

Weiss, G. H., and Ferretti, J. A. (1985). The choice of optimal parameters for measurement of spin-lattice relaxation times. III. Mathematical preliminaries for nonideal pulses. *J. Magn. Reson.* **61,** 490–498.

Wilson, C., Mace, J. E., and Agard, D. A. (1991). A computational method for the design of enzymes with altered substrate specificity. *J. Mol. Biol.* **220,** 495–506.

Wilson, R. J. (1972). "Introduction to Graph Theory." Longman, Harlow, UK.

Wold, S. (1991). Chemometrics. *J. Pharm. Biomed. Anal.* **9,** 589–596.

Word, J. M., Lovell, S. C., LaBean, T. H., Taylor, H. C., Zalis, M. E., Presley, B. K., Richardson, J. S., and Richardson, D. C. (1999). Visualizing and quantifying molecular goodness-of-fit. *J. Mol. Biol.* **285,** 1711–1733.

Xue, L. L., Wang, Y. H., Xie, Y., Yao, P., Wang, W. H., Qian, W., Huang, Z. X., Wu, J., and Xia, Z. X. (1999). Effect of mutation at valine 61 on the three-dimensional structure, stability and redox potential of cytochrome b5. *Biochem.* **38,** 11961–11972.

Young, M. M., Tang, N., Hempel, J. C., Oshiro, C. M., Taylor, E. W., Kuntz, I. D., Gibson, B. W., and Dollinger, G. (2000). High throughput protein fold identification by using experimental constraints derived from intramolecular cross-links and mass spectrometry. *Proc. Nat. Acad. Sci. USA* **97,** 5802–5806.

Zhou, R., Freisner, R. A., Ghosh, A., Rizzo, R. C., Jorgensen, W. L., and Levy, R. M. (2001). New linear interaction method for binding affinity calculations using a continuum solvent model. *J. Phys. Chem. B* **105,** 10388.

Zou, X., Sun, Y., and Kuntz, I. D. (1999). Inclusion of solvation in ligand binding free energy calculations using the generalized-Born model. *J. Amer. Chem. Soc.* **121,** 8033–8043.

Zucker, M. (1989). On finding all suboptimal foldings of an RNA molecule. *Science* **244,** 48–52.

FORCE FIELDS FOR PROTEIN SIMULATIONS

By JAY W. PONDER* AND DAVID A. CASE†

*Department of Biochemistry and Molecular Biophysics, Washington University School of Medicine, St. Louis, Missouri 63110, and †Department of Molecular Biology, The Scripps Research Institute, La Jolla, California 92037

I. Introduction 27
II. Protein Force Fields, 1980 to the Present 30
A. The Amber Force Fields 30
B. The CHARMM Force Fields 35
C. The OPLS Force Fields 38
D. Other Protein Force Fields 39
E. Comparisons Among Protein Force Fields 41
III. Beyond Fixed Atomic Point-Charge Electrostatics 45
A. Limitations of Fixed Atomic Point-Charges 46
B. Flexible Models for Static Charge Distributions 48
C. Including Environmental Effects via Polarization 50
D. Consistent Treatment of Electrostatics 52
E. Current Status of Polarizable Force Fields 57
IV. Modeling the Solvent Environment 62
A. Explicit Water Models 62
B. Continuum Solvent Models 64
C. Molecular Dynamics Simulations with the Generalized Born Model 75
V. Conclusions 77
References 78

I. Introduction

Molecular dynamics and Monte Carlo simulations of proteins, which began about 25 years ago, are by now widely used as tools to investigate their structure and dynamics under a wide variety of conditions, ranging from studies of ligand binding and enzyme-reaction mechanisms to problems of denaturation and protein re-folding. Fundamental to such simulations is the representation of the energy of the protein as a function of its atomic coordinates. The states expected to be populated at thermal equilibrium are the low-energy regions of this (potential) energy function, and forces on individual atoms are related to the gradient of this function, which is why such functions are also commonly referred to as "force fields."

Atomistic simulations of the properties of proteins commonly consider an average over the much-faster electronic motions, so that the energy surface on which the atoms move is the Born-Oppenheimer ground-state

ADVANCES IN PROTEIN CHEMISTRY, Vol. 66

0065-3233/03 $35.00

energy. It is not yet feasible to calculate directly such surfaces for macromolecules with high accuracy by means of quantum chemistry electronic structure calculations, so most practical simulations use a set of simple classical functions to represent the energy, adjusting a large number of parameters to optimize agreement with experimental data and with quantum calculations on smaller molecules. The design and parameterization of force fields for use in protein simulations is a complex task, involving many decisions concerning which data to emphasize in the fits, expectations of transferability to areas outside the "fit set," and computational efficiency. Our goal here is to provide an overview of the thinking that goes into the choices that must be made, a general description of the sorts of force fields that are most commonly used at present, and an indication of the directions of current research that may yield better functions in the near future. Although we hope the discussion can be read with profit by non-specialists, we will assume a general familiarity with molecular mechanics and dynamics and with their applications to proteins. Simulation methods for proteins are by now well-enough established that there are many good textbooks and monographs that cover the basics (Burkert and Allinger, 1982; Harvey and McCammon, 1987; Leach, 2001; Becker *et al.*, 2001; Cramer, 2002; Schlick, 2002).

It is impossible for a short review like this one to be comprehensive, and we have chosen to concentrate on a few areas, necessarily at the expense of others. We will restrict our discussion to simulations of proteins in water, leaving aside the interesting questions of how to deal with the large variety of small molecules that interact with proteins or with non-aqueous solvents. With perhaps less justification, we have chosen to focus on a few force fields that are very widely used, and even here it is not possible to give any real account of the differences among potentials. With regard to newer developments, we have chosen to concentrate on two main areas: the use of continuum methods to model the electrostatic effects of hydration, and the introduction of polarizability to model the electronic response to changes in the environment. These are not the only recent developments worthy of note, but we do expect them to be keys areas of research for the next few years.

There is much current development, and other good reviews of force fields (MacKerell, 2001; Hünenberger and van Gunsteren, 1997; Damm and Halgren, 2001). Hence, after a brief survey of current models, mostly generated during the 1990s, our review will focus on the general directions the field is taking in developing new models.

The most commonly used protein force fields incorporate a relatively simple (sometimes called "Class I") potential energy function:

$$V(r) = \sum_{bonds} k_b(b - b_0)^2 + \sum_{angles} k_\theta(\theta - \theta_0)^2 + \sum_{torsions} k_\phi[\cos(n\phi + \delta) + 1] + \sum_{\substack{nonbond\\pairs}} \left[\frac{q_i q_j}{r_{ij}} + \frac{A_{ij}}{r_{ij}^{12}} - \frac{C_{ij}}{r_{ij}^{6}} \right] \tag{1}$$

The first three summations are over bonds (1-2 interactions), angles (1-3 interactions), and torsions (1-4 interactions). These are illustrated in Fig. 1. The torsion term can also include so-called "improper" torsions, where the four atoms defining the angle are not all connected by covalent bonds; such terms serve primarily to enforce planarity around sp^2 central atoms, and other functional forms can be used for the same purpose. The final sum (over pairs of atoms i and j) excludes 1-2 and 1-3 interactions and often uses separate parameters for 1-4 interactions as compared with those used for atoms separated by more than three covalent bonds. It describes electrostatics that use partial charges q_i on each atom that interact via Coulomb's law. The combination of dispersion and exchange repulsion forces are represented by a Lennard-Jones 6-12 potential; this is often called the "van der Waals" term.

Equation 1 is about the simplest potential energy function that can reproduce the basic features of protein energy landscapes at an atomic level of detail, and it has proved to give insight into a remarkably broad range of properties. The combination of a potential energy function (as in equation 1) and all the parameters that go into it (k_b, b_0, k_θ, θ_0, *etc.*) constitutes a "force field." In practice, there is a close connection between the force fields and the computer codes that implement them, although this is becoming less true now than it used to be.

In section II, we review some of the history and performance of widely used protein force fields based on equation 1, or closely related equations. Sections III and IV outline some promising developments that go beyond this, primarily by altering the way electrostatics interactions are treated.

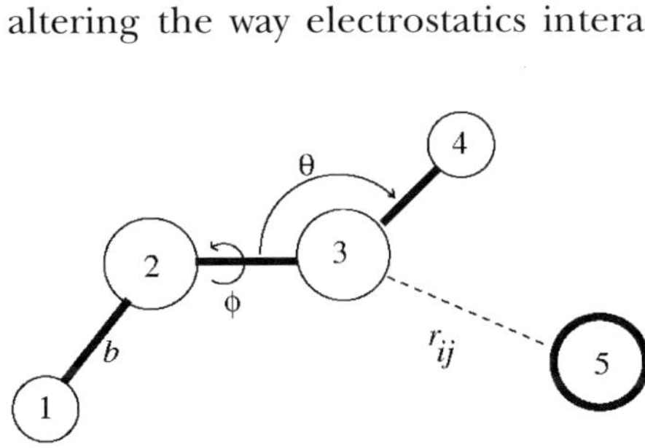

FIG. 1. Schematic view of force field interactions. Covalent bonds are indicated by heavy solid lines, nonbonded interactions by a light, dashed line.

Section III considers the use of atomic multipoles and off-center charge distributions, as well as attempts to incorporate electronic polarizability. Rather than making the electrostatic model more complex, Section IV considers continuum solvent models that simplify the problem by accounting for all solvent (water plus mobile counterions) effects in an averaged fashion.

II. Protein Force Fields, 1980 to the Present

There have been a large number of force fields used over the years for simulations of proteins, and it is not our intent to try to provide any sort of comprehensive review. Sections A through C provide an overview of three sets of parameters that have been widely used. The discussion emphasizes the decisions that went into their development. A very short summary of other important protein force fields is presented in section D, and a brief comparison of the results from different force fields is given in section E. Many of the more recent developments in protein force fields address electronic polarizability or the effects of solvation; these are discussed in sections III and IV, respectively.

Our historical discussion of protein force fields begins roughly in 1980, when molecular dynamics and Monte Carlo simulations of proteins were in their infancy. It is important to note, however, that the developments outlined below did not arise *sui generis* but rather were built on earlier force field developments in organic chemistry. Of particular importance were the ECEPP potentials from Scheraga and co-workers (Momany *et al.*, 1975; Némethy *et al.*, 1983) and the consistent force field (CFF) developments from the Lifson group (Lifson and Warshel, 1969; Hagler *et al.*, 1974; Hagler and Lifson, 1974; Niketic and Rasmussen, 1977). These provided important starting points for the work described here, as did efforts to develop potential energy functions in the general area of organic chemistry (Allinger, 1976; Burkert and Allinger, 1982).

We hope that our introduction below to three popular force fields will help to prepare readers to appreciate a literature that is much broader than we can cover here. Other good reviews exist (MacKerell, 2001; Hüenberger and van Gunsteren, 1997) that complement our presentation.

A. *The Amber Force Fields*

By the first half of the 1980s, enough experience had accumulated with earlier parameterizations for several groups to begin fairly systematic projects to develop a new generation of force fields. The earliest of these efforts were still done at a time when the limited power of computers made it attractive to not include all hydrogen atoms as explicit force centers. The importance of hydrogen bonding, however, led many investigators to adopt a compromise whereby polar hydrogens were explicitly represented but hydrogens bonded to carbon were combined into united atoms. A widely used force field at this level was developed in 1984 in the Kollman group (Weiner *et al.*, 1984) and incorporated into the Amber molecular mechanics package, which was at an early stage of development as well (Weiner and Kollman, 1981). The key ideas in this initial work were to be used repeatedly in later efforts by this group. Charges were derived from quantum chemistry calculations at the Hartree-Fock STO-3G level, via fitting of partial atomic charges to the quantum electrostatic potential; these are generally called ESP (for electrostatic potential) charges. The van der Waals terms were adapted from fits to amide crystal data by Lifson's group (Hagler *et al.*, 1974; Hagler and Lifson, 1974) and from liquid-state simulations pioneered by Jorgensen (Jorgensen, 1981). Force constants and idealized bond lengths and angles were taken from crystal structures and adapted to match normal mode frequencies for a number of peptide fragments. Finally, torsion force constants were adjusted to match torsional barriers extracted from experiment or from quantum chemistry calculations. Since it is only the total potential energy, as a function of torsion angle, that needs to agree with the target values, and since (in this potential energy function) these barriers have significant electrostatic and van der Waals interactions between the end atoms (the so-called '1-4' interactions), the k_ϕ values are closely coupled to the nonbonded potentials used and are hardly transferable from one force field to another.

Three problems with this "polar hydrogen only" approach, along with improvements in the speed of available computers, led many groups to move to an all-atom approach. First, aromatic rings such as benzene have a significant quadrupolar charge distribution, with an effective positive charge near the hydrogens and an effective negative charge nearer to the middle of the ring. This effect can be crucial in determining the ways in which aromatic side chains in proteins interact with other groups. For example, "T-shaped" geometries between rings are stabilized relative to "stacked" geometries that optimize van der Waals interactions (Williams, 1991, 2001). Also important are π-cation interactions, where positive groups are found directly above the centers of aromatic rings (Dougherty, 1995; Zacharias and Dougherty, 2002). Second, the forces that affect the pseudorotation between conformations, or "pucker," of five-member aliphatic rings (Tomimoto and Go, 1995) are difficult to describe when only the heavy atoms are available as force centers. This affects only proline residues in proteins, but analogous problems involving ribose and deoxyribose in nucleic acids (whose force fields were being parameterized at the same time) led momentum toward all-atom force fields. Finally, it is difficult with united atom models to make comparisons between computed and observed vibrational frequencies. An extension of the 1984 force field to an all-atom model was published in 1986, as a collaboration between the Kollman and Case groups (Weiner *et al.*, 1986). Both the 1984 and 1986 parameter sets were primarily developed based on experience with gas phase simulations.

The continued increase in the speed of computers led the Kollman group to decide in the early 1990s that a new round of force-field development was warranted; this came to be known as the ''Cornell *et al.*'' or *ff94* force field (Cornell *et al.*, 1995). In addition to improvements in the parameters, a more serious attempt was made to explicitly describe the algorithm by which the parameters were derived, so that consistent extensions could be made to molecules other than proteins (Fox and Kollman, 1998). This goal was not really achieved until the development almost a decade later of the *antechamber* program that completely automates all of the steps in the creation of an Amber-like force field for an arbitrary molecule or fragment.

A key motivation for this development was a desire to produce potentials suitable for condensed phase simulations, since the earlier work had concentrated in large part on gas phase behavior. In particular, the ways in which the OPLS potentials (discussed below) had been parameterized to reproduce the densities and heats of vaporization of neat organic liquids was very influential, along with recognition of the importance of having a balanced description of solute-solvent versus solvent-solvent interactions. A second point arose from the ability to use larger basis sets and fragment sizes to determine atomic charges that mimic the electrostatic potentials outside the molecule found from quantum mechanical calculations. Earlier work had established that fitting charges to the potentials at the Hartree-Fock 6-31G* level tended to overestimate bond-dipoles (compared with observed gas phase values) by amounts comparable to that in empirical water models such as SPC/E or TIP3P; such ''overpolarization'' is an expected consequence of electronic polarization in liquids. Hence, the use of fitted charges at the HF/6-31G* level appeared to offer a general procedure for quickly developing charges for all 20 amino acids in a way that would be roughly consistent with the water models that were expected to be used. Tests of this idea, with liquid-state simulations of amides and simple hydrocarbons, gave encouraging results.

The actual implementation of this scheme for developing charges had to deal with two complications, which continue to plague force field

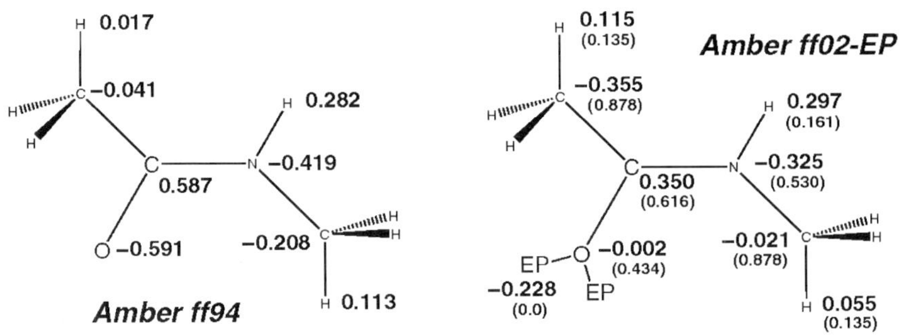

FIG. 2. Charge models for the Amber potentials. (Left) HF-6-31G*RESP charges in the style of *Amber ff94*. (Right) polarizable, extra-point charge model in the style of *Amber ff02-EP* with the atomic polarizabilities in Å^3 given in parentheses.

developers to the present day. First, the effective charges of the more-buried atoms are often underdetermined, so that charges for atoms in similar environments in different molecules might vary significantly. In effect, there are many combinations of atomic charges that will fit the electrostatic potential almost equally well. There are a variety of ways to overcome this problem, often involving statistical techniques based on singular-value decomposition, but Bayly *et al.* (Bayly *et al.*, 1993; Cornell *et al.*, 1993) chose to use a hyperbolic restraint term to limit the absolute magnitude of charges on non-hydrogen atoms. This is called RESP (for restrained electrostatic potential fit) and weakly favors solutions with smaller charges for buried atoms, yielding fairly consistent charge sets with little degradation in the quality of the fit to the electrostatic potential outside the molecule. As an example, the left-hand side of Fig. 2 shows the charges determined in this way for *N*-methylacetamide, modeling the peptide bond; the right-hand side of this figure is for a more complex electrostatic model, described below.

A second and more fundamental problem with the RESP procedure is that the resulting charges depend on molecular conformation, often in significant ways. This is a manifestation of electronic polarizability, which can only be described in a very averaged way if fixed atomic charges are to be used. Any real solution to this problem must involve a more complex model, such as those described in section III, below. The compromise chosen for the *ff94* force field was to fit charges simultaneously to several conformations, in the hopes of achieving optimal averaged behavior.

Once the charges and the "stiff" internal parameters for bonds and angles were available (the latter estimated in the same way as outlined above), the Lennard-Jones parameters could be established primarily by reference to densities and heats of vaporization in liquid-state simulations. Only a small number of sets of 6-12 parameters were necessary to achieve reasonable agreement with experiment. A key expansion from earlier work was the notion that parameters for hydrogens should depend in an important way on the electronegativity of the atoms they are bonded to (Gough *et al.*, 1992; Veenstra *et al.*, 1992).

As with many other force-field projects, the final parameters to be fit were the "soft" torsional potentials about single bonds. It makes some sense to address these after the charges and Lennard-Jones parameters have been developed, since the energy profile for rotation about torsion angles depends importantly on the nonbonded interactions between the moving groups at the ends, as well as on whatever intrinsic torsional potential is assigned. The question of how best to partition torsional barriers into "bonded" versus "nonbonded" interactions is a thorny one, and many developers of force fields have adopted a strictly empirical approach, fitting k_ϕ, n, and δ so that the total profile (including the nonbonded terms) matches some target extracted from quantum mechanics or from experiment.

A key set of torsional parameters are those for the ϕ and ψ backbone angles, since these affect every amino acid residue and heavily influence the relative energies of helices, sheets, and turns in proteins. The *ff94* parameters were fit to representative points on the dipeptide maps for glycine and alanine, computed at the MP2 level with a triple-ζ + polarization (TZP) basis set. This is not an unreasonable choice for a target function, but it has a number of intrinsic difficulties. First, the α-helix region near $\phi,\psi = -60,-40$ is not a minimum for a gas-phase dipeptide, so fitting just a representative point (as was done) can lead to errors in the surface as a whole, compared to the full MP2/TZP surface. More importantly, the use of a gas phase dipeptide model as a target ignores both the non-local electronic structure contributions that would be seen in larger fragments (Beachy *et al.*, 1997) and the polarization effects inherent in a condensed phase environment (Ösapay *et al.*, 1996). Some account of the longer-range effects was provided in subsequent parameterizations, referred to as *ff96* (Kollman *et al.*, 1997) and *ff99* (Wang *et al.*, 2000), in which the ϕ and ψ potentials were fit to tetrapeptide as well as dipeptide quantum mechanical conformational energies. These later fits provided potential surfaces that were significantly different from those in *ff94*, but it was hard to tell if physical realism was really being improved.

In recent years it has become computationally feasible to test protein potentials (and especially their backbone torsion angle behavior) by carrying out converged or nearly converged simulations on short peptides and comparing the resulting conformational populations to those derived from experiment (Damm and van Gunsteren, 2000; Mitsutake *et al.*, 2001; García and Sanbonmatsu, 2001). The experimental estimates, obtained mainly from circular dichroism or from NMR, are often only qualitative, but this can be enough to identify obvious errors in computed ensembles. For example, the *ff94* parameters appear to over-stabilize helical peptide conformers in many if not all instances. Computed melting temperatures for polyalanine helices are too high (García and Sanbonmatsu, 2002), and helical conformers can predominate in simulations of sequences that experimentally form other structures, such as β-hairpins. At least two modifications of the *ff94* ϕ and ψ potentials have been proposed and tested on large-scale peptide simulations (García and Sanbonmatsu, 2002; Simmerling *et al.*, 2002). It will be of interest to see how these ideas develop as a new generation of long time-scale peptide simulations becomes feasible.

B. The CHARMM Force Fields

As with Amber, the CHARMM program (Chemistry at HARvard using Molecular Mechanics) (Brooks *et al.*, 1983) was originally developed in the early 1980s and initially used an extended atom force field with no explicit hydrogens. By 1985, this had been replaced by the *CHARMM19* parameters, in which hydrogen atoms bonded to nitrogen and oxygen are explicitly represented, while hydrogens bonded to carbon or sulfur are treated as part of extended atoms (Reiher, 1985; Neria *et al.*, 1996). Key to the parameterization of this model were fits to quantum calculations at the HF/6-31G level of hydrogen bonded complexes between water and the H-bond donors or acceptors of the amino acids or fragments. This involves a series of supermolecular calculations of the model compound, such as formamide or *N*-methylacetamide and a single water molecule at each of several interaction sites. Before making the fits, the interaction energies are scaled by a factor of 1.16, which is the ratio of the water dimerization energy predicted by the TIP3P model to that predicted at the HF/6-31G level. As in the Amber parameterizations described above, the goal here was to obtain a balanced interaction between solute-water and water-water energies when the latter are represented by TIP3P. For peptides, it was found that fitting the peptide-water interactions in this way led to peptide-peptide hydrogen bonds that were also larger than HF/6-31G values by a factor very close to 1.16; in other cases, explicit fitting to solute-solute hydrogen bonded dimers may be needed for parameter generation (MacKerell, 2001).

As with the contemporaneous Amber 1984 united-atom parameterization, the *CHARMM19* values were developed and tested primarily on gas-phase simulations. However, the *CHARMM19* potential seems to do well (perhaps fortuitously) in solvated simulations and continues to be used for peptide and protein simulations; this is in contrast to the 1984 Amber force field, which is no longer widely used. In addition, the *CHARMM19* values have often been used in conjunction with a distance-dependent dielectric constant as a rough continuum solvation model.

In the early 1990s, the CHARMM development group also recognized the need to refine parameters more explicitly pointed to obtaining a good balance of interaction energies in explicit solvent simulations. The resulting *CHARMM22* protein force field was first included in the corresponding version of CHARMM, released in 1992, and was fully described a few years later (MacKerell *et al.*, 1998; MacKerell, 2001). The key approach from *CHARMM19* was carried over by deriving charge models primarily from fits to solute-water dimer energetics (now calculated at the HF/6-31G* level). In addition to fitting the dimer interaction energies, charges for model compounds were adjusted to obtain dipole moments somewhat larger than experimental or *ab initio* values. This has the same goal as the RESP procedure described above: bonds are expected to be more polarized in condensed phases than in the gas phase. The use of empirical charges that yield enhanced dipoles both reflects this behavior and allows a reasonably balanced set of interactions with the TIP3P water model, which has a similarly enhanced dipole moment.

Once the charges were determined by these dimer studies, the Lennard-Jones parameters were refined to reproduce densities and heats of vaporization of liquids as well as unit cell parameters and heats of sublimation for crystals. As with the Amber parameterization, generally only small adjustments from earlier values were required to fit the empirical data. Nevertheless, because of the steep dependence of these forces, such adjustments may be crucial for a well-balanced and successful set of parameters.

As with the Amber *ff94* force field, the torsional parameters were finally adjusted to target data derived from vibrational spectra and from *ab initio* calculations. The torsional potentials for the ϕ and ψ torsions were initially fit to HF/6-31+G* calculations on an analog of the alanine dipeptide in which the terminal methyl groups are replaced by hydrogen. These were then refined in an iterative procedure to improve the agreement with experiment of the backbone angles in simulations of myoglobin. In principle at least, this latter adjustment provides a way of correcting the *ab initio* dipeptide energy map for effects caused by the protein environment.

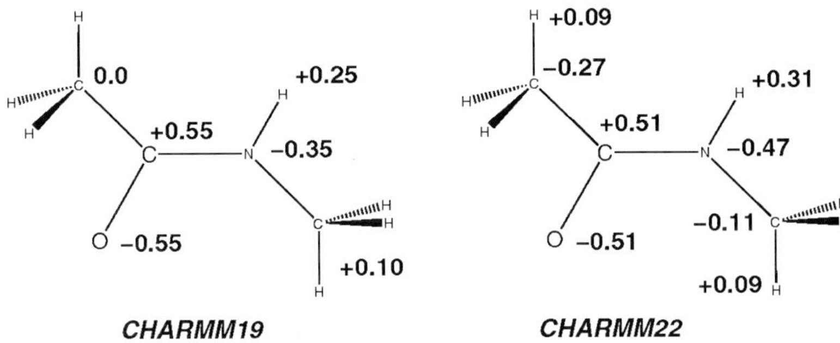

FIG. 3. Charge models for *N*-methylacetamide for two of the CHARMM force field parameterizations. In *CHARMM19*, the methyl groups are treated as united atoms, so that there is no breakdown into separate C and H charges.

As with the Amber parameterization, the question of how best to obtain good backbone torsional potentials is a vexing one, and studies are continuing, both at the dipeptide level and with solvated simulations of oligopeptides. Most recently, an extensive reworking of the nucleic acid parameters has resulted in the *CHARMM27* force field (Foloppe and MacKerell, 2000). However, the *CHARMM27* protein parameters are essentially identical to those from the *CHARMM22* force field.

One feature of the CHARMM parameterizations is the enforcement of *neutral groups*, which are small sets of contiguous atoms whose atomic charges are constrained to sum to zero. This is illustrated in Fig. 3, which shows partial atomic charges for *N*-methylacetamide in the *CHARMM19* and *CHARMM22* force fields. For example, charges for the C and O atoms of the peptide group form a small neutral group. These groups can be useful when truncating long-range electrostatic interactions: if an entire group is either included or ignored (or scaled by a smoothing factor), then there is never any splitting of dipoles. Ignoring charged side chains, each atom would then feel the electrostatic effects of a net neutral environment. The same behavior occurs with solvent molecules, if the interactions of a given water molecule are always treated as a group. Although it was long deemed plausible that such a group-based truncation scheme would yield better results than an atom-based scheme (wherein some members of a group might interact with a given atom, while others would not), this is probably not the case for most biomolecular simulations in water (Steinbach and Brooks, 1994; Darden *et al.*, 1998; Darden, 2001; Mark and Nilsson, 2002). In any event, such considerations are now much less important than in earlier times, since many current simulations use Ewald or fast multipole schemes to handle long-range electrostatics, where nothing is gained by having small neutral groups.

Comparison with the corresponding Amber charges from *ff94* (Fig. 2) shows that the peptide carbonyl group is somewhat less polar in *CHARMM22* than in Amber, whereas the opposite is true for the NH dipole. Still, the similarities in the charge models for these force fields are more striking than the differences.

C. *The OPLS Force Fields*

A third main development in the early 1980s (already alluded to above) involved potentials developed by Jorgensen and co-workers to simulate liquid state properties, initially for water and for more than 40 organic liquids. These were called OPLS (Optimized Potentials for Liquid Simulations) and placed a strong emphasis on deriving nonbonded interactions by comparison to liquid-state thermodynamics (Jorgensen, 1998). Indeed, the earliest applications of OPLS potentials were to rigid-molecule Monte Carlo simulations of the structure and thermodynamics of liquid hydrogen fluoride (Jorgensen, 1981). The reproduction of densities and heats of vaporization provides some confidence in both the size of the molecules and in the strengths of their intermolecular interactions. These early models (now called OPLS-UA) treated hydrogens bonded to aliphatic carbons as part of an extended atom but represented all other hydrogens explicitly.

The initial applications to proteins (Jorgensen and Swenson, 1985; Jorgensen and Tirado-Rives, 1988; Tirado-Rives and Jorgensen, 1990) used a polar-hydrogen–only representation, taking the atom types and the valence (bond, angle, dihedral) parameters from the 1984 Amber force field. This was called the AMBER/OPLS force field, and for some time was reasonably popular. As with Amber and CHARMM, an all-atom version (OPLS-AA) was developed later, but with much the same philosophy for derivation of charges and van der Waals parameters from simulations on pure liquids (Jorgensen *et al.*, 1996; Rizzo and Jorgensen, 1999; Kaminski *et al.*, 2001). Torsional parameters were developed in a consistent way by fits to HF/6-31G* energy profiles (Maxwell *et al.*, 1995), along with some recent modifications, especially for charged side chains (Kaminski *et al.*, 2001). Bond stretching and angle bending terms were standardized but were largely taken from the 1986 Amber all-atom force field. The parameter choices were intended to be "functional group friendly," so that they could be easily transferred to other molecules with similar chemical groupings. Although the parameters were principally derived with reference to condensed phase simulations, comparisons to gas-phase peptide energetics also show good results (Beachy *et al.*, 1997).

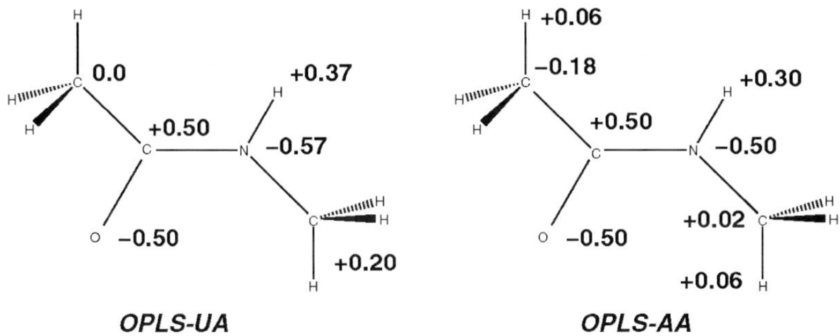

FIG. 4. Charge models for *N*-methylacetamide for two OPLS force field parameterizations. In OPLS-UA, the methyl groups are treated as united atoms, so that there is no breakdown into separate C and H charges.

Fig. 4 shows the fitted atomic charge models for the OPLS-UA and OPLS-AA force fields. Again, the polarity of the CO and NH bonds is similar to that in the Amber and CHARMM force fields. As with CHARMM, the OPLS charge fitting procedures yield small neutral groups.

D. *Other Protein Force Fields*

Our abbreviated discussion neglects many important advances in the development of protein force fields. In particular, there are several other protein potentials that have been widely used but will not be assessed in detail here. The GROMOS force fields (van Gunsteren *et al.*, 1998) were developed in conjunction with the program package of the same name (van Gunsteren and Berendsen, 1987; Scott *et al.*, 1999). The all-atom CEDAR and GROMACS force fields are largely derived from GROMOS. The Merck Molecular Force Field (MMFF) was developed by Halgren (Halgren, 1996a,b,c,d; Halgren and Nachbar, 1996; Halgren, 1999a,b), and has been aimed more at drug-like organic compounds than at proteins. MMFF was not derived for use in bulk phase simulations and performs poorly when used to model organic liquids (Kaminski and Jorgensen, 1996). This deficiency is not inherent in the buffered 14-7 function (Halgren, 1992) used in MMFF's van der Waals term, because this same functional form can be reparameterized to fit liquid data (Ren and Ponder, 2003). The DISCOVER force field (Maple *et al.*, 1998) has seen use primarily in conjunction with the commercial INSIGHT modeling package. The MM3 and MM4 potentials for amides (Lii and Allinger, 1991; Langley and Allinger, 2002) are an offshoot of

TABLE I
Other Force Fields for Peptide and Protein Modeling

Force field	Potential type	Key references
BUFF	All Atom	Carlson, 2000
CEDAR	All Atom	Hermans *et al.*, 1984; Hu *et al.*, 2003
CVFF	All Atom	Kitson and Hagler, 1988
DISCOVER	All Atom	Maple *et al.*, 1998
ECEPP/3	All Atom, Torsional	Némethy *et al.*, 1993
ENCAD	All Atom	Daggett and Levitt, 1993; Levitt *et al.*, 1995
GROMOS87	United Atom	van Gunsteren and Berendsen, 1987
GROMOS96	United Atom	Scott *et al.*, 1999
MM2	All Atom	Lii *et al.*, 1989
MM3	All Atom	Lii *et al.*, 1991
MM4	All Atom	Langley and Allinger, 2002
MMFF	All Atom	Halgren, 1996a,b,c,d
NEMO	Polarizable	Hermida-Ramón *et al.*, 2003
PROSA	Polarizable	Stern *et al.*, 1999
SCHRODINGER	Polarizable	Kaminski *et al.*, 2002
SDFF	Polarizable	Palmo *et al.*, 2003
SIBFA	Polarizable	Gresh, 1997; Guo *et al.*, 2000
SPASIBA	All Atom	Derreumaux and Vergoten, 1995
TRIPOS	All Atom	Clark *et al.*, 1989
UCSD-WILSON	All Atom	Mackay *et al.*, 1984
UFF	All Atom	Rappé *et al.*, 1992
UPJOHN	All Atom	Oie *et al.*, 1981
YETI	United, Torsional	Vedani, 1988

Allinger's highly respected molecular mechanics parameterizations and have been applied primarily to peptides. These MM methods use atomic charges only at formally charged groups, and rely on bond dipole moments to provide for most electrostatic interactions. A series of potentials refined over many years in Levitt's group (Levitt, 1983; Levitt and Sharon, 1988; Levitt *et al.*, 1995, 1997) are incorporated in the ENCAD (*EN*ergy *C*alculation *A*nd *D*ynamics) program and have been notably used to study protein folding and unfolding (Daggett, 2002). The ENCAD potential is unique in its use of group-based, rather than atom-based, neighbor exclusion of short-range electrostatic interactions. It also uses pairwise nonbonded potentials shifted to zero energy at short range, and specifically parameterized to reflect these small cutoff distances. The polarizable force fields listed in Table I will be discussed in section III of this review.

E. Comparisons Among Protein Force Fields

One might imagine that more than a quarter century of experience with force field methods for proteins would have led to some secure conclusions about their relative quality, at least for certain types of applications. It is very difficult, however, to generate "gold standards" by which different force fields may be compared. For one thing, experimental measurements reflect a great deal of thermal averaging of conformers in a complex condensed-phase environment. Until recently, few interesting simulations on peptides or proteins have been converged sufficiently well in their conformational sampling to allow deviations from experiment to be ascribed solely to deficiencies in the force field. Quantum chemistry calculations offer the attractive possibility of much simpler comparisons for fixed conformations or families of conformations depending on only a few degrees of freedom. For example, Halgren (1999b) has compared the predictions of inter- and intramolecular conformational energies for a variety of force fields to experiment and to quantum results. These comparisons offer useful insights into how force fields work but are not directly relevant to peptides and proteins. Furthermore, it is not clear that a force field optimized to fit such data for isolated molecules will work best in liquid-state simulations. This is especially true if electronic polarizability is ignored, as in the force fields described above, since the charge distributions appropriate for the gas phase will change significantly in a high-dielectric liquid such as water. Hence, simple class I–type force fields involve rather severe compromises between accuracy and simplicity, so that efforts to judge their "quality" are likely to depend strongly on what sorts of simulations are being considered and on the details of what is judged to be "correct" behavior.

With the passage of time, the fixed charge force fields appear to be converging toward a common electrostatic model. In Table II, it is seen that the most recent parameterizations of the Amber, CHARMM, and OPLS force fields use charge values that are more similar than those from previous sets—that is, compare the older (Amber *ff84, CHARMM19*, OPLS-UA) charges against the newer (*CHARMM22/27*, Amber *ff94/99*, OPLS-AA) values. Some of the convergence of these values represents movement toward an optimal fixed-charge model, but it also reflects the increasingly similar parameterization protocols and test sets (Beachy *et al.*, 1997) used by the Amber, CHARMM and OPLS developers. As the table indicates, other current force fields such as GROMOS96 and BUFF use very different charge values for amino acid residues.

After a lengthy period of refinement of partial charge values during the 1980s and 1990s, much effort has shifted toward improvement of the

TABLE II

Comparison of Fixed Partial Charge Models for Serine Taken from Current and Previous Generation Protein Force Fields

	Amber 84	Amber 94/99	CHARMM19	CHARMM22/27	OPLS-UA	OPLS-AA	GROMOS96	BUFF
N	−0.463	−0.4157	−0.35	−0.47	−0.57	−0.50	−0.28	−0.749
HN	0.252	0.2719	0.25	0.31	0.37	0.30	0.28	0.328
CA	0.035	−0.0249	0.10	0.07	0.20	0.14	0.00	0.189
HA	0.048	0.0843		0.09		0.06		0.048
C	0.616	0.5973	0.55	0.51	0.50	0.50	0.38	0.828
O	−0.504	−0.5679	−0.55	−0.51	−0.50	−0.50	−0.38	−0.679
CB	0.018	0.2117	0.25	0.05	0.265	0.145	0.15	0.296
HB	0.119	0.0352		0.09		0.06		0.006
OG	−0.55	−0.6546	−0.65	−0.66	−0.70	−0.683	−0.548	−0.764
HO	0.31	0.4275	0.40	0.43	0.435	0.418	0.398	0.491

protein torsional potentials. Each of the three major potentials discussed above has recently undergone revision of torsional parameters. The torsional parameters are traditionally the last values to be determined in the generation of a new protein force field, and these values have less theoretical underpinning than other force field terms. With the convergence of many of the other functional forms and parameters, the torsions are often used as a more general "error function" to correct the final force field results to agree with a desired set of conformational energy differences. For example, the recent OPLS-AA/L force field consists of a series of modifications of torsional parameters fit to *ab initio* results (Kaminski *et al.*, 2001). One important application for protein force fields with highly optimized torsional parameters is the prediction of side-chain rotamer preferences (Jacobson *et al.*, 2002), which is a critical component of accurate protein homology modeling protocols.

Even with the use of similar functional forms and parameter values that seem to be converging with the passage of time, recent versions of the traditional force fields still exhibit significant differences. Shown in Fig. 5 are Ramachandran free energy maps for solvated alanine dipeptide computed with the *CHARMM27*, Amber *ff94*, and OPLS-AA force fields. The simulation system consisted of one dipeptide molecule and 206 waters in a cubic box with periodic boundaries and with Ewald summation used to account for long-range electrostatics. Each map was determined by running 288 separate umbrella sampling MD trajectories restrained to small regions of ϕ-ψ space and totaling about 70 ns of simulation for each force field. The separate trajectories were then stitched together to generate full ϕ-ψ maps with a 2-D WHAM procedure (Kumar *et al.*, 1995). A similar comparison based on analysis of single lengthy MD trajectories has recently been published by Hermans' group (Hu *et al.*, 2003). In addition, this same group performed tight-binding DFT-based QM-MM simulations of a QM alanine dipeptide molecule solvated by MM water. Integrated conformer populations given in Table III show that all three force fields exhibit large differences from the DFT map: AMBER *ff94* strongly favors α-helical structures, OPLS-AA overpredicts β-sheet structure, and *CHARMM27* does not connect the helix and sheet structures via the expected bridge region (across $\psi = 0$) of the map.

One should not infer from these difficulties that there is a lack of interest in comparing one force field to another or in improving the existing parameterizations. It has been recognized for several years that many features seen in simulations are insensitive to the details of the force field parameterization. These are expected to include most qualitative features of the average structure of folded proteins and the nature and extent of fluctuations about the average structure. This expectation has

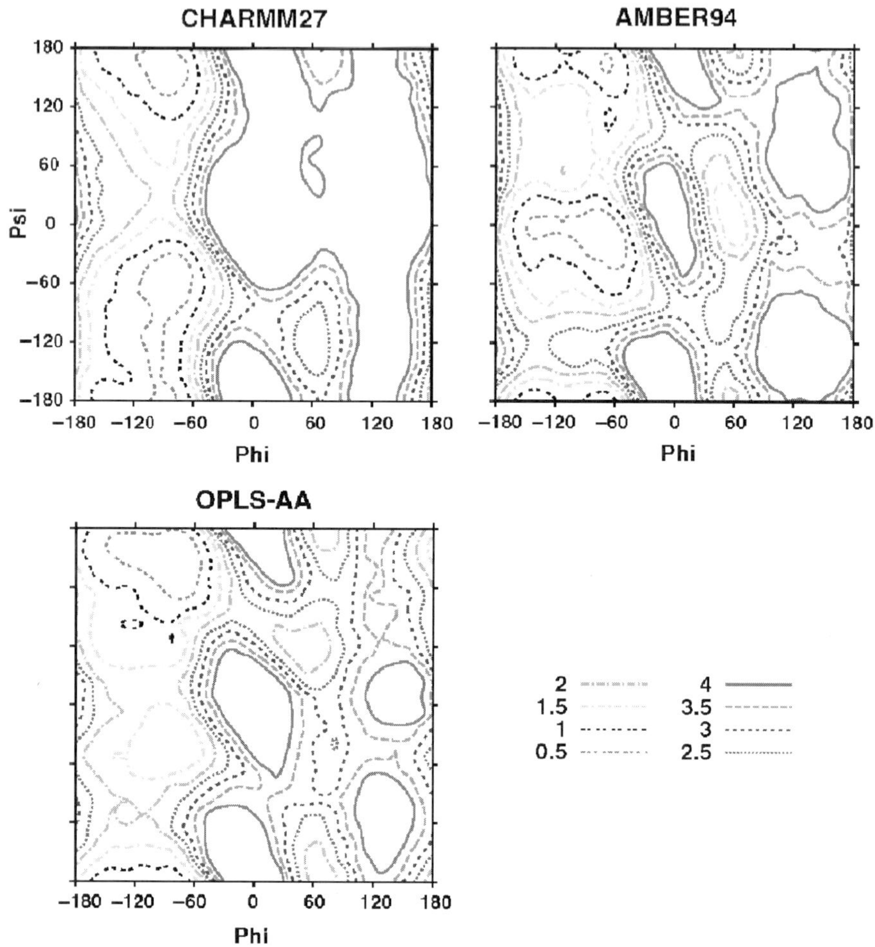

FIG. 5. Comparison of ϕ-ψ free energy plots from 2-D WHAM analysis of umbrella sampling trajectories for alanine dipeptide as computed with the AMBER *ff94*, CHARMM27 and OPLS-AA pairwise force fields. The contours shown are at 0.5 kcal/mol intervals from the global minimum of each Ramachandran map. (See Color Insert.)

been borne out in explicit comparisons of protein simulations with different force fields, concentrating on structure details (Roterman *et al.*, 1989; Ceccarelli and Marchi, 1997) or on fluctuation behavior (Teeter and Case, 1990; Price and Brooks, 2002). These results are often highly dependent on the strongly repulsive portions of the force field, which tend to be quite similar in different parameterizations, especially when

TABLE III
Alanine Dipeptide Population Percentages from QM-MM and Force Fields

	Alpha R	Bridge	Beta
DFT QM-MM	27	16	48
AMBER *ff94*	57	6	29
CHARMM27	46	3	49
OPLS-AA	14	10	70

The force field values were obtained by integrating data from Fig. 5 over the alpha R, bridge ("pass") and beta regions defined by Hu *et al.*, 2003.

liquid-state densities are a part of the fitting process. More variable are results that depend also on the softer, longer-range portions of the potential, where differences in the description of dispersion and electrostatic interactions become more important. Comparative tests on organic liquids (Kaminski and Jorgensen, 1996) show reasonable accord between popular force fields for those properties considered during the parameterization process. However, it is far from clear that this agreement would carry over, for example, to estimates of ligand-binding affinities or to helix-coil transition temperatures in peptides, which can be quite sensitive to energetic details.

III. Beyond Fixed Atomic Point-Charge Electrostatics

Unlike their biopolymer counterparts, empirical potentials for gas-phase, nonpolar organic molecules are exquisitely accurate. The widely respected MM3 force field for hydrocarbons has been used to compute the heat of formation of a set of 57 molecules, many of them very strained. The average estimated experimental error in ΔH_f was 0.40 kcal/mol for the full set of molecules. The standard deviation of the difference between the MM3 computed ΔH_f and the experimental values was 0.42 kcal/mol (Allinger *et al.*, 1989). The more recent MM4 (Allinger *et al.*, 1996) and Class II QMFF (Maple *et al.*, 1994) force fields give results at least as good as those for the original MM3. The quality of computed structures from any of these hydrocarbon force fields is equally high. We can conclude that, at least for hydrocarbons, a molecular mechanics computation is as trustworthy as the corresponding experimental result. For proteins and other biopolymers, the situation is currently much less satisfactory. Most of the effort to improve protein force fields has gone toward more accurate treatment of electrostatic and solvation effects, because these are thought to be the largest sources of error for these polar molecules in aqueous environments.

A. *Limitations of Fixed Atomic Point-Charges*

It is widely recognized that the use of an electrostatic model based on fixed atom-centered charges has two significant shortcomings. First, the restriction to only partial charges and to only the nuclear sites results in a model insufficiently flexible to describe certain features of molecular charge distributions. Second, the use of fixed charges means that the model is unable to respond directly to the molecular environment. For example, a carbonyl oxygen typically carries the same charge whether it is buried in the center of a folded protein or exposed to water on the protein surface. Both of these problems can be addressed via relatively simple extensions of the traditional fixed atomic point-charge model, but these modified force fields have yet to make their way into routine use.

Simple partial charge representations are intrinsically unable to accurately model the electrostatic potential around polar molecules. Algorithms to fit partial charges to the electrostatic potential defined by an *ab initio* wave function have been reported (PDM from Williams, 1988; CHELPG from Breneman and Wiberg, 1990). Williams' results indicate that optimally fit partial charges produce a potential with a relative RMS error of 5% to 15% from a target potential derived from quantum electronic structure calculations. Since partial charge electrostatic energies are usually expressed as a sum over pairs of interaction sites, the relative error in interaction energies will be even larger between two neighboring sets of sites. The key result is that atomic partial charge models lack the mathematical flexibility to describe the static, permanent electrostatics of general polar molecules to within "chemical accuracy." No amount of reparameterization can change this basic fact. A number of publications (Bayly *et al.*, 1993; Chipot *et al.*, 1993; Winn *et al.*, 1997) have proposed detailed schemes for arriving at atomic partial charges for biopolymers. In particular, Kollman's RESP model (Bayly *et al.*, 1993) solves some problems by applying penalty function restraints during the fitting process. Simultaneous fitting to multiple conformations of a flexible molecule can provide charges that are better determined statistically (Reynolds *et al.*, 1992). Stouch and Williams (1993) found similar improvement could be made by fixing ill-defined charges at chemically reasonable values. However, the inherent inflexibility of a nonpolarizable, fixed-charge model is not overcome by these methods.

Following on the work of Buckingham and Fowler (1985), atom- or bond-centered multipole expansions have been used to try to achieve a better fit to quantum-derived electrostatic potentials (for example, Sokalski *et al.*, 1993 and Colonna *et al.*, 1992). The relative RMS error in an optimal atomic multipole-based potential truncated after the quadrupole

TABLE IV
Energy and Structure of the K^+-Benzene Dimer

	ΔE_0	K^+-Centroid	ΔH_{298}
OPLS-AA	−9.32	2.90	
CHARMM27	−11.06	2.81	
Amber *ff94*	−12.55	2.74	
Amber *ff02*	−15.87	2.63	
AMOEBA	−19.27	2.81	−18.15
MP2/6-311+G(2d,2p)	−18.4	2.81	
MP2/aVQZ	−19.9	2.79	
CCSD(T)/CBS	−20.6	2.79	−20.1
Expt (HPMS)			−18.3
Expt (CID)			−17.7

Minimum energies and binding enthalples are given in kcal/mol, the K^+ to benzene centroid distance is given in Ångstoms. MP2/aVQZ and extrapolated CCSD(T)/CBS theoretical results are from Feller *et al.*, 2000; MP2/6-311+G(2d,2p) and CID experimental values are from Amicangelo and Armentrout (2000); the HPMS value, corrected for unimolecular dissociation, is from Sunner *et al.* (1981).

moments is usually less than 0.1% from the corresponding quantum-based target potential (Williams, 1988). The advantages of multipole methods are their functional flexibility and implicit inclusion of conformational effects (Dykstra, 1993; Price, 2000).

The need for explicit inclusion of polarization effects in protein force fields is a matter of current discussion in the literature (van der Vaart, 2000; Roux and Bernèche, 2002). A large body of work suggests that explicit polarization is required if a single set of parameters is to correctly describe both gas-phase cluster and bulk environments. However, the increased complexity and expense of polarizable force fields has seemed to require strong justification of their use in protein modeling..

A striking example of the importance of polarization in force-field calculations is provided by the interaction of a potassium cation with benzene. The benzene molecule is highly polarizable, and polarization effects are responsible for much of the large binding energy. As shown in Table IV, the simple partial charge force fields grossly underestimate the strength of this interaction. It is impossible to find reasonable sets of parameters that repair the simple force fields, especially in light of the concurrent need to fit water-ion and liquid benzene properties.

Table IV also shows results from two polarizable force fields, the Amber *ff02* parameter set and a polarizable force field being developed by one of us for the TINKER molecular modeling package (Ren and Ponder, unpublished). Both of these models yield interactions closer to experimental and high-level quantum results, though the *ff02* binding energy is not tight enough.

Since truly accurate quantum calculations are finally becoming available for a variety of small molecule interactions (Dunning, 2000; Huang and MacKerell, 2002), polarizable protein force fields that are directly validated against such data will be parameterized in the near future. Optimal simple partial charge force fields for biopolymer modeling in water must implicitly include the effects of the aqueous environment. Without explicit polarization, current force fields should not be parameterized directly against high-level gas-phase quantum calculations. The ability to transfer quantum-derived electrostatics to bulk-phase modeling is a major practical advantage of polarizable force fields.

B. *Flexible Models for Static Charge Distributions*

Better static electrostatic models for polar molecules necessitate going beyond the atomic partial charge paradigm. This can be done by adding additional partial charges at sites other than nuclei or by placing a more elaborate model at the atomic centers. There is significant experience with both types of extensions to the basic atomic charge model.

Lone pairs sites have been used by various force fields on selected nitrogen, oxygen, or sulfur atoms to fine-tune hydrogen bond interactions. Both the MM methods of Allinger and some of the Amber parameterizations (Dixon and Kollman, 1997) have adopted lone pairs for some protein atoms. An example of off-center charge modeling in recent Amber force fields is given in the right-hand side of Fig. 2. A generalization of this method would be to use distributed off-site point charges that would in principle achieve any level of accuracy in reproducing a target potential (Brobjer and Murrell, 1982). The XED force field uses a charge on the nucleus and up to five nearby partial charges on "orbital points" to represent each atom, and it has advantages over traditional force fields in modeling aromatic interactions (Chessari *et al.*, 2002). Simple application of Coulomb's law then leads to an $O(N^2)$ CPU-time dependence on the total number of charges, which is inefficient when compared with atom-centered multipole approaches. In addition, off-site charges may result in "site fusion" during procedures that can generate large conformational changes between energy evaluations such as random or Monte Carlo searches. Finally, it is possible to replace simple point charges with isotropic atom-centered charge distributions. A method using a point charge and counterbalancing Gaussian charge distribution on each atom yields a novel water model claimed to mimic some of the effects of the environment without the need for an explicit polarization calculation (Guillot and Guissani, 2001).

Another means of increasing the mathematical flexibility of a static electrostatic description beyond the limitations of partial atomic charges is to use atomic multipole moments. The derivation, use, and manipulation of higher-order moments in the study of intermolecular interactions are thoroughly covered in the important monograph by Stone (1996). Given the density matrix or electrostatic potential from a high-level electronic structure calculation, it is possible to derive atomic multipoles that provide an optimal least-squares fit to the target density. The PDM programs of Williams (1988) have been used to show that an atom-centered expansion through quadrupole moments can provide a potential within 0.1% of an MO-derived target. Alternatively, multipole expansions can be built by moving elements of overlap density from a quantum calculation to a distributed set of sites such as atomic nuclei. The similar distributed multipole analysis (DMA; Stone, 1981) and cumulative atomic multipole moments (CAMM; Sokalski and Poirier, 1983) protocols are examples of this approach. Spatial decomposition of the density, such as the atoms-in-molecules (AIM) procedure from Bader's lab (Bader, 1990; Popelier, 2000), has also been used to construct multipoles for protein atom types (Matta and Bader, 2000). It appears that atomic DMA values converge slightly more rapidly than AIM expansions (Popelier *et al.*, 2001), but the more important question for use in protein force fields relates to transferability of the multipole values.

If a limited set of lone pair partial charges are used, their motion relative to the rest of the structure can be governed by standard molecular mechanics bond and angle terms involving the lone pair sites. More generally, the use of off-site charges or atomic multipole moments requires definition of a local coordinate frame at each atom and motion of each atom's electrostatic model relative to its frame (Koch and Egert, 1995). There are two formulations available for computation of the electrostatic interactions between point multipoles: a polytensor scheme (Applequist, 1983; Dykstra, 1988) that becomes very efficient for higher-order derivatives and a spherical harmonic expansion (Price *et al.*, 1984; Popelier and Stone, 1994) leading to the energy, forces, and torques. In brief, the atomic multipoles at a site i can be represented as $M_i = [q_i, \mu_{i,1}, \mu_{i,2}, \mu_{i,3}, Q_{i,11}, Q_{i,12}, Q_{i,13}, \ldots Q_{i,33}]$ where q, μ and Q are the monopole, dipole, and quadrupole components, respectively. Then the interaction energy between two sites i and j is given by $M_j^T \cdot T \cdot M_i$ where the T tensors are constructed from extended derivatives of the inverse distance between the sites, $1/R$. Following Stone's notation (Stone, 1996), the multipole interaction T elements, with $4\pi\varepsilon_0$ omitted for clarity, are expressed as:

$$
\begin{aligned}
T &= \frac{1}{R} \\
T_\alpha &= \nabla_a T = -\frac{R_\alpha}{R^3} \\
T_{\alpha\beta} &= \nabla_\alpha T_\beta \\
T_{\alpha\beta\gamma} &= \nabla_\alpha T_{\beta\gamma} \\
&\ldots\ldots \\
(\alpha, \beta, \gamma, \ldots &= 1, 2, 3)
\end{aligned}
\tag{2}
$$

A concise description of the Ewald summation technique for multipole interactions is also available (Smith, 1998).

C. *Including Environmental Effects via Polarization*

As with the static electrostatic model, there are several possible methods available for including polarization in force-field calculations. An excellent recent review by Rick and Stuart (2002) covers the current methodologies for incorporating polarizability in simulations. An earlier summary of the status of polarizable force fields was provided by Halgren and Damm (2001). We will not repeat their treatments of the various polarization formalisms but focus instead on problems common to all polarizable force fields as they are extended from small, relatively rigid molecules such as water to large, flexible protein structures.

Three basic methods for including polarization have been studied: fluctuating charge, Drude oscillator, and induced dipole models. Fluctuating charge models use the principle of electronegativity equalization to produce a set of point charges that optimize the total electrostatic energy. In theory, intermolecular charge transfer is then handled by requiring conservation of charge for the whole system. In actual application, charge conservation is often enforced for each individual molecule. The original force field model of this class was the charge equilibration (QEq) method of Rappé and Goddard (1991) that is still used in their UFF force field (Rappé *et al.*, 1992) and its BUFF biopolymer counterpart (Carlson, 2000). Problems in computing analytical derivatives of QEq have led to the recent formulation of a "consistent" variation, CQEq (Kitao and Ogawa, 2003). The more recent dynamical fluctuating charge (FQ) method (Rick *et al.*, 1994) can be viewed as a perturbation formulation of QEq. In the typical use of FQ, the electrostatic forces required for dynamics simulation are propagated via an extended Lagrangian (Liu *et al.*, 1998), thus avoiding the need for a closed analytic form of the derivative.

Drude oscillator (DO) methods, also referred to as shell models, use a harmonic restraint potential to tether a mobile point charge to an interaction site. In a general Drude oscillator model, an atom will carry a charge fixed at the nucleus and a second restrained charge of variable position. The charge magnitudes and harmonic force constants are fit to atomic and molecular polarizability data and experimental energies. As with FQ, the DO model is most often used with an extended Lagrangian treatment of the variable charges during a molecular dynamics simulation (Mitchell and Fincham, 1993).

Perhaps the best-studied method for handling polarization is use of induced multipole moments. While higher-order multipole polarization and hyperpolarization can be included in force fields (Dykstra, 1989), only induced dipoles are usually considered. The use of induced dipoles in molecular dynamics dates back at least to Vesely's treatment of polarizable Stockmayer-type systems (Vesely, 1977). Point polarizabilities are generally assigned to either molecular centers of mass or distributed over some or all atomic sites. The direct, permanent electrostatic field felt at each polarizable site then results in a site of "direct" induced dipoles. But since the induced dipoles alter the field at each site, the procedure must be iterated to generate a self-consistent set of "mutual" induced dipoles arising from the mutual polarization.

Three methods are available for calculation of the induced dipole moments. The first is the iteration alluded to above. The induced dipole at each atomic site is computed as $\mu_{\mathrm{i},\alpha}{}^{ind} = \alpha_i E_{i,\alpha}$, where α_i is the atomic polarizability and $E_{i,\alpha}$ is the sum of the fields generated by both permanent multipoles and induced dipoles:

$$\mu_{i,\alpha}^{ind} = \alpha_i \left(\sum_{\{j\}} T_{\alpha}^{ij} M_j + \sum_{\{j'\}} T_{\alpha\beta}^{ij'} \mu_{j',\beta}^{ind} \right) \quad \text{for } \alpha, \beta = 1, 2, 3 \tag{3}$$

where $M_j = [q_j, \mu_{j,1}, \mu_{j,2}, \mu_{j,3}, \ldots]^T$ contains the permanent multipole components and $T_{\alpha}{}^{ij} = [T_{\alpha}, T_{\alpha 1}, T_{\alpha 2}, T_{\alpha 3}, \ldots]$ is the interaction tensor between site i and j introduced in the previous section. The sets $\{j\}$ and $\{j'\}$ in equation 3 consist of all atomic sites except for separate small lists of omitted near-neighbor sites that may differ based on the details of the model. In fully interactive models (see below) the set $\{j'\}$ includes all atomic sites other than i itself. It can be shown that the solution of the above self-consistent equation can be written as:

$$\mu_{i,\alpha}^{ind}(n+1) = \mu_{i,\alpha}^{ind}(n) + \alpha_i \sum_{\{j'\}} T_{\alpha\beta}^{ij'} \mu_{j',\beta}^{ind}(n) \quad \text{for } n = 0, 1, 2, \ldots \tag{4}$$

where $\mu_{i,\alpha}^{ind}(0) = \alpha_i \sum_{\{j\}} T_{\alpha}^{ij} M_j$ is the "direct" induced dipole on site i due to the electric field from permanent multipoles of other molecules, and $\alpha_i \sum_{\{j'\}} T_{\alpha\beta}^{ij'} \mu_{j',\beta}^{ind}(n)$ is the "mutual" induced dipole further induced by dipoles on all the other sites. The iterative solution can generally be accelerated by well-known successive overrelaxation (SOR) techniques (Young, 1971) by using a value of $\omega = 0.75$ in equation 5, which converges rapidly to any reasonable level of accuracy.

$$\mu_{j,\alpha}^{ind}(n+1) = (1-\omega)\mu_{j,\alpha}^{ind}(n) + \omega\left[\mu_{j,\alpha}^{ind}(n) + \alpha_i \sum_{\{j'\}} T_{\alpha\beta}^{ij'} \mu_{j',\beta}^{ind}(n)\right] \tag{5}$$

A second method for finding the induced dipoles is full-matrix direct solution of the set of coupled linear equations, avoiding the iterative procedure. Also, in the context of molecular dynamics, the dipoles can again be updated via an extended Lagrangian scheme. Ewald summation has been described for various polarizable multipole schemes (Nymand and Linse, 2000; Ren and Ponder, 2003), and particle mesh Ewald (PME) has been derived for induced dipole interactions (Toukmaji *et al.*, 2000).

D. *Consistent Treatment of Electrostatics*

In this section, we discuss the features necessary for a consistent and effective treatment of polarization that is intended for use in modeling flexible, polar molecules such as proteins. Discussion will be limited to induced dipole models, since these are the most commonly used type at present, but most of the points below also apply and can be adapted for fluctuating charge or Drude oscillator formulations.

At the most basic level, the polarization method used in a force field should be able to compute molecular polarizabilites with reasonable fidelity. A number of atomic polarizability models, including additive (Miller, 1990; Stout and Dykstra, 1998) and interactive models (Applequist *et al.*, 1972; Bode ande Applequist, 1996; Thole, 1981; van Duijnen *et al.*, 1998), have been proposed for treatment of molecular polarizability. Additive models allow the polarizable sites to respond to an external field but not to other sites within the molecule. Interactive models include the mutual effects of polarizable sites within a molecule on each other, as well as polarization induced by an external field. Simple additive models are sufficient for use with a force field if intramolecular polarization can be neglected, as is often the case for very small molecules. However, for larger peptides and proteins the distinction between intra- and intermolecular polarization is obviously blurred, and an interactive model that treats both on the same basis is required.

The successful Dang-Chang polarizable model for water (Dang and Chang, 1997) uses only a single polarizable site that carries the experimental molecular polarizability of 1.444 $Å^3$. Other models have used distributed polarizabilities on oxygen and the hydrogens that linearly sum to the correct water value. In these models, intramolecular polarization is neglected and is implicitly included within the overall parameterization scheme.

Alternatively, the POL3 water model (Caldwell and Kollman, 1995) and the Amber *ff02* protein force field (Cieplak *et al.*, 2001) use a modified form of Applequist's interactive polarization. The Applequist model uses generally smaller distributed atomic polarizabilities than most additive models. When the sites interact via mutual polarization, the total polarizability is increased and correct molecular polarizabilities are generated. For example, the POL3 model for water contains point polarizabilities of 0.528 $Å^3$ on oxygen and 0.17 $Å^3$ on hydrogen. When these sites interact via full mutual induction, a molecular polarizability of 1.46 $Å^3$ is generated in good agreement with the experimental water value. However, the Amber *ff02* energy model calls for neglect of mutual induction between atoms that are 1-2 or 1-3 bonded in analogy with the molecular mechanics tradition of omitting electrostatic interactions between near neighbor atoms. The result is that the atomic polarizabilities in the POL3 model do not interact, and the molecular polarizability of POL3 water is reduced to the additive value of 0.87 $Å^3$. Based on this analysis, we might predict that the POL3 model will be underpolarized, and indeed this appears to be the case. The POL3 water dimer interaction energy and O-O separation distance are 5.44 kcal/mol and 2.79 Å, respectively. These values are intermediate between "fully polarized" high-level quantum results of 5.0 kcal/mol and 2.91 Å and a typical nonpolarizable pairwise model such as TIP3P with dimer values of 6.54 kcal/mol and 2.75 Å. The analysis extends to proteins as well. Neglect of neighboring atom mutual polarization leads to a systematic underestimation of polarization effects in *ff02*. The remaining polarization can, of course, be implicitly included in the overall parameterization as is the case with the simple pairwise protein force fields.

Unfortunately, it is not possible to simply "turn on" the 1-2 and 1-3 mutual polarization to regain the full molecular polarizability. Attempts to include full mutual polarization within a modified POL3 model lead to a "polarization catastrophe" in the liquid wherein the induced dipole moments fail to converge and become infinite in magnitude. The origin of the catastrophe resides in the non-physical approximation of using a point polarizability model. The diffuse nature of the true charge distribution and the fact that polarizability is more realistically spread over corresponding regions of space suggest that polarization should be damped at short range. The lack of damping in the Applequist model leads to failure of energetic models derived from it. The inability of the original Applequist scheme to adequately model the molecular polarizability of very polar, anisotropic molecules such as benzene is another symptom of the same problem (Applequist, 1993).

Among the interactive models, the damped induction suggested by Thole (1981) exhibits several advantages: (1) it avoids the polarization catastrophe at short range by replacing point dipole interactions with interactions between smeared dipoles, (2) it produces anisotropic responses to an external field by using only isotropic atomic polarizabilities, and (3) atomic polarizabilities derived within the model are highly transferable. For example, a single atomic polarizability value for each of the elements C, N, O, and H gives an excellent fit to a large number of experimental molecular polarizabilities (van Duijnen and Swart, 1998).

To avoid a "polarization catastrophe" at very short range, Thole introduced a modification scheme in which dipole interactions are damped as though one of the point dipoles in each pairwise interaction is replaced by a smeared charge distribution. As a result, the dipole interaction energy approaches a finite value instead of becoming infinite as the separation distance approaches zero. Several charge distributions yield roughly similar results, but the one most commonly used to date has the form

$$\rho = \frac{3a}{4\pi} \exp(-au^3) \tag{6}$$

where $u = R_{ij}/(\alpha_i\alpha_j)^{116}$ is the effective distance as a function of atomic polarizabilities of sites i (α_{i}) and j (α_j). The factor a is a dimensionless width parameter of the smeared charge distribution and controls the strength of damping. Using the charge distribution given by equation 6, the damped T matrix elements corresponding to equation 3 can be derived. It has been shown that the damped first-order T element is (Kong, 1997)

$$T_{\alpha}^{D} = -[1 - \exp(-au^3)]\frac{R_{\alpha}}{R^3} \tag{7}$$

The modified higher-order T matrix elements can be obtained successively by taking the derivative of the preceding lower rank elements:

$$
\begin{aligned}
T^D_{\alpha\beta} &= \lambda_5 \frac{3R_\alpha R_\beta}{R^5} - \lambda_3 \frac{\delta_{\alpha\beta}}{R^3} \\
T^D_{\alpha\beta\gamma} &= -\lambda_7 \frac{15 R_\alpha R_\beta R_\gamma}{R^7} - \lambda_5 \frac{3(R_\alpha \delta_{\beta\gamma} + R_\beta \delta_{\alpha\gamma} + R_\gamma \delta_{\alpha\beta})}{R^5} \\
T^D_{\alpha\beta\gamma\eta} &= \lambda_9 \frac{105 R_\alpha R_\beta R_\gamma R_\eta}{R^9} \\
&\quad -\lambda_7 \frac{15(R_\alpha R_\beta \delta_{\gamma\eta} + R_\alpha R_\gamma \delta_{\beta\eta} + R_\alpha R_\eta \delta_{\beta\gamma} + R_\beta R_\gamma \delta_{\alpha\eta} + R_\beta R_\eta \delta_{\alpha\gamma} + R_\gamma R_\eta \delta_{\alpha\beta})}{R^7} \\
&\quad +\lambda_5 \frac{3(\delta_{\alpha\beta}\delta_{\gamma\eta} + \delta_{\alpha\gamma}\delta_{\beta\eta} + \delta_{\alpha\eta}\delta_{\beta\gamma})}{R^5}
\end{aligned}
\tag{8}
$$

where the λ_i for $i \in \{3,5,7,9\}$ are the damping coefficients that modify the standard interactions

$$
\begin{aligned}
\lambda_3 &= 1 - \exp(-au^3) \\
\lambda_5 &= 1 - (1 + au^3)\exp(-au^3) \\
\lambda_7 &= 1 - (1 + au^3 + \frac{3}{5}a^2u^6)\exp(-au^3) \\
\lambda_9 &= (1 - [1 + au^3 + (18a^2u^6 + 9a^3u^9)/35]\exp(-au^3)
\end{aligned}
\tag{9}
$$

After replacing the original interaction matrix elements with the above damped ones, the energy, force, and electric field are computed in the usual fashion. Some groups apply damping to all electrostatic interactions (Burnham *et al.*, 1999). Other workers choose to damp only those energy terms involving induced moments, leaving the interactions between permanent multipoles unmodified (Ren and Ponder, 2003).

Under a Thole-style scheme, atomic polarizabilities are generally larger than those from simple additive fits because of the reduction in polarization caused by the short-range damping. For example, a modified Thole water model devised by the Ponder group (Ren and Ponder, 2003) uses atomic polarizabilities of 0.837 Å^3 on oxygen and 0.496 Å^3 on each hydrogen but yields a molecular polarizability of only 1.41 Å^3. The Thole interaction model was introduced into energy calculations by the Groningen group (de Vries *et al.*, 1997) and later adopted by Burnham *et al.* (1999) for their initial "Thole-type" model (TTM) for water. It has also recently been used by Karlstrom's group for their NEMO polarizable potential (Brdarski *et al.*, 2000).

Once a polarization model gives satisfactory results for molecular polarizabilities, the next test is generation of accurate electrostatic potentials in the region surrounding a polarizable molecule. A key point is that permanent electrostatics taken from a quantum calculation should not be used directly as the permanent model in a polarizable force field. If mutual induction is allowed as in the Applequist or Thole models, then some intramolecular polarization must be removed from the quantum results, where it is already implicitly included, to avoid ''double counting'' of induction effects. Various groups have devised procedures to modify molecular orbital-derived charges or multipoles for use with their intramolecular polarization schemes (Cieplak *et al.*, 2001; Ren and Ponder, 2002). The multipoles on each atom taken directly from a quantum calculation may be considered as a sum of intrinsic ''permanent'' and ''induced'' moments:

$$M_i = M_i^{perm} + M_i^{ind}, \tag{10}$$

where M_i^{ind} is produced by direct and mutual induction from all sites in the absence of an external field:

$$M_{i,\alpha}^{ind} = \alpha_i \left(\sum_{\{j\}} T_\alpha^{ij} M_j^{perm} + \sum_{\{j'\}} T_\alpha^{ij'} M_{j'}^{ind} \right). \tag{11}$$

This is the same relation as in equation 3, except that the induced dipole is replaced by generalized induced moments. Substitution of equation 10 into the above expression yields

$$M_{i,\alpha}^{ind} = \alpha_i \left(\sum_{\{j\}} T_\alpha^{ij} (M_j - M_j^{ind}) + \sum_{\{j'\}} T_\alpha^{ij'} M_{j'}^{ind} \right). \tag{12}$$

If the same scaling factors are employed for mutual and direct induction using identical groups $\{j\}$ and $\{j'\}$, then equation 12 reduces to a surprisingly simple expression:

$$M_{i,\alpha}^{ind} = \alpha_i \sum_{\{j\}} T_\alpha^{ij} M_j. \tag{13}$$

In the case where $\{j\}$ and $\{j'\}$ are not the same, or when different fractional scaling factors are desired for short-range direct and mutual interactions, a more complex formulation requiring iterative solution is necessitated. The underlying ''permanent'' moments, M_i^{perm}, calculated via equation 10 will return the full quantum mechanical result when combined with the force field polarization model defined by equation 11. Following this protocol, ''permanent'' moments from small quantum calculated fragments can be combined in a consistent fashion to build electrostatic models for larger structures. Finally, we should point out that it may be convenient (or necessary) to differentiate the short-range treatment of direct induction caused by the permanent field from the treatment of mutual induction. A number of empirical models for inclusion of direct induction have been tested (Ren and Ponder, 2002), and the best choice will depend on other details of the particular polarization implementation being considered.

Finally, a polarizable force field has to achieve a balance between inter- and intramolecular energetics. All molecular mechanics methods face the challenge of combining a description of nonbonded interactions (vdW, electrostatics, polarization) with short-range valence interactions (bonds, angles, torsions). As noted above, force fields usually ignore or scale intramolecular vdW and permanent electrostatic interactions between atoms separated by three or fewer bonds. Such interactions are thought to be handled by the valence terms. Polarizable force fields developed to date apply similar schemes for scaling of short-range intramolecular polarization. However, there is no reason that the scaling scheme used for the induction calculation will be optimal, or even appropriate, for computing energetic terms arising from polarization. Use of separate scaling schemes provides a useful flexibility during parameterization of a polarizable force field but at the cost of a somewhat more complex calculation. The felicitous merging of valence and nonbonded terms is a critical consideration in any successful force field for flexible proteins, and polarizable models are no exception.

E. *Current Status of Polarizable Force Fields*

The systems studied via polarizable potentials fall into two separate classes, each exposing an additional layer of challenges for the force-field developer. The simpler class contains water and most other small molecules. A large number of water models have been proposed, many of which use potentials that would be difficult to extend to more general molecular systems. As noted above, several workers have found that polarization in water is treated adequately by a single isotropic polarizable site. Force fields for small organics require potentials applicable for several elements or atom types and a distributed polarization model. However, as long as the systems considered are monofunctional or lacking rotatable bonds, it is usually sufficient to ignore explicit intramolecular polarization by including it implicitly in the parameterization. The second class would include polarizable force fields for arbitrarily large, flexible molecules such as proteins. In this case, the force field developer has the additional problem of maintaining consistency and balance in the treatment of intra- vs. intermolecular polarization. Also, there is the hope that the final model will account for at least some of the conformational effects of polarization in flexible systems.

1. Polarizable Models for Small Molecules

There have undoubtedly been more polarizable simulations reported for liquid water than for any other substance. Previous attempts to parameterize a water model applicable across multiple phases date back to the "polarizable electropole" model of Barnes *et al.* (1979) that treated electrostatics via a single site carrying the experimental dipole and quadrupole moments and isotropic dipole polarizability. An interesting polarizable and dissociable potential was suggested by Stillinger and David (1978) and applied to water clusters and ion monohydrates. The original MCY model (Matsuoka *et al.*, 1976) based on a fit to points on an *ab initio* dimer surface has evolved into the NCC-vib potential (Corongiu, 1992) that includes induced dipole polarization and vibrational flexibility. Sprik and Klein (1988) made an early attempt at modifying the existing TIP4P potential to include polarization within a Drude oscillator framework particularly suitable for molecular dynamics simulation. Another early water model from the Levy group (Bernardo *et al.*, 1994) uses partial charges and induced dipoles on each atomic center, along with a Thole-like polarization damping scheme.

Recent work on polarizable water includes a simple, rigid, non-iterative 3-site model (Yu *et al.*, 2003) parameterized to include only direct induction and an elaborate Drude oscillator model (MCDHO; Saint-Martin *et al.*, 2000) that was compared with high-quality *ab initio* calculations. The POL5/TZ and POL5/QZ water models (Stern *et al.*, 2001), which combine fluctuating charge and induced dipole polarization, yield good agreement with a range of structural and thermodynamic properties. The AMOEBA water model (Ren and Ponder, 2003) also gives excellent cluster and liquid phase results and has been designed for compatibility with a protein force field built around a polarizable atomic multipole electrostatic description.

The family of second-generation TTM models, TTM2-R and TTM2-F (Burnham and Xantheas, 2002a,b), exhibit excellent agreement with *ab initio* structures and energies for water clusters through $(H_2O)_6$ but perform less well in liquid simulations. The TTM2-F model points up an interesting problem not solved by including polarization in the potential. It is known experimentally that the bond angle in water increases from 104.52° in the gas phase to a value near 106° in the liquid. Essentially all flexible water models, pairwise or polarizable, exhibit the opposite trend in that the average angle value decreases on moving from gas to liquid. As Burnham and Xantheas show with TTM2-F, the problem is that the electrostatic model does not respond correctly to changes in valence geometry. In particular, the dipole moment derivative vector with respect to bond stretching lies in the wrong direction in all empirical potentials, including those with explicit polarization. The solution lies in explicitly coupling the electrostatic model to the valence terms. The complex TTM2-F model uses a numerical correction based on a theoretical dipole moment surface for the water molecule. A potentially similar, but simpler, method based on bond charge fluxes has been proposed by Palmo and Krimm (1998). It seems likely that analogous problems will be exposed for the backbone amide group of proteins. Many studies suggest that typical protein force fields are too stiff with respect to pyramidalization at the amide nitrogen. A very recent publication, also from the Krimm group (Mannifors *et al.*, 2003), provides an explanation and fix based again on electrostatic-valence coupling and charge flux.

A major impetus for "next generation" empirical potentials is the hope that accurate prediction of ligand and drug-binding energies with proteins will become a reality. The specific electrostatic interactions encoded in a polarizable force field are certainly capable of describing effects completely missing in current generation force fields. For example, Fig. 6 shows a comparison of the OPLS-AA and AMOEBA force fields for

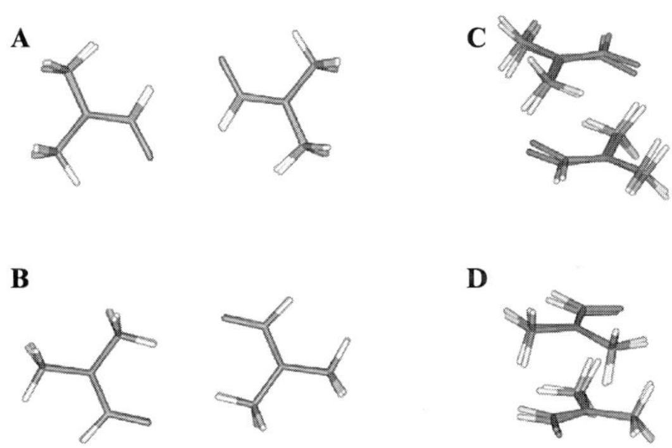

Energy Model	A	B	C	D
MP2/aug-cc-pVTZ (BSSE)	−5.35	−4.14	−8.34	−8.90
OPLS-AA	−3.45 (0.25)	−3.54 (0.64)	−5.42 (0.41)	−5.18 (0.69)
AMOEBA	−4.97 (0.08)	−5.37 (0.09)	−7.80 (0.24)	−8.79 (0.15)

FIG. 6. Comparison of *ab initio*, OPLS-AA and AMOEBA structures and energies for dimer configurations of dimethylformamide (DMF). The MP2 values are taken from Vargas *et al.* (2000), who suggested the DMF dimer as a model for possible C—H to O=C hydrogen bonding interactions in proteins. All energies are in kcal/mol. RMS values in Å from the *ab initio* minima are given in parentheses. The upper panel shows the superposition of the AMOEBA minima on the Vargas *et al.* structures.

four configurations of the dimethylformamide (DMF) dimer. According to correlated large-basis molecular orbital calculations (Vargas *et al.*, 2000), configurations C and D exhibit much tighter binding than either A or B because of the presence of non-traditional C-H to O=C hydrogen bonds. This structural feature is not included in the parameterization of current generation force fields, and OPLS-AA fares poorly on all of the DMF configurations, especially C and D. Such specific, but subtle, energetic effects almost certainly play an important role in regulating binding at protein receptor sites.

2. Polarizable Protein Force Fields

As of early 2003, there is no polarizable force field that sees wide use in protein modeling. Several efforts are under way, but parameter sets are still under development, and published applications are limited to those from the development groups. This is partly a reflection of the difficulty of the problems involved and the fact that the behavior of polarizable force fields for flexible molecules is not yet completely understood. Also, the methods and software used to treat polarization are not as standardized as for the current generation of pairwise protein potentials.

The Amber *ff02* potential (Cieplak *et al.*, 2001) has been mentioned above. It represents an initial polarizable member of the Amber family of force fields and is the result of Peter Kollman's long-standing interest (Lybrand and Kollman, 1985) in non-additive force fields. Unlike other developing polarizable protein models, *ff02* was intentionally derived via perturbation of an existing pairwise force field, a fact that should make it easier to judge the overall effect of including *ff02*-style polarization in protein simulations.

The Columbia groups of Friesner and Berne, in collaboration with researchers at Schrödinger, have produced an evolving series of polarizable protein models. The first, sometimes referred to as the PROSA force field, is based on a fluctuating charge formalism (Banks *et al.*, 1999). This model was later extended via the addition of induced dipoles on some sites (Stern *et al.*, 1999). Recently, many of these same workers have published a new "first generation" protein potential that abandons the FQ methodology in favor of exclusive use of induced atomic dipoles. The latest model claims to reproduce quantum mechanical energies of di- and tetrapeptide test cases to within an average RMSD of 0.5 kcal/mol. The same torsional fitting techniques currently being used for the OPLS-AA parameter set were applied by the Columbia and Schrödinger workers in an attempt to improve side chain rotamer prediction.

The SIBFA force field developed by Gresh (1997) has been applied to a series of protein structure and molecular recognition problems.

The energy function is based on a rather direct decomposition of *ab initio* SCF computations, including multipole electrostatics, repulsion, and dispersion-like terms and separate terms for polarization and charge transfer. The latter term is considered critical for ion interactions but also contributes a substantial fraction of the energy of polar hydrogen bonds as computed by SIBFA (Gresh, 1997; Masella *et al.*, 1998). The SIBFA potential was combined with a continuum reaction field solvation model to analyze the conformational preferences of alanine dipeptide and other oligopeptides (Gresh *et al.*, 1998).

Very recently, a NEMO potential for a capped glycine residue has been reported (Hermida-Ramón *et al.*, 2003). This study devoted considerable attention to the intramolecular potential, and presented comparisons with several nonpolarizable protein force fields. The authors conclude that their model accounts for most of the intramolecular polarization, but that errors on the order of 1-2 kcal/mol should be expected in conformational energies.

The SDFF (spectroscopically determined force field) from Krimm's group at Michigan (Palmo *et al.*, 2003) is unique among current polarizable protein force-field efforts in its extensive use of cross-terms and emphasis on valence potential functions. In particular the use of charge flux methods, as described earlier, appears to provide an important next step with its flexible coupling of electrostatics to bond stretching (Palmo and Krimm, 1998; Mannfors *et al.*, 2003). At present the SDFF has been described for amides and peptide backbone components, but it remains to be elaborated into a full protein force field.

In 1997 Beachy *et al.* reported an extensive comparison of nonpolarizable protein force fields and uncovered a systematic difference in alternative hydrogen bonding arrangements between fixed charge potentials and *ab initio* quantum calculations. Fig. 7 highlights the difference in dimerization energy between the simple amide *cis*-NMA and a larger protein β-sheet model based on alanine dipeptide. These systems were subsequently modeled with the SIBFA force field (Gresh *et al.*, 1999). The SIBFA results are shown in the figure and compared with some nonpolarizable models, a preliminary AMOEBA protein force field, and quantum calculations. The current generation of fixed-charge force fields predicts the β-sheet model to be more stable than the *cis*-NMA structure, whereas more elaborate force fields such as AMOEBA and SIBFA predict the opposite, in agreement with the quantum results. The most reasonable conclusion is that fixed charge force fields are missing cooperative effects in hydrogen bonding networks that can lead to significant energetic errors.

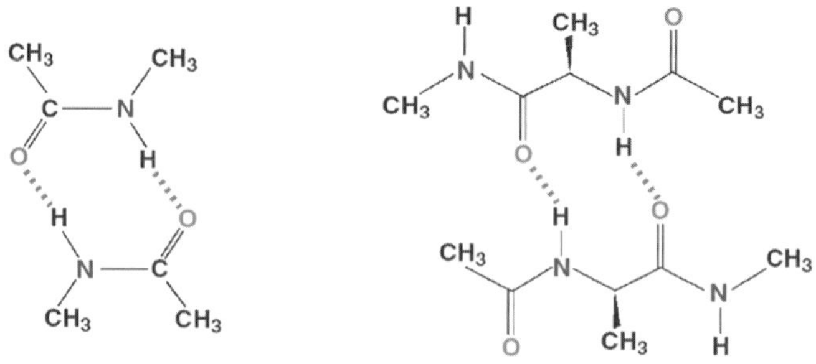

Energy Model	***cis*-NMA**	***β*-sheet**	**ΔE**
OPLS-AA	–11.5	–16.9	–5.4
CHARMM27	–11.6	–16.9	–5.3
AMBER *ff94*	–11.3	–14.8	–3.5
AMBER *ff02*	–13.5	–14.8	–1.3
AMOEBA	–18.5	–12.6	+5.9
SIBFA	–18.7	–17.1	+1.6
MP2/(CEP)4-31G+(2d)	–20.5	–17.5	+3.0
BP/DZVP *(BSSE corrected)*	–16.2	–8.4	+7.8

FIG. 7. Comparison of interaction energies for dimers of *cis*-N-methylacetamide (NMA) and alanine dipeptide (β-Sheet) as computed via quantum and force field methods. All energies are in kcal/mol. Force field values for the NMA dimer are fully optimized geometries, while the β-sheet structures were restrained to $\phi = -139^\circ$ and $\psi = +135^\circ$. The SIBFA and quantum values are taken from Gresh *et al.* (1999).

IV. Modeling the Solvent Environment

A. *Explicit Water Models*

The original atomic-scale computational model for liquid water was proposed by Bernal and Fowler (1933). At the time of the development of the earliest protein force fields, the ST2 model of Stillinger and Rahman (1974) was in wide use. The ST2 model was adopted for some of the first protein simulations done in the Karplus group at Harvard. The early 1980s saw the introduction of the SPC (Berendsen *et al.*, 1981) and TIP3P (Jorgensen *et al.*, 1983) potentials, two similar, rigid 3-site water models parameterized to reproduce the basic bulk phase structure and thermodynamics of liquid water. At present, these two potentials are still the most commonly cited solvent models for use in protein modeling and simulation. Still other water potentials have added intramolecular flexibility and interpolate the parameter space between the SPC and TIP3P models (for example, the F3C model of Levitt *et al.*, 1997 and a model from Ferguson, 1995). The addition of a fourth site along the H-O-H bisector leads to the TIP4P model (Jorgensen *et al.*, 1983). This potential yields improvements in the gas-phase dimer structure and liquid O-O radial distribution function and has been generalized to the flexible TIP4F model (Mahoney and Jorgensen, 2001).

The recently developed TIP5P potential (Mahoney and Jorgensen, 2000) exhibits excellent agreement with the experimental internal energy, density, and O-O radial distribution at room temperature. This 5-site model is essentially a modernization of the ST2 potential. It uses only fixed partial charges to model electrostatics and includes polarization response to the environment only in an averaged, mean-field sense. As a result, TIP5P provides an excellent description of the homogeneous bulk phase but is a poor model for gas-phase clusters and for nonpolar solutes in polar solvents. For example, the gas-phase binding energy of the water dimer is overestimated by more than 30% in the TIP5P model.

In application to large biomolecular systems, there is concern that such models cannot correctly account for situations where the same nonpolarizable moiety is exposed to different electrostatic environments, either within a single large static structure or during a course of simulation. In addition, there is an inherent inconsistency in most nonpolarizable models related to their static inclusion of average bulk polarization within the potential. This results in internal energies and other properties that are derived against a gas-phase reference state that is already "pre-polarized" for the liquid phase. While it is possible to correct for the resulting self-energy of the reference state, as in the SPC/E water model (Berendsen *et al.*, 1987), such corrections are not routinely used for heterogeneous systems. This suggests that use of SPC/E with uncorrected protein potentials will lead to a mismatch in the balance of water-water and protein-water electrostatics. Of course, this reference state problem disappears if fully polarizable potentials are used for both the protein and water components.

Protein simulations containing explicit water have obvious advantages over continuum solvent treatments in some respects. For example, detailed bound solvent motifs, such as water bridges, require solvent molecules as an integral part of the structural model. A recent report of potentials of mean force (PMFs) for side chain pairs in explicit and various continuum solvents (Masunov and Lazaridis, 2003) shows a dramatic lack of fine structure in the continuum versus explicit PMFs because of continuum averaging of specific water interactions.

B. *Continuum Solvent Models*

There are many circumstances in molecular modeling studies in which a simplified description of solvent effects has advantages over the explicit modeling of each solvent molecule. One of the most popular models, especially for water, treats the solvent as a high dielectric continuum, interacting with charges that are embedded in solute molecules of lower dielectric. The solute charge distribution, and its response to the reaction field of the solvent dielectric, can be modeled either by quantum mechanics or by partial atomic charges in a molecular mechanics description. In spite of the severity of the approximation, this model often gives a good account of equilibrium solvation energetics and is widely used to estimate pKs, redox potentials, and the electrostatic contributions to molecular solvation energies (Tomasi and Persico, 1994; Cramer and Truhlar, 1999; Beroza and Case, 1998; Schaefer *et al.*, 1998; Roux and Simonson, 1999). For molecules of arbitrary shape, the Poisson-Boltzmann equations that describe electrostatic interactions in a multiple-dielectric environment are typically solved by finite-difference or boundary element numerical methods. These can be efficiently solved for small molecules but may become expensive for proteins or nucleic acids. Although progress continues to be made in numerical solutions, there is a clear interest in exploring more efficient, if approximate, approaches to this problem.

Most of the simulation techniques to be used here rest on models in which the protein degrees of freedom are treated explicitly but the solvent degrees of freedom are not. This requires that the energy surface used for the protein degrees of freedom be a potential of mean force (PMF) in which the solvent degrees of freedom are implicitly averaged over (Roux and Simonson, 1999). Assuming that the full potential energy function consists of a term, U_{vac} for the interactions within the protein, depending only on the protein degrees of freedom, $\boldsymbol{r}$, and additional terms for the protein-solvent and solvent-solvent interactions, the PMF is ideally,

$$U_{pmf}(\boldsymbol{r}) = U_{vac}(\boldsymbol{r}) + \Delta G_{sol}(\boldsymbol{r}) \tag{14}$$

where $\Delta G_{sol}(\boldsymbol{r})$ is the free energy of transferring the protein from vacuum to the solvent with its internal degrees of freedom fixed at $\boldsymbol{r}$. A very common model divides $\Delta G_{sol}(\boldsymbol{r})$ into electrostatic and non-polar contributions, which are defined in terms of a hypothetical thermodynamic cycle in which the process of vacuum to solvent transfer is carried out in three steps: in vacuum, set to zero all atomic partial charges; then transfer the resulting apolar molecule to solvent; and finally, in solvent, restore the atomic partial charges to their original values. ΔG_{np} is then the free energy change for the apolar transfer step, and ΔG_{elec} is the difference of the electrostatic work of charging the system in solvent versus vacuum.

The non-polar term should be very similar to the free energy of transferring from vacuum to solvent an aliphatic compound with the same steric form as the solute molecule of interest. For small, fairly rigid molecules, this can be taken from experimental measurements on hydrocarbons (Chen *et al.*, 1994) or from a variety of theories for solvation of non-polar molecules. A common estimate is that aliphatic solvation energies in water have linear dependence on solvent accessible surface area,

$$\Delta G_{np} = \gamma(SA) + b \tag{15}$$

where SA is the solvent accessible surface area and γ and b are empirical parameters fit to the solvation free energy of alkanes (Sitkoff *et al.*, 1994; Simonson and Brünger, 1994). This simple model is certainly not fully correct, and there have been a number of attempts to extend it (Sharp *et al.*, 1996; Gallichio *et al.*, 2002), but a full discussion of this topic is beyond the scope of this review. For the rest of this section, we will consider models for the electrostatic contribution to hydration.

1. *The COSMO Model*

The COSMO approach (Klammt and Schüürmann, 1993) is a very simple yet effective model for continuum solvents that allows the basic ideas to be developed in a few lines of math. COSMO stands for Conductor-like Screening Model and treats the situation illustrated in Fig. 8, where the solvent dielectric becomes infinite (as in a conductor). As with other models, we imagine the solute to be a low-dielectric object (with embedded charges) that is immersed in this conductor-like solvent environment.

Upon immersion, charge carriers in the solvent can move freely to the surface of the solute, creating a non-uniform charge distribution at the molecular surface. The electrostatic energy of this set of charges will be:

$$E = E_{gas} + \sum_i \int z_i \frac{1}{|r_i - r_q|} q dS + \frac{1}{2}\int q \frac{1}{|r_q - r_{q'}|} q' dS dS' \tag{16}$$

Here E_{gas} is the electrostatic energy of the molecular charge distribution in the absence of solvent. The z_i are partial atomic charges in the molecule, at positions r_i, and q represents the surface charge in an element of surface area, dS. Hence, the second term in equation 16 represents the interaction energy between the molecular charge distribution and the induced surface charges, whereas the final term represents the mutual repulsion among these induced charges. Anticipating the fact that the

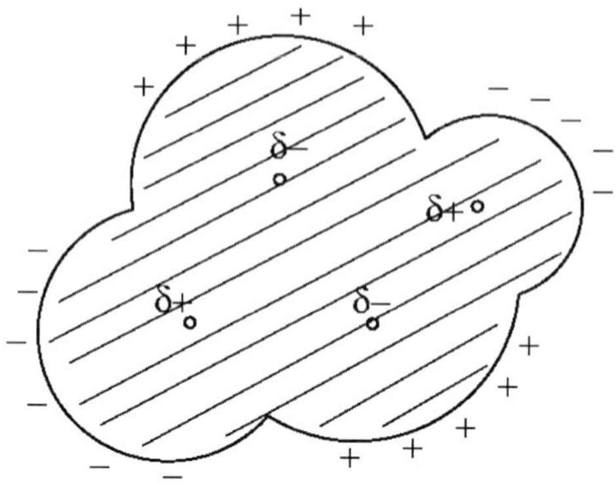

FIG. 8. Schematic diagram of a solute embedded in a medium of infinite dielectric. The shaded area represents the molecule, which consists of low dielectric material with embedded partial atomic charges. The "+" and "−" signs along the molecular surface represent the accumulation of charges to balance the atomic charges. This surface charge distribution can be determined by minimizing as the electrostatic energy of the system, as described in the text.

integrals above eventually will be estimated by some sort of quadrature, we can re-write this as a matrix equation:

$$E = E_{gas} + \mathbf{z}^T\mathbf{Bq} + \frac{1}{2}\mathbf{q}^T\mathbf{Aq} \tag{17}$$

Here **z** and **q** are vectors representing the molecular and surface charges, and **A** and **B** are matrices of inverse inter-charge distances. For the present discussion, the details of how the discretization is carried out are not relevant.

To determine the induced charge distribution, we simply minimize the energy expression in equation 17 by setting $(\partial E/\partial \mathbf{q}) = 0$, which yields

$$\mathbf{Aq} = -\mathbf{Bz} \tag{18}$$

or

$$\mathbf{q} = -\mathbf{A}^{-1}\mathbf{Bz} \tag{19}$$

Substituting **q** from equation 19 into equation 17 gives expressions for the solute-solvent interaction energy:

$$E_{solute\text{-}solvent} = -\mathbf{z}^T\mathbf{BA}^{-1}\mathbf{Bz} \tag{20}$$

and for the solvent-solvent repulsion energy:

$$E_{solvent\text{-}solvent} = \frac{1}{2}\mathbf{z}^T\mathbf{BA}^{-1}\mathbf{AA}^{-1}\mathbf{Bz} = \frac{1}{2}\mathbf{z}^T\mathbf{BA}^{-1}\mathbf{Bz} \tag{21}$$

There are two very useful features of these models that can be seen from this very simple, symbolic, approach. First, the solvent-solvent energy is proportional to the squares of the atomic charges, since **z** appears twice in equation 20. The is known to be approximately true for simple ions (e.g., the hydration free energy of divalent ions such as Mg^{2+} is about four times that of monovalent ions such as Na^{+}), and the same ideas hold for *partial* atomic charges as well: solvation effects for partial charges near ± 0.5 (such as the C or O atoms in a ketone or amide group) are much more important (by something like a factor of 25) than are solvation effects for most CH groups, which have partial atomic charges near ± 0.1. Furthermore, in this model, the sign of the charge is not important, so that both $+0.5$ and -0.5 charges have the same (favorable) solvation free energy; this aspect of the model is only approximately true for water, where negative charges (of equivalent size) are somewhat better solvated than the corresponding positive charges.

The second general feature of continuum solvent models that is evident from the above derivation is that the solvent-solvent repulsion or penalty is exactly minus one-half of the (favorable) solute-solvent interaction. This is a general feature of linear-response theories, and again appears to be approximately valid for real solutes as well, at least as judged by microscopic simulations (Simonson, 2002).

Of course, biological solvents are not immersed in infinite dielectrics, and so the model as presented above would be of limited usefulness: for relatively high dielectrics like water, it gives solvation energies that are roughly correct, but electrostatic interactions between solute molecules would be completely screened by the surface charges, completely removing all long-range electrostatic interactions. To a remarkable extent, these deficiencies can be ameliorated by increasing the electrostatic interactions between the surface charge sites (i.e., the second term in equation 16, or equivalently the **A** matrix) by a factor $\varepsilon/(\varepsilon - 1)$. This has the effect of making the screening charge opposite in sign, but slightly smaller in magnitude, than the corresponding atomic charge, so that charge screening by the solvent is not complete. These ideas were originally developed in the context of a quantum mechanical description of the solute charge density and are still widely used for such calculations (Andzelm *et al.*, 1995; Truong and Stefanovich, 1995; Baldridge and Klammt, 1997; Barone and Cossi, 1998; Dolney *et al.*, 2000), primarily because energy gradients can be expressed in a simple fashion. The model can also be used to describe the electrostatic effects of solvation in a force field context (York and Karplus, 1999); here, however, the COSMO model has mostly been overshadowed by Poisson-Boltzmann models (which are more general) and generalized Born models (which are faster). Nevertheless, the basic physics present in the conductor-like model are present as well in these alternative approaches, which we describe next.

2. *The Poisson-Boltzmann Models*

For an approach that is more general and fundamental than COSMO, but which preserves its continuum character, one can continue to treat with the two-dielectric model illustrated in Fig. 5 but then formulate and solve the corresponding electrostatic model without resorting to any approximations concerning the magnitude of the exterior dielectric constant. An additional advantage of this model is that the effects of mobile co- and counterions in the solvent can be included as in Debye-Hückel theory. For this model (Sharp and Honig, 1990; Tomasi and Persico, 1994; Cramer and Truhlar, 1999), the electrostatic potential ϕ is determined by the Poisson or linearized Poisson-Boltzmann (PB) equation,

$$\nabla\varepsilon(r)\,\nabla\phi(r) - \kappa^2(r)\varepsilon(r)\phi(r) = -4\pi\rho(r) \tag{22}$$

where $\varepsilon(r)$ has the interior or exterior value depending on whether r lies inside or outside the molecular surface (Connolly, 1983); $\kappa(r)$ is the usual Debye-Huckel parameter in the solvent region but is zero in the interior and within an ion exclusion radius from the molecular surface, and ρ is the solute charge distribution. Because of the linearity of equation 22, the work of creating a charge distribution, ρ is simply $\int\rho\phi dV/2$, and the difference in the work of charging required for the calculation of ΔG_{es} is

$$\Delta G_{es} = \frac{1}{2}\sum_i q_i[\phi_{sol}(r_i) - \phi_{vac}(r_i)] \tag{23}$$

where the q_i and r_i are the solute atomic partial charges and atomic coordinates, respectively, and ϕ_{sol} and ϕ_{vac} are the solutions of equation 22 for exterior conditions corresponding to solvent and vacuum, respectively.

Because of the complex geometry of the dielectric and ion-exclusion boundaries, equation 22 must generally be solved numerically. A variety of programs have been developed to carry out such calculations that are beyond the scope of this review. There have been several efforts to develop charge and radius parameters for use in these models based on fitting small molecule solvation energies (Lim *et al.*, 1991; Sitkoff *et al.*, 1994; Nina *et al.*, 1997, 1999). There have also been a number of studies comparing potential energy profiles and surfaces computed with the continuum solvent models to potentials of mean force calculated with explicit solvent for both hydrogen bonding and ion-ion interactions (Ösapay *et al.*, 1996, Luo *et al.*, 1999; Masunov and Lazaridis, 2003). The qualitative features of the continuum results are generally in good agreement with the explicit solvent results, but the details and fine structure can depend critically on parameterization. When a continuum solvent model is being used in conjunction with a molecular mechanics potential energy function, it is important that the continuum model use the same atomic partial charges and dielectrics as the molecular mechanics potential, or artifacts related to the over-screening or under-screening of charge-charge interactions can occur (Ösapay *et al.*, 1996). Since almost all protein force field charge distributions were designed to be used in a $\varepsilon = 1$ environment, troublesome complications arise when these are combined with continuum solvent models that other dielectrics (such as 2, 4, or 20) for the molecular interior. These difficulties are not insuperable, but a discussion of them is beyond the scope of this review.

Although the Poisson-Boltzmann model has a much lower computational cost than an analogous explicit solvent model, the direct use of the model in molecular dynamics simulations has only recently been shown to be feasible. First, the cost of solving equation 22 at each time step is significantly greater than the cost of evaluating U_{vac}. Second, it is difficult, although not impossible, to calculate the energy gradients that are needed for molecular mechanics (Sharp, 1991; Gilson *et al.*, 1995; Cortis *et al.*, 1996; Ripoll *et al.*, 1996; Luo *et al.*, 2002). Finally, the possibility that solvent-sized and thus high-dielectric-filled voids can appear in the protein interior during a simulation of unfolding creates discontinuities of the energy surface that would be very problematic in dynamics. These considerations have led many workers to consider approximate models, such as the generalized Born theories considered next.

3. *The Generalized Born Model*

The Born (1920) model computes the electrostatic work required to move a charged sphere from a vacuum environment into a continuous dielectric region. The result is proportional to the square of the charge (as in the COSMO model discussed above), and is inversely proportional to the size of the ion. The basic *ansatz* of generalized Born theory (Still *et al.*, 1990; Bashford and Case, 2000) is to extend these ideas to non-spherical molecules by casting the electrostatic contribution to solvation into the following form:

$$\Delta G_{es} = -\frac{1}{2}\sum_i \sum_j \frac{q_i q_j}{f_{GB}}\left(1 - \frac{1}{\varepsilon_{out}}\right) \tag{24}$$

A simple plausibility argument for the generalized Born model is the following. Imagine a molecule consisting of charges q_i embedded in spheres of radii, a_i; if the separation r_{ij} between any two spheres is sufficiently large in comparison to the radii, then the solvation free energy can be given by a sum of individual Born terms, and pairwise Coulombic terms:

$$\Delta G_{es} = -\sum_i \frac{q_i^2}{2a_i}\left(1 - \frac{1}{\varepsilon_{out}}\right) - \frac{1}{2}\sum_i \sum_{j \neq i} \frac{q_i q_j}{r_{ij}}\left(1 - \frac{1}{\varepsilon_{out}}\right) \tag{25}$$

where the factor $(1 - 1/\varepsilon_{out})$ appears in the pairwise terms because the Coulombic interactions are re-scaled by the change of dielectric constant on going from vacuum to solvent.

The goal of generalized Born theory can be thought of as an effort to find a relatively simple analytical formula resembling equation 25 that for real molecular geometries will capture as much as possible the physics of the Poisson equation. Hence, to obtain a formula such as equation 24, we seek a function f_{GB}, such that in the self terms (the first summation in equation 25), f_{GB} acts as an "effective Born radius" while in the pairwise terms (second summation in equation 25) it becomes an effective interaction distance. A variety of functional forms for f_{GB} have been tested (Onufriev *et al.*, 2002), but the most common form chosen is (Still *et al.*, 1990):

$$f_{GB}(r_{ij}, R_i, R_j) = \left[r_{ij}^2 + R_i R_j \exp\left(-\frac{r_{ij}^2}{4R_i R_j}\right)\right]^{1/2} \tag{26}$$

Here, the R_i are the effective Born radii of the atoms, which generally depend not only on a_i, the intrinsic atomic radii, but also on the radii and relative positions of all other atoms. Ideally, R_i should be chosen so that if one were to solve the Poisson equation for a single charge placed at the position of atom i, and a dielectric boundary determined by all of the molecule's atoms and their radii, then the self-energy of charge q_i in its reaction field, $q_i\phi(r_i)/2$, would be equal to

$$\frac{q_i^2}{2R_i}\left(1 - \frac{1}{\varepsilon_{out}}\right)$$

Such a scheme leads to what have been called "perfect" radii (Onufriev *et al.*, 2002). These, in conjunction with the functional form for the cross terms, equation 26, then yields very accurate cross terms when compared with numerical solutions of the Poisson equation. Thus the key to accurate GB models is the estimation of the effective radii (Onufriev *et al.*, 2002; Lee *et al.*, 2002).

In practice, of course, obtaining "perfect" radii requires an expensive solution to the Poisson-Boltzmann equation, and a faster algorithm for computing the R_i values is needed for efficient molecular dynamics simulations. In the classical electrostatics of a linearly polarizable medium (Jackson, 1999), the work required to assemble a charge distribution can be formulated either in terms of a product of the charge distribution with the electric potential, as in equation 23 above, or in terms of the scalar product of the electric field **E** and the electric displacement **D**:

$$W = \frac{1}{8\pi}\int \mathbf{E}\cdot\mathbf{D}dV \tag{27}$$

We now introduce the essential approximation used in most forms of generalized Born theory: that the electric displacement is Coulombic in form and remains so even as the exterior dielectric is altered from 1 to ε_{out} in the solvation process. In other words, the displacement that is due to the charge of atom i (which for convenience is here presumed to lie on the origin) is,

$$\mathbf{D}_i \approx \frac{q_i\mathbf{r}}{r^3}$$

This is called the *Coulomb field approximation.* In the spherically symmetric case (as in the Born formula) it is exact, but in more complex geometries, there can be substantial deviations. The work of placing a charge q_i at the origin within a molecule whose interior dielectric constant is ε_{in}, surrounded by a medium of dielectric constant ε_{out} and in which no other charges have yet been placed is then,

$$W = \frac{1}{8\pi}\int (\mathbf{D}/\varepsilon)\cdot\mathbf{D}dV \approx \frac{1}{8\pi}\int_{in}\frac{q_i^2}{r^4\varepsilon_{in}}dV + \frac{1}{8\pi}\int_{out}\frac{q_i^2}{r^4\varepsilon_{out}}dV \tag{28}$$

The electrostatic component of the solvation energy is found by taking the difference in W when ε_{out} is changed from 1 to 80:

$$\Delta G_{es} = \frac{-1}{8\pi}\left(1-\frac{1}{80}\right)\int_{out}\frac{q_i^2}{r^4}dV \tag{29}$$

where the contribution that is due to the interior region has canceled in the subtraction.

Comparing equation 29 to the Born formula (equation 16), we conclude that the effective Born radius should be,

$$R_i^{-1} = \frac{1}{4\pi}\int_{out} r^{-4}dV \tag{30}$$

It is often convenient to re-write this in terms of an integration over the interior region, excluding a radius a_i around the origin,

$$R_i^{-1} = a_i^{-1} - \frac{1}{4\pi}\int^{\prime} r^{-4}dV \tag{31}$$

Here the integral is over the interior of the molecule, excluding regions within a sphere of radius a_i of atom i, and we have used the fact that the integration of r^{-4} over all space outside radius a is simply $4\pi a^{-1}$. In the case of a monatomic ion, where the molecular boundary is simply the sphere of radius a_i, the Born formula is recovered exactly.

The original formulation (Still *et al.*, 1990), as well as some more recent methods (Lee *et al.*, 2002), use a numerical evaluation of this integral. In alternative ''pairwise'' approaches, the integral is estimated via a sum over pairs of atoms. There are a variety of ways to implement a pairwise formula. In the ACE (analytical continuum electrostatics) model (Schaefer and Karplus, 1996; Schaefer *et al.*, 1998; Calimet *et al.*, 2001), Voronoi volumes for various atom types are determined from a database of protein structures, and these are used to estimate the contribution of each atom to the integral in equation 30. An alternative approach considers the molecule as a set of (overlapping) spheres. The integrals over spheres can then be obtained analytically, including the case that atom j overlaps atom i (Schaefer and Froemmel, 1990; Hawkins *et al.*, 1995, 1996). A straightforward pairwise summation using these ideas would over-count the solute region, since neighboring atoms j themselves overlap with each other. Hawkins *et al.* proposed scaling the neighboring values of a_j as an empirical correction to compensate for this neglect of overlap (Hawkins *et al.*, 1995, 1996); the original correction factors were subsequently optimized in the TINKER package to fit small molecule data. The expression for the generalized Born radii then takes the form,

$$R_i^{-1} = a_i^{-1} - \sum_j H(r_{ij}, S_j a_j) \tag{32}$$

where H is a fairly complex expression and the S_j scaling factors are the additional empirical parameters that account for overlaps. Several groups have adopted this idea, using different training sets to determine the intrinsic radii and scaling factors (Hawkins *et al.*, 1996; Jayaram *et al.*, 1998; Dudek *et al.*, 1998; Tsui and Case, 2000, 2001).

Ghosh *et al.* (1998) have proposed an alternative approach in which the Coulomb field is still used in place of the correct field and Green's theorem is used to convert the volume integral in equation 31 to a surface integral. At this level, the ''S-GB'' (surface-GB) model is formally identical to the model outlined above, although there are potential computational advantages in the surface integral approach, especially for large systems and for evaluating gradients. In practice, empirical short-range and long-range corrections are added to improve agreement with numerical Poisson theory. The combination of S-GB with novel models for the non-electrostatic component of solvation appears to offer excellent prospects for many applications, including distinguishing misfolded decoys from more native conformations (Gallicchio *et al.*, 2002; Felts *et al.*, 2002).

4. *Incorporation of Salt Effects*

Generalized Born models have not traditionally considered salt effects, but the model can be extended to low-salt concentrations at the Debye-Huckel level by the following arguments (Srinivasan *et al.*, 1999). The basic idea of the generalized Born approach can be viewed as an interpolation formula between analytical solutions for a single sphere and for widely separated spheres. For the latter, the solvation contribution becomes

$$\Delta G_{el} = -\left(1 - \frac{\exp[-\kappa r_{ij}]}{\varepsilon_{out}}\right)\frac{q_i q_j}{r_{ij}} \tag{33}$$

where κ is the Debye-Huckel screening parameter. The first term removes the gas-phase interaction energy, and the second term replaces it with a screened Coulomb potential. On the other hand, at short distances, the generalized Born formula is used to reduce to the result for a single spherical ion (Kirkwood, 1934; Tanford and Kirkwood, 1957):

$$\Delta G_{el} = -\frac{1}{2}\left(1 - \frac{1}{\varepsilon_{out}}\right)\frac{q^2}{a} - \frac{q^2\kappa}{2\varepsilon_{out}(1+\kappa b)} \tag{34}$$

where a is the radius of the sphere and b is the radial distance to which salt ions are excluded, so that b-a is the ion exclusion radius. To a close extent, these two limits can be obtained by the simple substitution

$$\left(1 - \frac{1}{\varepsilon_{out}}\right) \Rightarrow \left(1 - \frac{\exp[-\kappa f_{GB}]}{\varepsilon_{out}}\right) \tag{35}$$

in equation 24. This reduces directly to equation 33 for large distances and to the correct result as r_{ij} goes to zero to terms linear in κ. However, this model does not contain an ion exclusion radius and hence tends to overestimate salt effects (as compared with the usual PB model) by allowing counterions to approach the solute more closely than they should. A simple, *ad hoc* modification that leads to acceptable results can be obtained by scaling κ by 0.73 (Srinivasan *et al.*, 1999).

5. *"Second Generation" Generalized Born Models*

As outlined above, the key to making GB calculations more accurate, in the sense of agreement with PB calculations, is improved estimation of the effective Born radii. In the conventional models, two levels of approximation are involved, the Coulomb field approximation (CFA) used to derive equation 30 and approximations used to estimate the integral in that expression. Recent efforts to improve generalized Born models have looked at both of these issues.

Some work has gone beyond the CFA. Lee *et al.* (2002) have looked at replacing equation 30 with a formula that involves both volume integrals of r^{-4} and also r^{-5}, to provide a correction for deviations from spherical geometry. Requirements of dimensional consistency and reduction to the CFA form in the spherical case (where the CFA is exact) place strong constraints on the functional form. The most promising approach uses the formula

$$R_i^{-1} = P\left(\frac{1}{2a_i^2} - \frac{1}{4\pi}\int r^{-5}dV\right)^{1/2} - a_i^{-1} + \frac{1}{4\pi}\int r^{-4}dV \qquad (36)$$

which has one empirical parameter *P*, and where the integrals cover the same space as in equation 31. One can estimate the anticipated value for *P* from the asymptotic limit of a single spherical cavity, where $P = 2\sqrt{2}$ gives the exact Born result. Empirically, values slightly greater than 2.83 appear optimal. When the integrals are evaluated with an accurate grid-based numerical method, equation 36 provides remarkably good agreement with effective Born radii calculated from Poisson equation solutions. Further experimentation with the parameter *P*, for different dielectric environments, and molecular sizes and shapes is planned.

A second track for improving the GB model involves more accurate estimation of integrals such as that in equation 30. Onufriev *et al.* noted that models that estimate the integral based on a fused-spheres picture (and parameterized for small molecules) tend to understimate the effective radii of atoms deeply buried in proteins (Onufriev *et al.*, 2002). They argued that this happens because interstices between spheres are not counted as part of the low-dielectric interior, even when they are much too small to accommodate a water molecule. In their new method, an adjustable parameter λ is introduced into equation 32:

$$R_i^{-1} = a_i^{-1} - \lambda \sum_j H(r_{ij}, S_j a_j) \qquad (37)$$

This has the effect of keeping the effective radii for atoms near the surface the same as in conventional GB models (where the approximations are quite good) and to increase the R_i values for more buried atoms. Tests on protein structures showed substantially improved agreement with PB solutions not only for the total solvation energies but for the individual interaction terms that go into a pK_a calculation. Further elaborations of this basic idea appear to give considerable improvement in protein molecular dynamics simulations and folding landscape characterization (Onufriev, Bashford and Case, unpublished).

Another "second track" method (Lee *et al.*, 2002) takes a more systematic approach to improving the accuracy of these integrals. Developing some ideas from the ACE model (Schaefer *et al.*, 1998, 2001), this approach approximates the molecular volume by a set of atom-based, overlapping weight functions; these are chosen so that their sum is nearly constant inside the molecular surface and decays to zero very quickly outside it. In this way, the effective Born radii (and their derivatives) can be computed analytically in a sum-over-atoms approach.

C. *Molecular Dynamics Simulations with the Generalized Born Model*

It should be emphasized that adding a GB/SA solvation model to an existing protein molecular mechanics potential creates a new combination whose characteristics and quality can be difficult to predict. In principle, all of the parameters of the combined model should be reoptimized, following a protocol similar to that used for force fields designed for explicit solvent simulations. This time-consuming task has not yet been carried out, and early reports of results of molecular dynamics simulations are mostly still in a mode of testing the accuracy of the models.

Because GB simulations can be faster than explicit water simulations, and because the lack of solvent friction allows conformational space to be sampled more quickly, continuum solvent simulations are very attractive in studies of peptide and protein folding, where the qualitative features of the landscapes (including likely folding pathways and intermediates) are still unclear. Models with continuum solvation have reasonably good performance in distinguishing native folds from misfolded "decoys" (Vorobjev *et al.*, 1998; Dominy and Brooks, 2002; Felts *et al.*, 2002; Feig and Brooks, 2002), and searches for native conformations and folding pathways have also yielded promising results (Pande and Rokhsar, 1999; Bursulaya and Brooks, 2000; Mitsutake *et al.*, 2001; Zagrovic *et al.*, 2001, 2002a,b; Simmerling *et al.*, 2002; García and Sanbonmatsu, 2002). Simulations of folded proteins have also been reported (Wagner and Simonson, 1999; Dominy and Brooks, 1999; Calimet *et al.*, 2001; Cornell, *et al.*, 2001; Tsui and Case, 2001). The general aspects of the dynamics, as measured by NMR order parameters or by fluctuations about the mean,

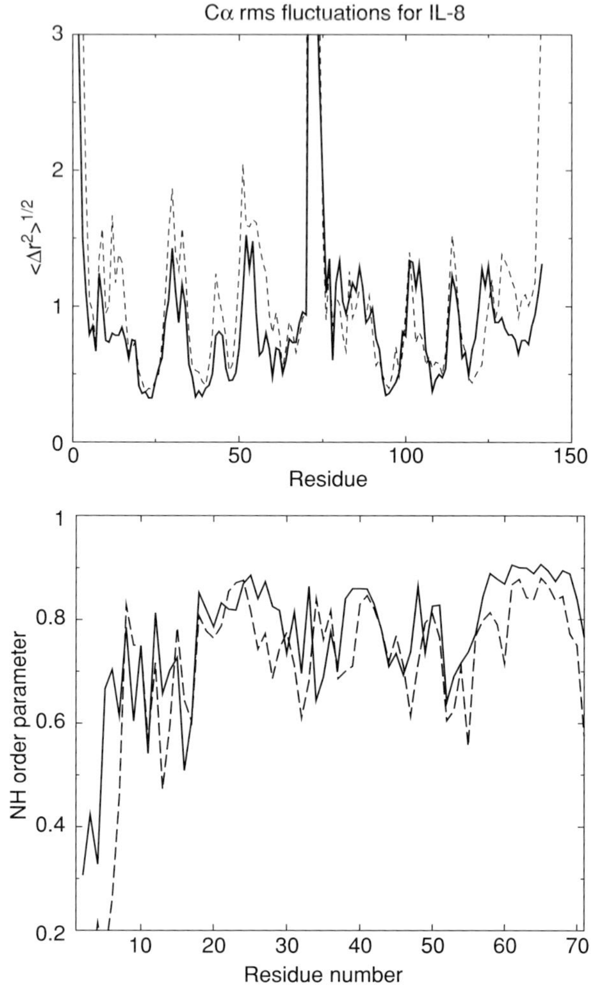

FIG. 9. (Top) Atomic fluctuations of the Cα atoms of interleukin-8 about the mean. Solid line: explicit water simulation; dashed line: generalized Born simulation. Residues 1-71 are from the first monomer, 72-142 from the second monomer. (Bottom) Order parameters (S^2) for N-H motion in interleukin-8, with the same meaning for solid and dashed lines. Adapted from Cornell *et al.* (2001).

are nearly the same as in more conventional, explicit water simulations. This is illustrated in Fig. 9 for interleukin-8. However, deviations from the crystallographic or NMR starting structures have sometimes been noticeably larger than in explicit solvent simulations of comparable length.

To some extent, this might be a consequence of the improved conformational sampling, which lets the simulation more quickly find non-native structures that are energetically favored by the particular force field. But it also seems likely that the current generation of GB models do not have as good a balance between protein-protein and protein-solvent interactions as do the more widely tested explicit solvent models. Further experience and reparameterization will clearly be very useful in sorting this out, but users of the current generation of GB models should be aware that they have definite limitations, many of which are not yet fully understood.

V. Conclusions

An increase in computer power of at least two orders of magnitude should occur over the next decade. Without further research into the accuracy of force-field potentials, future macromolecular modeling may well be limited more by validity of the energy functions, particularly electrostatic terms, than by technical ability to perform the computations. For many calculations related to ligand binding, drug design, and protein structure prediction, accuracy of the underlying potential functions is critical. For example, imagine a hypothetical protein of $N = 100$ residues for which the lowest potential energy basin, representing the family of "native" folds, lies $\Delta E = 5$ kcal/mol below either a compact "misfolded" structure or the massive regime of unfolded states. Then the average acceptable error per residue if we are to be able to distinguish the correct fold is approximately $\Delta E/N^{1/2}$, or 0.5 kcal/mol. The above example has been discussed by Dill (1997) and Abagyan (1997), with a similar analysis advanced by many others. Note that the 0.5 kcal/mol value assumes that error is randomly distributed. Systematic errors in the potential function can further reduce the acceptable error per residue. However, contrary to the pessimistic dogma held by some in the modeling community, this simple error analysis indicates that further work on potential function accuracy is valuable.

Calculation of ligand and drug binding energetics would benefit greatly from potentials of "chemical accuracy"—errors of less than 0.5 kcal/mol in relative binding energies. The promise of structure-based drug design remains largely unmet as a direct result of the inability of current modeling protocols to generate reliable answers. In contrast to gas-phase hydrocarbon potentials, protein potentials have not reached the required level of accuracy, but they are close enough to suggest that improved treatment of the major sources of error will lead to success. In addition, the argument should be made that we do not adequately understand the basics of protein structure and energetics until quantitative force-field modeling methods become truly predictive tools.

In this article we have sought to provide a description of the current standard practice of protein simulations, providing some of the details of how three of the currently popular protein force fields were developed and tested. Along with this, we have highlighted two directions in which force fields are currently moving, incorporating models of electronic polarizability and simplifying solvent interactions with continuum ideas. This selection of topics is of course heavily influenced by our interests and experience and necessarily leaves out a discussion of other important aspects of protein force fields. Nevertheless, we hope that even this restricted view of the successes and limitations of current force fields will provide a useful perspective on ways in which "chemical accuracy" may be more closely approached in practical protein simulations in the near future.

Acknowledgments

This work was supported by NIH grant GM56531 (to DAC) and NSF grant 9808317 (to JWP). We thank Dr. Alan Grossfield for the WHAM calculations presented in Fig. 5 and Table III and Dr. Pengyu Ren for the AMOEBA polarizable force field results reported in Table IV and Figs. 6 and 7.

References

Abagyan, R. A. (1997). *In* "Computer Simulation of Biomolecular Systems" Vol. 3, (W. F. van Gunsteren, P. K. Weiner, and A. J. Wilkerson, Eds.), pp. 363–394. Kluwer Academic Publishers, Dordrecht, The Netherlands.

Allinger, N. L. (1976). *Adv. Phys. Org. Chem.* **13,** 1.

Allinger, N. L., Yuh, Y. H., and Lii, J.-H. (1989). *J. Am. Chem. Soc.* **111,** 8551–8566.

Allinger, N. L., Chen, K., and Lii, J.-H. (1996). *J. Comput. Chem.* **17,** 642–668.

Amicangelo, J. C., and Armentrout, P. B. (2000). *J. Phys. Chem. A* **104,** 11420–11432.

Andzelm, J., Kölmel, C., and Klamt, A. (1995). *J. Chem. Phys.* **103,** 9312–9320.

Applequist, J. (1983). *J. Math. Phys.* **24,** 736–741.

Applequist, J. (1993). *J. Phys. Chem.* **97,** 6016–6023.

Applequist, J., Carl, J. R., and Fung, K.-K. (1972). *J. Am. Chem. Soc.* **94,** 2952–2960.

Bader, R. F. W. (1990). "Atoms in Molecules: A Quantum Theory." Oxford University Press, New York.

Baldridge, K., and Klamt, A. (1997). *J. Chem. Phys.* **106,** 6622–6633.

Banks, J. L., Kaminski, G. A., Zhou, R., Mainz, D. T., Berne, B. J., and Friesner, R. A. (1999). *J. Chem. Phys.* **110,** 741–754.

Barnes, P., Finney, J. L., Nicholas, J. D., and Quinn, J. E. (1979). *Nature* **282,** 459–464.

Barone, V., and Cossi, M. (1998). *J. Phys. Chem. A* **102,** 1995–2001.

Bashford, D., and Case, D. A. (2000). *Annu. Rev. Phys. Chem.* **51,** 129–152.

Bayly, C., Cieplak, P., Cornell, W., and Kollman, P. A. (1993). *J. Phys. Chem.* **97,** 10269–10280.

Beachy, M. D., Chapman, D., Murphy, R. B., Halgren, T. A., and Friesner, R. A. (1997). *J. Am. Chem. Soc.* **119,** 5908–5920.

Becker, O., MacKerell, A. D., Jr., Roux, B., and Watanabe, M. (2001). "Computational Biochemistry and Biophysics." Marcel Dekker, New York.

Berendsen, H. J. C., Grigera, J. R., and Straatsma, T. P. (1987). *J. Phys. Chem.* **91,** 6269–6271.

Berendsen, H. J. C., Postma, J. P. M., van Gunsteren, W. F., and Hermans, J. (1981). *In* "Intermolecular Forces." (B. Pullmann, Ed.), pp. 331–342. D. Reidel Publishing Company, Dordrecht, The Netherlands.

Bernal, J. D., and Fowler, R. H. (1933). *J. Chem. Phys.* **1,** 515–548.

Bernardo, D. N., Ding, Y., Krogh-Jespersen, K., and Levy, R. M. (1994). *J. Phys. Chem.* **98,** 4180–4187.

Beroza, P., and Case, D. A. (1998). *Methods Enzymol.* **295,** 170–189.

Bode, K. A., and Applequist, J. (1996). *J. Phys. Chem.* **100,** 17820–17824.

Born, M. (1920). *Z. Phys.* **1,** 45–48.

Brdarski, S., Astrand, P. O., and Karlström, G. (2000). *Theoret. Chem. Acc.* **105,** 7–14.

Breneman, C. M., and Wiberg, K. B. (1990). *J. Comput. Chem.* **11,** 361–373.

Brobjer, J. T., and Murrell, J. N. (1982). *J. Chem. Soc. Faraday Trans. 2* **78,** 1853–1870.

Brooks, B. R., Bruccoleri, R. E., Olafson, B. D., States, D. J., Swaminathan, S., and Karplus, M. (1983). *J. Comput. Chem.* **4,** 187–217.

Buckingham, A. D., and Fowler, P. W. (1985). *Can. J. Chem.* **63,** 2018–2025.

Burkert, U., and Allinger, N. L. (1982). "Molecular Mechanics." American Chemical Society, Washington, D.C.

Burnham, C. J., Li, J., Xantheas, S. S., and Leslie, M. (1999). *J. Chem. Phys.* **110,** 4566–4581.

Burnham, C. J., and Xantheas, S. S. (2002a). *J. Chem. Phys.* **116,** 1479–1510.

Burnham, C. J., and Xantheas, S. S. (2002b). *J. Chem. Phys.* **116,** 5115–5124.

Bursulaya, B. D., and Brooks, C. L., III. (2000). *J. Phys. Chem. B* **104,** 12378–12383.

Caldwell, J. W., and Kollman, P. A. (1995). *J. Phys. Chem.* **99,** 6208–6219.

Calimet, N., Schaefer, M., and Simonson, T. (2001). *Proteins* **45,** 144–158.

Carlson, M. J. (2000). "BUFF: A Biological Universal Forcefield Derived from Quantum Mechanics." Ph.D. thesis, California Institute of Technology.

Ceccarelli, M., and Marchi, M. (1997). *J. Phys. Chem. B* **101,** 2105–2108.

Chen, J. L., Noodleman, L., Case, D. A., and Bashford, D. (1994). *J. Phys. Chem.* **98,** 11059–11068.

Chessari, G., Hunter, C. A., Low, C. M. R., Packer, M. J., Vinter, J. G., and Zonta, C. (2002). *Chem. Eur. J.* **8,** 2860–2867.

Chipot, C., Angyan, J. G., Maigret, B., and Scheraga, H. A. (1993). *J. Phys. Chem.* **97,** 9788–9796.

Cieplak, P., Caldwell, J., and Kollman, P. (2001). *J. Comput. Chem.* **22,** 1048–1057.

Clark, M., Cramer, R. D., and Van Opdenbosch, N. (1989). *J. Comput. Chem.* **10,** 982–1012.

Colonna, F., and Evleth, E. (1992). *J. Comput. Chem.* **13,** 1234–1245.

Connolly, M. L. (1983). *Science* **221,** 709–713.

Cornell, W., Abseher, R., Nilges, M., and Case, D. A. (2001). *J. Mol. Graph. Model.* **19,** 136–145.

Cornell, W. D., Cieplak, P., Bayly, C. I., Gould, I. R., Merz, K. M., Jr., Ferguson, D. M., Spellmeyer, D. C., Fox, T., Caldwell, J. W., and Kollman, P. A. (1995). *J. Am. Chem. Soc.* **117,** 5179–5197.

Cornell, W. D., Cieplak, P., Bayly, C. I., and Kollman, P. A. (1993). *J. Am. Chem. Soc.* **115,** 9620–9631.

Corongiu, G. (1992). *Int. J. Quant. Chem.* **42,** 1209–1235.

Cortis, C. M., Langlois, J.-M., Beachy, M. D., and Friesner, R. A. (1996). *J. Chem. Phys.* **105,** 5472–5484.

Cramer, C. J. (2002). "Essentials of Computational Chemistry: Theory and Models." John Wiley and Sons, New York.

Cramer, C. J., and Truhlar, D. G. (1999). *Chem. Rev.* **99,** 2161–2200.

Daggett, V. (2002). *Acc. Chem. Res.* **35,** 422–429.

Daggett, V., and Levitt, M. (1993). *Annu. Rev. Biophys. Biomol. Struct.* **22,** 353–380.

Damm, W., and van Gunsteren, W. F. (2000). *J. Comput. Chem.* **21,** 774–787.

Dang, L. X., and Chang, T.-M. (1997). *J. Chem. Phys.* **106,** 8149–8159.

Darden, T. A. (2001). *In* "Computational Biochemistry and Biophysics." (O. Becker, A. D. MacKerell, Jr., B. Roux, and M. Watanabe, Eds.), pp. 91–114. Marcel Dekker, New York.

Darden, T., Pearlman, D., and Pedersen, L. G. (1998). *J. Chem. Phys.* **109,** 10921–10935.

Derreumaux, P., and Vergoten, G. (1995). *J. Chem. Phys.* **102,** 8586–8605.

de Vries, A. H., van Duijnen, P. T., Zijlstra, R. W. J., and Swart, M. (1997). *J. Electron Spect. Related Phenom.* **86,** 49–55.

Dill, K. A. (1997). *J. Biol. Chem.* **272,** 701–704.

Dixon, R. W., and Kollman, P. A. (1997). *J. Comput. Chem.* **18,** 1632–1646.

Dolney, D. M., Hawkins, G. D., Winget, P., Liotard, D. A., Cramer, C. J., and Truhlar, D. G. (2000). *J. Comput. Chem.* **21,** 340–366.

Dominy, B. N., and Brooks, C. L., III. (1999). *J. Phys. Chem. B* **103,** 3765–3773.

Dominy, B. N., and Brooks, C. L., III. (2002). *J. Comput. Chem.* **23,** 147–160.

Dougherty, D. A. (1995). *Science* **271,** 163–168.

Dudek, M. J., Ramnarayan, K., and Ponder, J. W. (1998). *J. Comput. Chem.* **19,** 548.

Dunning, T. H. (2000). *J. Phys. Chem. A* **104,** 9062–9080.

Dykstra, C. E. (1988). *J. Comput. Chem.* **9,** 476–487.

Dykstra, C. E. (1989). *J. Am. Chem. Soc.* **111,** 6168–6174.

Dykstra, C. E. (1993). *Chem. Rev.* **93,** 2339–2353.

Feig, M., and Brooks, C. L., III. (2002). *Proteins* **49,** 232–245.

Felts, A. K., Gallicchio, E., Wallqvist, A., and Levy, R. M. (2002). *Proteins* **48,** 404–422.

Ferguson, D. M. (1995). *J. Comput. Chem.* **16,** 501–511.

Foloppe, N., and MacKerell, Jr., A. D. (2000). *J. Comput. Chem.* **21,** 86–104.

Fox, T., and Kollman, P. A. (1998). *J. Phys. Chem. B* **102,** 8070–8079.

Gallicchio, E., Zhang, L. Y., and Levy, R. M. (2002). *J. Comput. Chem.* **23,** 517–529.

García, A. E., and Sanbonmatsu, K. Y. (2001). *Proteins* **42,** 345–354.

García, A. E., and Sanbonmatsu, K. Y. (2002). *Proc. Natl. Acad. Sci. USA* **99,** 2782–2787.

Ghosh, A., Rapp, C. S., and Friesner, R. A. (1998). *J. Phys. Chem. B* **102,** 10983–10990.

Gilson, M. K., McCammon, J. A., and Madura, J. D. (1995). *J. Comput. Chem.* **16,** 1081–1095.

Gough, C., DeBolt, S., and Kollman, P. A. (1992). *J. Comput. Chem.* **13,** 963–970.

Gresh, N. (1997a). *J. Chim. Phys. PCB* **94,** 1365–1416.

Gresh, N. (1997b). *J. Phys. Chem. A* **101,** 8680–8694.

Gresh, N., Guo, H., Salahub, D. R., Roques, B. P., and Kafafi, S. A. (1999). *J. Am. Chem. Soc.* **121,** 7885–7894.

Gresh, N., Tiraboschi, G., and Salahub, D. R. (1998). *Biopolymers* **45,** 405–425.
Guillot, B., and Guissani, Y. (2001). *J. Chem. Phys.* **114,** 6720–6733.
Guo, H., Gresh, N., Roques, B. P., and Salahub, D. R. (2000). *J. Phys. Chem. B* **104,** 9746–9754.
Hagler, A., Euler, E., and Lifson, S. (1974). *J. Am. Chem. Soc.* **96,** 5319–5327.
Hagler, A., and Lifson, S. (1974). *J. Am. Chem. Soc.* **96,** 5327–5335.
Halgren, T. A. (1992). *J. Am. Chem. Soc.* **114,** 7827–7843.
Halgren, T. A. (1996a). *J. Comput. Chem.* **17,** 490–519.
Halgren, T. A. (1996b). *J. Comput. Chem.* **17,** 520–552.
Halgren, T. A. (1996c). *J. Comput. Chem.* **17,** 553–586.
Halgren, T. A. (1996d). *J. Comput. Chem.* **17,** 616–641.
Halgren, T. A. (1999a). *J. Comput. Chem.* **20,** 720–729.
Halgren, T. A. (1999b). *J. Comput. Chem.* **20,** 730–748.
Halgren, T. A., and Damm, W. (2001). *Curr. Opin. Struct. Biol.* **11,** 236–242.
Halgren, T. A., and Nachbar, R. B. (1996). *J. Comput. Chem.* **17,** 587–615.
Harvey, S., and McCammon, J. A. (1987). "Dynamics of Proteins and Nucleic Acids." Cambridge University Press, Cambridge.
Hawkins, G. D., Cramer, C. J., and Truhlar, D. G. (1995). *Chem. Phys. Lett.* **246,** 122–129.
Hawkins, G. D., Cramer, C. J., and Truhlar, D. G. (1996). *J. Phys. Chem.* **100,** 19824–19839.
Hermans, J., Berendsen, H. J. C., van Gunsteren, W. F., and Postma, J. P. M. (1984). *Biopolymers* **23,** 1513–1518.
Hermida-Ramón, J. M., Brdarski, S., Karlström, G., and Berg, U. (2003). *J. Comput. Chem.* **24,** 161–176.
Hu, H., Elstner, M., and Hermans, J. (2003). *Proteins* **50,** 451–463.
Huang, N., and MacKerell, A. D., Jr. (2002). *J. Phys. Chem. A* **106,** 7820–7827.
Hünenberger, P. H., and van Gunsteren, W. F. (1997). *In* "Computer Simulations of Biomolecular Systems" Vol. 3, (W. F. van Gunsteren, P. K. Weiner, and A. J. Wilkinson, Eds.), pp. 3–82. Kluwer Academic Publishers, Dordrecht, The Netherlands.
Jackson, J. D. (1999). "Classical Electrodynamics." 3rd edition. John Wiley and Sons, New York.
Jacobson, M. P., Kaminski, G. A., Friesner, R. A., and Rapp, C. S. (2002). *J. Phys. Chem. B* **106,** 11673–11680.
Jayaram, B., Liu, Y., and Beveridge, D. L. (1998). *J. Phys. Chem.* **109,** 1465–1471.
Jorgensen, W. L. (1981). *J. Am. Chem. Soc.* **103,** 335–340.
Jorgensen, W. L. (1998). *In* "Encyclopedia of Computational Chemistry" (P. von R. Schleyer, N. L. Allinger, T. Clark, J. Gasteiger, P. A. Kollman, and H. F. Schaefer III, Eds.), pp. 1986–1989. John Wiley and Sons, Chichester.
Jorgensen, W. L., Chandrasekhar, J., Madura, J. D., Impey, R. W., and Klein, M. L. (1983). *J. Chem. Phys.* **79,** 926–935.
Jorgensen, W. L., Maxwell, D. S., and Tirado-Rives, J. (1996). *J. Am. Chem. Soc.* **118,** 11225–11236.
Jorgensen, W. L., and Swenson, C. J. (1985). *J. Am. Chem. Soc.* **107,** 569–578.
Jorgensen, W. L., and Tirado-Rives, J. (1988). *J. Am. Chem. Soc.* **110,** 1657–1671.
Kaminski, G. A., Friesner, R. A., Tirado-Rives, J., and Jorgensen, W. L. (2001). *J. Phys. Chem. B* **105,** 6474–6487.
Kaminski, G., and Jorgensen, W. L. (1996). *J. Phys. Chem.* **100,** 18010–18013.
Kaminski, G. A., Stern, H. A., Berne, B. J., Friesner, R. A., Cao, Y. X., Murphy, R. B., Zhou, R., and Halgren, T. A. (2002). *J. Comput. Chem.* **23,** 1515–1531.
Kirkwood, J. G. (1934). *J. Chem. Phys.* **2,** 351–361.

Kitao, O., and Ogawa, T. (2003). *Mol. Phys.* **101,** 3–17.
Kitson, D. H., and Hagler, A. T. (1988). *Biochemistry* **27,** 5246–5257.
Klamt, A., and Schüürmann, G. (1993). *J. Chem. Soc. Perkin Trans.* **2,** 799–805.
Koch, U., and Egert, E. (1995). *J. Comput. Chem.* **16,** 937–944.
Kollman, P., Dixon, R., Cornell, W., Fox, T., Chipot, C., and Pohorille, A. (1997). *In* "Computer Simulations of Biomolecular Systems," Vol. 3 (W. F. van Gunsteren, P. K. Weiner, and A. J. Wilkinson, Eds.), pp. 83–96. Kluwer Academic Publishers, Dordrecht, The Netherlands.
Kong, Y. (1997). "Multipole Electrostatic Methods for Protein Modeling with Reaction Field Treatment." Ph.D. thesis, Washington University School of Medicine.
Kumar, S., Rosenberg, J. M., Bouzida, D., Swendsen, R. H., and Kollman, P. A. (1995). *J. Comput. Chem.* **16,** 1339–1350.
Langley, C. H., and Allinger, N. L. (2002). *J. Phys. Chem. A* **106,** 5638–5652.
Leach, A. R. (2001). "Molecular Modelling. Principles and Applications." Second Edition. Prentice-Hall, Harlow, England.
Lee, M. S., Salsbury, F. R., Jr., and Brooks, C. L. III. (2002). *J. Chem. Phys.* **116,** 10606–10614.
Levitt, M. (1983). *J. Mol. Biol.* **168,** 595–620.
Levitt, M., Hirshberg, M., Sharon, R., and Daggett, V. (1995). *Comp. Phys. Commun.* **91,** 215–231.
Levitt, M., Hirschberg, M., Sharon, R., Laidig, K. E., and Daggett, V. (1997). *J. Phys. Chem. B* **101,** 5051–5061.
Levitt, M., and Sharon, R. (1988). *Proc. Natl. Acad. Sci.* **85,** 7557–7561.
Lifson, S., and Warshel, A. (1969). *J. Chem. Phys.* **49,** 5116–5129.
Lii, J.-H., and Allinger, N. (1991). *J. Comput. Chem.* **12,** 186–199.
Lim, C., Bashford, D., and Karplus, M. (1991). *J. Phys. Chem.* **95,** 5610–5620.
Liu, Y.-P., Kim, K., Berne, B. J., Friesner, R. A., and Rick, S. W. (1998). *J. Chem. Phys.* **108,** 4739–4755.
Luo, R., David, L., and Gilson, M. K. (2002). *J. Comput. Chem.* **23,** 1244–1253.
Luo, R., David, L., Hung, H., Devaney, J., and Gilson, M. K. (1999). *J. Phys. Chem. B* **103,** 727–736.
Lybrand, T. P., and Kollman, P. A. (1985). *J. Chem. Phys.* **83,** 2923–2933.
Mackay, D. H. J., Berens, P. H., Wilson, K. R., and Hagler, A. T. (1984). *Biophys. J.* **46,** 229–248.
MacKerell, A. D., Jr. (2001). *In* "Computational Biochemistry and Biophysics" (O. Becker, A. D. MacKerell, Jr., B. Roux, and M. Watanabe, Eds.), pp. 7–38. Marcel Dekker, New York.
MacKerell, A. D., Jr., Bashford, D., Bellott, M., Dunback, R. L., Jr., Evanseck, J. D., Field, M. J., Fischer, S., Gao, J., Guo, H., Ha, S., Joseph-McCarthy, D., Kuchnir, L., Lau, F. T. K., Mattos, C., Michnick, S., Ngo, T., Nguyen, D. T., Prodhom, B., Reiher, W. E. III, Roux, B., Schlenkrich, M., Smith, J. C., Stote, R., Straub, J., Watanabe, M., Wiórkiewicz-Kuczera, J., Yin, D., and Karplus, M. (1998). *J. Phys. Chem. B* **102,** 3586–3616.
Mahoney, M. W., and Jorgensen, W. L. (2000). *J. Chem. Phys.* **112,** 8910–8922.
Mahoney, M. W., and Jorgensen, W. L. (2001). *J. Chem. Phys.* **115,** 10758–10768.
Mannfors, B. E., Mirkin, N. G., Palmo, K., and Krimm, S. (2003). *J. Phys. Chem. A.* **107,** 1825–1832.
Maple, J. R., Hwang, M.-J., Jalkanen, K. J., Stockfisch, T. P., and Hagler, A. T. (1998). *J. Comput. Chem.* **18,** 430–458.

Maple, J. R., Hwang, M.-J., Stockfisch, T. P., Dinur, U., and Hagler, A. T. (1994). *J. Comput. Chem.* **15,** 162–182.
Mark, P., and Nilsson, L. (2002). *J. Comput. Chem.* **23,** 1211–1219.
Masella, M., Gresh, N., and Flament, J.-P. (1998). *J. Chem. Soc. Faraday Trans.* **94,** 2745–2753.
Masunov, A., and Lazaridis, T. (2003). *J. Am. Chem. Soc.* **125,** 1722–1730.
Matsuoka, O., Clementi, E., and Yoshimine, M. (1976). *J. Chem. Phys.* **64,** 1351–1361.
Matta, C. F., and Bader, R. F. W. (2000). *Proteins* **40,** 310–329.
Maxwell, D. W., Tirado-Rives, J., and Jorgensen, W. L. (1995). *J. Comput. Chem.* **16,** 984–1010.
Miller, K. J. (1990). *J. Am. Chem. Soc.* **112,** 8533–8542.
Mitchell, P. J., and Fincham, D. (1993). *J. Phys.: Condens. Matter* **5,** 1031–1038.
Mitsutake, A., Sugita, Y., and Okamoto, Y. (2001). *Biopolymers* **60,** 96–123.
Momany, F. A., McGuire, R. F., Burgess, A. W., and Scheraga, H. A. (1975). *J. Phys. Chem.* **79,** 2361–2381.
Némethy, G., Pottle, M. S., and Scheraga, H. A. (1983). *J. Phys. Chem.* **87,** 1883–1887.
Neria, E., Fischer, S., and Karplus, M. (1996). *J. Chem. Phys.* **105,** 1902–1921.
Niketic, S. R., and Rasmussen, K. (1997). "The Consistent Force Field: A Documentation." Springer-Verlag, New York.
Nina, M., Beglov, D., and Roux, B. (1997). *J. Phys. Chem. B* **101,** 5239–5248.
Nina, M., Im, W., and Roux, B. (1999). *Biophys. Chem.* **78,** 89–96.
Nymand, T. M., and Linse, P. (2000). *J. Chem. Phys.* **112,** 6152–6160.
Oie, T., Maggiora, G. M., and Christoffersen, R. E. (1981). *Int. J. Quant. Chem., Quant. Biol. Symp.* **8,** 1–47.
Onufriev, A., Bashford, D., and Case, D. A. (2000). *J. Phys. Chem. B* **104,** 3712–3720.
Onufriev, A., Case, D. A., and Bashford, D. (2002). *J. Comput. Chem.* **23,** 1297–1304.
Ösapay, K., Young, W. S., Bashford, D., Brooks, C. L. III, and Case, D. A. (1996). *J. Phys. Chem.* **100,** 2698–2705.
Palmo, K., and Krimm, S. (1998). *J. Comput. Chem.* **19,** 754–768.
Palmo, K., Mannfors, B., Mirkin, N. G., and Krimm, S. (2003). *Biopolymers* **68,** 383–394.
Pande, V. S., and Rokhsar, D. S. (1999). *Proc. Natl. Acad. Sci. USA* **96,** 9062–9067.
Popelier, P. (2000). "Atoms in Molecules: An Introduction." Pearson Education Ltd., Essex, England.
Popelier, P. L. A., Joubert, L., and Kosov, D. S. (2001). *J. Phys. Chem. A* **105,** 8254–8261.
Popelier, P. L. A., and Stone, A. J. (1994). *Mol. Phys.* **82,** 411–425.
Price, D. J., and Brooks, C. L. III. (2002). *J. Comput. Chem.* **23,** 1045–1057.
Price, S. L. (2000). *Rev. Comput. Chem.* **14,** 225–289.
Price, S. L., Stone, A. J., and Alderton, M. (1984). *Mol. Phys.* **52,** 987–1001.
Rappé, A. K., Casewit, C. J., Colwell, K. S., Goddard, W. A., and Skiff, W. M. (1992). *J. Am. Chem. Soc.* **114,** 10024–10035.
Rappé, A. K., and Goddard, W. A. (1991). *J. Phys. Chem.* **95,** 3358–3363.
Reiher, W. E., III, (1985). "Theoretical Studies of Hydrogen Bonding." Ph.D. thesis, Harvard University.
Ren, P., and Ponder, J. W. (2002). *J. Comput. Chem.* **23,** 1497–1506.
Ren, P., and Ponder, J. W. (2003). *J. Phys. Chem. B* **107,** 5933–5947.
Reynolds, C. A., Essex, J. W., and Richards, W. G. (1992). *J. Am. Chem. Soc.* **114,** 9075–9079.
Rick, S. W., and Stuart, S. J. (2002). *Rev. Comput. Chem.* **18,** 89–146.
Rick, S. W., Stuart, S. J., and Berne, B. J. (1994). *J. Chem. Phys.* **101,** 6141–6156.

Ripoll, D. R., Vorbjev, Y. N., Liwo, A., Vila, J. A., and Scheraga, H. A. (1996). *J. Mol. Biol.* **264,** 770–783.

Rizzo, R. C., and Jorgensen, W. L. (1999). *J. Am. Chem. Soc.* **121,** 4827–4836.

Roterman, I. K., Lambert, M. H., Gibson, K. D., and Scheraga, H. A. (1989). *J. Biomol. Struct. Dyn.* **7,** 421–453.

Roux, B., and Bernèche, S. (2002). *Biophys. J.* **82,** 1681–1684.

Roux, B., and Simonson, T. (1999). *Biophys. Chem.* **78,** 1–20.

Saint-Martin, H., Hernández-Cobos, J., Bernal-Uruchurtu, M. I., Ortega-Blake, I., and Berendsen, H. J. C. (2000). *J. Chem. Phys.* **113,** 10899–10912.

Schaefer, M., Bartels, C., and Karplus, M. (1998). *J. Mol. Biol.* **284,** 835.

Schaefer, M., Bartels, C., Leclerc, F., and Karplus, M. (2001). *J. Comput. Chem.* **22,** 1857–1879.

Schaefer, M., and Froemmel, C. (1990). *J. Mol. Biol.* **216,** 1045–1066.

Schaefer, M., and Karplus, M. (1996). *J. Phys. Chem.* **100,** 1578–1599.

Schaefer, M., van Vlijmen, H. W. T., and Karplus, M. (1998). *Adv. Protein Chem.* **51,** 1–57.

Schlick, T. (2002). "Molecular Modeling and Simulation: An Interdisciplinary Guide." Springer-Verlag, New York.

Scott, W. R., Hünenberger, P. H., Tironi, I. G., Mark, A. E., Billeter, S. R., Fennen, J., Torda, A. E., Huber, T., Krüger, P., and van Gunsteren, W. F. (1999). *J. Phys. Chem. A* **103,** 3596–3607.

Sharp, K. (1991). *J. Comput. Chem.* **12,** 454–468.

Sharp, K. A., and Honig, B. (1990). *Annu. Rev. Biophys. Biophys. Chem.* **19,** 301–332.

Sharp, K. A., Kumar, S., Rossky, P. J., Friedman, R. A., and Honig, B. (1996). *J. Phys. Chem.* **100,** 14166–14177.

Simmerling, C., Strockbine, B., and Roitberg, A. E. (2002). *J. Am. Chem. Soc.* **124,** 11258–11259.

Simonson, T. (2002). *Proc. Natl. Acad. Sci. USA* **99,** 6544–6549.

Simonson, T., and Brünger, A. T. (1994). *J. Phys. Chem.* **98,** 4683–4694.

Sitkoff, D., Sharp, K. A., and Honig, B. (1994). *J. Phys. Chem.* **98,** 1978–1988.

Smith, W. (1998). *CCP5 Newsletter* **46,** 18–30.

Sokalski, W. A., Keller, D. A., Ornstein, R. L., and Rein, R. (1993). *J. Comput. Chem.* **14,** 970–976.

Sokalski, W. A., and Poirier, R. A. (1983). *Chem. Phys. Lett.* **98,** 86–92.

Sprik, M., and Klein, M. L. (1988). *J. Chem. Phys.* **89,** 7556–7560.

Srinivasan, J., Trevathan, M. W., Beroza, P., and Case, D. A. (1999). *Theor. Chem. Acc.* **101,** 426–434.

Steinbach, P. J., and Brooks, B. R. (1994). *J. Comput. Chem.* **15,** 667–683.

Stern, H. A., Kaminski, G. A., Banks, J. L., Zhou, R., Berne, B. J., and Friesner, R. A. (1999). *J. Phys. Chem. B* **103,** 4730–4737.

Stern, H. A., Rittner, F., Berne, B. J., and Friesner, R. A. (2001). *J. Am. Chem. Phys.* **115,** 2237–2251.

Still, W. C., Tempczyk, A., Hawley, R. C., and Hendrickson, T. (1990). *J. Chem. Phys.* **115,** 6127–6129.

Stillinger, F. H., and David, C. W. (1978). *J. Chem. Phys.* **69,** 1473–1484.

Stillinger, F. H., and Rahman, A. (1974). *J. Chem. Phys.* **60,** 1545–1557.

Stone, A. J. (1981). *Chem. Phys. Lett.* **83,** 233–238.

Stone, A. J. (1996). "The Theory of Intermolecular Forces." Oxford University Press, New York.

Stouch, T. R., and Williams, D. E. (1993). *J. Comput. Chem.* **14,** 858–866.

Stout, J. M., and Dykstra, C. E. (1998). *J. Phys. Chem. A* **102,** 1576–1582.

Sunner, J., Nishizawa, K., and Kebarle, P. (1981). *J. Phys. Chem.* **85,** 1814–1820.

Tanford, C., and Kirkwood, J. G. (1957). *J. Am. Chem. Soc.* **79,** 5333–5339.

Teeter, M. M., and Case, D. A. (1990). *J. Phys. Chem.* **94,** 8091–8097.

Thole, B. T. (1981). *Chem. Phys.* **59,** 341–350.

Tirado-Rives, J., and Jorgensen, W. L. (1990). *J. Am. Chem. Soc.* **112,** 2773–2781.

Tomasi, J., and Persico, M. (1994). *Chem. Rev.* **94,** 2027–2094.

Tomimoto, M., and Go, N. (1995). *J. Phys. Chem.* **99,** 563–577.

Toukmaji, A., Sagui, C., Board, J., and Darden, T. (2000). *J. Chem. Phys.* **113,** 10913–10927.

Truong, T. H., and Stefanovich, E. V. (1995). *J. Chem. Phys.* **103,** 3709–3717.

Tsui, V., and Case, D. A. (2000). *J. Am. Chem. Soc.* **122,** 2489–2498.

Tsui, V., and Case, D. A. (2001). *Biopolymers (Nucl. Acid. Sci.),* **56,** 275–291.

van der Vaart, A., Bursulaya, B. D., Brooks III, C. L., and Merz, K. M. J. (2000). *J. Phys. Chem. B* **104,** 9554–9563.

van Duijnen, P. T., and Swart, M. (1998). *J. Phys. Chem. A* **102,** 2399–2407.

van Gunsteren, W. F., and Berendsen, H. J. C. (1987). "Groningen Molecular Simulation (GROMOS) Library Manual." University of Groningen, The Netherlands.

van Gunsteren, W. F., Daura, X., and Mark, A. E. (1998). *In* "Encyclopedia of Computational Chemistry" (R. von, P. Schleyer, N. L. Allinger, T. Clark, J. Gasteiger, P. A. Kollman, and H. F. Schaefer III, Eds.), pp. 1211–1216. John Wiley, Chichester.

Vargas, R., Garza, J., Dixon, D. A., and Hay, B. P. (2000). *J. Am. Chem. Soc.* **122,** 4750–4755.

Vedani, A. (1988). *J. Comput. Chem.* **9,** 269–280.

Veenstra, D., Ferguson, D., and Kollman, P. A. (1992). *J. Comput. Chem.* **13,** 971–978.

Vesely, F. J. (1977). *J. Comput. Phys.* **24,** 361–371.

Vorobjev, Y. N., Almagro, J. C., and Hermans, J. (1998). *Proteins* **32,** 399–413.

Wagner, F., and Simonson, T. (1999). *J. Comput. Chem.* **20,** 322–335.

Wang, J., Cieplak, P., and Kollman, P. A. (2000). *J. Comput. Chem.* **21,** 1049–1074.

Weiner, P. K., and Kollman, P. A. (1981). *J. Comput. Chem.* **2,** 287–303.

Weiner, S. J., Kollman, P. A., Case, D. A., Singh, U. C., Ghio, C., Alagona, G., Profeta, S. Jr., and Weiner, P. (1984). *J. Am. Chem. Soc.* **106,** 765–784.

Weiner, S. J., Kollman, P. A., Nguyen, D. T., and Case, D. A. (1986). *J. Comput. Chem.* **7,** 230–252.

Williams, D. E. (1988). *J. Comput. Chem.* **9,** 745–763.

Williams, D. E. (1991). *Rev. Comput. Chem.* **2,** 219–271.

Williams, D. E. (2001). *J. Comput. Chem.* **22,** 1154–1166.

Winn, P. J., Ferenczy, G. G., and Reynolds, C. A. (1997). *J. Phys. Chem. A* **101,** 5437–5445.

York, D. M., and Karplus, M. (1999). *J. Phys. Chem. A* **103,** 11060–11079.

Young, D. M. (1971). "Iterative Solution of Large Linear Systems." Academic Press, New York.

Yu, H., Hansson, T., and van Gunsteren, W. F. (2003). *J. Chem. Phys.* **118,** 221–234.

Zacharias, N., and Dougherty, D. A. (2002). *Trends Pharm. Sci.* **23,** 281–287.

Zagrovic, B., Snow, C. D., Khaliq, S., Shirts, M. R., and Pande, V. S. (2002a). *J. Mol. Biol.* **323,** 153–164.

Zagrovic, B., Snow, C. D., Shirts, M. R., and Pande, V. S. (2002b). *J. Mol. Biol.* **323,** 927–937.

Zagrovic, B., Sorin, E. J., and Pande, V. (2001). *J. Mol. Biol.* **313,** 151–169.

PROTEIN SIMULATION AND DRUG DESIGN

By CHUNG F. WONG* AND J. ANDREW McCAMMON†

*Howard Hughes Medical Institute and Department of Pharmacology, University of California–San Diego, La Jolla, California 92093-0365, and †Howard Hughes Medical Institute, Department of Pharmacology, and Department of Chemistry and Biochemistry, University of California–San Diego, La Jolla, California 92093-0365

I. Introduction 87
II. Fixed-Conformation Models 88
A. Force-Field Energy Calculations 88
B. Poisson Model 90
C. Poisson-Boltzmann Model: Including Salt Effects 91
D. Generalized Born Model 92
E. Protonation States of Proteins and Ligands 94
III. Including Protein and Ligand Flexibility in Modeling Molecular Recognition 95
A. Free Energy Calculations 96
B. Kirkwood Integration Formula 97
C. Locally Enhanced Sampling 98
D. λ Dynamics Method 99
E. Parallel Fast-Growth Method 100
F. Translational and Rotational Entropy 101
IV. Fast Approximate Simulation Methods for Calculating Binding Free Energies 103
A. Systematic Sensitivity Analysis 103
B. Single-Step Zwanzig's Perturbation Method 105
C. Combining Explicit and Implicit Solvent Models 107
D. Chemical-Scanning Computational Experiments 109
E. Use of Explicit-Solvent Models to Parametrize Implicit-Solvent Models 110
F. Semi-Empirical Linear Response Theory 110
V. Including Protein Flexibility in Virtual Screening 111
A. Relaxed Complex Methods 112
B. Dynamic Pharmacophore Method 114
VI. Enhanced Sampling via the Replica-Exchange Method 115
VII. Conclusions 117
References 118

I. Introduction

The rapid advance of computer technology, force-field developments, and simulation methods and algorithms has gradually made force-field–based methods more easily and reliably useful in computer-aided drug design. In some applications, simple molecular mechanics geometry optimization and energy calculations can already suggest compounds that

ADVANCES IN PROTEIN CHEMISTRY, Vol. 66

are worthwhile to make and/or screen. Such calculations can be improved by using modern continuum solvent models in improving the description of solvation effects. Proper account of protein flexibility—via molecular dynamics simulations, for example—can further enhance the predictive reliability of simulation models. Combined with rigorous statistical mechanical theories, these simulations can help predict binding free energy (Tembe and McCammon, 1984; Wong and McCammon, 1986b; Wong and McCammon, 1986a; Mezei and Beveridge, 1986; Bash *et al.*, 1987; Jorgensen, 1989; Beveridge and DiCapua, 1989). While these simulations were limited to several tens of picoseconds in the early stages, they had moved a step forward beyond the fixed-conformation picture that dominated many scientists' thinking at the time. Recent developments in simulation methods and computer hardware have made it possible to extend the time scale of these simulations so that more reliable predictions can be made and so that results can be obtained more quickly. Methods that improve conformational sampling should further facilitate these simulations. Various approximate methods with different degrees of sophistication have also been introduced to facilitate the modeling of molecular recognition. In practical drug design, it should be beneficial to use a hierarchical approach in which simple but fast methods are first used for preliminary screening of real or virtual compound libraries and for suggesting potentially useful drug candidates to make/screen, followed by further evaluations by models with increasing levels of sophistication. In this article, we will review various approximate and rigorous force-field-based methods that can fit into such a hierarchical approach.

II. Fixed-Conformation Models

A. *Force-Field Energy Calculations*

In practical drug-design applications, a modeler often needs to come up with suggestions of compounds to make or screen within hours or days and it is difficult to use complicated models such as explicit-solvent molecular dynamics simulations on a day-to-day basis. As a result, simple molecular mechanics models have been used more frequently than molecular dynamics simulation models. In favorable cases, molecular mechanics studies can already be useful in drug design. For example, a ligand can be docked into the binding site of its target via constrained energy minimization in which the constraints are obtained from experimental data or a pharmacophore model. Once the structural model of a protein-ligand complex is built, one can examine what analogs of the ligand should be made to target specific features of its receptor. Simple geometric, electrostatics, and hydrogen donor/acceptor considerations can often suggest suitable compounds that are worthwhile to make or screen. These suggestions can be further refined by quantitative calculations. A simple way to estimate the binding energy between a ligand and its receptor can be obtained from

$$\Delta E = E_{\text{complex}} - E_{\text{protein}} - E_{\text{ligand}} \tag{1}$$

where E_{complex}, E_{protein}, and E_{ligand} are the force-field energies of the complex, the protein, and the ligand, respectively. The fixed-conformation approximation is often used for speedy calculations. For example, an energy-minimized structure of the complex can provide initial conformational models for the complex, the protein, and the ligand. Often, simple distance-dependent dielectric functions are used in the energy minimization and the subsequent energy calculations to model solvent-corrected electrostatic interactions. Increasing computing power and new theoretical and algorithmic developments have also made it easier to use more sophisticated continuum solvent models in energy calculations. A popular model includes two terms—one describes electrostatics and the other hydrophobic contributions—to account for the effects of solvation on binding. A common way to describe the hydrophobic contributions to binding free energy is via

$$G_{\phi} = \gamma SA + c \tag{2}$$

in which γ is a microscopic surface tension, SA is the solvent-accessible surface area of a solute, and c is a constant. The use of this relationship can be traced back to 1974, when Chothia (Chothia, 1974) analyzed the experimental data of Nozaki and Tanford (Nozaki and Tanford, 1971). On the other hand, several electrostatic models are becoming popular in biomolecular modeling. Although distance-dependent dielectric models are still often used for quick calculations, Poisson-Boltzmann and Generalized Born models have received increasing attention and have been reviewed recently (see, for example, Honig and Nicholls, 1995; Warshel and Papazyan, 1998; Orozco and Luque, 2000; Bashford and Case, 2000; Simonson, 2001). These models have not yet been used extensively in drug discovery but are likely to be used because these calculations can now be done within hours or minutes on common desktop computers by using the fixed-conformation approximation.

B. *Poisson Model*

The Poisson equation can provide a reasonably accurate description of the electrostatic properties of molecules. It reads

$$-\nabla \cdot \varepsilon(\vec{r})\nabla\phi(\vec{r}) = \rho(\vec{r}) \tag{3}$$

where $\varepsilon(\vec{r})$ is the dielectric function, $\phi(\vec{r})$ is the electrostatic potential, and $\rho(\vec{r})$ is the charge density at position $\vec{r}$. Analytical solutions are possible only for simple systems, but one can solve this equation numerically for complicated molecular models. For example, DelPhi (Rocchia *et al.*, 2002) and UHBD (Madura *et al.*, 1995) are two packages that have been developed to solve the Poisson equation for complicated systems by using the finite-difference method. The APBS package (Baker *et al.*, 2001) makes it possible to solve the Poisson equation for very large molecular systems. Typically, a solvated molecular system is represented by treating the molecule as a low dielectric surrounded by solvent with a different dielectric. The detailed shape of the molecule can be taken into account in these numerical calculations. Although the use of a single dielectric constant to describe the dielectric property of a protein, ligand, or protein-ligand complex is an oversimplification, such models have provided useful semi-quantitative insights into molecular recognition. For example, Schapira *et al.* (Schapira *et al.*, 1999) showed recently that the Poisson model, combined with suitable descriptions of hydrophobic effects and entropy changes, could predict the relative binding free energy between protein and ligand, protein and peptide, and protein and protein to within about 2.5 kcal/mol. Hünenberger *et al.* (Hünenberger *et al.*, 1999) calculated the binding free energy of a number of small-molecule inhibitors to protein kinase A, and the results agreed with the experimental trend. Wong *et al.* (Wong *et al.*, 2001) suggested that the predictive ability of such models may be improved by introducing a few empirical parameters determined by fitting them to existing experimental data, as in semi-empirical linear-response theories described in section IV, F. Our recent applications of similar continuum solvent models to study the binding of a number of pyrido[2,3-d]pyrimidine tyrosine kinase inhibitors to the fibroblast growth factor receptor 1, and the binding of flavonoids to the protein kinases CDK1 and CDK2, have yielded good correlation with measured log IC_{50}, having correlation coefficients ranging from 0.6 to over 0.9. Thus, although there are uncertainties in choosing model parameters—such as the solute dielectric, the atomic partial charges, and van der Waals radii—and in defining the dielectric boundary, the use of a two dielectric approximation, atomic partial charges from popular force fields or quantum calculations, and van der Waals radii from typical force-fields appears to yield useful semi-quantitative results in some applications. In these calculations, we used the van der Waals surface, rather than the solvent-accessible surface, to define the dielectric boundary between a solute and its surrounding solvent. The results can be quite different depending on which surface is used. But Vijayakumar and Zhou (Vijayakumar and Zhou, 2001) showed in studying the stabilization of barnase by salt bridges that results obtained by using the van der Waals surface agreed better with experimental results than those obtained by using the solvent-accessible surface area. The use of the solvent-accessible surface area with a small solute dielectric constant around 2–4 as commonly used in continuum electrostatics calculations tends to overestimate the desolvation costs of charged groups. This finding was reiterated in the study of the electrostatic stabilization of T4 lysozyme (Dong and Zhou, 2002). Since these calculations can be done relatively quickly, it should be worthwhile to apply such models to more systems relevant to drug design to gain further insights into the predictive reliability of these models.

C. *Poisson-Boltzmann Model: Including Salt Effects*

Experimental measurements are usually done in salt solutions. The Poisson equation can be generalized to account for salt effects. If one assumes the number density of the i-th type of ion n_i at different points in space to be given by

$$n_i = n_i^0 \exp\left(-\frac{q\phi}{RT}\right) \tag{4}$$

where n_i^0 is the number density of the i-th type of ion in a pure salt solution, R is the gas constant, and T is the absolute temperature, the charge density due to the i-th type of ion becomes

$$\rho_i = n_i q_i \tag{5}$$

where q_i is the charge of this type of ion. This charge density that is due to mobile ions can be added to the Poisson equation to obtain the well-known Poisson-Boltzmann equation

$$-\nabla \cdot \varepsilon(\vec{r})\ \nabla\phi(\vec{r}) = \rho(\vec{r}) + \sum_i q_i n_i^0 \exp\left(-\frac{q_i\phi}{RT}\right) \tag{6}$$

in which the sum is over all types of mobile ions. Programs such as APBS (Baker *et al.*, 2001), DelPhi (Rocchia *et al.*, 2002) and UHBD (Madura *et al.*, 1995) can now solve this nonlinear Poisson-Boltzmann equation directly. Before nonlinear solvers were developed, this equation was solved in its linearized form. The linearized equation is valid when $q_i\phi/RT$ is much smaller than one—which can be satisfied for low univalent salt concentrations and for moderately charged systems—and has the form

$$-\nabla \cdot \varepsilon(\vec{r})\ \nabla\phi(\vec{r}) + \varepsilon\kappa^2\phi\ (\vec{r}) = \rho(\vec{r}) \tag{7}$$

where κ, the inverse of the Debye-Hückel screening length, is given by

$$\kappa^2 = 2Ie^2/RT\varepsilon \tag{8}$$

in which the ionic strength I is defined as

$$I = \sum_i \frac{1}{2}\left(q_i^2/e^2\right)n_i \tag{9}$$

and ε is the dielectric constant of the solvent.

The solution of the Poisson-Boltzmann equation allows the effects of salts to be taken into account in a mean-field sense. This mean-field approximation should be adequate for many drug-design applications, although it breaks down for highly charged molecules or concentrated multivalent salt solutions. If necessary, Monte Carlo or Brownian dynamics simulations that simulate ions explicitly can be carried out, although with a significant increase in computational costs. For example, Monte Carlo simulations were carried out to study the binding of calcium ions to bovine calbindin (Svensson *et al.*, 1991). In these simulations, counter ions and salts surrounded a fixed protein within a large sphere and a uniform dielectric model was used in calculating electrostatic interactions. The calculated binding constants agreed well with experimental results for binding constants ranging over six orders of magnitude. Yu *et al.* (Yu *et al.*, 1996) developed a similar algorithm based on Brownian dynamics rather than Monte Carlo simulations, and periodic boundary conditions were used. These simulation models can be further improved by using inhomogeneous rather than uniform dielectric models. For example, the Poisson equation for the solute can first be solved to obtain the electrostatic potential surrounding the solute and the interactions between an ion and the solute can be estimated from the product of the potential and the charge of the ion.

D. *Generalized Born Model*

A somewhat simpler but faster model is the Generalized Born model (Still *et al.*, 1990). There are now different variants that perform better than the original model. For example, a recent model by Qiu *et al.* (Qiu *et al.*, 1997) utilizes the following formula to estimate the electrostatic contributions to the solvation energy

$$-\frac{1}{2}\left(1-\frac{1}{\varepsilon}\right)\sum_{i=l}^{N}\sum_{j=i}^{N}\frac{q_i q_j}{\sqrt{r_{ij}^2 + a_{ij}^2 e^{-D}}} \tag{10}$$

where

$$a_{ij}^2 = a_i a_j \tag{11}$$

and

$$D = r_{ij}^2/(2a_{ij})^2 \tag{12}$$

in which a_i is the Born radius of atom i and r_{ij} is the distance between atoms i and j. The Generalized Born equation (equation 10) includes a Born-like term

$$-\frac{1}{2}\left(1-\frac{1}{\varepsilon}\right)\frac{q_i^2}{a_i} \tag{13}$$

when $i = j$ and a Coulombic-like interaction term when $i \neq j$ except that the distance between two charges is replaced by an effective distance $\sqrt{r_{ij}^2 + a_{ij}^2 e^{-D}}$ that is dependent upon the Born radii. Qiu *et al.* (Qiu *et al.*, 1997) determine the effective Born radius of an atom in a molecule or molecular complex by using the equation

$$G_{pol} = -166.0\left(1-\frac{1}{\varepsilon}\right)\frac{q_i^2}{a_i} = \left(1-\frac{1}{\varepsilon}\right) \left(\frac{-166.0}{R_{vdw}+R_{off}+P_1} + \sum_{j\varepsilon\text{stretch}}\frac{P_2 V_j}{r_{ij}^4} + \sum_{j\varepsilon\text{bend}}\frac{P_3 V_j}{r_{ij}^4} + \sum_{j\varepsilon\text{nonbonded}}\frac{f_{\text{contact}} P_4 V_j}{r_{ij}^4}\right) \tag{14}$$

where R_{vdw} is the van der Waals radius of the ion, R_{off} is a dielectric offset distance, V_j is the volume of a neighboring atom j of atom i in the molecule or molecular complex, and P_1, P_2, P_3, and P_4 are four parameters of the model. The use of this equation is based on the assumption that the presence of another atom j at a distance r_{ij} from atom i displaces a high dielectric region of the solvent with volume V_j and changes the solvation energy by an amount proportional to the volume and inversely proportional to the fourth power of r_{ij} (based on charge-induced dipole arguments). Different scaling parameters P_2, P_3, and P_4 are used for stretching, bending, and nonbonded terms because atomic overlaps differ for these terms. For the nonbonded terms, an extra close contact function $f_{contact}$ is used to reduce the effective volume when atom *j* gets too close to the atom for which an effective Born radius is sought. Another parameter P_5 is used to fine-tune the contact function $f_{contact}$. The parameters P_1, P_2, P_3, P_4, and P_5 are determined by requiring equation 14 to give the same polarization energy as that obtained by solving the Poisson equation for a "representative" set of molecules. This analytical equation allows a Born radius to be determined quickly. David *et al.* (David *et al.*, 2000) found that this Generalized Born model did not work as well for proteins as for smaller molecules. However, they suggested that a special set of parameters could be developed for a specific protein if one intends to study the protein in detail. Methods (Ghosh *et al.*, 1998; Lee *et al.*, 2002) adding correction terms to the classical charge-induced dipole interactions have also been introduced to further improve the performance of the Generalized Born model.

E. *Protonation States of Proteins and Ligands*

An important issue to consider in reliably modeling protein-ligand interactions is the description of the protonation states of ionizable groups in the proteins and the ligands. The protonation state of acidic and basic amino acids of a protein and that of its ligand can be influenced by their mutual interactions, by their interactions with their environment, and by protein-ligand binding. It is necessary to estimate well the preferred protonation states of these groups to reliably estimate binding affinity. For example, small molecules containing amines could be protonated in an aqueous solution but deprotonated in a protein environment that does not favor charged groups. One can evaluate whether this change of protonation state on binding can occur by carrying out numerical calculations. For example, continuum electrostatics models have already been used to study the pH-dependent properties of proteins (Bashford and Karplus, 1990; Yang *et al.*, 1993; Yang and Honig, 1993; Antosiewicz *et al.*, 1994; Schaefer *et al.*, 1998), and they can be used to assign the most likely protonation states of a ligand and a protein and examine whether their protonation states change on binding. For a molecule or molecular complex containing N sites that can be either protonated or unprotonated, there are 2^N different possible protonation states for the molecule or molecular complex. Continuum electrostatics calculations within the fixed-conformation approximation can provide a reasonably fast identification of the most probable protonation state(s). One can use the Poisson-Boltzmann model, for example, to calculate the energies of structures with different combinations of the protonation states of different ionizable groups and identify the combinations that are most suitable for modeling the protein, the ligand, and the complex in a binding energy calculation. However, even within the single-conformation approximation, the number of calculations can grow rapidly as N increases. Different approximations have been introduced to facilitate these calculations. For example, one can base on the pK_a's of simple compounds having ionizable groups similar to those in the protein or in the ligand to predetermine the protonation states of amino acids or chemical functional groups in the protein and in the ligand. If the pK_a's are very different from the pH at which a binding free energy is desired, it is less likely that the protein environment can significantly shift these pK_a's to alter their preferred protonation states in solution. One can then remove these groups from the N possible ionizable groups to reduce the number of 2^N combinations one needs to consider in determining the most probable protonation states of a molecule or molecular complex. A cluster method may also be used in which ionizable groups are clustered in such a way that intercluster interactions are negligible (Gilson, 1993). One then only needs to do $2^{N'}$ calculations for each cluster but now N' is smaller than N so that the number of calculations can be significantly reduced. A Monte Carlo simulation (Beroza *et al.*, 1991) can also be performed to search for the most probable protonation states of a protein, ligand, and protein-ligand complex.

III. Including Protein and Ligand Flexibility in Modeling Molecular Recognition

Protein and ligand flexibility, particularly the former, is hard to deal with computationally because of the large amount of computer time required to carry out thorough conformational sampling. It is necessary to have a reasonable description of intra- and intermolecular interactions and solvation effects to simulate conformational distributions reliably. Explicit-solvent simulations have been used for sampling protein conformation since about the mid 1980s, when computers began to be powerful enough to permit practical simulations lasting tens of picoseconds. With the rapid increase in computing power and the introduction of novel simulation algorithms, simulations now routinely cover over several hundred picoseconds and even tens of nanoseconds for proteins containing several hundred residues. These advances will make it easier to account for the role of protein and ligand flexibility in drug-design applications, at least for systems that warrant detailed examination.

A. *Free Energy Calculations*

There exist rigorous theories for simulating binding free energy, taking into account molecular flexibility and going beyond the continuum description of solvation effects. For example, in 1984 (Tembe and McCammon, 1984), the thermodynamic cycle-perturbation method was introduced to help compare the binding affinities of a group of similar inhibitors to a receptor. This method is based on constructing a thermodynamic cycle relating two binding processes:

$$
\begin{array}{ccccccc}
 & \mathrm{P}+\mathrm{L} & \longrightarrow & \mathrm{P{:}L} & & & \Delta G_{\mathrm{PL}} \\
\Delta G_{\mathrm{L}\to\mathrm{L}'} & \downarrow & & \downarrow & \Delta G_{\mathrm{PL}\to\mathrm{PL}'} & & \\
 & \mathrm{P}+\mathrm{L}' & \longrightarrow & \mathrm{P{:}L} & & & \Delta G_{\mathrm{PL}'}
\end{array}
$$

Because free energy is a state function

$$\Delta\Delta G = \Delta G_{\mathrm{PL}'} - \Delta G_{\mathrm{PL}}, \tag{15}$$

which compares the binding affinity of two ligands, L and L′, to a protein P, can also be obtained from

$$\Delta\Delta G = \Delta G_{\mathrm{PL}\to\mathrm{PL}'} - \Delta G_{\mathrm{L}\to\mathrm{L}'}. \tag{16}$$

This trick converts the difficult problem of calculating $\Delta G_{\mathrm{PL}'}$ and ΔG_{PL} into evaluating $\Delta G_{\mathrm{L}\to\mathrm{L}'}$ and $\Delta G_{\mathrm{PL}\to\mathrm{PL}'}$. $\Delta G_{\mathrm{PL}'}$ and ΔG_{PL} are usually more difficult to calculate directly because they involve simulating the displacement of a large number of water molecules before P and L/L′ can approach each other and both the protein and the ligands may need to adjust their conformation appropriately to allow binding to occur. $\Delta G_{\mathrm{L}\to\mathrm{L}'}$ and $\Delta G_{\mathrm{PL}\to\mathrm{PL}'}$ are often much easier to calculate, especially when the two ligands are similar, because they involve relatively small changes in chemical functional groups. One can use special methods, such as Zwanzig's perturbation theory (Zwanzig, 1954), to calculate the free energy differences, $\Delta G_{\mathrm{L}\to\mathrm{L}'}$ and $\Delta G_{\mathrm{PL}\to\mathrm{PL}'}$. For the Helmholtz free energy, Zwanzig's perturbation theory reads

$$\Delta A = -RT \ln\langle \exp(-\Delta H/RT)\rangle_r \tag{17}$$

where $\Delta H = H_p - H_r$, in which H_p and H_r are the classical Hamiltonians of the perturbed and reference systems, respectively; and $\langle\cdots\rangle_r$ represents an ensemble average over the reference state. (The quantities ΔG and ΔA are virtually the same in many cases. If not, pressure-volume contributions can be explicitly accounted for in calculating ΔG.) This formula is exact, although it is hard to obtain a reliable estimate of ΔA when the difference between the reference and the perturbed systems is large. One way to alleviate this problem is to change the reference system into the perturbed one in N steps, rather than one single step, and use the perturbation formula N times to estimate ΔA from

$$\Delta A = -RT\sum_{i=l}^{N} \ln\langle \exp(-\Delta H_i/RT)\rangle_i \tag{18}$$

where $i = 1$ refers to the reference state and $i = N+1$ corresponds to the final state, and $\Delta H_i = H_{i+1} - H_i$. This type of so-called "windowing" calculation can be very expensive to do when large modifications are made because many windows are required, and each window may require a long simulation to obtain sufficiently good statistics. Consequently, these calculations are usually practical to use only for a small number of ligands that warrant a close look. However, methods to further speed up free energy calculations are constantly being introduced, and some recent development are discussed here.

B. *Kirkwood Integration Formula*

The Kirkwood integration formula (Kirkwood, 1935) can also be used to calculate the free energy difference between two states. Assuming there is a coupling parameter λ that takes a classical Hamiltonian $H(\lambda)$ from one state to another state when λ goes from 0 to 1, the free energy difference between the two end states can be obtained from (Kirkwood, 1935)

$$\Delta G = \int_0^1 \left\langle \frac{\partial H(\lambda)}{\partial \lambda} \right\rangle_\lambda d\lambda \tag{19}$$

One way to take advantage of this formula is to use the polynomial path method (Mezei, 1992; Resat and Mezei, 1993) to calculate the integral. This essentially amounts to using a quadrature to approximate the integral

$$\Delta G = \int_0^1 \left\langle \frac{\partial H(\lambda)}{\partial \lambda} \right\rangle_\lambda d\lambda \approx \sum_k \left\langle \frac{\partial H(\lambda)}{\partial \lambda} \right\rangle_{\lambda_k} \omega_k \tag{20}$$

in which the integrand is evaluated at the quadrature points λ_k and the ω_k are the corresponding quadrature weights. Rather than using equally spaced abscissas for calculating the integrand, numerical quadrature employs an "optimal" distribution of unequally spaced abscissas to reduce the number of points at which the integrand needs to be computed to obtain a good estimate of the integral. This can save much computer time because the integrands are expensive to calculate, involving a long simulation at each $\lambda = \lambda_k$. It is harder to do this with the "windowing technique" described above because each window attempts to calculate the free energy difference between two neighboring states directly, and the results from adjacent windows need to be matched at their boundaries. This polynomial path technique has recently been used to study the conformational properties of gonadotropin-releasing hormone (Mezei and Guarnieri, 1998; Watts *et al.*, 2001).

The choice of path in taking one system into another can also significantly affect the ease of numerically calculating the integral in the Kirkwood formula. Mezei and Beveridge (Mezei and Beveridge, 1986) suggested the use of nonlinear scaling with the following formula:

$$H(\lambda) = \lambda^k H_1 + (1 - \lambda)^k H_o \tag{21}$$

where H_0 and H_1 are the classical Hamiltonians corresponding to the two end states when λ equals to 0 and 1, respectively. Linear scaling is obtained when k is set to 1 and was found to present problems when atoms were created or removed when singularity of the integrand occurs at the end points (Mruzik *et al.*, 1976; Postma *et al.*, 1982). By using nonlinear scaling separately for the attractive and repulsive terms of the Lennard-Jones potential and for the Coulombic potential, Resat and Mezei (Resat and Mezei, 1993) found that the Kirkwood integration formula provided good estimates for the conformational free energy differences of simple peptides.

C. *Locally Enhanced Sampling*

Protein side chains or ligand functional groups can adopt multiple conformations, and the barriers separating these conformations may be so high that transitions among them are difficult. Thus it may require a long simulation to reliably estimate a free energy change if more than one conformer contributes significantly to binding. To improve the sampling of these conformations in free energy calculations, Verkhivker *et al.* (Verkhivker *et al.*, 1992) utilized the locally enhanced sampling (LES) method (Elber and Karplus, 1990). The essence of this method is to replace a side chain or a functional group by N copies that do not interact with each other, and each copy only interacts with itself and its surrounding with $1/N^{th}$ of the original strength. The use of multiple copies and the reduction of the interaction potentials could significantly enhance the sampling of the conformation of amino acid side-chain or functional groups of the ligand. Fig. 1 depicts how a thermodynamic cycle

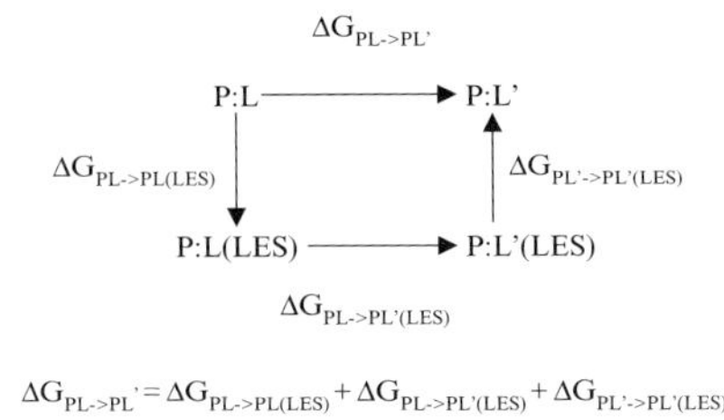

FIG. 1. Connection between normal and locally enhanced sampling (LES) results in free energy calculations.

can be constructed to connect the LES results to the desired free energy change.

As illustrated by the figure, two additional perturbation calculations, $\Delta G_{\mathrm{PL}\rightarrow\mathrm{PL(LES)}}$ and $\Delta G_{\mathrm{PL}'\rightarrow\mathrm{PL}'(\mathrm{LES})}$, relating the single-copy to the multiple-copy representation, are required. But Verkhivker *et al.* (Verkhivker *et al.*, 1992) demonstrated that the benefits of adopting a multiple-copy representation outweighed the additional costs of introducing two additional perturbation calculations. More recently, Simmerling *et al.* (Simmerling *et al.*, 1998) applied this method to study the $\alpha \rightarrow \beta$ anomerization of glucose and found that the free energy calculations converged an order of magnitude faster than with the single-copy method.

D. λ Dynamics Method

The λ dynamics method (Banba *et al.*, 2000; Damodaran *et al.*, 2001) is another technique introduced to speed up free energy calculations. In this method, multiple ligands are placed in the binding site of their receptor all together with the interaction potential of each ligand with its target reduced from its full strength. The fraction, λ_i^2, of the interaction potential of each ligand is treated as a fictitious particle with a fictitious mass. During a molecular dynamics simulation, λ_i is allowed to evolve just as a coordinate of an atom. Because the interaction potential of each ligand is reduced, the barriers for conformational transitions are lower. The reduced barriers enable a ligand to explore different orientations and conformations more easily. Also, the ranking of the ligands can emerge quickly during the simulation, because λ_i^2 can increase rapidly for the winners at the expense of the losers. The identification of the strong binders can therefore be quicker than by doing many free energy perturbation calculations including only one ligand at a time. This method was able to quickly distinguish strong benzamidine inhibitors of trypsin from weaker ones (Guo and Brooks, 1998).

E. Parallel Fast-Growth Method

Another way to calculate the free energy difference between two states is the use of the slow-growth method in which one evaluates the free energy difference between two states by calculating the work in taking one state into the other. This was once obtained by doing a molecular dynamics simulation in which the classical Hamiltonian $H(\lambda)$ is slowly changed from one state to another when λ goes from 0 to 1 and the Helmholtz free energy difference ΔA was estimated by

$$\Delta A = \int_0^1 \frac{\partial H(\lambda)}{\partial \lambda} d\lambda \tag{22}$$

Jarzynski (Jarzynski, 1997) recently derived an interesting, more general relationship in which nonequilibrium measurements can be used to calculate free energy differences. For Markovian processes that are microscopically reversible, it can be shown that the Helmholtz free energy difference ΔA between two states (Jarzynski, 1997; Crooks, 1998) can be obtained by the relation

$$\overline{\exp(-\beta W)} \equiv \int dW \rho(W,t) \exp(-\beta W) = \exp(-\beta \Delta A) \tag{23}$$

where W is the work associated with switching one system into another in time t, β is the inverse of the product between the gas constant R and the absolute temperature T, and $\rho(W,t)\ dW$ is the probability of work found within the range W and $W + dW$. If one switches from one state into the other in one step, $W = \Delta H = H_1 - H_0$, this relation reduced to the Zwanzig perturbation formula (Zwanzig, 1954)

$$\begin{aligned} \exp(-\beta\Delta A) = \overline{\exp(-\beta W)} &= \overline{\exp(-\beta \Delta H)} \Rightarrow \Delta A \\ &= -RT < \exp(-\beta\Delta H) >_0 \end{aligned} \tag{24}$$

If one switches states infinitely slowly,

$$\rho(W, t) = \delta(W - W_{\mathrm{inf}}) \tag{25}$$

so that

$$\exp(-\beta\Delta A) = \int dW \rho(W,t) \exp(-\beta W) = \exp(-\beta W_{\mathrm{inf}}) \Rightarrow A = W_{\mathrm{inf}} \tag{26}$$

where W_{inf} can be obtained from the slow-growth formula, equation 22, for example. If one considers a single mutation, the fast-growth formulation may not be able to take advantage of parallel computing well. Although the different W's can be calculated independently of each other, they can be calculated only after an equilibrium ensemble of a reference system has been generated, and it takes time to generate the equilibrium ensemble for large systems, especially when explicit-solvent models are used. In contrast, the "windowing" and thermodynamic integration formalisms allow multiple equilibrium simulations to be carried out simultaneously on different computers or processors. However, once an equilibrium ensemble for a reference system has been generated, multiple alchemical calculations can be performed in parallel by using the fast-growth method.

F. *Translational and Rotational Entropy*

It is usually difficult to directly obtain the translational and rotational contributions to the entropy from practical simulations. An approximate way to estimate these contributions is to assume them to be uncoupled from the contributions from the other degrees of freedom so that they can be estimated by ideal gas-type formulae. The classical Hamiltonian of a system can be written as

$$H(p_{tr}, q_{tr}, p, q) = H(p_{tr}, q_{tr}) + H(p, q) + V(p_{tr}, q_{tr}, p, q) \tag{27}$$

where q_{tr} and p_{tr} are respectively the translational-rotational coordinates and their conjugate momenta, q and p denote respectively all the other coordinates and their corresponding conjugate momenta, and $V(p_{tr}, q_{tr}, p, q)$ represents the coupling potential between the Hamiltonian $H\ (p_{tr}, q_{tr})$ describing the translational-rotational motion and the Hamiltonian $H\ (p, q)$ that describes the remaining motion. If $V(p_{tr}, q_{tr}, p, q)$ is negligible (this is not true if there are significant couplings between rotational and vibrational motion and between translational-rotational motion and the solvent dynamics), the entropy contributions from the translational-rotational motion and the remaining motion can be calculated separately—the latter one from free energy simulations and the former from approximate analytical models. For example, one can use ideal gas-type formulae described in statistical mechanics textbooks to estimate the contributions from the translational and rotational motion. The translational entropy, S_{trans}, has the form

$$S_{\text{trans}} = Nk \ln\left[\left(\frac{2\pi MkT}{h^2}\right)^{3/2} \frac{Ve^{5/2}}{N}\right] \tag{28}$$

where N is the number of molecules in a volume of V, M is the mass of the molecule, and h is the Planck constant. There are usually three such terms corresponding to the protein, ligand and their complex for each binding process and the entropy change corresponding to such a binding process is proportional to

$$\ln\left[\left(\frac{M_{\text{complex}}}{M_{\text{protein}}M_{\text{ligand}}}\right)^{3/2}\right] \tag{29}$$

where M_{protein}, M_{ligand}, and M_{complex} denote the mass of the protein, the ligand, and the complex, respectively. When the protein is much larger than the ligand, this can be approximated by

$$\ln\left[\left(\frac{1}{M_{\text{ligand}}}\right)^{3/2}\right] \tag{30}$$

When comparing the binding energies between two ligands to a receptor, the translational entropy is proportional to

$$\ln\left[\left(\frac{M_{\text{ligand1}}}{M_{\text{ligand0}}}\right)^{3/2}\right] \tag{31}$$

For the type of free energy difference calculations discussed in the previous sections, the two ligands, ligand0 and ligand1, usually differ only by small functional groups that contribute only a small fraction to the total mass of the ligands. Under this situation, the contributions of translational entropy to a free energy difference are often negligible. Similar arguments suggest that the contributions of rotational entropy are also small. In this case, ratios of moments of inertia instead of masses are involved. Therefore, when comparing a number of similar ligands, the entropy contributions from translational and rotational motion may be neglected in many cases.

On the other hand, loss of entropy caused by rigidifying rotatable bonds on binding may not be negligible, especially when the different ligands considered contain different numbers of rotatable bonds. When it is difficult to use short free energy calculations to estimate these effects, one way to estimate such entropy contributions is to assume each sp^3 hybridized rotatable bond to lose a constant value (e.g., $T\Delta S = \sim 0.3$ kcal/mol [Bohm, 1994; Morris *et al.*, 1998]) on binding. The entropy contributions from other rotatable bonds—for example, hydroxyl groups attached to aromatic rings—can also be significant, as demonstrated by Gidofalvi *et al.* (Gidofalvi *et al.*, 2002). They examined this issue by using potential energy curves obtained from quantum and continuum electrostatic calculations to estimate such entropy contributions. In studying the interactions between balanol and protein kinase A, they estimated that the entropy loss associated with each hydroxyl group could contribute from about 0 to 0.6 kcal/mol to the binding free energy, depending on which hydroxyl group of the ligand was considered. More extensive discussions of entropy changes on binding can be found in Gilson *et al.* 1997, and Luo and Sharp, 2002.

IV. Fast Approximate Simulation Methods for Calculating Binding Free Energies

A. *Systematic Sensitivity Analysis*

One does not always need to aim for highly accurate binding constants to be productive in drug design. It is already useful to generate rules or constraints from computational studies to guide the design of chemical libraries for high throughput screening, to direct the optimization of a drug lead, and to help the construction of a pharmacophore model for mining new drug leads with different chemical scaffolds from small-molecule libraries. To this end, rather than exploring physical modifications, it may be useful to explore nonphysical ones that probe the relative significance of different features of functional groups in affecting binding. For example, by sequentially turning off the atomic partial charge or dipole moment of every relevant functional group in a lead compound, one can determine which charges or dipole moments are truly significant and which are not. This understanding can then turn into rules that can help determine how functional groups of lead compounds should be modified to improve the odds of finding stronger binders, that can help generate constraints on designing focused chemical libraries targeting towards specific targets, and that can help construct pharmacophore models. For these more modest goals, one can use mathematical tricks to carry out many approximate free energy difference calculations by using simulation results for only one reference system.

One way to do this is to adopt the sensitivity analysis approach that has been used by engineers for a long time (Franck, 1987; Tomovick and Vukobratovic, 1972) and has recently been extended to study molecular and biomolecular systems (Thacher *et al.*, 1991; Susnow *et al.*, 1991; Wong, 1991; Wong and Rabitz, 1991; Zhu and Wong, 1993a; Zhu and Wong, 1993b; Susnow *et al.*, 1994; Zhu and Wong, 1994; Bleil *et al.*, 1995; Zhang *et al.*, 1995; Wong *et al.*, 1998). To identify the key model parameters determining system properties, engineers calculate the derivatives $\partial O/\partial \lambda_i$ of a property O of the system with respect to the model parameters λ_i

to measure the sensitivity of the observable to parameter changes. (The dimensionless logarithmic derivatives $\partial \ln O / \partial \ln \lambda_i = (\partial O / \partial \lambda_i)(\lambda_i / O)$ are also often calculated.) Parameters that do not affect system properties yield negligible $\partial O / \partial \lambda_i$. On the other hand, important parameters would yield large $\partial O / \partial \lambda_i$. Analytical expressions can be worked out for calculating these derivatives from the dynamical behavior of a single reference system. Formulas for mechanical and thermodynamic properties of molecular systems have been worked out (Zhu and Wong, 1993b; Bleil et al., 1995; Wong et al., 1998). Those for the Helmholtz free energy are particularly simple. The first order derivative is simply

$$\frac{\partial A}{\partial \lambda_i} = \left\langle \frac{\partial V}{\partial \lambda_i} \right\rangle \tag{32}$$

where V is the potential energy of the system. Because the calculation of the derivatives of a number of observables with respect to all the parameters of a model adds only a small fraction to the costs of doing dynamical simulations on a reference system, it is easier to use this approach to systematically compare the role of all the parameters in a model so that no important parameters are overlooked. This idea can be generalized to include higher-order derivatives so that a Taylor's series expansion

$$\Delta O = \sum_i \frac{\partial O}{\partial \lambda_i} \Delta \lambda_i + \frac{1}{2} \sum_{i,j} \frac{\partial^2 O}{\partial \lambda_i \partial \lambda_j} \Delta \lambda_i \Delta \lambda_j + \ldots . \tag{33}$$

can be used to predict the influence of larger parameter changes on system properties. The Taylor's series expansion also permits the effects of many different combinations of parameter changes to be examined after the expensive derivatives have already been calculated. Moreover, the second order derivatives yield useful information on the significance of the interactions between two parameters on affecting system properties. For example, in studying protein-ligand interactions, it is useful to find out the parts of the protein with which a particular functional group of a ligand are favorably interacting. This information can be obtained by identifying the large second order derivatives having one parameter belonging to the ligand and the other to the protein and examining which parameters are involved. The second order parametric derivatives of the Helmholtz free energy is given by the form (Bleil *et al.*, 1995; Wong *et al.*, 1998)

$$\frac{\partial^2 A}{\partial \lambda_i \partial \lambda_j} = \left\langle \frac{\partial^2 V}{\partial \lambda_i \partial \lambda_j} \right\rangle - \beta \left[\left\langle \left(\frac{\partial V}{\partial \lambda_i} \right) \left(\frac{\partial V}{\partial \lambda_j} \right) \right\rangle - \left\langle \left(\frac{\partial V}{\partial \lambda_i} \right) \right\rangle \left\langle \left(\frac{\partial V}{\partial \lambda_j} \right) \right\rangle \right] \tag{34}$$

Although calculating a large number of first and higher-order derivatives provides an efficient and systematic way to identify the determinants of molecular recognition, using the Taylor's series expansion, equation 33, to predict actual free energy changes may not be most effective, because the series may converge slowly when large perturbations are made and higher-order derivatives are harder to calculate well from short simulations. Other strategies can be used to predict the free energy changes resulting from somewhat larger modifications.

B. Single-Step Zwanzig's Perturbation Method

The single-window Zwanzig perturbation theory (Zwanzig, 1954) described above can provide quick estimates of free energy changes when parameter perturbations are sufficiently small. In fact, this strategy was used in earlier free energy calculations when computers were orders of magnitude slower. For example, an early free energy perturbation study focused on examining the effects of making conservative modifications on free energy changes (Wong and McCammon, 1986b). In one case, benzamidine was modified into parafluorobenzamidine and the effects on trypsin binding were examined. In this calculation, only simulations on the reference systems, benzamidine and the trypsin-benzamidine complex, were performed, and the single-window Zwanzig perturbation formula was used to calculate $\Delta G_{\mathrm{L}\rightarrow\mathrm{L}'}$ and $\Delta G_{\mathrm{PL}\rightarrow\mathrm{PL}'}$ directly, in which the parahydrogen in the benzamidine ring was changed into a fluorine with the appropriate change in force field parameters and bond length.

By focusing on small perturbations, a single-window Zwanzig perturbation formula (Zwanzig, 1954) can be used to provide initial estimates of the effects of making many physical or nonphysical changes to the ligand or the protein on binding affinity without carrying out expensive molecular dynamics simulations for all the perturbed systems. Only simulations of the reference system are needed. This technique, or its close cousin in which the Helmholtz free energy change is obtained by expanding ΔA in terms of ΔH and keeping up to second order term

$$\Delta A = \langle \Delta H \rangle_r + \frac{1}{2RT} \left\langle (\Delta H \; - \langle \Delta H \rangle_r)^2 \right\rangle_r \tag{35}$$

has been used to examine the effects of adding or removing protons (Del Buono *et al.*, 1994; Levy *et al.*, 1991) and the effects of changing molecular charge distribution on free energy changes (Simonson *et al.*, 1997). In drug design applications—focusing on a small chemical subspace for which a single-window perturbation formula can be used to study many modifications—should already be useful for finding better derivatives of a drug lead. One can then use a few of the promising derivatives for further single-window perturbation calculations to enlarge the chemical subspace for identifying other drug candidates. Although this full-blown molecular-dynamics–based method has not yet been applied extensively to drug-design applications, an implicit solvent model has already been used to carry out proof-of-concept–type calculations on protein kinases to help generate rules for guiding the development of pharmacophore models to aid the design of focused chemical libraries and to guide the optimization of a drug lead (Wong *et al.*, 2001; Gould and Wong, 2002).

The range of application of this idea can be extended by using soft-core potentials in reference simulations (Liu *et al.*, 1996). A single perturbation formula does not work well when larger atoms or atomic groups are added or deleted, because the reference simulation does not adequately sample the configuration states relevant to the modified systems. If a large atom or atomic group is going to be deleted, the reference simulation may not have sampled well the space allowing the surrounding solute or solvent atoms to get closer. If a large atom or atomic group is going to be created, the reference simulations may have many configurations that create unfavorable steric clashes at the modified sites. To alleviate this problem, Liu *et al.* (Liu *et al.*, 1996) utilized soft-core potentials. For example, a modified Lennard-Jones potential of the form

$$V(r_{ij}) = 4\varepsilon_{ij}\left[\frac{\sigma_{ij}^{12}}{\left(\alpha\sigma_{ij}^{6} + r_{ij}^{6}\right)^{2}} - \frac{\sigma_{ij}^{6}}{\alpha\sigma_{ij}^{6} + r_{ij}^{6}}\right] \tag{36}$$

can be used at selected sites in a reference simulation so as to create more space where atoms are going to be added and to allow atoms to get closer to sites where atoms are going to be deleted. In the above equation, ε_{ij} and σ_{ij} are the Lennard-Jones parameters between atoms i and j, r_{ij} is the distance between the two atoms, and α is a softening parameter that prevents the potential from diverging as $r_{ij} \to 0$. This approach was able to predict well the free energy differences among a number of para-substituted phenols in water solvent (Liu *et al.*, 1996). There is a limit to which this approach works well. Mordasini and McCammon (Mordasini and McCammon, 2000) later examined the range of applicability of this single-reference approximation by introducing increasingly larger modifications. They found that this model could still yield reasonable qualitative scoring when functional groups involving up to three atoms were deleted.

C. *Combining Explicit and Implicit Solvent Models*

Recently, explicit and implicit solvent models have been combined to facilitate free energy calculations (Vorobjev *et al.*, 1998; Vorobjev and Hermans, 1999; Kuhn and Kollman, 2000a; Lee *et al.*, 2000; Massova and Kollman, 2000). This approach uses explicit solvent molecular dynamics simulations to relax crystal structures to solution ones and then uses the simulated solution structures in implicit-solvent calculations to obtain free energy. Using implicit-solvent models eliminates the extensive simulation time required for sampling solvent configurations. This approach assumes that the free energy of a system can be obtained by averaging the potential of mean force, obtained from an implicit solvent model, of dynamics snapshots generated from explicit-solvent models. Configurational entropy contributions of the solute can be estimated from the harmonic or quasiharmonic model. These approximations appear to work well. Vorobjev *et al.* (Vorobjev *et al.*, 1998) used one such approach successfully in distinguishing correctly folded protein conformations from misfolded ones. In their study, Vorobjev *et al.* (Vorobjev *et al.*, 1998) ran a quick molecular dynamics simulation of a protein for about 50 to 100 ps and used an implicit-solvent model to calculate the free energy of the correctly and incorrectly folded protein. The free energy was found to be lower for the correctly folded protein. The implicit solvent model included contributions from the gas-phase energy of the solute, the energy of cavity formation, the solute-solvent interaction energy, and the solvent electrostatic polarization energy. The energy of cavity formation was assumed to be proportional to the solvent accessible surface area of the solute. The solvent electrostatic polarization energy was obtained by solving the Poisson-Boltzmann equation, and the other terms were obtained from a molecular mechanics force field.

A similar MM/PBSA approach was used to study protein-ligand interactions. For example, Kuhn and Kollman (Kuhn and Kollman, 2000a) obtained encouraging results by applying this method to predict the binding affinity of seven ligands to avidin and streptavidin. They obtained a correlation coefficient of 0.92 between the calculated and the experimental binding affinities. The root-mean-square difference between calculated and experimental results, which covered a range of about 16 kcal/mol, was on the order of 1.7 kcal/mol. In these calculations, the length of the molecular dynamics simulation used for the averaging was 300 ps. No simulations on the separated protein and ligands were done. Instead, the protein and ligands were assumed to adopt the same conformation as that in the molecular dynamics simulations of the complexes. Kuhn and Kollman (Kuhn and Kollman, 2000a) also used the harmonic approximation to calculate the entropy change on binding by using six quenched dynamics snapshots. This approximation introduced a relatively large uncertainty in calculating entropy changes; the discrepancy among results from the six snapshots amounted to 5 kcal/mol in the worst case that they studied. Therefore the solute entropy contributions remain a challenge to calculate reliably.

To further speed this approach, one can replace the expensive explicit-solvent simulations with implicit ones. Statistical mechanical theory gives the Helmholtz free energy A, apart from the scaling constant of the classical partition function that cancels out in binding energy calculations, as

$$A = -RT \ln \iint \exp(-\beta H(u, v))\,du\,dv \tag{37}$$

where R is the gas constant, T is the absolute temperature, $\beta = 1/RT$, and $H(u, v)$ is the classical Hamiltonian expressed in terms of the solute coordinates u and the solvent coordinates v. Integrating over the solvent coordinates gives

$$A = -RT \ln \int \exp(-\beta W(u))\,du \tag{38}$$

where $W(u)$ is the potential of mean force of the solute with conformation defined by u, and the kinetic energy term for the solute is ignored because it cancels out in binding energy calculations. $W(u)$ can be estimated by a continuum solvent model, such as the Poisson-Boltzmann and the Generalized Born models—described in sections II, C, and II, D—plus accessible surface area–dependent terms. If one can calculate the atomic forces resulting from $W(u)$, one can carry out a molecular dynamics simulation to generate an ensemble of structures for calculating the Helmholtz free energy according to equation 38. However, it is difficult to obtain a free energy from this equation directly in practical simulations with limited simulation lengths. If one adopts similar approximations as in previous MD/PBSA calculations (Vorobjev *et al.*, 1998; Vorobjev and Hermans, 1999; Kuhn and Kollman, 2000a; Lee *et al.*, 2000; Massova and Kollman, 2000), one can calculate the internal energy E of a system via

$$E = \langle W(u) \rangle \tag{39}$$

where $\langle \cdots \rangle$ represents an ensemble average over snapshots obtained from an implicit solvent simulation and the Helmholtz free energy can be obtained from

$$A = E - TS \tag{40}$$

Again, one can use the harmonic or quasiharmonic approximation to estimate the entropy term. For a normal mode analysis, one can also use a more realistic implicit solvent model to calculate the Hessian matrix. Most of the previous MM/PBSA calculations employed more approximate distance-dependent dielectric models in normal mode calculations.

Several methods have already been introduced to use the relatively sophisticated Poisson-Boltzmann model to calculate electrostatic forces during molecular dynamics simulations (Gilson *et al.*, 1993; Im *et al.*, 1998). Methods for calculating forces resulting from solvent-accessible surface area-dependent hydrophobic terms have also been developed (e.g., see Sridharan *et al.*, 1995). However, it is still expensive to use these methods to calculate Poisson-Boltzmann forces on the fly during molecular dynamics simulations. An alternative is to use the cheaper Generalized Born model. Dominy and Brooks (Dominy and Brooks, 1999) have parameterized the Generalized Born model by Qui *et al.* (Qiu *et al.*, 1997) for the CHARMM force field and found that the model performed quite well in reproducing molecular solvation energy and conformational free energy. Use of this model in molecular dynamics simulations of a 56-residue protein yielded results that agree well with corresponding explicit-solvent simulations. These results are encouraging because they demonstrate that this type of model has the potential of making it easier to include protein flexibility, via molecular simulations, in modeling protein-drug interactions. Qiu *et al.*'s Generalized Born model (Qiu *et al.*, 1997) has also been incorporated into the UHBD program (Madura *et al.*, 1995; Davis *et al.*, 1991) for molecular dynamics simulations of HIV protease (David *et al.*, 2000) and for constrained Brownian dynamics simulations of peptides (Shen *et al.*, 2001). The constrained Brownian dynamics simulation algorithm can use a larger time step than molecular dynamics simulation algorithms to speed up the conformational sampling of small molecules. This will be useful for improving the calculation of the free energy of floppy ligands in solution. As mentioned before, many earlier MM/PBSA simulations assumed the ligand conformational distribution in solution to be the same as that in a protein-ligand complex. This approximation may not serve well for floppy ligands and for ligands that adopt very different conformations in the bound and unbound states.

D. Chemical-Scanning Computational Experiments

The explicit/implicit solvent approach just described requires doing at least one simulation for each protein-ligand complex. Therefore it is still difficult to examine the binding of a large number of compounds to a receptor. However, if one focuses on a small subset of chemical space around a lead compound, one can adopt the same approximations as described earlier in free energy calculations so that simulations on the reference systems alone can be used to predict the effects of making many modifications on a lead compound. In these calculations, no molecular dynamics simulation simulations are performed on the derivatives of a lead compound. Instead, snapshots of the reference simulations are modified to change different functional groups of the lead compound into new ones. For example, Kuhn and Kollman (Kuhn and Kollman, 2000b) were able to predict a derivative that binds stronger than biotin to avidin by changing different C-H groups of biotin into C-F groups. This approach has also been applied to cases where larger modifications are made. For example, Massova and Kollman (Massova and Kollman, 1999) performed a computational alanine scanning experiment and were still able to obtain qualitative agreement with experiments when studying protein-protein interactions even though the chemical modifications were rather large (they involved changing non-alanine amino acids into alanines).

E. *Use of Explicit-Solvent Models to Parametrize Implicit-Solvent Models*

Another way to combine explicit and implicit solvent calculations is to use the former to help parametrize the latter. In one study, Archontis *et al.* (Archontis *et al.* 1998) first carried out rigorous free energy perturbation calculations by using an explicit-solvent model. They then calibrated the parameters, especially the internal dielectric of the protein, of a Poisson-Boltzmann model by using the results from the explicit-solvent free energy calculations. Once the cheaper Poisson-Boltzmann model was calibrated, they used it to facilitate more mutagenesis studies. In principle, this approach can also be used in drug-design applications. Zhang *et al.* (Zhang *et al.*, 2001), on the other hand, recalibrated the OPLS (Jorgensen *et al.*, 1996) Lennard-Jones radii for use in their Surface Generalized Born model by using results from explicit solvent free energy perturbation calculations on 40 organic compounds.

F. *Semi-Empirical Linear Response Theory*

Another way to facilitate the comparison of the binding affinity among a number of rather different ligands is the semi-empirical linear response approach (Åqvist *et al.*, 1994; Carlson and Jorgensen, 1995; Paulsen and

Ornstein, 1996; McDonald *et al.*, 1997; Smith *et al.*, 1998; Hansson *et al.*, 1998). This method approximates the binding affinity ΔG between a ligand-receptor pair by the relation:

$$\Delta G = \alpha\Big(\Big\langle U^{\text{Bound}}_{\text{electrostatic}}\Big\rangle - \Big\langle U^{\text{Unbound}}_{\text{electrostatic}}\Big\rangle\Big) + \beta\Big(\Big\langle U^{\text{Bound}}_{\text{Lennard-Jones}}\Big\rangle - \Big\langle U^{\text{Bound}}_{\text{Lennard-Jones}}\Big\rangle\Big) + \gamma\Big(\Big\langle \Omega^{\text{Bound}}\Big\rangle - \Big\langle \Omega^{\text{Unbound}}\Big\rangle\Big) \tag{41}$$

where $\langle U^{\text{Bound,Unbound}}_{\text{electrostatic}}\rangle$ is the averaged ligand-surrounding electrostatic interaction energy, $\langle U^{\text{Bound,Unbound}}_{\text{Lennard-Jones}}\rangle$ is the averaged ligand-surrounding Lennard-Jones interaction energy, and $\langle \Omega^{\text{Bound,Unbound}}\rangle$ is the averaged solvent-accessible surface area of the complex or the uncomplexed molecule(s). α, β, and γ are empirical parameters determined by least-square fit of the experimental binding free energy of a number of inhibitors to the ensemble averaged quantities of equation 41 obtained from molecular simulations. Once α, β, and γ are determined, they can be used to predict the binding affinity of inhibitors whose binding affinity has not been measured. Encouraging results have appeared in a number of applications. For example, a recent application of this approach to study the binding of the tetrahydroimidazo[4,5,1-jk][1,4]benzodiazepine-2(1H)-thione and one class of compounds to HIV reverse transcriptase (Smith *et al.*, 1998) yielded a root-mean-square deviation of less than 1 kcal/mol from experimental results when the observed range of binding affinity was ~4 kcal/mol. The Åqvist group did not use the solvent-accessible surface area term in their semi-empirical model but still obtained good correlation with experimental data in a number of applications (Marelius *et al.*, 1998).

V. Including Protein Flexibility in Virtual Screening

Protein flexibility has often been ignored in docking experiments but can be important in improving the reliability of virtual screening experiments. Rigorous methods involve simulating the docking between a protein and a ligand directly, taking into account the flexibility of the protein and ligand simultaneously. It is still very difficult to carry out such brute-force simulations, although some work has been done along this line. For example, Mangoni *et al.* (Mangoni *et al.*, 1999) directly simulated the docking of a flexible ligand to a flexible protein by using molecular dynamics simulations in which a higher temperature was applied to the translational motion of the center-of-mass of the ligand. Nakajima *et al.*

(Nakajima *et al.*, 1997), on the other hand, applied the multicanonical molecular dynamics simulation method to dock a flexible peptide into the SH3 domain of Abl tyrosine kinase. Near term, it seems more computationally tractable to decouple the conformational dynamics of the protein from that of the ligand. In other words, a flexible model of the protein is first generated without the ligand, and then the ligand is docked to the flexible receptor model. There are different levels of sophistication that one can adopt to generate a flexible receptor model. Since side chains are typically more disordered than main chains, one model is to simply generate multiple side-chain conformations and keep the main chain fixed; see, for example, the work by Leach and coworkers (Leach and Lemon, 1998; Leach, 1994). When different experimental structures for the same target are available, one can also use these structures to construct docking templates that take multiple conformations into account. This strategy has been used by Knegtel *et al.* (Knegtel *et al.*, 1997) in DOCK (Meng *et al.*, 1992) and by Claussen (Claussen *et al.*, 2001) in FLEXX (Kramer *et al.*, 1999). Force-field methods provide another tool for generating snapshots for docking experiments. One way to do this is to perform a normal mode analysis and use the resulting low-frequency (or high-amplitude) modes to generate alternate conformations for docking (Zacharias and Sklenar, 1999; Keseru and Kolossvary, 2001). As described below, relaxed complex methods make use of detailed molecular dynamics simulations that take explicit-solvent and anharmonic motion into account.

A. *Relaxed Complex Methods*

The relaxed complex methods were inspired by two successful experimental methods for rapid discovery of ligands that bind strongly to a target, namely the ''SAR by NMR'' method (Shuker *et al.*, 1996) and the ''tether method'' (Erlanson *et al.*, 2000). These methods all recognize that ligands may bind to conformations that occur only rarely in the dynamics of the target and that strong binding often reflects multivalent attachment of the ligand to the target. Variants of the relaxed complex approach include a single-ligand method and a double-ligand method.

The basic element of these new methods is the automated docking of small libraries of compounds to a diverse selection of target conformations. The first phase of the approach involves generating the target conformations. The initial publications on the relaxed complex methods made use of long molecular dynamics simulations of the unliganded target molecule (Lin *et al.*, 2002a; Lin *et al.*, 2002b). Other procedures that could provide more extensive sampling of the target configurations include replica exchange sampling, as discussed in section VI.

The second phase of the relaxed complex methods involves the rapid docking of minilibraries of candidate inhibitors to the conformational snapshots of the target. In this phase, a relatively simple scoring algorithm is used to allow fast docking. In the initial applications (Lin *et al.*, 2002a; Lin *et al.*, 2002b), AutoDock (Morris *et al.*, 1998) has been used for the docking, but the use of other docking algorithms is certainly possible. Favorable complexes are recorded for further analysis.

The third phase attempts to improve the scoring of the best complexes found in the docking calculations by use of a slower but more accurate algorithm for estimating the standard free energies of binding. This phase was introduced in the second publication on the relaxed complex methods (Lin *et al.*, 2002b), in which the MM/PBSA approach, described in section IV, C, was used for rescoring. As described earlier, the MM/PBSA approach combines a standard molecular mechanics evaluation of the intramolecular interactions of the solute molecules with a continuum evaluation of the free energy of solvation (by using the Poisson-Boltzmann model for the electrostatic part and a surface tension model for the apolar part). In the usual MM/PBSA approach, certain approximations are used for changes in the entropy on binding (viz., the translational, rotational, conformational, and vibrational entropies). It should be possible to improve on some of these approximations by placing the calculations on a more rigorous statistical mechanical foundation (Gilson *et al.*, 1997).

The scheme described above represents the single-ligand method. The double-ligand variant recognizes that two ligands with relatively low binding affinities to the target can be linked to form a high-affinity ligand. Because the binding of the first ligand could introduce unfavorable interactions for the binding of the second ligand, the combination of the best-ranked ligands for respective binding sites does not necessarily produce the best composite compound. Continuing from the previous single-ligand studies, the first ligand is therefore treated as part of the target, and the docking simulations of the second ligand are repeated in a limited search space, based on the allowable lengths of linkers. Again, the binding of the second ligand is subsequently rescored by more accurate approaches.

The first applications of the relaxed complex methods focused on an experimentally well-characterized system, the FK506 binding protein FKBP-12 (Lin *et al.*, 2002a; Lin *et al.*, 2002b). A 2 ns molecular dynamics calculation was used to sample the FKBP-12 conformations, and the AutoDock software (Morris *et al.*, 1998) was used for the initial docking. The first article (Lin *et al.*, 2002a) considered the binding of compounds 2 and 9 from the "SAR by NMR" article by Shuker *et al.* (Shuker *et al.*, 1996). It was shown that the binding of the ligands is quite sensitive to conformational fluctuations of the target protein FKBP-12, even though the latter is a relatively rigid protein. In particular, with the AutoDock 3.0.5 scoring function (Morris *et al.*, 1998), the binding energies of compound 2 covered a range of 3 to 4 kcal/mol, corresponding to a 100- to 1000-fold difference in binding affinities.

In the second article (Lin *et al.*, 2002b), rescoring was done by using the MM/PBSA approach (Vorobjev *et al.*, 1998; Vorobjev and Hermans, 1999; Kuhn and Kollman, 2000a; Lee *et al.*, 2000; Massova and Kollman, 2000). The solutions of the Poisson-Boltzmann equation were obtained from the new APBS software (Baker *et al.*, 2001). As in the first article, significant ranges of binding energies were found for the ligands (dimethylsulfoxide, 4-hydroxy-2-butanone, and tetrahydrothiophene-1-oxide, in this case). These variations result in part from steric effects, since the difference between the largest and smallest solvent-accessible molecular surface of the FKBP-12 binding site is found to be about 187 A^2. For these ligands, use of the MM/PBSA rescoring allowed the correct prediction of the binding modes in comparison with the crystallographic structures.

B. Dynamic Pharmacophore Method

Many receptor-based pharmacophore models have been developed by using one crystal, NMR, or model structure. Pharmacophore models based on a single receptor structure could fail to identify inhibitors that bind to structures that are somewhat different from the experimental or model structure but that are still readily accessible at physiological temperatures. To address this issue, Carlson *et al.* (Carlson *et al.*, 1999; Carlson and McCammon, 2000; Carlson *et al.*, 2000) developed a dynamic pharmacophore model in which a number of snapshots from molecular dynamics simulations were used to construct the model. For each snapshot, they determined components of a pharmacophore model by identifying favorable binding sites of chemical functional groups with the multi-unit search for interacting conformers (MUSIC) program available in the BOSS program (Jorgensen, 1997). The MUSIC procedure identifies favorable binding sites of probe molecules by simultaneously energy-refining a large number of probe molecules, which do not interact with others, in the potential field of a drug target. Strong binding sites tend to cluster many probe molecules in well-defined orientations and locations. By carrying out MUSIC calculations on a number of dynamic snapshots, one can identify strong binding sites consistently appearing in many rather than only one or a few snapshots. These sites can form important components in a pharmacophore model. This approach can also uncover useful binding sites that are not presented by the initial starting structure. By using methanol as a probe molecule, Carlson *et al.* (Carlson *et al.*, 1999; Carlson and McCammon, 2000; Carlson *et al.*, 2000) developed dynamic pharmacophore models that perform better than the single conformation model in identifying potent inhibitors of HIV-1 integrase. Unfortunately, the dynamic model also increased the number of false positives.

VI. Enhanced Sampling via the Replica-Exchange Method

Obviously, thorough conformational sampling requires efficient algorithms. When taking multiple structures of the same receptor from experimental data, conformations that are separated by high-energy barriers can sometimes be obtained. On the other hand, it can be very slow to use a brute-force molecular dynamics simulation to find conformations separated by large energy barriers. More sophisticated algorithms will be helpful. One method that has become popular recently is the replica-exchange method (Mitsutake *et al.*, 2001; Sanbonmatsu and Garcia, 2002). In this method, multiple molecular dynamics or Monte Carlo runs are carried out at a range of temperatures, and configurations in different runs are allowed to exchange every specified number of simulation steps according to Metropolis-like criteria for the transition probability between two replicas i and j at different temperatures:

$$\begin{aligned} \omega(X_i -> X_j) &= 1 \qquad\qquad\quad \textit{for} \quad \Delta \leq 0 \\ &= \exp(-\Delta) \quad \textit{for} \quad \Delta > 0 \end{aligned} \tag{42}$$

where

$$\Delta = (\beta_j - \beta_i)(E_i - E_j) \tag{43}$$

and E_i and E_j are the energy of replica i (at temperature $(k\beta_i)^{-1}$ where k is the Boltzmann constant) and j (at temperature $(k\beta_j)^{-1}$), respectively, at a given simulation step. Usually, one only considers exchanging two replicas at adjacent temperatures because the transition probability decreases rapidly with temperature difference. Initially, this method was implemented with Monte Carlo simulation, but molecular dynamics is becoming a popular tool for doing the configurational sampling. In molecular dynamics, the exchange between two replicas involves changing the instantaneous temperature of each replica. One way to implement this is to scale the velocity of every atom in replica i by the factor $\sqrt{T_j/T_i}$ where T_i and T_j are the temperatures of the replica before and after exchange. As a result, the replicas involved in the exchange are no longer in equilibrium states right after the exchange. However, if replica exchange is only carried out at an interval longer than the momentum relaxation time and snapshots right after the exchange are not used, the ensembles of configurations generated from a replica-exchange simulation can be used to calculate useful equilibrium and thermodynamic properties. In simulating a small peptide Met-enkephalin in explicit solvent, Sanbonmatsu and Garcia (Sanbonmatsu and Garcia, 2002) found that replica exchange simulations lasting 32 ns sampled about 5 times more space than a constant temperature simulation for the same length.

An important practical issue to consider in extending the replica-exchange method to large proteins is the choice of temperature spacing between adjacent replicas. A thorough sampling of configurational space requires the temperature spacing between adjacent replicas to be close enough so that adequate replica exchange can occur during the simulation to allow a system to escape from local energy minima. The energy gap between adjacent replicas needs to be small compared to the fluctuation of energy to facilitate replica exchange. To aid the choice of temperatures in a replica-exchange simulation, one can enforce a relation like (Mitsutake *et al.*, 2001)

$$U_j - U_i \approx \delta U_i \tag{44}$$

where U_i and U_j are the internal energies of replicas simulated at two temperatures T_i and T_j respectively, and δU_i is the fluctuation of U_i. Since U_i is an extensive variable, one can assume it to be proportional to nkT_i, where n is the number of degrees of freedom (this needs to be refined near phase-transition temperatures). The fluctuation of U_i is proportional to $\sqrt{nkT_i}$. Substituting these expressions into equation 44 yields

$$\Delta T_i = T_j - T_j \propto \frac{T_i}{\sqrt{n}} \tag{45}$$

This implies the temperature gap is directly proportional to the absolute temperature but inversely proportional to the square root of the number of degrees of freedom. Equation 45 can be satisfied by choosing

$$T_i \propto \exp\left(\frac{i}{\sqrt{n}}\right) \quad \forall i = 1, 2, \ldots, N_{\max} \tag{46}$$

and

$$T_1 = T_{\min}, T_2, \ldots\ldots, T_{N_{\max}} = T_{\max} \tag{47}$$

where T_{min} and T_{max} are the minimum and maximum temperatures used in the replica-exchange simulation. Equation 46 suggests that the temperature spacing needs to be smaller for bigger systems. For large proteins, one needs to choose small temperature spacings, and therefore many replicas, to improve sampling. This can be very expensive to do. However, if the goal is to extend coverage of the conformational space of a folded protein rather than to study processes involving very large conformational change, such as protein folding, the use of a T_{max} only somewhat larger than the desired temperature might already be useful for obtaining conformations that are not readily obtained by a brute-force molecular dynamics simulation at a single temperature. Related generalized-ensemble simulation methods have also been developed and a useful review has recently appeared (Mitsutake *et al.*, 2001).

VII. Conclusions

Force-field methods have a relatively firm theoretical basis but unfortunately have been rather expensive to use. However, with the rapid advance of high-performance computer hardware and software, it has become easier to use force-field techniques in drug discovery. One can also adopt a hierarchical approach in which approximate but faster methods can first be used to prescreen a large number of compounds or to generate rules for aiding drug design, followed by further exploration by more sophisticated methods. Fixed-conformation approximations can sometimes be adequate for generating quick suggestions on what compounds are worthwhile to make and/or screen. These calculations can further benefit from recent developments of more accurate solvation models that are now cheap enough to use on a day-to-day basis. On the other extreme are rigorous explicit-solvent simulation methods such as free energy perturbation calculations. Although these approaches are expensive to use, they provide a rigorous basis for evaluating more approximate methods. For example, they have helped evaluate and calibrate continuum solvent models. With the growing availability of inexpensive computer clusters, it should also be increasingly easier to carry out free energy perturbation calculations on at least a small number of potential drug candidates that deserve more thorough investigation. Intermediate models varying between the extremes of elaborate explicit solvent models and simple fixed-conformation molecular mechanics models have also been developed to fit into the hierarchical approach to drug design. These intermediate models include those that aim at deciphering the determinants of molecular recognition and using them to construct rules to aid drug design (e.g., sensitivity analysis), those that focus on first exploring a smaller chemical subspace (e.g., single-step Zwanzig perturbation method), those that mix empirical parameters with simulation data (e.g., semi-empirical linear response theories), and those that use a combination of explicit and implicit solvent models (e.g., the MM/PBSA method). Force-field methods have also begun to play a larger role in accounting for protein flexibility in virtual screening (e.g., the dynamic pharmacophore method and the relaxed complex methods). Thorough conformational sampling remains a challenging issue to tackle in the realistic treatment of protein and ligand flexibility in modeling drug-receptor interactions. However, recent developments such as the introduction of the replica-exchange and related methods may offer some hope in relieving these problems. Recent developments on using implicit solvent models in molecular simulations should also make it easier to incorporate protein flexibility in drug design.

Acknowledgments

The authors acknowledge support from the Howard Hughes Medical Institute, the National Institutes of Health, the National Science Foundation, the Keck Foundation, the National Biomedical Computing Resource, the San Diego Supercomputer Center, and Accelrys Inc. Some of the material was previously published in the Annual Review of Pharmacology and Toxicology (2003) **43,** 31–45.

References

Antosiewicz, J., McCammon, J. A., and Gilson, M. K. (1994). *J. Mol. Biol.* **238,** 415–436.

Åqvist, J., Medina, C., and Samuelsson, J.-E. (1994). *Protein Eng.* **7,** 385–391.

Archontis, G., Simonson, T., Moras, D., and Karplus, M. (1998). *J. Mol. Biol.* **275,** 823–846.

Baker, N. A., Sept, D., Joseph, S., Holst, M. J., and McCammon, J. A. (2001). *Proc. Natl. Acad. Sci. USA* **98,** 10037–10041.

Banba, S., Guo, Z. Y., and Brooks, C. L. (2000). *J. Phys. Chem. B* **104,** 6903–6910.

Bash, P. A., Singh, U. C., Langridge, R., and Kollman, P. A. (1987). *Science* **236,** 564–568.

Bashford, D., and Case, D. A. (2000). *Ann. Rev. Phys. Chem.* **51,** 129–152.

Bashford, D., and Karplus, M. (1990). *Biochemistry* **29,** 10219–10225.

Beroza, P., Fredkin, D. R., Okamura, M. Y., and Feher, G. (1991). *Proc. Natl. Acad. Sci. USA* **88,** 5804–5808.

Beveridge, D. L., and DiCapua, F. M. (1989). *Annu. Rev. Biophys. Biophys. Chem.* **18,** 431–492.

Bleil, R. E., Wong, C. F., and Rabitz, H. (1995). *J. Phys. Chem.* **99,** 3379–3386.

Bohm, H. J. (1994). *J. Comput.-Aided Mol. Design* **8,** 243–256.

Carlson, H. A., and Jorgensen, W. L. (1995). *J. Phys. Chem.* **99,** 10667–10673.

Carlson, H. A., Masukawa, K. M., and McCammon, J. A. (1999). *J. Phys. Chem. A* **103,** 10213–10219.

Carlson, H. A., Masukawa, K. M., Rubins, K., Bushman, F. D., Jorgensen, W. L., Lins, R. D., Briggs, J. M., and McCammon, J. A. (2000). *J. Med. Chem.* **43,** 2100–2114.

Carlson, H. A., and McCammon, J. A. (2000). *Mol. Pharmacol.* **57,** 213–218.

Chothia, C. (1974). *Nature* **248,** 338.

Claussen, H., Buning, C., Rarey, M., and Lengauer, T. (2001). *J. Mol. Biol.* **308,** 377–395.

Crooks, G. E. (1998). *J. Stat. Phys.* **90,** 1481–1487.

Damodaran, K. V., Banba, S., and Brooks, C. L. (2001). *J. Phys. Chem. B* **105,** 9316–9322.

David, L., Luo, R., and Gilson, M. K. (2000). *J. Comput. Chem.* **21,** 295–309.

Davis, M. E., Madura, J. D., Luty, B. A., and McCammon, J. A. (1991). *Comput. Phys. Commun.* **62,** 187–197.

Del Buono, G. S., Freire, E., and Levy, R. M. (1994). *Proteins* **20,** 85–94.

Dominy, B. N., and Brooks, C. L. (1999). *J. Phys. Chem. B* **103,** 3765–3773.

Dong, F., and Zhou, H.-X. (2002). *Biophys. J.* In press.

Elber, R., and Karplus, M. (1990). *J. Amer. Chem. Soc.* **112,** 9161–9175.

Erlanson, D. A., Braisted, A. C., Raphael, D. R., Randal, M., Stroud, R. M., Gordon, E. M., and Wells, J. A. (2000). *Proc. Natl. Acad. Sci. USA* **97,** 9367–9372.

Franck, P. (1987). *"Introduction to System Sensitivity Theory."* Academic Press, New York.

Ghosh, A., Rapp, C. S., and Friesner, R. A. (1998). *J. Phys. Chem. B* **102,** 10983–10990.

Gidofalvi, G., Wong, C. F., and McCammon, J. A. (2002). *J. Chem. Ed.* **79,** 1122–1126.

Gilson, M. K. (1993). *Proteins* **15,** 266–282.

Gilson, M. K., Davis, M. E., Luty, B. A., and McCammon, J. A. (1993). *J. Phys. Chem.* **97,** 3591–3600.

Gilson, M. K., Given, J. A., Bush, B. L., and McCammon, J. A. (1997). *Biophys. J.* **72,** 1047–1069.

Gould, C., and Wong, C. F. (2002). *Pharmacol. Ther.* **93,** 169–178.

Guo, Z. Y., and Brooks, C. L. (1998). *J. Amer. Chem. Soc.* **120,** 1920–1921.

Hansson, T., Marelius, J., and Aqvist, J. (1998). *Journal of Computer-Aided Molecular Design* **12,** 27–35.

Honig, B., and Nicholls, A. (1995). *Science* **268,** 1144–1149.

Hünenberger, P. H., Helms, V., Narayana, N., Taylor, S. S., and McCammon, J. A. (1999). *Biochemistry* **38,** 2358–2366.

Im, W., Beglov, D., and Roux, B. (1998). *Comput. Phys. Commu.* **111,** 59–75.

Jarzynski, C. (1997). *Phys. Rev. E.* **56,** 5018–5035.

Jorgensen, W. L. (1989). *Acc. Chem. Res.* **22,** 184–189.

Jorgensen, W. L. (1997). *"BOSS."* Yale University, New Haven.

Jorgensen, W. L., Maxwell, D. S., and Tirado-Rives, J. (1996). *J. Amer. Chem. Soc.* **118,** 11225–11236.

Keseru, G. M., and Kolossvary, I. (2001). *J. Amer. Chem. Soc.* **123,** 12708–12709.

Kirkwood, J. G. (1935). *J. Chem. Phys.* **3,** 300–313.

Knegtel, R. M., Kuntz, I. D., and Oshiro, C. M. (1997). *J. Mol. Biol.* **266,** 424–440.

Kramer, B., Rarey, M., and Lengauer, T. (1999). *Proteins* **37,** 228–241.

Kuhn, B., and Kollman, P. A. (2000a). *J. Med. Chem.* **43,** 3786–3791.

Kuhn, B., and Kollman, P. A. (2000b). *J. Amer. Chem. Soc.* **122,** 3909–3916.

Leach, A. R. (1994). *J. Mol. Biol.* **235,** 345–56.

Leach, A. R., and Lemon, A. P. (1998). *Proteins* **33,** 227–239.

Lee, M. R., Duan, Y., and Kollman, P. A. (2000). *Proteins* **39,** 309–316.

Lee, M. S., Salsbury, F. R., and Brooks, C. L. (2002). *J. Chem. Phys.* **116,** 10606–10614.

Levy, R. M., Belhadj, M., and Kitchen, D. B. (1991). *J. Chem. Phys.* **95,** 3627–3633.

Lin, J. H., Perryman, A. L., Schames, J. R., and McCammon, J. A. (2002a). *J. Am. Chem. Soc.* **124,** 5632–3.

Lin, J. H., Perryman, A., Schames, J., and McCammon, J. A. (2002b). *Biopolymers* In press.
Liu, H. Y., Mark, A. E., and van Gunsteren, W. F. (1996). *J. Phys. Chem.* **100,** 9485–9494.
Luo, H., and Sharp, K. (2002). *Proc. Natl. Acad. Sci. USA* **99,** 10399–10404.
Madura, J. D., Briggs, J. M., Wade, R. C., Davis, M. E., Luty, B. A., Ilin, A., Antosiewicz, J., Gilson, M. K., Bagheri, B., Scott, L. R., and McCammon, J. A. (1995). *Comput. Phys. Commu.* **91,** 57–95.
Mangoni, R., Roccatano, D., and Di Nola, A. (1999). *Proteins* **35,** 153–162.
Marelius, J., GraffnerNordberg, M., Hansson, T., Hallberg, A., and Aqvist, J. (1998). *J. Comput.-Aided Mol. Design* **12,** 119–131.
Massova, I., and Kollman, P. A. (1999). *J. Amer. Chem. Soc.* **121,** 8133–8143.
Massova, I., and Kollman, P. A. (2000). *Perspectives in Drug Discovery and Design* **18,** 113–135.
McDonald, N. A., Carlson, H. A., and Jorgensen, W. L. (1997). *J. Phys. Org. Chem.* **10,** 563–576.
Meng, E. C., Shoichet, B. K., and Kuntz, I. D. (1992). *J. Comput. Chem.* **13,** 505–524.
Mezei, M. (1992). *J. Comp. Chem.* **13,** 651–656.
Mezei, M., and Beveridge, D. L. (1986). *Ann. Acad. Sci. NY* **482,** 1–23.
Mezei, M., and Guarnieri, F. (1998). *J. Biomol. Struct. Dyn.* **16,** 723–732.
Mitsutake, A., Sugita, Y., and Okamoto, Y. (2001). *Biopolymers* **60,** 96–123.
Mordasini, T. Z., and McCammon, J. A. (2000). *J. Phys. Chem. B* **104,** 360–367.
Morris, G. M., Goodsell, D. S., Halliday, R. S., Huey, R., Hart, W. E., Belew, R. K., and Olson, A. J. (1998). *J. Comput. Chem.* **19,** 1639–1662.
Mruzik, M. R., Farid, F. F., Abraham, F., Screiber, D. E., and Pound, G. M. (1976). *J. Chem. Phys.* **64,** 481.
Nakajima, N., Higo, J., Kidera, A., and Nakamura, H. (1997). *Chem. Phys. Lett.* **278,** 297–301.
Nozaki, Y., and Tanford, C. (1971). *J. Biol. Chem.* **246,** 2211.
Orozco, M., and Luque, F. J. (2000). *Chem. Rev.* **100,** 4187–4225.
Paulsen, M. D., and Ornstein, R. L. (1996). *Protein Eng.* **9,** 567–571.
Postma, J. P. M., Berendsen, H. J. C., and Haak, J. R. (1982). *Faraday Symp. Chem. Soc.* **17,** 481.
Qiu, D., Shenkin, P. S., Hollinger, F. P., and Still, W. C. (1997). *J. Phys. Chem. A* **101,** 3005–3014.
Resat, H., and Mezei, M. (1993). *J. Chem. Phys.* **99,** 6052–6061.
Rocchia, W., Sridharan, S., Nicholls, A., Alexov, E., Chiabrera, A., and Honig, B. (2002). *J. Comput. Chem.* **23,** 128–137.
Sanbonmatsu, K. Y., and Garcia, A. E. (2002). *Proteins* **46,** 225–234.
Schaefer, M., van Vlijmen, H. W., and Karplus, M. (1998). *Adv. Protein. Chem.* **51,** 1–57.
Schapira, M., Totrov, M., and Abagyan, R. (1999). *J. Mol. Recognit.* **12,** 177–190.
Shen, T. Y., Wong, C. F., and McCammon, J. A. (2001). *J. Amer. Chem. Soc.* **123,** 9107–9111.
Shuker, S. B., Hajduk, P. J., Meadows, R. P., and Fesik, S. W. (1996). *Science* **274,** 1531–1534.
Simmerling, C., Fox, T., and Kollman, P. A. (1998). *J. Amer. Chem. Soc.* **120,** 5771–5782.
Simonson, T. (2001). *Curr. Opin. Struct. Biol.* **11,** 243–252.
Simonson, T., Wong, C. F., and Brünger, A. T. (1997). *J. Phys. Chem. A* **101,** 1935–1945.
Smith, R. H., Jr., Jorgensen, W. L., Tirado-Rives, J., Lamb, M. L., Janssen, P. A., Michejda, C. J., and Kroeger Smith, M. B. (1998). *J. Med. Chem.* **41,** 5272–5286.
Sridharan, S., Nicholls, A., and Sharp, K. A. (1995). *J. Comput. Chem.* **16,** 1038–1044.

Still, W. C., Tempczyk, A., Hawley, R. C., and Hendrickson, T. (1990). *J. Amer. Chem. Soc.* **112,** 6127–6129.

Susnow, R., Nachbar, R. B., Jr., Schutt, C., and Rabitz, H. (1991). *J. Phys. Chem.* **95,** 8585–8597.

Susnow, R., Schutt, C., Rabitz, H., and Subramaniam, S. (1994). *J. Comp. Chem.* **15,** 947–962.

Svensson, B., Jonsson, B., Woodward, C. E., and Linse, S. (1991). *Biochemistry* **30,** 5209–5217.

Tembe, B. L., and McCammon, J. A. (1984). *Computers and Chemistry* **8,** 281.

Thacher, T. S., Hagler, A. T., and Rabitz, H. (1991). *J. Am. Chem. Soc.* **113,** 2020–2033.

Tomovick, R., and Vukobratovic, M. (1972). *"General Sensitivity Theory."* American Elsevier, New York.

Verkhivker, G., Elber, R., and Nowak, W. (1992). *J. Chem. Phys.* **97,** 7838–7841.

Vijayakumar, M., and Zhou, H. X. (2001). *J. Phys. Chem. B* **105,** 7334–7340.

Vorobjev, Y. N., Almagro, J. C., and Hermans, J. (1998). *Proteins* **32,** 399–413.

Vorobjev, Y. N., and Hermans, J. (1999). *Biophys. Chem.* **78,** 195–205.

Warshel, A., and Papazyan, A. (1998). *Curr. Opin. Struct. Biol.* **8,** 211–217.

Watts, C. R., Mezei, M., Murphy, R. F., and Lovas, S. (2001). *J. Biomol. Struct. Dyn.* **18,** 733–748.

Wong, C. F. (1991). *J. Am. Chem. Soc.* **113,** 3208–3209.

Wong, C. F., Hünenberger, P. H., Akamine, P., Narayana, N., Diller, T., McCammon, J. A., Taylor, S., and Xuong, N. H. (2001). *J. Med. Chem.* **44,** 1530–1539.

Wong, C. F., and McCammon, J. A. (1986a). *Isr. J. Chem.* **27,** 211–215.

Wong, C. F., and McCammon, J. A. (1986b). *J. Am. Chem. Soc.* **108,** 3830–3832.

Wong, C. F., and Rabitz, H. (1991). *J. Phys. Chem.* **95,** 9628–9630.

Wong, C. F., Thacher, T., and Rabitz, H. (1998). *In* "Reviews in Computational Chemistry," (K. B. Lipkowitz and D. B. Boyd, Eds.), pp. 281–326. Wiley-VCH, New York.

Yang, A.-S., Gunner, M. R., Sampogna, R., Sharp, K., and Honig, B. (1993). *Proteins* **15,** 252–265.

Yang, A.-S., and Honig, B. (1993). *J. Mol. Biol.* **231,** 459–474.

Yu, W., Wong, C. F., and Zhang, J. (1996). *J. Phys. Chem.* **100,** 15280–15289.

Zacharias, M., and Sklenar, H. (1999). *J. Comput. Chem.* **20,** 287–300.

Zhang, H., Wong, C. F., Thacher, T., and Rabitz, H. (1995). *Proteins* **23,** 218–232.

Zhang, L. Y., Gallicchio, E., Friesner, R. A., and Levy, R. M. (2001). *J. Comput. Chem.* **22,** 591–607.

Zhu, S.-B., and Wong, C. F. (1993a). *J. Chem. Phys.* **99,** 9047–9053.

Zhu, S.-B., and Wong, C. F. (1993b). *J. Chem. Phys.* **98,** 8892–8899.

Zhu, S.-B., and Wong, C. F. (1994). *J. Phys. Chem.* **98,** 4695–4701.

Zwanzig, R. W. (1954). *J. Chem. Phys.* **22,** 1420.

FREE ENERGY CALCULATIONS AND LIGAND BINDING

By BJØRN O. BRANDSDAL, FREDRIK ÖSTERBERG, MARTIN ALMLÖF, ISABELLA FEIERBERG, VICTOR B. LUZHKOV, AND JOHAN ÅQVIST

Department of Cell and Molecular Biology, Biomedical Center, Uppsala University, Uppsala, Sweden, SE-75124

I. Introduction 123
II. Free Energy Perturbation and Thermodynamic Integration 125
A. Applications to Ligand Binding 129
III. Extrapolation of Free Energies 135
IV. Linear Interaction Energy Approaches 137
A. Plasmepsin II 141
B. Trypsin-BPTI 143
C. P450 cam 145
V. MM-PBSA 147
VI. PROFEC 149
VII. λ-Dynamics and Chemical MC/MD 151
VIII. Conclusions 154
References 155

I. Introduction

The ability of proteins to bind to one another and to different ligands in a highly specific manner is an important feature of many biological processes. The characterization of the structure and the energetics of molecular complexes is thus a key factor for understanding biological functions, and the energetics often provides the most important and useful link between structure and function of biomolecular systems. Furthermore, the prediction and design of ligands that can reversibly bind to pharmaceutical targets (enzyme inhibitors, receptor agonists and antagonists, etc.) is at the heart of structure-based drug design. The ability to predict the strength of noncovalent associations, as well as the structures of molecular complexes, has therefore been an important objective in computational chemistry.

A number of different types of computational approaches have been developed over the years for predicting binding constants. These range from purely empirical or statistical ones, such as quantitative structure activity relationships (QSAR), to more or less rigorous methods based on evaluation of the actual physical energies involved in the binding process.

ADVANCES IN PROTEIN CHEMISTRY, Vol. 66

0065-3233/03 $35.00

A few general points should be kept in mind when deciding which computational strategy to use for estimating binding energies. One of the most important aspects when trying to predict the potency of a set of ligands is the time required for calculating the affinity (or score) of a typical ligand. Screening of large virtual libraries demands a high throughput of ligands, and the time spent on evaluating a single compound must be short. However, if the three-dimensional (3D) structure of a "lead" compound in complex with the receptor is known it may be affordable and desirable to carry out more accurate and time-consuming calculations, particularly when the objective is to explore a limited number of chemical modifications of this lead.

The most rapid methods for estimating binding free energies are so-called empirical or knowledge-based (statistical) scoring approaches. These are based on simple energy functions (Böhm, 1994; Jain, 1996; Eldridge *et al.*, 1997) or on the frequency of occurrence of different atom-atom contact pairs in complexes of known structure (Muegge and Martin, 1999; Gohlke *et al.*, 2000), respectively. The simplicity of the energy function along with the lack of conformational sampling and explicit water treatment makes these approaches very fast, but usually at the cost of accuracy. The most time-consuming and rigorous methods are based on molecular force fields and involve slow, gradual transformations between the states of interest using either molecular dynamics (MD) or Monte Carlo (MC) simulations for generating ensemble averages. Extensive conformational sampling and the fact that a large number of pair-wise interactions must be calculated at each MD or MC step make such methods time-consuming. In this chapter, we will mainly focus on the latter category of techniques that can be characterized by the use of molecular force fields for obtaining the relevant energy components.

Although force field calculations on molecular complexes have a rather long history, the emergence of free energy calculation methods (Torrie and Valleau, 1974, 1977; Mezei *et al.*, 1978; Warshel, 1982; Postma *et al.*, 1982) was of major importance because it provided the crucial link to experimental work, whereby calculated and observed energetics could be compared directly. The statistical mechanical framework for free energy calculations had been available for quite some time (Kirkwood, 1935; Zwanzig, 1954), but it was not applied to chemical and biological problems until the early 1980s when more powerful computers became available to the research community. At this time the first application of the methodology to proteins was reported, dealing with free energy profiles for proton transfer in lysozyme (Warshel, 1984). Another important step was the combination of free energy simulations with thermodynamic cycles describing the binding of different ligands to a given receptor (Tembe and

McCammon, 1984). This type of approach in which the calculations focus on non-physical "mutations" between different ligands became the standard procedure for obtaining relative-binding free energies (Wong and McCammon, 1986; Bash *et al.*, 1987; Hwang and Warshel, 1987).

Many of these studies of protein-ligand binding in the mid-1980s showed a remarkable agreement between theory and experiment, which led to an explosion of activity in the field of free energy calculations. More recent investigations, however, have demonstrated that significantly longer simulations than those used in the original reports are often required to obtain reliable results in protein-ligand binding studies. The increasing number of applications of free energy calculations also showed that the use of these methods was not as straightforward as expected; therefore much effort was spent on improving the methodology. It became clear that such simulations often could not be performed routinely, as "black box" jobs, but rather required careful attention to the computational setup and to the interpretation of results.

II. Free Energy Perturbation and Thermodynamic Integration

We will begin our discussion of free energy calculations with a short review of some of the important aspects of what can be called the rigorous approaches. Several excellent accounts of these methods have been published elsewhere (e.g., Beveridge and DiCapua, 1989; Straatsma and McCammon, 1992; Kollman, 1993; Lamb and Jorgensen, 1997). Most free energy calculations are generally formulated in terms of estimating the relative free energy differences, ΔG, between two equilibrium states. This is of great importance in many applications, because it is normally the difference in the thermodynamic properties between two such states that is of interest. Estimation of the difference in binding free energy for two similar ligands to a common receptor is one example in which such calculations play a central role.

The free energy difference between two states A and B can be formally obtained from Zwanzig's formula (Zwanzig, 1954):

$$\Delta G = G_B - G_A = -\beta^{-1}\ln\langle\exp(-\beta\Delta V)\rangle_A \tag{1}$$

where $\beta = 1/kT$ and $\langle\ \rangle_A$ denotes an MD- or MC-generated ensemble average of $\Delta V = V_B - V_A$ that is sampled using the V_A potential. Equation (1) assumes that the configurational sampling is carried out under constant temperature and pressure conditions (isothermal-isobaric ensemble), whereas (N, V, T)-simulations instead yield the corresponding Helmholtz free energy. Furthermore, a kinetic contribution to the free energy difference (e.g., due to a possible change in atomic masses) is not considered because it will always cancel out by virtue of the equipartioning theorem and the relevant thermodynamic cycle (e.g., if one considers the absolute solvation free energy of an ion, the kinetic contribution will be the same as in the gas-phase and therefore vanish in the solvation cycle).

The main criterion for Equation (1) to be useful is that the configurations sampled on the potential V_A should have a reasonable (at least non-vanishing) probability of also ocurring on V_B. This essentially means that thermally accessible regions of the two potentials should have a significant degree of overlap. If not, the result will be a very slow convergence of the average. That convergence can be assessed, for example, by interchanging the labels A and B and changing the sign of ΔG in Equation (1), thus applying the formula "backwards."

In order to solve the above convergence problem associated with the implementation of Equation (1), a multistep approach is normally adopted. A path between the states A and B is defined by introducing a set of intermediate potential energy functions that are usually constructed as linear combinations of the initial *(A)* and final *(B)* state potentials:

$$V_m = (1 - \lambda_m) V_A + \lambda_m V_B \tag{2}$$

where λ_m varies from 0 to 1. In practice this path is thus discretized into a number of points ($m = 1, \ldots, n$), each represented by a separate potential energy function that corresponds to a given value of λ. This coupling parameter approach rests on the fact that the free energy difference is uniquely defined by the initial and final states (i.e., a state function) and can be computed along any reversible path connecting those states. Now the total free energy change can be obtained by summing over the intermediate states along the λ variable:

$$\Delta G = G_B - G_A = -\beta^{-1} \sum_{m=1}^{n-1} \ln\langle \exp[-\beta(V_{m+1} - V_m)] \rangle_m \tag{3}$$

This approach is generally referred to as the free energy perturbation (FEP) method. If we denote $\Delta\lambda_m = \lambda_{m+1} - \lambda_m$ and combine Equations (2) and (3), we find that:

$$\Delta G = -\beta^{-1} \sum_{m=1}^{n-1} \ln\langle \exp[-\beta \Delta V \Delta\lambda_m] \rangle_m \tag{4}$$

with $\Delta V = V_B - V_A$, as above. With our definition of V_m, one also finds that $\partial V_m / \partial \lambda_m = \Delta V$. Even if the intermediate potentials are not defined as a simple linear combination of the states A and B, we can still write the potential difference in the exponent of Equation (3) as $V_{m+1} - V_m = \frac{\partial V_m}{\partial \lambda_m} \Delta\lambda_m$, provided that the λ-steps are sufficiently small. This means that Equation (3) takes the following form:

$$\Delta G = -\beta^{-1} \sum_{m=1}^{n-1} \ln \left\langle \exp\left[-\beta \frac{\partial V_m}{\partial \lambda_m} \Delta\lambda_m \right] \right\rangle_m \tag{5}$$

For small steps in λ, this equation can be linearized by retaining only the leading terms in the Taylor expansion of the exponent and logarithm to yield:

$$\Delta G = \sum_{m=1}^{n-1} \left\langle \frac{\partial V_m}{\partial \lambda_m} \right\rangle_m \Delta\lambda_m \tag{6}$$

which with $\lambda \rightarrow 0$ can instead be written as an integral over λ:

$$\Delta G = \int_0^1 \left\langle \frac{\partial V(\lambda)}{\partial \lambda} \right\rangle_\lambda d\lambda \tag{7}$$

This equation, which is, in fact, exact and can be derived directly from the configuration integral, just as Equation (1), is usually referred to as the *thermodynamic integration (TI)* formula for the free energy. For the case in which the intermediate potentials, V_m or $V(\lambda)$, are defined as linear combinations of V_A and V_B, it also takes the simpler form:

$$\Delta G = \int_0^1 \langle \Delta V \rangle_\lambda d\lambda \tag{8}$$

From the above considerations it is evident that it is very convenient to use the linear combination of Equation (2) because it means that for each value of λ, only the two endpoint potentials need to be used and both energies and forces are then simply scaled with the appropriate coefficients to yield the desired trajectories. In this respect it seems unnecessarily complicated to instead scale all the parameters of the potential energy function separately as is done in some programs (van Gunsteren and Berendsen, 1987).

In the early days of free energy simulations the TI approach was synonymous with what has later become called *slow growth.* This meant that the value of λ actually was changed in each time step of an MD simulation. Although this method was claimed to be more efficient than the discrete FEP formulation (van Gunsteren, 1988), the consensus today is that a "non-continuous" change in λ is the better choice (we usually recommend 50 to 100 discrete points). The primary reasons are that equilibration can be allowed at each point, extra points can be added at any time, and any pattern of spacing between the λ-points can be used in order to optimize the efficiency. Fig. 1 shows an example of an FEP calculation of the free energy of charging a Na^+ ion in water, where a non-uniform spacing of λ-points can be seen to yield a nearly constant free energy contribution at each step. This minimizes the convergence error, for a given total number of MD/MC steps, because the error in Equation (1) and in each term of Equation (3) is generally proportional to the magnitude of the free energy change.

The so-called "double-wide sampling" approach (Mezei, 1987) involves the insertion of fictitious λ-points midway between the real ones so that each λ-interval can be made effectively half as big. One then makes use of both forward and backward application of the FEP formula in order to "connect" the virtual points with the real ones. For sufficiently closely spaced λ-points, however, this approach is equivalent to averaging the free energies over the forward and backward direction for each perturbation step. The advantage of the latter approach is that an estimate of the

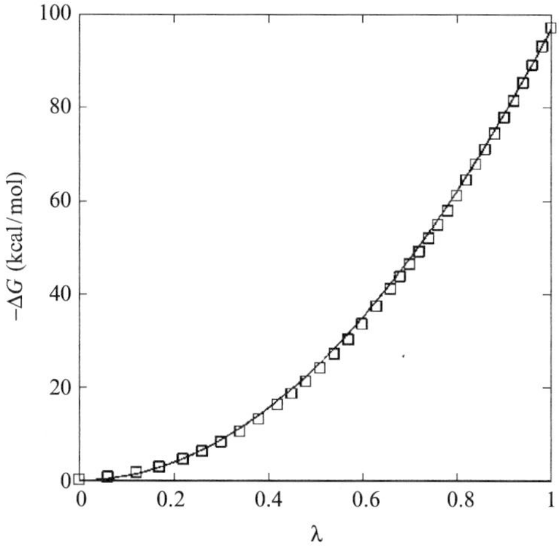

FIG. 1. Result from an FEP/MD simulation (140 ps) of the free energy of charging a sodium ion (including the Born correction) in a 40 Å radius sphere of water that contains around 27,000 atoms. This calculation uses the local reaction field method (Lee and Warshel, 1992), which accurately reproduces the results with an infinite cutoff, so that only about 20 M pairwise interactions need to be calculated at each step (whereas the full ~350 M interactions are only evaluated directly every 50 MD steps). The simulation demonstrates the size consistency of the SCAAS boundary model (King and Warshel, 1989) with respect to earlier results (Åqvist, 1990, 1994) and illustrates the validity of linear response in this case, because the free-energy function closely fits a quadratic behavior (solid line).

convergence or summation error (this is just one out of several quality indicators) can be obtained from the two directions.

Much has been said in the literature about FEP versus TI, but once one has decided on a simulation protocol (length of trajectories, equilibration, number of λ-points and their spacing, etc.), the difference mainly pertains to which formula is to be used for evaluating the free energy. As shown previously, the formulas are, in principle, equivalent. However, although the FEP formulation (e.g., Equations [1], [3], and [4]) is exact, the discretized version of TI usually used today is an approximation to the exact integral formula. Equation (7) is then approximated by Equation (6), or perhaps a trapezoidal version thereof. Therefore, Equation (6) is only valid in the limit of small λ-steps, whereas higher order fluctuation terms in ΔV (if we use the linear combination rule), or $\partial V_m/\partial\lambda_m$, otherwise contribute to the free energy (see Åqvist and Hansson, 1996, for example). Therefore, discretized TI with only a few λ-steps (say, <20) is generally not recommended in our opinion, unless a more sophisticated approximation of the integral (Mezei, 1987) is used.

Some say that an advantage with TI over FEP is that the former allows for an analysis of the contribution of different potential energy components to the free energy, because the total derivative of the potential in Equation (7) is the sum of derivatives of the individual force field terms. However, as is evident from the linearized FEP formula (Equation [6]), such a component analysis can be carried out equally well within the FEP scheme provided that the λ-steps are sufficiently small. On the other hand, it is now well recognized that such a component analysis will depend on the path chosen to connect the end states (the components of the free energy are not state functions), therefore care should be taken in interpreting the results of such analyses. A point in case is the interpretation of contributions from bonded terms in free energy calculations (Boresch and Karplus, 1999a,b).

A. *Applications to Ligand Binding*

Most studies of ligand binding are carried out in the context of a thermodynamic cycle as depicted in Fig. 2 (Tembe and McCammon, 1984), which is used to calculate the relative binding free energy between two ligands (L′ and L) and a given receptor. ΔG^{p}_{mut} and ΔG^{w}_{mut} in Fig. 2 denote the difference in free energy between L and L′ when bound to the solvated receptor binding site and in water, respectively, while ΔG_{bind} are the corresponding binding free energies. The relative free energy of binding hence becomes:

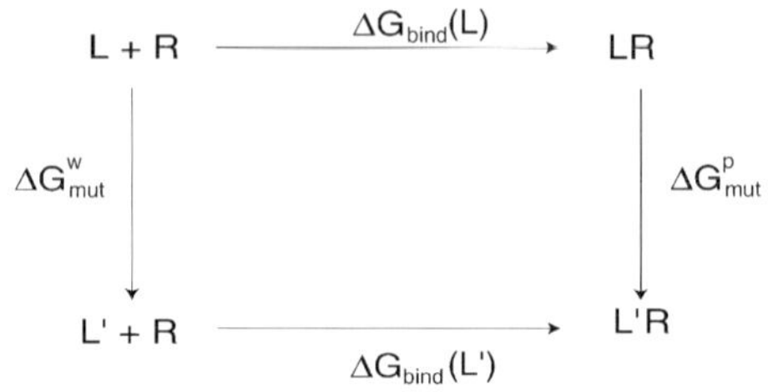

FIG. 2. Thermodynamic cycle used in FEP/TI calculations of the relative binding free energies of two ligands, L and L′, to a receptor molecule R. The absolute binding free energy can be obtained, in principle, by treating L as a dummy ligand and taking into account the relevant standard state in terms of volume/concentration (Hermans and Wang, 1997).

$$\Delta\Delta G_{bind} = \Delta G_{bind}(L') - \Delta G_{bind}(L) = \Delta G^{p}_{mut} - \Delta G^{w}_{mut} \tag{9}$$

With the free energy perturbation approach, the free energy associated with the two unphysical paths, L→L′ and LR→L′R, corresponding to a mutation of L into L′ in the free and bound state, respectively, is calculated. The nice aspect about this type of procedure is that it is usually much more efficient than trying to calculate ΔG_{bind} directly from the potential of mean force along an assumed "binding coordinate."

A much discussed problem in free energy calculations is the so-called end-point catastrophe associated with vanishing or appearing atoms (Pitera and van Gunsteren, 2002). This is particularly pertinent to applications of the above type in which different ligands are transformed into one another and atoms need to appear and disappear during the process. The problem is basically two-fold. First, there is a possible numerical instability of the free energy formulas associated primarily with the infinite repulsive ($1/r^{12}$) Lennard-Jones term for $r = 0$. That is, in the state in which a certain atom has no interactions (a dummy atom), other atoms can lie on top of it, so that the energy of the state in which this atom is present with its full interaction would become infinite, as well as its derivative. In the FEP approach this causes no numerical problem in the direction where the real atom is being created, because the exponent in Equation (1) then becomes minus infinity ($V_B > V_A$) and the contribution of such configurations to the free energy difference is 0. Furthermore, in the FEP approach the very end-points ($\lambda = 0$ or $\lambda = 1$) are formally used in only one summation direction (i.e., as the initial point of a summation) and never as the final point. In TI, on the other hand, the end-point derivatives are often used for the numerical integration of Equation (7), and an infinite integrand can obviously cause a severe instability. This might be considered an advantage of FEP over TI, but, unfortunately, this issue is not only a numerical one.

The second problem is perhaps more serious and has to do with a deficient configurational sampling introduced by appearing/disappearing atoms. It can be understood from the fact that the repulsive Lennard-Jones potential of a very small atom (i.e., one whose interactions are scaled by a very small value of λ) is still infinite for $r = 0$. This means that sampling of the positions occupied by vanishing atoms cannot be accomplished until these atoms have completely disappeared (i.e., at the end-point of the λ-range). This is particularly problematic in confined geometries, such as a protein-binding site, where it is possible that the space occupied by vanishing atoms cannot be properly filled until the very last end-point simulation. Furthermore, when MD (as opposed to MC) is used for sampling, the dynamics can become unstable near the end-points of vanishing atoms. This is due to the fact that for very small atoms, interacting with δ-function–like Lennard-Jones spikes, the associated forces change too rapidly with distance and would thus require successively smaller timesteps as the end-point is approached.

The main remedies for the problems just described are (1) to transform the Lennard-Jones potential into a softer one (that is not infinite for $r = 0$) (Beutler *et al.*, 1994; Zacharias *et al.*, 1994), (2) to shrink bonds to vanishing atoms so that they are effectively ''pulled'' within the van der Waals spheres of the atoms to which they are connected, and (3) to use a denser distribution of λ-points near problematic end-points. The different approaches have been examined recently by Pitera and van Gunsteren (2002), who advocate the potential-softening method as the most reliable one. That study, however, still appears somewhat inconclusive because FEP (that does not involve infinite derivatives) was not examined and a TI protocol with very few λ-points was used. It is clear, however, that the use of so-called soft-core nonbonded potentials can improve sampling considerably. As far as bond-shrinking is concerned, a note of caution may be appropriate: the possible application of constraints for changing bond-lengths in this context (and otherwise) suppresses some contribution to the free energy (Pearlman and Kollman, 1991; Boresch and Karplus, 1999a,b). This type of contribution may vanish along a thermodynamic cycle, but no guarantee exists for this in general.

As an example of the end-point effect and the use of a nonuniform λ-spacing, Fig. 3A shows the results from a FEP calculation of the conversion of hexachloroethane to 1,1,1-trichloroethane in water. The illustration shows that the slope of the free energy near $\lambda = 0$ (hexachloroethane) is not very steep and that the curve is practically linear in this region. So, on the left-hand side of Fig. 3A, a dense λ-spacing is actually not necessary

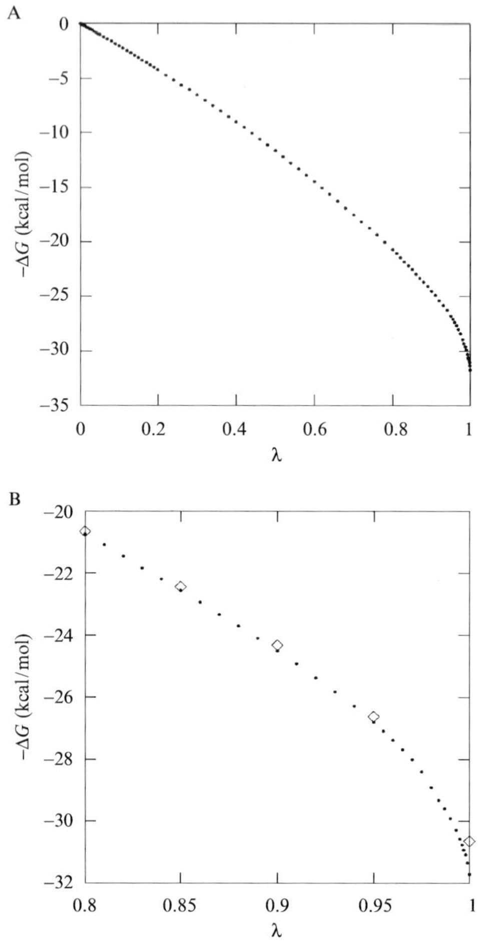

FIG. 3. Results from FEP/MD simulations of transforming hexachloroethane to 1,1,1-trichloroethane in water. (A) An example of having more closely spaced λ-values near the end-points is shown, where it can be seen that this is only needed near $\lambda = 1$ where the curve becomes steep. (B) Close-up view of the problematic end-point region, where the error caused by using evenly and sparsely distributed λ-points is also shown (*diamonds*).

and does not improve the accuracy significantly. Near $\lambda = 1$ (trichloroethane), on the other hand, where the larger chlorine substituents are vanishing, the slope becomes very steep and a dense spacing is required. Fig. 3B shows the effect of using a constant λ-spacing of 0.05 (in fact, a rather common value), where an error of >1 kcal/mol is immediately introduced.

As an application of the FEP/TI methodology to the typical problem of predicting relative-binding free energies of protein-ligand complexes, we will discuss a useful example that illustrates several aspects of this type of calculation. This example deals with the prediction of relative-binding free energies for complexes of quaternary ammonium ions with the KcsA potassium channel. The well-studied extracellular potassium channel binding site is formed in this case by a cage of four tyrosine residues Y82 near the entry to the permeation pore (Doyle *et al.*, 1998) of the channel (Fig. 4). Stabilization of the alkylammonium cations in the binding site is achieved by electrostatic interactions with carbonyl groups lining the channel surface near the entry and by nonpolar interactions with aromatic rings of the tyrosine residues (Luzhkov and Åqvist, 2001). Experimental studies show a well-defined selectivity for this site where the optimum affinity (K_D in the mM range) is observed for the complex of

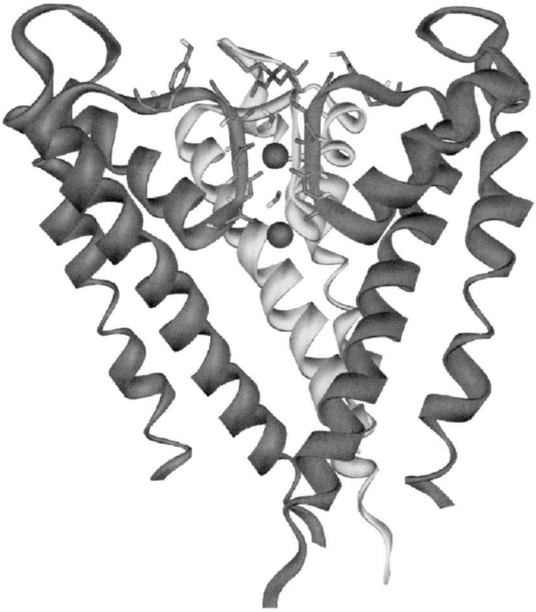

FIG. 4. Snapshot from an MD simulation of the KcsA potassium channel in complex with tetrapropylammonium ion at the extracellular binding site (one channel subunit has been removed for clarity). Two K^+ ions in the selectivity filter are also shown. (See Color Insert.)

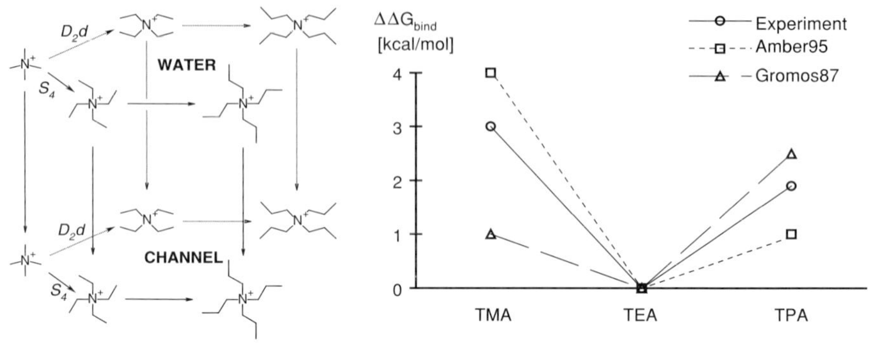

FIG. 5. Thermodynamic cycles involved in the calculations of quaternary ammonium ion binding, in different conformations, to the K^+ channel (*left*), and binding energetics for Me_4N^+ (TMA), Et_4N^+ (TEA), and *n*-Pr_4N^+ (TPA) obtained with different force fields (*right*). In these calculations, changing bonds were described by harmonic potentials and no bond-length constraints were applied.

tetraethylammonium ion, whereas smaller and larger alkylammonium ions bind less tightly (Meuser *et al.*, 1999).

In principle, a rigorous evaluation of the binding energetics should consider both of the stable conformations of the quarternary centre of the ligands, namely S_4 and D_{2d}, that are separated by a too-high energy barrier to allow interconversion between the conformers on the sub-μs timescale (Luzhkov *et al.*, 2002). In other words, as one goes from Me_4N^+ to Et_4N^+, two conformational "valleys" appear that need to be sampled separately. Fig. 5 shows the thermodynamic cycles corresponding to this situation. We have recently examined the binding of the three ions Me_4N^+, Et_4N^+, and *n*-Pr_4N^+ to the external binding site of the KcsA ion channel by means of automated docking and FEP/MD simulations (Luzhkov *et al.*, 2003). As an illustration, the calculated and experimental results are shown in Fig. 5 for the D_{2d} conformation, which has been shown both experimentally and theoretically to be about 0.6 to 1.0 kcal/mol more stable than S_4 in water (Naudin *et al.*, 2000; Luzhkov *et al.*, 2002). In these calculations the Gromos87 (van Gunsteren and Berendsen, 1987) and Amber95 (Cornell *et al.*, 1995) force fields were employed. Fig. 5 shows that both of these force fields reproduce the experimentally observed binding optimum for Et_4N^+ with Amber95 giving the closest agreement with experiment and the correct ranking of blockers. This is an interesting case because it shows how an "ergodicity" problem can appear as a consequence of different regions of the conformational space effectively becoming "disconnected" along the perturbation path on a given time scale. This type of situation, in fact, may not be so uncommon when one considers perturbations between ligands with different number of torsional degrees of freedom.

III. Extrapolation of Free Energies

Even though the FEP and TI calculations have been shown to give accurate results in a number of cases, a substantial amount of computer time is still often required to obtain the relative binding free energy between two ligands and their receptor. This will, of course, render the methodology less tractable when the goal is to screen a large number of compounds or when guiding experimental design of drug candidates. Methods that rapidly predict changes in the binding constant of a ligand associated with specific modifications are much more suitable in such cases, but there is usually a trade-off between speed and accuracy. Approaches that estimate the free energy difference between two states based on a single simulation of just one reference state nevertheless may provide a useful alternative to TI or FEP in this respect.

A starting point for such approaches is to expand the free energy difference between two states in a power series around one of the states (Jayaram and Beveridge, 1990; Åqvist *et al.*, 1994; Smith and van Gunsteren, 1994):

$$\Delta G_{AB} = \langle \Delta V \rangle_A - \frac{\beta}{2}\left\langle (\Delta V - \langle \Delta V \rangle_A)^2 \right\rangle_A + \frac{\beta^2}{6}\left\langle (\Delta V - \langle \Delta V \rangle_A)^3 \right\rangle_A - \cdots \quad (10)$$

where ΔV is defined by Equation (2). While the regular FEP formula is recovered as the infinite limit of this series, it may be interesting to try to truncate Equation (10) after a finite (small) number of terms. In particular, truncation after the second term corresponds to the assumption of a linear response of the state A to a perturbation, which is equivalent to a Gaussian distribution of ΔV. This approach was examined by Levy *et al.* (1991) in the context of ion-pair hydration and found to yield rather accurate results, although the calculated trajectories were somewhat short. That is, the higher (≥ 2) order terms in Equation (10) converge slower and slower, and require long simulations for convergence (Smith and van Gunsteren, 1994). An early attempt by Gerber and co-workers (1993) even used a truncation after the first term for estimating binding free energies between different ligands and dihydrofolate reductase, but that type of approximation is not expected to work in general.

In cases where the system obeys linear response, it turns out to be much more efficient to use two simulations, of both the initial and final states, to calculate the free energy from a truncated expansion (Lee *et al.*, 1992). For example, if we consider the response to electric fields as linear and characterized by a single polarization force constant, the second-order (and higher) fluctuation terms in Equation (10) will formally cancel when we take the average of the free energy estimates from the two simulations (Åqvist *et al.*, 1994; Åqvist and Hansson, 1996):

$$\Delta G_{AB} = \frac{1}{2}\{\langle \Delta V \rangle_A + \langle \Delta V \rangle_B\} \tag{11}$$

This equation converges much faster than Equation (10) simply because only the linear terms in ΔV are needed. When two states, A and B, denote the end-points of a charging process in which a charge distribution is created from nothing or annihilated, Equation (11) often simplifies into a single state formula. That is, if we consider, for example, the charging of a (rigid) molecule then the term $\langle \Delta V \rangle$ will tend to zero (Åqvist *et al.*, 1994; Åqvist and Hansson, 1996). Hence, the simple formula $\Delta G_{AB} = 1/2 \langle \Delta V \rangle_B$ can work well for estimating the electrostatic contribution to solvation energies. Another way to turn this linear response equation into a single state formula is to use an intermediate state (e.g., $\lambda = 0.5$) (King and Barford, 1993) and apply Equation (11) from that state to both end-points, in which case we get:

$$\Delta G_{AB} = \langle \Delta V \rangle_{\lambda = 0.5} \tag{12}$$

with $\Delta V = V_B - V_A$ as above. An interesting feature of this approach (see also Gerber *et al.*, 1993) is that it prescribes the use of an arbitrary reference state for calculating the free energy differences between other states.

The type of procedure just described that used simulations of a single reference state for free energy calculations has been further examined by van Gunsteren and co-workers (Liu *et al.*, 1996; Schäfer *et al.*, 1999) and by Radmer and Kollman (1998). Liu *et al.* (1996) have shown that the free energy resulting from the charge rearrangement studied earlier (Smith and van Gunsteren, 1994) could be computed using a single-step perturbation approach, and they obtained essentially the same results as with the series expansion. van Gunsteren and co-workers have also shown that it is possible to study mutations that involve creation or removal of atoms with the single perturbation approach by choosing an appropriate reference state (Liu *et al.*, 1996; Schäfer *et al.*, 1999). The reliability of the results obtained with the perturbation formula (Equation 1) will be highly dependent of course on whether the configurations sampled at the reference state are representative of the ensemble as a whole. Using the perturbation formula implicitly includes all higher-order derivatives, and is thus formally exact while the series expansion must be considered as approximate with truncation of higher-order terms. However, the range of potential modifications can be extended by biasing the sampling of the reference state with, for example, soft interactions sites at the positions where atoms are created or removed. This strategy has been used recently to study the affinity of natural ligands and xenoestrogens to the estrogen receptor (Oostenbrink *et al.*, 2000). In order to sample the configurations, a nonphysical reference state was constructed with several soft interaction sites. The use of a ''soft'' ligand allows occasional overlaps of the ligand atoms with its surroundings, and thereby samples configurations that are favorable for real ligands of different shapes and sizes. It is important to emphasize that only relative energies can be obtained with such extrapolation methods, but they do reduce the overall time needed compared with FEP or TI significantly.

Although considerable effort has thus been devoted to examining the performance of various truncated formulas of the Taylor expansion (Equation 10), the single-step perturbation formula (Equation 1) seems generally more reliable when just one simulation of a single reference state is used. In cases where the linear response approximation holds (i.e., typically for charging processes (electrostatics)), the single-step perturbation formula reduces to the first two terms of Equation (10). However, in such cases Equation (11) is valid and then provides a more efficient approach than either one-step perturbation or truncation after the second term of the series, because Equation (11) only requires convergence of the plain ΔV averages and often only one of them. Åqvist and Hansson (1996) has examined the validity of Equation (11) in detail for a number of different types of compounds in different solvents. The main advantage with the single reference state approach is, of course, that a number of different end-points (usually different molecules) can be considered simultaneously. However, the sampling problems associated with creation and annihilation of atoms, even using soft-core potentials, still severely limit the use of this type of method for examining molecules that differ significantly from each other.

IV. Linear Interaction Energy Approaches

The linear interaction energy (LIE) approach (Åqvist *et al.*, 1994) is another type of method that relies on molecular dynamics or Monte Carlo simulations to generate ensemble averages. This method has gained considerable attention in past years, particularly with respect to the estimation of absolute or relative binding free energies for widely different compounds. It has been reviewed earlier (Åqvist and Marelius, 2001; Åqvist *et al.*, 2002), and only the basic concepts of the method will be presented here.

The initial idea was to consider the absolute binding free energy of a ligand—the change in free energy when transferred from solution to the solvated receptor binding site—as composed of a polar and a nonpolar contribution. The main point is to consider only the physically relevant states, the corners of the thermodynamics cycle (Fig. 2), and not to spend time on sampling uninteresting intermediate states as traditional FEP or TI does. Furthermore, the linear response approximation is used to determine the electrostatic contribution to the binding free energy, whereas the nonpolar contribution is estimated using an empirically derived parameter that scales the intermolecular van der Waals (Lennard–Jones) interaction energies from the MD simulations. This was motivated by the fact that solute-solvent van der Waals energies are found to be correlated with the same variables as hydrophobic solvation–free energies (e.g., accessible surface area), and that average van der Waals energies also scale approximately linearly with solute size (Åqvist *et al.*, 1994). This led to an approximate equation for the binding free energy of the following type:

$$\Delta G_{\text{bind}} = \alpha\Delta\langle V_{l-s}^{vdw}\rangle + \beta\Delta\langle V_{l-s}^{el}\rangle + \gamma \tag{13}$$

where $\langle\ \rangle$ denotes MD or MC averages of the nonbonded van der Waals (*vdw*) and electrostatic (*el*) interaction of the ligand and its surrounding environment (*l–s*). The Δ's denote the change in these averages when transferring the ligand from solution (free state) to the receptor binding site (bound state). The response of intramolecular energy terms in the ligand and receptor, just as the solvent energy, is thus embedded in the coefficients on Equation (13) (e.g., the classical linear response factor of $\frac{1}{2}$ for electrostatic solvation expresses precisely this) (Åqvist and Marelius, 2001). Hence, two simulations are required to determine the absolute binding free energy of a ligand to a receptor: one of the ligand free in solution and one when it is bound to the solvated receptor binding site. The parameters are the weight coefficients α and β for the nonpolar and the electrostatic contribution, respectively, and possibly an additional constant γ that also can be referred to the nonpolar component (Åqvist *et al.*, 1994; Hansson *et al.*, 1998; Wang, W. *et al.*, 1999).

The first applications of the LIE method used a β coefficient of $\frac{1}{2}$ as predicted by the linear response approximation, along with a nonpolar coefficient (α) of 0.16 and the additional constant γ set to zero. With this parameterization the LIE method was able to reproduce the experimental binding data with good accuracy for a number of systems (Åqvist *et al.*, 1994; Hansson and Åqvist, 1995; Åqvist and Mowbray, 1995; Åqvist, 1996; Hultén *et al.*, 1997). Further investigation of the validity of the linear response approximation led to the implementation of a ligand-dependent

β that can take on a few different values in the range of 0.33 to 0.50. These different values reflect deviations from electrostatic linear response and were directly taken from comparing FEP calculations to Equation (11), for a set of test compounds (Åqvist and Hansson, 1996). When taking these deviations from $\beta = \frac{1}{2}$ into account, the optimal nonpolar coefficient was found to be $\alpha = 0.18$ for a calibration set of 18 ligand-receptor complexes, with the optimal value of γ being very close to zero (Hansson *et al.*, 1998).

This revised LIE model has subsequently been used in studies of dihydrofolate reductase (DHFR) (Marelius *et al.*, 1998a; Graffner-Nordberg *et al.*, 2001) and human thrombin inhibitors (Ljungberg *et al.*, 2001), as well as a number of complexes with ligand recognition and transport proteins, namely arabinose, lysine, fatty acid, and retinol binding protein (Åqvist and Marelius, 2001). The work on DHFR inhibitor binding involved both calculations on analogues of the classical antifolate methotrexate (Marelius *et al.*, 1998a) and on newly designed lipophilic ester soft drugs against the *Pneumocystis carinii* enzyme (Graffner-Nordberg *et al.*, 2001). An essential aspect of these studies was to examine not only the ranking of different inhibitors, but also the selectivity of a given inhibitor for different DHFR enzymes. Hence, Marelius *et al.* (1998a) addressed the effects of point mutations of the human enzyme on methotrexate affinity, whereas the calculations on the nonclassical ester inhibitors focused on the selectivity between the human and the *P. carinii* enzyme (Graffner-Nordberg *et al.*, 2001). In these studies, as well as others (Luzhkov and Åqvist, 2001), automated docking methods (Morris *et al.*, 1998) were used as a first stage in order to generate a set of starting models for subsequent LIE calculations, and this type of hierarchical approach seems to be a viable strategy for lead optimization.

The work on thrombin inhibitors demonstrated the capability of this approach in predicting the relative affinities of chemically very different ligands, as well as the possibility of estimating stereoselectivity. However, in the case of thrombin, it was found that Equation (13) requires a constant term ($\gamma = -2.9$ kcal/mol) in order to reproduce the absolute binding free energies. The revised LIE model discussed in the previous section with such an additional constant gives a mean unsigned error of 0.6 kcal/mol for the data set of eight thrombin inhibitors. Interestingly, it was also found that a free parameterization of all three coefficients in Equation (13) yielded essentially the same values of α and β as before. In our view, this suggests that the possible system dependence of the parameterization of Equation (13) might be reducible to different constant terms (γ) for different types of receptor sites.

The above conclusion is also supported by simulations of complexes of the previously mentioned recognition and transport proteins (Åqvist and

Marelius, 2001). That is, ligand binding to the polar binding sites of arabinose, lysine, and muscular fatty acid binding protein appears well described by the revised LIE model with $\gamma = 0$. On the other hand, the absolute binding free energies of four examined complexes with the entirely hydrophobic cavity of retinol binding protein require a constant term of about -7 kcal/mol in Equation (13) in order to reproduce the experimental data (Åqvist and Marelius, 2001). These results, as well as those for the thrombin complexes, seem to indicate that the hydrophobicity of the receptor site may be a source of system dependency that can be alleviated by including a specific constant γ. A similar idea has also been put forward by W. Wang *et al.* (1999) who suggested an interesting method based on desolvated nonpolar surface areas in the complex as means to distinguish between different types of binding sites.

The possibility of introducing a constant term γ in Equation (13) was suggested already in the original description of the LIE method (Åqvist *et al.*, 1994), where it was also noted that such a term is needed in general if the approach is to be used for estimating solvation free energies. Jorgensen and co-workers have instead used a third term in Equation (13) containing the difference in solvent accessible surface area (SASA) of the ligand, scaled by an empirical coefficient (Carlson and Jorgensen, 1995; Jones-Hertzog and Jorgensen, 1997; Lamb *et al.*, 1999). We have

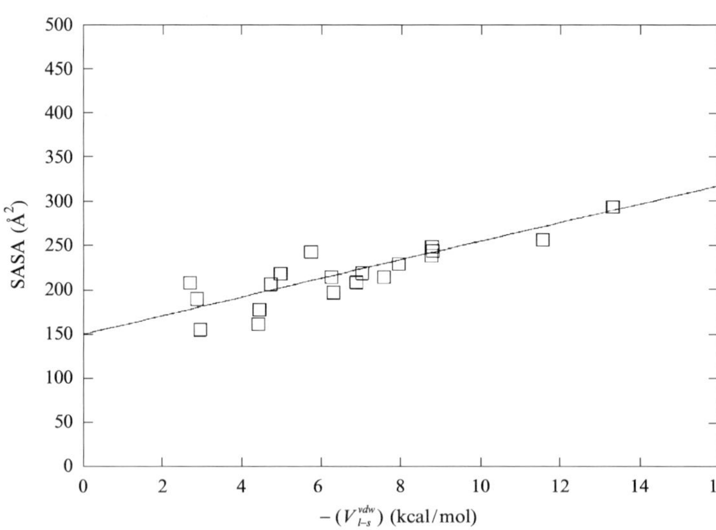

FIG. 6. Illustration of the correlation between solvent accessible surface area and the average solute-solvent van der Waals energy for a number of organic compounds in water (data from McDonald *et al.*, 1997).

argued earlier (Åqvist *et al.*, 2002; Hansson *et al.*, 1998) that this is basically equivalent to using a constant γ because the SASA value is also strongly correlated with the intermolecular van der Waals energy (Fig. 6). Another difference in the approach of Jorgensen (Carlson and Jorgensen, 1995) and others (Wall *et al.*, 1999) is that the electrostatic coefficient is treated as a free parameter in the optimization of Equation (13). In such cases, the method really does not have much to do with linear response, and, in our opinion, it is rather indicative of a problem with the electrostatic treatment when β becomes close to zero or even negative (Åqvist and Marelius, 2001).

The differing parameterizations of Equation (13) reported in several works (Paulsen and Ornstein, 1996; Jones-Hertzog *et al.*, 1997; Lamb *et al.*, 1999; Wall *et al.*, 1999) has sometimes led to the notion that coefficients in the LIE method are strongly dependent on the system under study and the force field being used. As discussed previously, we have found examples in which the absolute (but not the relative) binding free energies do need a systematic correction to fit the experimental data. This correction, however, only appeared as a constant term (γ) while the previously optimized values of α and β seemed to be robust. We will briefly discuss these issues below in the light of a few new examples, but first it may be useful to illustrate the fact that the parameterization of the LIE method of Hansson *et al.* (1998) is quite predictive for complexes with an aspartic protease not included in the earlier calibration set.

A. *Plasmepsin II*

The aspartic protease plasmepsin II (Plm II) of the malaria parasite *Plasmodium falciparium* is a key enzyme in the degradation of host hemoglobin, which takes place inside acidic vacuoles when the parasite is in its intraerythrocytic stage. Because inhibition of the hemoglobin degradation pathway is lethal for the parasite, the enzymes in this pathway are putative drug targets (Werbovetz, 2000). A C_2-symmetric scaffold previously used against HIV-protease (Alterman *et al.*, 1999) was investigated for Plm II inhibition in terms of stereochemistry and sidechain identity (Ersmark *et al.*, 2003) using LIE calculations and *in vitro* inhibition assays. No x-ray structure was available of Plm II complexed with any of these compounds, so the 2.7 Å resolution crystal structure of *P. falciparium* Plm II in complex with the inhibitor pepstatin A was used as a starting point (Silva *et al.*, 1996). Nine ligands were docked manually into the active site, guided by the position of the pepstatin A molecule. The docking was also assisted by an x-ray structure of HIV-1 protease in complex with a compound similar to the C_2-symmetric ligands (Alterman *et al.*, 1998). In the LIE calculations the standard parameters of Hansonn *et al.* (1998) were used and MD simulations were carried out with the Gromos87 force field (van Gunsteren and Berendsen, 1987) as implemented in the program Q (Marelius *et al.*, 1998b). The results of the study, shown in Fig. 7, showed an excellent agreement between the LIE and experimental binding energies, correctly predicting that the *SRRRRS* isomers are the only active Plm II inhibitors. Fig. 8 shows the predicted conformation of one of the active inhibitors superimposed on the crystal structure of the complex with pepstatin. Replacing the terminal valine and methylamide groups of the *SRRRRS* isomer with (1*S*,2*R*)-1-amino-2-indanol yielded an increase in affinity, which was quantitatively predicted by the LIE calculations. As a comparison, the binding affinities were also estimated using an empirical scoring function (Eldridge *et al.*, 1997). One hundred snapshots from the MD trajectory of each complex were minimized and scored, and the average results are also given

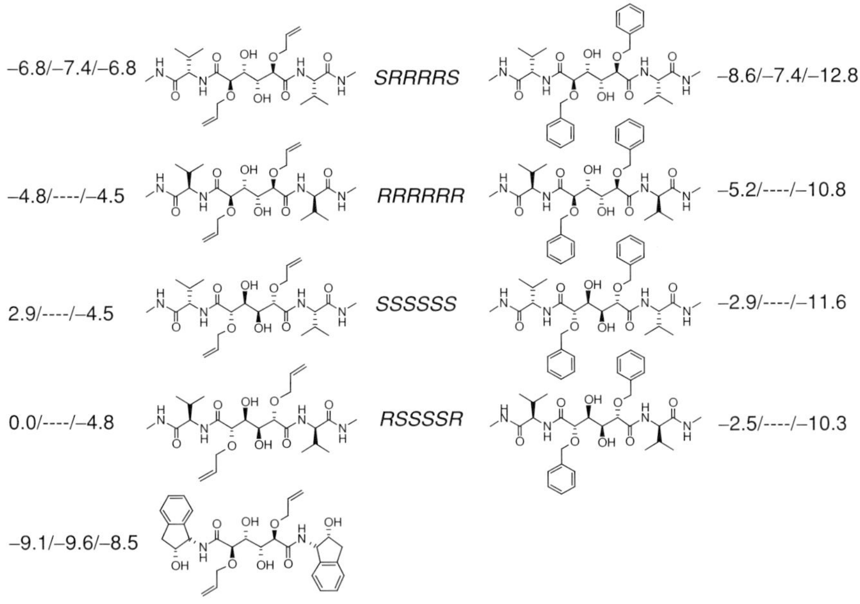

FIG. 7. The ligands used in the plasmepsin II inhibition study. The structures of the two stereoisomer series are shown with their stereochemistry indicated, together with the compound with modified end groups (*leftmost, bottom*). The numbers in the left, middle, and right columns are the binding free energies obtained from LIE calculations, experimental studies, and an empirical scoring function (Eldridge *et al.*, 1997), respectively. The symbols "---" denote no observed activity in the enzyme assay, which was sensitive up to an inhibitor concentration of 10 μM.

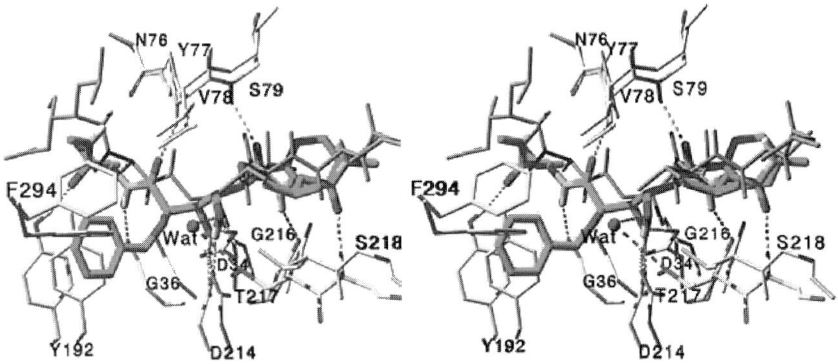

FIG. 8. Stereo view of the predicted (MD average) binding conformation of an active inhibitor (*top, rightmost in Fig. 7;* green) in complex with Plm II (yellow), superimposed on the x-ray structure of pepstatin A (purple) complexed with Plm II (light blue) (Silva *et al.*, 1996). (See Color Insert.)

in Fig. 7. The affinities of the ligands containing the allyloxy moiety were fairly well predicted, whereas those of the benzyloxy compounds were highly overestimated, with the predicted affinity of the active stereoisomer being 5.4 kcal/mol too high. The overprediction can be attributed to the ligand-size dependent lipophilic term of the scoring function, which clearly overestimates the hydrophobic binding contribution. Furthermore, conformational averaging was found to clearly improve the results of the scoring function in the sense that the correct ranking of stereoisomers for each compound was then obtained, which was not the case if only an initial minimized structure was used.

B. *Trypsin-BPTI*

Evaluation of absolute or even relative binding free energies of protein-protein complexes is a difficult task to address with computer simulations approaches. The interaction energies can be on the order of several thousand kcal/mol, and extremely long simulations would then be required in order to get stable energies. Protein-protein interfaces are, however, generally composed of a cluster of "hot spot" residues at the center of the interface surrounded by energetically less important residues (Fig. 9). For example, the primary (P1) binding residue of the bovine pancreatic trypsin inhibitor (BPTI) has been found to be responsible for almost 70% of the interaction free energy in the binding of BPTI to trypsin (Krowarsch *et al.*, 1999). Instead of trying to predict the absolute binding free energy, one can then try to calculate

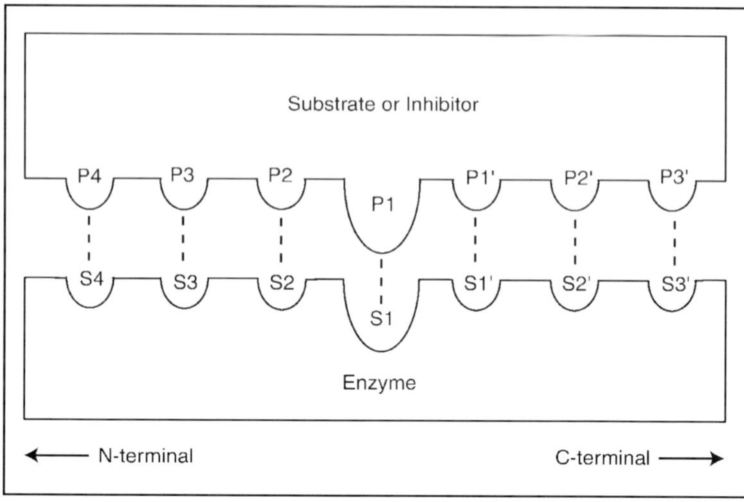

FIG. 9. Schematic representation of the specificity sub-sites on serine proteases and their inhibitors/substrates as introduced by Schechter and Berger (1967).

the effect of single or multiple point mutations on the association energy. In the present case, BPTI with the P1 residue mutated from the native lysine to glycine was used as a reference state for analyzing the effects of P1-mutations on the trypsin-BPTI binding affinity (Brandsdal *et al.*, 2001a). The P1-Gly mutant does not have any side-chain that enters the substrate specificity pocket of trypsin and its association constant thus only reflects contributions from secondary interactions. The idea here was simply to treat each residue at the primary binding position as a ''ligand'' in the LIE framework, whereas the rest of the inhibitor was considered as part of the surroundings. Out of the 20 possible trypsin-BPTI complexes differing only at the P1-position, 13 were selected such that most of the binding range was covered. This strategy was found to give very good results with respect to the experimental association energies, and a correlation coefficient of 0.99 was obtained (Fig. 10) excluding the P1-Asp and Glu variants that are associated with uncertainties regarding their protonation state and possible counterions (Brandsdal *et al.*, 2001a). A subsequent LIE study of cold-active trypsin from Atlantic salmon revealed that its enhanced binding affinity for positively charged ligands is entirely caused by electrostatic effects (Brandsdal *et al.*, 2001b).

Besides pointing to a useful approach for examining the energetics of protein-protein recognition interfaces, the aforementioned study also indicates that the LIE method may, in fact, not be as sensitive to the

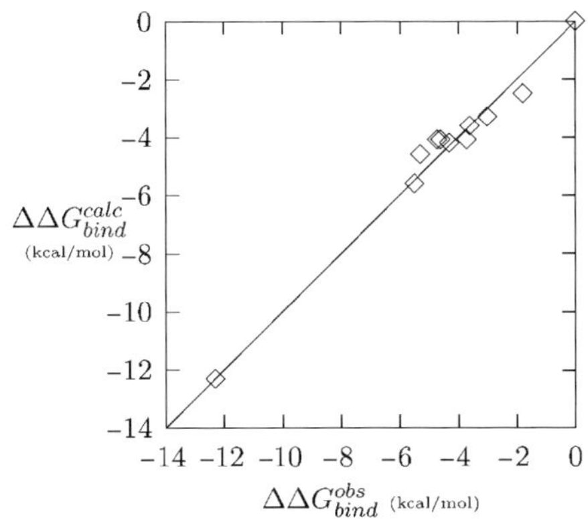

FIG. 10. Scatter diagram of calculated versus experimental binding free energies (kcal/mol) of trypsin-BPTI complexes relative to P1-Gly (Brandsdal et al, 2001a). The correlation coefficient obtained was 0.99 with a mean unsigned error of 0.38 kcal/mol.

choice of force field as might be expected. That is, the work of Brandsdal *et al.* (2001a) used the Amber95 (Cornell *et al.*, 1995) force field with exactly the same parameterization of Equation (13) as that used with the Gromos87 (van Gunsteren and Berendsen, 1987) potential. It has also been shown by Kollman and co-workers that this parameterization worked well with Amber95 for the trypsin-benzamidine complex (J. Wang *et al.*, 1999).

C. *P450cam*

As mentioned previously, force field dependence has sometimes been invoked as an explanation for differing parameterizations of the LIE method in the past years. Ideally, the coefficients of Equation (13) should be independent of the force field used to study the energetics of ligand binding. However, obvious errors or imbalances of a given force field (including the water model used) are bound to affect the parameters that are calibrated against experimental data. Paulsen and Ornstein (1996) studied a series of cytochrome P450-camphor analogue complexes with the LIE method using the CVFF force field (Dauber-Osguthorpe *et al.*, 1988), and excellent agreement with experimental binding data was obtained using $\alpha = 1.043$, $\beta = 0.5$, and $\gamma = 0$. This specific parameterization was proposed to arise from the use of CVFF as opposed to the Gromos87 force field that was used in the initial calibration of the LIE method (Åqvist *et al.*, 1994). However, as noted previously, the calculations on the different trypsin-ligand complexes using

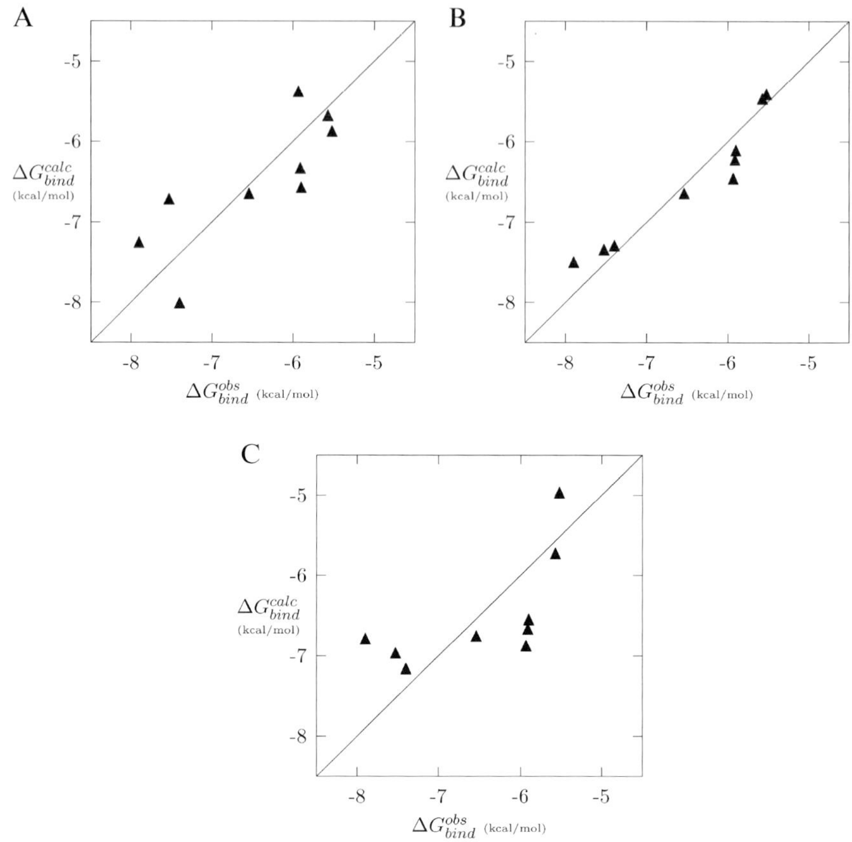

FIG. 11. Calculated versus observed binding free energies for seven ligand complexes with P450cam obtained from LIE calculations (Almlöf *et al.*, 2003), using the earlier parametrization of Equation (13) (Hansson *et al.*, 1998). Three different force fields were used: (A) Gromos87, (B) OPLS-AA, and (C) Amber95. A constant term of $\gamma = -4.2, -4.4$, and -4.9 *kcal*/mol, respectively, is used.

the Gromos87 and Amber95 force fields (Åqvist, 1996; J. Wang *et al.*, 1999; Brandsdal *et al.*, 2001b), suggests that force field dependence may not be a major issue. Furthermore, the results for thrombin and some other systems discussed previously indicate that the addition of a constant term (γ), which depends on the hydrophobicity of the binding site, may sometimes be required for getting the absolute binding energies right. For systems that need such a constant, the nonpolar coefficient (α) will acquire a higher value upon calibration if the γ is omitted from Equation (13).

In order to elucidate the issue of force field dependence, we have carried out LIE calculations on seven P450cam-ligand complexes (Almlöf *et al.*, 2003) that were also considered by Paulsen and Ornstein (1996), using three different force fields: Gromos87, Amber95, and OPLS-AA (Jorgensen *et al.*, 1996). The results for these three force fields are presented in Fig. 11. Using the earlier LIE parameterization (Hansson *et al.*, 1998) (with $\gamma = 0$) gives relative binding free energies that agree reasonably well with the experimental binding data, but for all three force fields the absolute binding free energy is significantly too positive. However, this appears as a systematic offset by approximately 4 to 5 kcal/mol that is practically independent of the force field. In particular, we note that Gromos87 and OPLS-AA yields virtually the same value of γ (-4.2 and -4.4) with very small resulting mean unsigned errors of 0.47 and 0.23 kcal/mol, respectively, using the earlier parametrization of α and β (Hansson *et al.*, 1998). In fact, in these cases free optimization of all three parameters in Equation (13) again returns very similar values of α and β to those found earlier (Hansson *et al.*, 1998). For Amber95 the optimal value of γ is somewhat more negative (-4.9 kcal/mol), and the overall quality of the results is worse with an average error of 0.57 kcal/mol. It is also worth noting that these binding free energies are dominated by hydrophobic interactions, and for such systems it thus seems necessary to include an additional constant, as previously discussed. Our belief is therefore that the LIE parameterization should not be particularly dependent upon the force field, but that it is more sensitive to the simulation protocols being used (e.g., cutoffs, treatment of electrostatics, sampling time) (Åqvist and Marelius, 2001). Also, because the energy differences between the bound and free states are considered, there is likely to be some cancellation of errors even if a given force field does not reproduce absolute solvation energies exactly. An interesting extension of the LIE method that uses the surface-generalized Born model of Still *et al.* (1990) for the solvent has also been reported recently (Zhou *et al.*, 2001).

V. MM-PBSA

Another approach that has gained considerable attention in the last few years for estimating association free energies of molecular complexes is the so-called MM-PBSA method (Molecular Mechanics/Poisson-Boltzmann/Surface Area) (Srinivasan *et al.*, 1998; Kollman *et al.*, 2000). This approach is based on an analysis of molecular dynamics trajectories using a continuum solvent approach and approximates the "average" free energy of a state as:

$$\langle G\rangle = \langle E_{MM}\rangle + \langle E_{PBSA}\rangle - T\langle S_{MM}\rangle \tag{14}$$

where $\langle E_{MM}\rangle$ is an average molecular mechanical energy that typically includes bond, angle, torsion, van der Waals, and electrostatic terms from a regular force field, and is evaluated with no nonbonded cutoff. Solvation-free energies are calculated using a numerical solution of the Poisson-Boltzmann equation (Warwicker and Watson, 1982; Gilson and Honig, 1988; Honig and Nicholls, 1995) and, together with a surface area–based estimate of the nonpolar free energy (Sitkoff *et al.*, 1994), constitute the $\langle G_{PBSA}\rangle$ term. Both $\langle E_{MM}\rangle$ and $\langle G_{PBSA}\rangle$ are obtained by averaging over a sample of representative geometries extracted from an MD trajectory of the system (typically around 100 snapshots). The last term, $-T\langle S_{MM}\rangle$, is the solute entropy, which can be estimated by quasi-harmonic analysis of the trajectory or by using normal mode analysis (Srinivasan *et al.*, 1998).

The MM-PBSA approach was initially used to study the stability of various DNA and RNA fragments (Srinivasan *et al.*, 1998), but has also been used to estimate ligand-binding free energies in the last years (Kollman *et al.*, 2000; Kuhn and Kollman, 2000a, 2000b; Wang *et al.*, 2001; Huo *et al.*, 2002). In order to calculate the binding free energy between a ligand and a receptor, two alternatives exist with this methodology. The first is to evaluate the terms in Equation (14) for the complex, receptor, and ligand based on separate trajectories with a subsequent determination of ΔG_{bind} according to:

$$\Delta G_{bind} = \left\langle G_{complex}\right\rangle - \left\langle G_{receptor}\right\rangle - \left\langle G_{ligand}\right\rangle \tag{15}$$

The second alternative is to determine each of the terms in Equation (15) based on snapshots from a trajectory of the complex only, in which case the two latter terms are estimated simply by "removing" one of the molecular partners from the trajectory. In practice, the first option does not ever seem to have been used in protein-ligand studies, which is understandable because there is no way to get the $\langle E_{MM}\rangle$ term to converge for the receptor or complex within reasonable computing time. Hence, the regular implementation of this method actually assumes that the structure of the receptor and the ligand does not change upon binding, because no intramolecular terms either in the receptor or ligand are taken into account. This is in contrast to the LIE method in which such terms are considered in terms of responses to the intermolecular interaction through the appropriate weight coefficients.

A fundamental question with the MM-PBSA approach is how to best determine the contribution from the entropy change upon binding. If the absolute binding free energy is to be estimated, the entropic contribution must be determined in a consistent fashion to yield meaningful results. This is in general a difficult task especially if the conformational fluctuations are significant, and even relative entropies are difficult to determine with high accuracy. In calculations of relative binding free energies for a series of ligands to a common protein receptor, the entropic contribution is often assumed to cancel when the ligands are of similar size (Kollman *et al.*, 2000). This would seem to be a rather questionable assumption, because different ligands may have different degrees of freedom that are affected by interactions with the receptor. An apparent example is provided by a recent analysis of inhibitor binding to cathepsin D (Huo *et al.*, 2002). On the other hand, to estimate, for example, ligand entropies in solution from normal mode analysis around conformations from a simulation of its complex with a receptor may be an equally drastic simplification. Nevertheless, researchers have obtained impressive results with these approximations by using the MM-PBSA method (Kuhn and Kollman, 2000b; J. Wang *et al.*, 2001; Huo *et al.*, 2002).

Our experience from many LIE calculations is that ligands (except for very small and rigid ones) often adopt and explore rather different conformations when free in solution, compared with the case when they are bound to a receptor. In such cases it would seem difficult to capture the correct binding energetics using only a simulation of the complex. For instance, large hydrophobic ligands that bind to their receptor in extended conformations are sometimes seen to undergo a "hydrophobic collapse" in solution, meaning that they arrange themselves in such a way as to minimize their water-exposed hydrophobic surface. This can lead to a decreased binding affinity, because the dissociated reference state in water then becomes more favorable, and such an effect is completely missed if no simulation of the ligand in water is carried out.

Nonetheless, it is clear that the MM-PBSA approach has several appealing features compared with the more rigorous approaches like FEP/TI, especially when dealing with diverse sets of ligands that differ significantly in their structural and chemical composition. Other variants of the MM-PBSA method have also been introduced, such as "computational alanine scanning" (Massova and Kollman, 1999) and "computational fluorine scanning" (Kuhn and Kollman, 2000a), which can be useful techniques when exploring the sensitivity of a receptor site to changes in composition.

VI. PROFEC

The ability to determine the effect on the binding energetics from certain modifications of a ligand prior to the experimental design is, of course, of great importance in a number of disciplines. Both the rigorous

FEP/TI and the more approximate methods such as LIE or MM-PBSA can be used to predict the potency of a set of ligands to a given receptor, but these methods are not well suited to directly suggest how to modify ligands to improve their binding capacities. That is, it is desirable to try to obtain information from a given simulation on how to modify a given ligand in order to improve its affinity, and this is, for example, the main appealing feature of the single reference state/one-step perturbation approaches discussed in Section III. Another interesting method in this respect is PROFEC (pictorial representation of free energy changes) (Radmer and Kollman, 1998), which considers the electrostatic and the van der Waals effect from inserting particles around a ligand. The basic idea is to define a grid centered at one of the ligand atoms and to calculate the cost of adding a Lennard-Jones particle for each point in the grid according to the traditional FEP equation:

$$\Delta G_{ins}(i,j,k) = -\beta^{-1}\ln\langle\exp(-\beta\Delta v(i,j,k))\rangle_0 \tag{16}$$

where $\Delta v(i, j, k)$ is the van der Waals interaction energy between the particle and the surrounding atoms. Based on contour surfaces around the ligand, suggestions of how to add new atoms that improve binding can be made. The electrostatic contribution can then be examined by calculating the derivative of the binding free energy with respect to the charge at each grid point, under the assumption that a particle has already been inserted. Again, the contour maps of the derivative can be displayed and might suggest how the charge distribution should be changed to improve binding. The maps are generated from two MD simulations: one of the protein-ligand complex and one of the ligand in solution. Thus, for each grid point the difference $\Delta\Delta G_{ins}$ of particles in the protein-ligand complex and the ligand in solution is calculated, and contour maps of $\Delta\Delta G_{ins}$ are constructed and visualized.

Typically, PROFEC would be used in combination with one of the more detailed approaches, such as traditional FEP, LIE, or MM-PBSA, to computationally validate the changes suggested prior to experimental design. For example, PROFEC was used to construct new TIBO-like inhibitors to HIV-1 reverse transcriptase with subsequent application of FEP/TI confirming the suggestions made by PROFEC (Eriksson *et al.*, 1999). In another recent study, inhibitor binding to cathepsin D was investigated using a combination of MM-PBSA and PROFEC (Huo *et al.*, 2002). Thus, these studies suggest another possible computational strategy for ligand design using complementary approaches. First, a crude but rapid method can be used to scan large virtual libraries to identify possible binding candidates, and then a more accurate estimation of binding free energies may be carried out. These studies can then be coupled to PROFEC in order to suggest possible modification that will enhance the binding capacity. However, it should be kept in mind that PROFEC suffers from some limitations. The main weakness of PROFEC is its inability to evaluate free energies when multiple sites are modified or when modifications induce large conformational changes.

VII. λ-Dynamics and Chemical MC/MD

Inspired by the work of Tidor (1993), Kong and Brooks (1996) have proposed a new approach to multiple-state free energy calculations. This method, λ-dynamics, treats the coupling parameter λ as a dynamic variable, and a set of variables $\{\lambda_i\}, i = 1, \ldots, n$, is used to scale different interactions terms instead of the traditional single coupling parameter as used in conventional free energy calculations. The methodology is based on the idea that multiple ligands will compete for a common receptor on the basis of their relative free energies, and that this can be explored using multiple copy simultaneous search approaches (Elber and Karplus, 1990). The hybrid potential energy function to perform such "competitive binding experiments" can be formulated as (Kong and Brooks, 1996):

$$V(\{\lambda\}) = V_{env} + \sum_{i=1}^{L} \lambda_i^2 (V_i - F_i) \tag{17}$$

where L is the total number of ligands, V_{env} is the interaction involving the surrounding atoms (e.g., solvent, protein, and the invariant atoms of the ligands), V_i is the interaction involving any of the atoms in ligand i, λ_i is the coupling parameter, and F_i is a reference energy. Atoms in different ligands are not allowed to interact with each other, and the ligands are thus invisible to one another. The dynamics of the system is described by an extended Hamiltonian:

$$H(\{\lambda\}) = T + T_{(\lambda)} + V_{env} + \sum_{i=1}^{L} \lambda_i^2 (V_i - F_i) \tag{18}$$

where each λ_i is treated as a fictitious particle with mass m_i and $T_{\{\lambda\}}$ is the kinetic energy associated with the λ-variables. From the configuration integral of the hybrid system:

$$Z(\{\lambda\}) = \int \exp\left[-\beta\left(V_{env} + \sum_{i=1}^{L} \lambda_i^2 (V_i - F_i)\right)\right] d\Gamma \tag{19}$$

the free energy difference between two molecules i and j can be calculated, with reference free energy F_i and F_j, respectively, based on the probability distribution of states dominated by $\lambda_i = 1$ and $\lambda_j = 1$ according to:

$$\Delta\Delta G_{i,j} = -\beta^{-1} \ln \frac{Z(\lambda_i = 1, \{\lambda_{m \neq i}\} = 0)}{Z(\lambda_j = 1, \{\lambda_{l \neq j}\} = 0)} \tag{20}$$

F_i in the above equations serves as both a reference free energy and a biasing potential. Most molecular mechanics-based free energy approaches consider the free energy of the solvated ligand and that of the complexed ligand-receptor state in order to estimate the free energy of binding. The solvation free energy of the ligand ("free state") can be calculated using conventional free energy methods (e.g., FEP) or more rapidly using continuum solvation approaches such as Poisson-Boltzmann and generalized Born models (Still *et al.*, 1990). The value of F_i is then taken as the relative solvation free energy for the ligands in the free state, and an iterative procedure is used in order to improve sampling chemical space (Kumar *et al.*, 1992).

The λ-dynamics method uses classical MD to propagate both the atomic coordinates and the chemical space (coupling parameter). It is, however, possible to use MC in order to sample the coupling parameter stochastically combined with MD for propagating the atomic coordinates. In principle, Monte Carlo methods can also be used for sampling of the configurational space, but when dealing with protein conformations, MD is still a better approach for sampling. Although the idea of using MC to sample the "chemical space" is not new (Bennett, 1976; Tidor, 1993), Pitera and Kollman (1998) have taken this approach further by applying mixed chemical MC/MD (CMC/MD) to the problem of multiple ligands.

If we represent the binding processes in terms of the thermodynamic perturbation cycle in Fig. 2, the relative binding free energy between two compounds can be written as:

$$\begin{aligned}\Delta\Delta G_{bind} &= \Delta G_{bind}(L') - \Delta G_{bind}(L) = \Delta G^{p}_{mut} - \Delta G^{w}_{mut} \\ &= -RT \ln\left\langle \exp[-\beta(\Delta V_{mut} - \Delta G^{w}_{mut})] \right\rangle_p\end{aligned} \tag{21}$$

where ΔV_{mut} denotes the energy difference between potentials describing the interactions of the two compounds. The $\Delta\Delta G_{bind}$ can thus be determined directly from the CMC/MD simulations by incorporating the relative mutation energies (ΔG^{w}_{mut}) as a "solvation offset" to the energy of each state, which can be considered as an umbrella sampling approach (Torrie and Valleau, 1977). ΔG^{w}_{mut} thus represents a biasing potential in the CMC/MD simulation of the bound state. In practice the system is simulated for a number of MD steps focusing on one ligand at a time, which generates a new configuration of the ligand (as well as the "ghost" ligands) and the surrounding environment (water and protein). From such a configuration the energetics of each ligand is evaluated, and a new ligand is chosen by using a random "trial move". Then the energy is evaluated and the move is either accepted or rejected according to the standard criterion:

$$\begin{cases} \Delta E_i \leq 0 \Rightarrow P_i = 1 \\ \Delta E_i > 0 \Rightarrow P_i = \exp(-\beta \Delta E_i) \end{cases} \tag{22}$$

where ΔE_i is the difference in the interaction energy between ligand i and the previous simulated ligand, and P_i is the acceptance probability. This method thus uses Monte Carlo steps to "jump" between different ligands and generates an ensemble of these. The relative free energy between two ligands is then calculated according to the same equation used in the λ-dynamics method (Equation 20). When applied to the TIBO derivatives in HIV-1 RT (Eriksson *et al.*, 1999), relative free energies calculated by CMC/MD were found to converge very slowly. In order to improve the convergence, biasing potentials of the form $\Delta V_{mut} - \Delta G^p_{mut}$ were instead introduced, where the offset energy (ΔG^p_{mut}) now reflects the relative free energy in the bound state. This procedure is applied in an iterative fashion, and the biasing offsets are initially set to zero and are subsequently calculated according to Equation (20), which is essentially the same as the WHAM procedure (Kumar *et al.*, 1992) used in the λ-dynamics method.

Both λ-dynamics and the CMC/MD have been used successfully to estimate relative binding free energies of similar compounds, but because of their complexity, the implementation of these methods is not so straightforward. In particular, if ligands differ considerably from each other, it does not seem easy to guess the offset free energies in Equations (17) and (21), for example, from relative solvation energies, because bonded energy terms will enter into these quantities. Another problem appears to be the control of the actual spatial coordinates for different ligands, because they cannot in practice be allowed to drift away from each other. A remedy for this is, of course, to restrain the various ligands to each other (Banba and Brooks, 2000), but that is bound to impose some limitations on the amount of conformational space that can be explored. As with most of the "FEP/TI-derived" methods, λ-dynamics and CMC/MD inevitably seem limited to "small perturbations," and this is probably the main reason for the restricted use of these two methods. In our opinion, the CMC/MD method appears to be the most promising of the two, because it avoids the complication of treating λ as a dynamic parameter. However, it is still early to rule out any of them, because both are still in their infancy. Extensive testing on real protein-ligand systems is required in order to address the full potential of the dynamic or MC treatment of the coupling parameter.

VIII. Conclusions

In this chapter we have tried to give an overview of some different methods for calculating ligand binding free energies, which are all based on force fields and conformational sampling. Although there are also a number of scoring approaches, based on both empirical functions and scaled molecular mechanics energies, that purport to estimate the binding affinity from a single (sometimes energy-minimized) conformation of a given complex, we have not considered such methods herein. Although they may be useful for rapid screening and docking, their scope in terms of accuracy still seems limited. We have also found that conformational averaging appears to improve the results of empirical scoring (Marelius *et al.*, 2001; Ersmark *et al.*, 2003).

Regarding our ability to obtain binding free energies from molecular simulations, a lot of progress clearly has been made since the earliest attempts to calculate relative affinities of closely related compounds in the 1980s by FEP/TI methods. However, we have still not quite reached the final goal of being able to "screen" a diverse set of ligands with high fidelity *in silico*, which would really benefit medicinal chemistry projects. As is evident from previous discussions, many of the problems still revolve around the sampling issue. That is, one often has to start with one particular 3D structural model of given complex and then somehow must try to extrapolate the structures and energetics of other molecular complexes from this model. To do this by MD or MC methods generally requires a considerable amount of sampling in order to reach sufficient numbers of configurations of the new systems so that the most relevant regions of their conformation spaces have been covered.

It is probably fair to say that the FEP/TI type of method has not really fulfilled its promise of being able to open a major new avenue to structure-based drug design. This is mainly due to slow convergence and sampling difficulties. In particular, in this type of extrapolation process in which one may want to look at 20 or so new (and different) ligands, arriving at the correct end-points by long perturbation paths sometimes seems hopeless. It appears that a better solution to this problem can often be provided by automated docking of individual compounds, at least when they differ significantly from each other, and then to try to evaluate the binding energetics by a method that does not require the unphysical transformations involved in FEP/TI and related methods. However, the general docking problem with flexible receptor and ligand, in our opinion, is still not solved, although "redocking" of experimental complexes with a rigid receptor might work well. The docking problem resembles the protein-folding one in many respects, and the only way to attack difficult cases seems to be by extensive conformational searching (with full flexibility) in combination with more reliable scoring methods (Halperin *et al.*, 2002). On the other hand, with experimental 3D data for some relevant complexes, the situation often looks much brighter. As far as the scoring or binding affinity prediction is concerned, a number of new methods that can provide alternatives to FEP/TI have been proposed. We have tried to elucidate some of them here, but it is still probably too early to elect one particular approach as the method of choice. In this respect, it is only their efficiency, reliability, and predictive power demonstrated in real medicinal chemistry projects that will eventually allow us to rank their usefulness.

Acknowledgments

Support from the Swedish Research Council (VR) and the Swedish Foundation for Strategic Research (SSF) to J.Å. and from the Norwegian Research Council to B.O.B. is gratefully acknowledged.

References

Almlöf, M., Brandsdal, B. O., and Åqvist, J. (2003). To be published.

Alterman, M., Andersson, H. O., Garg, N., Ahlsen, G., Lovgren, S., Classon, B., Danielson, U. H., Kvarnstrom, I., Vrang, L., Unge, T., Samuelsson, B., and Hallberg, A. (1999). *J. Med. Chem.* **42,** 3835–3844.

Alterman, M., Bjoersne, M., Muehlman, A., Classon, B., Kvarnstroem, I., Danielson, H., Markgren, P. O., Nillroth, U., Unge, T., Hallberg, A., and Samuelsson, B. (1998). *J. Med. Chem.* **41,** 3782–3792.

Åqvist, J. (1990). *J. Phys. Chem.* **94,** 8021–8024.

Åqvist, J. (1994). *J. Phys. Chem.* **98,** 8253–8255.

Åqvist, J. (1996). *J. Comp. Chem.* **17,** 1587–1597.

Åqvist, J., and Hansson, T. (1996). *J. Phys. Chem.* **100,** 9512–9521.

Åqvist, J., Luzhkov, V. B., and Brandsdal, B. O. (2002). *Acc. Chem. Res.* **35,** 358–365.

Åqvist, J., and Marelius, J. (2001). *Comb. Chem. High. T. Scr.* **4,** 613–626.

Åqvist, J., Medina, C., and Samuelsson, J. E. (1994). *Protein Eng.* **7,** 385–391.

Åqvist, J., and Mowbray, S. L. (1995). *J. Biol. Chem.* **270,** 9978–9981.

Banba, S., and Brooks, C. L. (2000). *J. Chem. Phys.* **113,** 3423–3433.

Bash, P. A., Singh, U. C., Brown, F. K., Langridge, R., and Kollman, P. A. (1987). *Science* **235,** 574–576.

Bennett, C. H. (1976). *J. Comp. Phys.* **22,** 245–268.

Beutler, T. C., Mark, A. E., van Schaik, R. C., Gerber, P. R., and van Gunsteren, W. F. (1994). *Chem. Phys. Lett.* **222,** 529–539.

Beveridge, D. L., and DiCapua, F. M. (1989). *Annu. Rev. Biophys. Biophys. Chem.* **18,** 431–492.
Boresch, S., and Karplus, M. (1999a). *J. Phys. Chem. A* **103,** 103–118.
Boresch, S., and Karplus, M. (1999b). *J. Phys. Chem. A* **103,** 119–136.
Brandsdal, B. O., Åqvist, J., and Smalås, A. O. (2001a). *Protein Sci.* **10,** 1584–1595.
Brandsdal, B. O., Smalås, A. O., and Åqvist, J. (2001b). *FEBS Lett.* **499,** 171–175.
Böhm, H. J. (1994). *J. Comput. Aided. Mol. Des.* **8,** 243–256.
Carlson, H. A., and Jorgensen, W. L. (1995). *J. Phys. Chem.* **99,** 10667–10673.
Cornell, W. D., Cieplak, P., Bayly, C. I., Gould, I. R., Merz Jr, K. M., Ferguson, D. M., Spellmeyer, D. C., Fox, T., Caldwell, J. W., and Kollman, P. A. (1995). *J. Am. Chem. Soc.* **117,** 5179–5197.
Dauber-Osguthorpe, P., Roberts, V. A., Osguthorpe, D. J., Wolff, J., Genest, M., and Hagler, A. T. (1988). *Proteins* **4,** 31–47.
Doyle, D. A., Cabral, J. M., Pfuetzner, R. A., Kuo, A. L., Gulbis, J. M., Cohen, S. L., Chait, B. T., and MacKinnon, R. (1998). *Science* **280,** 69–77.
Elber, R., and Karplus, M. (1990). *J. Am. Chem. Soc.* **112,** 9161–9175.
Eldridge, M. D., Murray, C. W., Auton, T. R., Paolini, G. V., and Mee, R. P. (1997). *J. Comput. Aided. Mol. Des.* **11,** 425–445.
Eriksson, M. A. L., Pitera, J., and Kollman, P. A. (1999). *J. Med. Chem.* **42,** 868–881.
Ersmark, K., Feierberg, I., Bjelic, S., Hultén, J., Samuelsson, B., Åqvist, J., and Hallberg, A. (2003). *Bioorg. Med. Chem.* **11,** 3723–3733.
Gerber, P. R., Mark, A. E., and van Gunsteren, W. F. (1993). *J. Comput. Aided. Mol. Des.* **7,** 305–323.
Gilson, M. K., and Honig, B. (1988). *Proteins* **4,** 7–18.
Gohlke, H., Hendlich, M., and Klebe, G. (2000). *J. Mol. Biol.* **295,** 337–356.
Graffner-Nordberg, M., Kolmodin, K., Åqvist, J., Queener, S. F., and Hallberg, A. (2001). *J. Med. Chem.* **44,** 2391–2402.
Halperin, I., Ma, B., Wolfson, H., and Nussinov, R. (2002). *Proteins* **47,** 409–433.
Hansson, T., and Åqvist, J. (1995). *Protein Eng.* **8,** 1137–1144.
Hansson, T., Marelius, J., and Åqvist, J. (1998). *J. Comput. Aided. Mol. Des.* **12,** 27–35.
Hermans, J., and Wang, L. (1997). *J. Am. Chem. Soc.* **119,** 2707–2714.
Honig, B., and Nicholls, A. (1995). *Science* **268,** 1144–1149.
Hultén, J., Bonham, N. M., Nillroth, U., Hansson, T., Zuccarello, G., Bouzide, A., Åqvist, J., Classon, B., Danielson, U. H., Karlen, A., Kvarnstrom, I., Samuelsson, B., and Hallberg, A. (1997). *J. Med. Chem.* **40,** 885–897.
Huo, S. H., Wang, J. M., Cieplak, P., Kollman, P. A., and Kuntz, I. D. (2002). *J. Med. Chem.* **45,** 1412–1419.
Hwang, J. K., and Warshel, A. (1987). *Biochemistry* **26,** 2669–2673.
Jain, A. N. (1996). *J. Comput. Aided. Mol. Des.* **10,** 427–440.
Jayaram, B., and Beveridge, D. L. (1990). *J. Phys. Chem.* **94,** 7288–7293.
Jones-Hertzog, D. K., and Jorgensen, W. L. (1997). *J. Med. Chem.* **40,** 1539–1549.
Jorgensen, W. L., Maxwell, D. S., and TiradoRives, J. (1996). *J. Am. Chem. Soc.* **118,** 11225–11236.
King, G., and Barford, R. A. (1993). *J. Phys. Chem.* **97,** 8798–8802.
King, G., and Warshel, A. (1989). *J. Chem. Phys.* **91,** 3647–3661.
Kirkwood, J. G. (1935). *J. Chem. Phys.* **3,** 300–313.
Kollman, P. (1993). *Chem. Rev.* **93,** 2395–2417.
Kollman, P. A., Massova, I., Reyes, C., Kuhn, B., Huo, S., Chong, L., Lee, M., Lee, T., Duan, Y., Wang, W., Donini, O., Cieplak, P., Srinivasan, J., Case, D. A., and Cheatham, T. E. (2000). *Acc. Chem. Res.* **33,** 889–897.

Kong, X. J., and Brooks, C. L. (1996). *J. Chem. Phys.* **105,** 2414–2423.
Krowarsch, D., Dadlez, M., Buczek, O., Krokoszynska, I., Smalås, A. O., and Otlewski, J. (1999). *J. Mol. Biol.* **289,** 175–186.
Kuhn, B., and Kollman, P. A. (2000a). *J. Am. Chem. Soc.* **122,** 3909–3916.
Kuhn, B., and Kollman, P. A. (2000b). *J. Med. Chem.* **43,** 3786–3791.
Kumar, S., Bouzida, D., Swendsen, R. H., Kollman, P. A., and Rosenberg, J. M. (1992). *J. Comp. Chem.* **13,** 1011–1021.
Lamb, M. L., and Jorgensen, W. L. (1997). *Curr. Opin. Chem. Biol.* **1,** 449–457.
Lamb, M. L., Tirado-Rives, J., and Jorgensen, W. L. (1999). *Bioorg. Med. Chem.* **7,** 851–860.
Lee, F. S., Chu, Z. T., Bolger, M. B., and Warshel, A. (1992). *Prot. Eng.* **5,** 215–228.
Lee, F. S., and Warshel, A. (1992). *J. Chem. Phys.* **97,** 3100–3107.
Levy, R. M., Belhadj, M., and Kitchen, D. B. (1991). *J. Chem. Phys.* **95,** 3627–3633.
Liu, H. Y., Mark, A. E., and van Gunsteren, W. F. (1996). *J. Phys. Chem.* **100,** 9485–9494.
Ljungberg, K. B., Marelius, J., Musil, D., Svensson, P., Norden, B., and Åqvist, J. (2001). *Eur. J. Pharm. Sci.* **12,** 441–446.
Luzhkov, V. B., and Åqvist, J. (2001). *FEBS Lett.* **495,** 191–196.
Luzhkov, V. B., Österberg, F., Acharya, P., Chattopadhyaya, J., and Åqvist, J. (2002). *Phys. Chem. Chem. Phys.* **4,** 4640–4647.
Luzhkov, V. B., Österberg, F., and Åqvist, J. (2003). To be published.
Marelius, J., Graffner-Nordberg, M., Hansson, T., Hallberg, A., and Åqvist, J. (1998a). *J. Comput. Aided. Mol. Des.* **12,** 119–131.
Marelius, J., Kolmodin, K., Feierberg, I., and Åqvist, J. (1998b). *J. Mol. Graphics Model.* **16,** 213–225, 261.
Marelius, J., Ljungberg, K. B., and Åqvist, J. (2001). *Eur. J. Pharm. Sci.* **14,** 87–95.
Massova, I., and Kollman, P. A. (1999). *J. Am. Chem. Soc.* **121,** 8133–8143.
McDonald, N. A., Carlson, H. A., and Jorgensen, W. L. (1997). *J. Phys. Org. Chem.* **10,** 563–576.
Meuser, D., Splitt, H., Wagner, R., and Schrempf, H. (1999). *FEBS Lett.* **462,** 447–452.
Mezei, M. (1987). *J. Chem. Phys.* **86,** 7084–7088.
Mezei, M., Swaminathan, S., and Beveridge, D. L. (1978). *J. Am. Chem. Soc.* **100,** 3255–3256.
Morris, G. M., Goodsell, D. S., Halliday, R. S., Huey, R., Hart, W. E., Belew, R. K., and Olson, A. J. (1998). *J. Comp. Chem.* **19,** 1639–1662.
Muegge, I., and Martin, Y. C. (1999). *J. Med. Chem.* **42,** 791–804.
Naudin, C., Bonhomme, F., Bruneel, J. L., Ducasse, L., Grondin, J., Lasségues, J. C., and Servant, L. (2000). *J. Raman Spectrosc.* **31,** 979–985.
Oostenbrink, B. C., Pitera, J. W., van Lipzig, M. M. H., Meerman, J. H. N., and van Gunsteren, W. F. (2000). *J. Med. Chem.* **43,** 4594–4605.
Paulsen, M. D., and Ornstein, R. L. (1996). *Protein Eng.* **9,** 567–571.
Pearlman, D. A., and Kollman, P. A. (1991). *J. Chem. Phys.* **94,** 4532–4545.
Pitera, J., and Kollman, P. (1998). *J. Am. Chem. Soc.* **120,** 7557–7567.
Pitera, J. W., and van Gunsteren, W. F. (2002). *Mol. Simul.* **28,** 45–65.
Postma, J. P. M., Berendsen, H. J. C., and Haak, J. R. (1982). *Faraday Symp. Chem. Soc.* **17,** 55–67.
Radmer, R. J., and Kollman, P. A. (1998). *J. Comput. Aided. Mol. Des.* **12,** 215–227.
Schäfer, H., van Gunsteren, W. F., and Mark, A. E. (1999). *J. Comp. Chem.* **20,** 1604–1617.
Schechter, I., and Berger, A. (1967). *Biochem. Biophys. Res. Commun.* **27,** 157–162.
Silva, A. M., Lee, A. Y., Gulnik, S. V., Majer, P., Collins, J., Bhat, T. N., Collins, P. J., Cachau, R. E., Luker, K. E., Gluzman, I. Y., Francis, S. E., Oksman, A., Goldberg, D. E., and Erickson, J. W. (1996). *Proc. Natl. Acad. Sci., USA* **93,** 10034–10039.

Sitkoff, D., Sharp, K. A., and Honig, B. (1994). *J. Phys. Chem.* **98,** 1978–1988.
Smith, P. E., and van Gunsteren, W. F. (1994). *J. Chem. Phys.* **100,** 577–585.
Srinivasan, J., Cheatham, T. E., Cieplak, P., Kollman, P. A., and Case, D. A. (1998). *J. Am. Chem. Soc.* **120,** 9401–9409.
Still, W. C., Tempczyk, A., Hawley, R. C., and Hendrickson, T. (1990). *J. Am. Chem. Soc.* **112,** 6127–6129.
Straatsma, T. P., and McCammon, J. A. (1992). *Annu. Rev. Phys. Chem.* **43,** 407–435.
Tembe, B. L., and McCammon, J. A. (1984). *Computers & Chemistry* **8,** 281–283.
Tidor, B. (1993). *J. Phys. Chem.* **97,** 1069–1073.
Torrie, G. M., and Valleau, J. P. (1974). *Chem. Phys. Lett.* **28,** 578–581.
Torrie, G. M., and Valleau, J. P. (1977). *J. Comp. Phys.* **23,** 187–199.
van Gunsteren, W. F., and Berendsen, H. J. C. (1987). "*Groningen Molecular Simulation (GROMOS) Library Manual.*" Biomos, B. V., Groningen, The Netherlands.
van Gunsteren, W. F. (1988). *Prot. Eng.* **2,** 5–13.
Wall, I. D., Leach, A. R., Salt, D. W., Ford, M. G., and Essex, J. W. (1999). *J. Med. Chem.* **42,** 5142–5152.
Wang, J., Dixon, R., and Kollman, P. A. (1999). *Proteins* **34,** 69–81.
Wang, J. M., Morin, P., Wang, W., and Kollman, P. A. (2001). *J. Am. Chem. Soc.* **123,** 5221–5230.
Wang, W., Wang, J., and Kollman, P. A. (1999). *Proteins* **34,** 395–402.
Warshel, A. (1982). *J. Phys. Chem.* **86,** 2218–2224.
Warshel, A. (1984). *Pontif. Acad. Sci. Scr. Varia.* **55,** 60–81.
Warwicker, J., and Watson, H. C. (1982). *J. Mol. Biol.* **157,** 671–679.
Werbovetz, K. A. (2000). *Curr. Med. Chem.* **7,** 835–860.
Wong, C. F., and McCammon, J. A. (1986). *J. Am. Chem. Soc.* **108,** 3830–3832.
Zacharias, M., Straatsma, T. P., and McCammon, J. A. (1994). *J. Chem. Phys.* **100,** 9025–9031.
Zhou, R., Friesner, R. A., Ghosh, A., Rizzo, R. C., Jorgensen, W. L., and Levy, R. M. (2001). *J. Phys. Chem. B* **105,** 10388–10397.
Zwanzig, R. W. (1954). *J. Chem. Phys.* **22,** 1420–1426.

MEMBRANE PROTEIN SIMULATIONS: ION CHANNELS AND BACTERIAL OUTER MEMBRANE PROTEINS

By CARMEN DOMENE, PETER J. BOND, AND MARK S. P. SANSOM

Laboratory of Molecular Biophysics (LMB), Department of Biochemistry, The University of Oxford, Oxford, United Kingdom

I. Introduction 159
A. Biological Membranes 160
B. Membrane Proteins 162
II. Simulation Methods 162
A. Classical Molecular Dynamics 163
III. Ion Channels 166
A. Ion Permeability and Selectivity in KcsA 167
B. Gating Models: Combining KcsA and MthK 170
C. Modeling Mammalian K Channels: Kir 172
D. Other Channels and Related Transporters 173
IV. Outer Membrane Proteins 174
A. Outer Membranes 174
B. Outer Membrane Proteins 174
C. Porins: Models for Transport, Selectivity, and Gating 175
D. OmpA: An Elusive Ion Channel 180
V. Future Prospects 183
A. A Virtual Outer Membrane? 183
B. *Ab Initio* Molecular Dynamics 184
References 185

I. Introduction

Membrane proteins are of great biological and biomedical significance. Understanding their structure and dynamics presents considerable challenges to experimenters and simulators alike. In particular, the complex membrane environment in which such proteins are embedded means that they present considerable challenges in terms of both the set-up and execution of biologically meaningful simulations.

Biomolecular simulation of membrane proteins is also playing an increasingly important role in structural biology. For example, studies of water transport in GlpF (Tajkhorshid *et al.*, 2002) reveal how the integration of structure determination by X-ray diffraction and MD simulation of protein dynamics can lead to further insights into biological function. Furthermore, the quality of simulations is such that we may start to use them to investigate how the dynamics of the protein in the experimental environment (protein/detergent co-crystal from X-ray diffraction;

ADVANCES IN PROTEIN CHEMISTRY, Vol. 66

0065-3233/03 $35.00

protein/detergent micelle for NMR) may be related to the dynamics of the protein when embedded in a membrane.

Such simulations are being extended to a wide range of membrane proteins including ion channels (Bernèche and Roux, 2001; Roux *et al.*, 2000; Sansom *et al.*, 2002; Sansom *et al.*, 2000), water and related transport proteins (De Groot and Grubmuller, 2001), proton pumps such as bacteriorhodopsin (Ulmschneider and Sansom, ms. in preparation) and bacterial outer membrane proteins (Bond *et al.*, 2002; Bond and Sansom, 2003; Faraldo-Gómez *et al.*, 2002; Faraldo-Gómez *et al.*, 2003). Current molecular dynamics simulations can be performed on systems of more than 10^6 atoms for periods of time of tens of nanoseconds, providing a detailed insight into the dynamics of these systems. This information helps us to bridge the gap between the essentially static images of membrane proteins provided by, for example, protein crystallography or NMR, and the dynamic aspects of their biological function. A combination of molecular modeling and simulation helps us to extrapolate from the structure of prokaryotic membrane proteins to the structure and dynamics of their human homologues, which may also aid in experimental structure determination. Ultimately, simulation offers the prospect of understanding the effects of mutations on physiological function in a variety of disease states (Ashcroft, 2000), as well as aiding the discovery and design of drugs to help combat such diseases. In this review, rather than attempt to cover all simulations of membrane proteins, we will focus on two classes that have been investigated by a number of research groups, namely ion channels and bacterial outer membrane proteins.

A. *Biological Membranes*

Biological membranes are composed principally of a lipid bilayer (ca. 40 Å thick) plus membrane proteins. Such membranes enable cells to create an internal environment separate from the outside world and also may provide different compartments within the cell. Protein molecules may be embedded within the lipid bilayer or bound to one or more surfaces. The lipid head group region of a bilayer is ca. 10 Å thick, providing a polar region in which both lipid headgroups and water molecules are present, with both of which a membrane protein may interact (Wiener and White, 1992). The central hydrophobic region of the bilayer creates a barrier to permeation of ions and polar solute molecules in the absence of specific channels and transporters.

It is therefore evident that a biological membrane provides a complex, anisotropic environment in which membrane proteins have to fold stably and function. The environment is anisotropic both in terms of its polarity and dielectric properties (ranging from an apolar, low dielectric environment in the centre of the membrane to a polar, higher dielectric environment in the vicinity of the headgroups) as well as its viscosity profile (Fig. 1). This will make attempts to approximate the environment of a membrane protein in simulations via, for example, generalized Born type models (Spassov *et al.*, 2002), that are more complex than for a 'simple' aqueous environment. Thus, to date, most membrane protein simulations have been performed with a fully atomistic model of the surrounding environment. Furthermore, one should remember that although *in vitro* experiments frequently use bilayers composed of a single lipid species or a relatively simple mixture of a few lipids, *in vivo* cell membranes are complex mixtures of many different species of lipids. This means that the chemical environment experienced by the membrane proteins may be laterally heterogeneous (see below).

In addition to the "simple" lipid bilayer present in the majority of cell membranes, more complex lipid bilayer environments are present in the outer membranes of Gram negative bacteria, for example (see below). These membranes have an inner leaflet composed mainly of phospholipid, but an outer leaflet containing large, complex lipopolysaccharide molecules. Thus, an additional degree of spatial anisotropy is introduced,

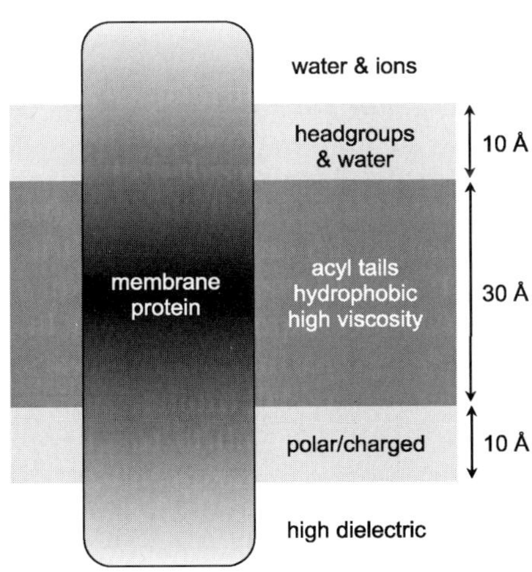

FIG. 1. Schematic of the anisotropic environment presented by a membrane to a membrane protein. The grey scale, for both membrane and protein, represents the polarity/dielectric of the environment from low (dark grey) to high (white). The two regions within the membrane (headgroup region and hydrophobic core) and their approximate dimensions are indicated.

which will eventually have to be included in simulations of this class of membranes.

B. Membrane Proteins

Membrane proteins constitute between 20 and 30% of genes in most organisms (Wallin and von Heijne, 1998). Thus, there may be ca. 10,000 membrane proteins encoded in the human genome. They play key roles in regulation of cellular activities. The function of membrane proteins is essential for a wide range of physiological events such as neurotransmitter transport, adhesion, regulation, and nerve impulse transmission. Also, it is suggested that ca. 50% of future drug targets will be membrane proteins (Terstappen and Reggiani, 2001). Two major classes of biomedically important membrane proteins are ion channels and solute transporters, both of which facilitate the movement of ions or other polar molecules across the impermeable core of the bilayer. To date, most simulation studies have focused on these families of membrane proteins.

Despite the large number of genes encoding membrane proteins, the number of membrane protein structures presently known at high resolution is rather small, just about 0.2% of them. Only around 50 structures of membrane proteins have been determined (see http://blanco.biomol.uci.edu/Membrane_Proteins_xtal.html), ranging from channels and transporters to membrane bound enzymes. Of these structure, only 15 are from eukaryotes. This is because bacterial membrane proteins are more easily over-expressed and crystallized (Werten *et al.*, 2002).

From those structures of membrane proteins that have been determined, and from a wide variety of indirect structural studies and from theoretical considerations (Popot and Engelman, 2000), it is evident that there are two basic folds for the transmembrane (TM) domain of a membrane protein: a bundle of TM α-helices or an antiparallel β-barrel (Fig. 2). The latter fold is only found in those membrane proteins found in bacterial outer membranes, or phylogenetically related membranes such as the mitochondrial outer membrane.

II. Simulation Methods

Due to the inherent experimental difficulties in obtaining structural data for membrane proteins, it is essential that we maximize our understanding of the structure/function relationships of those membrane proteins for which we have experimental structures. Computational approaches have proven to be useful and have become a standard tool

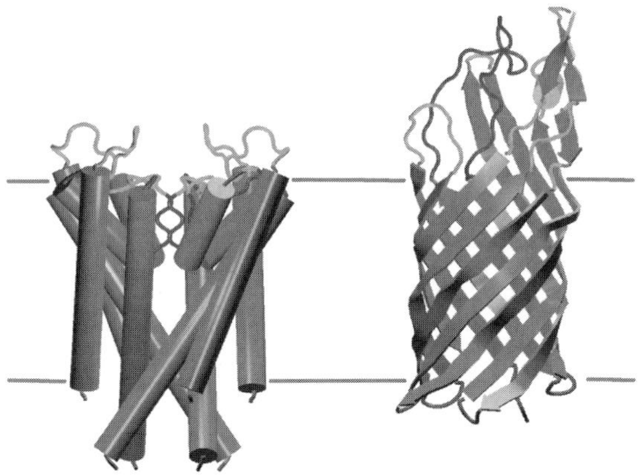

FIG. 2. The two classes of membrane protein: an α-helix bundle illustrated by KcsA (Zhou *et al.*, 2001); and a β-barrel, illustrated by OpcA (Prince *et al.*, 2002). The grey horizontal lines indicate the presumed location of the membrane. (See Color Insert.)

for investigations of membrane proteins (Chung and Kuyucak, 2002; Hansson *et al.*, 2002; Isralewitz *et al.*, 2001; Roux, 2002; Sansom *et al.*, 2002; Werten *et al.*, 2002).

A. *Classical Molecular Dynamics*

To date, most biomolecular simulations of membrane transport proteins have employed conventional molecular mechanic forcefields in which each atom is treated as a van der Waals sphere with a point charge. These methods have been successful for a wide range of problems in protein dynamics and have evolved considerably since their first application to proteins 25 years ago (Karplus and McCammon, 2002). Since then, advances in both computing power and algorithms have made it possible to study larger and more complex systems. There are a number of forcefields in use at present in the context of biomolecular systems and membrane proteins in particular such as CHARMM (Brooks *et al.*, 1983), GROMOS (van Gunsteren and Berendsen, 1987), OPLS (Jorgensen and Tirado-Rives, 1998; Kaminski *et al.*, 2001), and AMBER (Pearlman *et al.*, 1995). In all of them, the potential energy of the system is defined as a function of the positions of the particles that constitute it. The potential function is represented as the sum of several terms which describe the bond length and angle deformation energies, the energy associated with the torsion angles, the van der Waals non-bonded energy (dispersion and repulsion), and the non-bonded electrostatic energy. All these energies are calculated between pairs of atoms (pair-potentials), neglecting their many-body nature. The atomic partial charges and the Lennard-Jones radii that are the basic ingredients of the potential functions differ from one force field to another. A further consideration is whether to include all the hydrogen atoms in the calculation or whether to treat only potentially H-bonding hydrogens explicitly, with the hydrogens of apolar groups modelled via extended-atoms. A discussion of some aspects of potential functions related to MD studies of ion channels is provided by (Roux and Bernèche, 2002).

The first reported simulations of membrane were performed *in vacuo* due to computing limitations. Since then, there have been numerous simulation studies of lipid bilayers in which both lipid and water molecules are treated explicitly (Feller, 2000; Forrest and Sansom, 2000; Saiz and Klein, 2002; Tieleman *et al.*, 1997; Tobias, 2001; Tobias *et al.*, 1997). Simulations of solvated lipid bilayers are able to reproduce structural dynamic properties of lipid bilayers and reproduce experimental quantities such as order parameters and area per lipid with a good degree of accuracy (Husslein *et al.*, 1998; Koubi *et al.*, 2001; Moore *et al.*, 2001; Saiz and Klein, 2002; Tarek *et al.*, 2001; Tobias, *et al.*, 1997; Tu *et al.*, 1998a; Tu *et al.*, 1998b). There are also several MD calculations on zwitterionic detergents (Wendoloski *et al.*, 1989) and micelles for anionic, cationic, and mixed detergents (Bocker *et al.*, 1994; Bogusz *et al.*, 2000; Bogusz *et al.*, 2001; Marrink and Mark, 2002) which are of relevance with respect to understanding the structural dynamics of membrane proteins in the detergent environments used for solubilization, crystallization, and for NMR studies (see below).

Given the slow dynamics of lipid molecules in fluid-phase bilayers (with a diffusion coefficient D $\sim 10^{-3} - 10^{-4}$ Å^2 ps^{-1}) (Gennis, 1989), and the irregular shape of most membrane proteins, obtaining a correctly configured initial system is a non-trivial task; however, the reliability of the subsequent simulation may depend on how carefully this is performed. Two approaches have been employed. The first (Petrache *et al.*, 2000; Woolf and Roux, 1994; Woolf, 1996) consists of building a bilayer around the protein lipid by lipid, each individual molecule being selected from a library of lipid conformations. The second approach (Shen *et al.*, 1997; Tieleman and Berendsen, 1998) uses a previously equilibrated lipid bilayer, in which a cylindrical hole is created by the application of weak repulsive radial forces on the lipid atoms in order to accommodate the protein. An extension to the second approach enables the cavity in the bilayer to have an arbitrary shape (Fig. 3), thus generalizing the method to

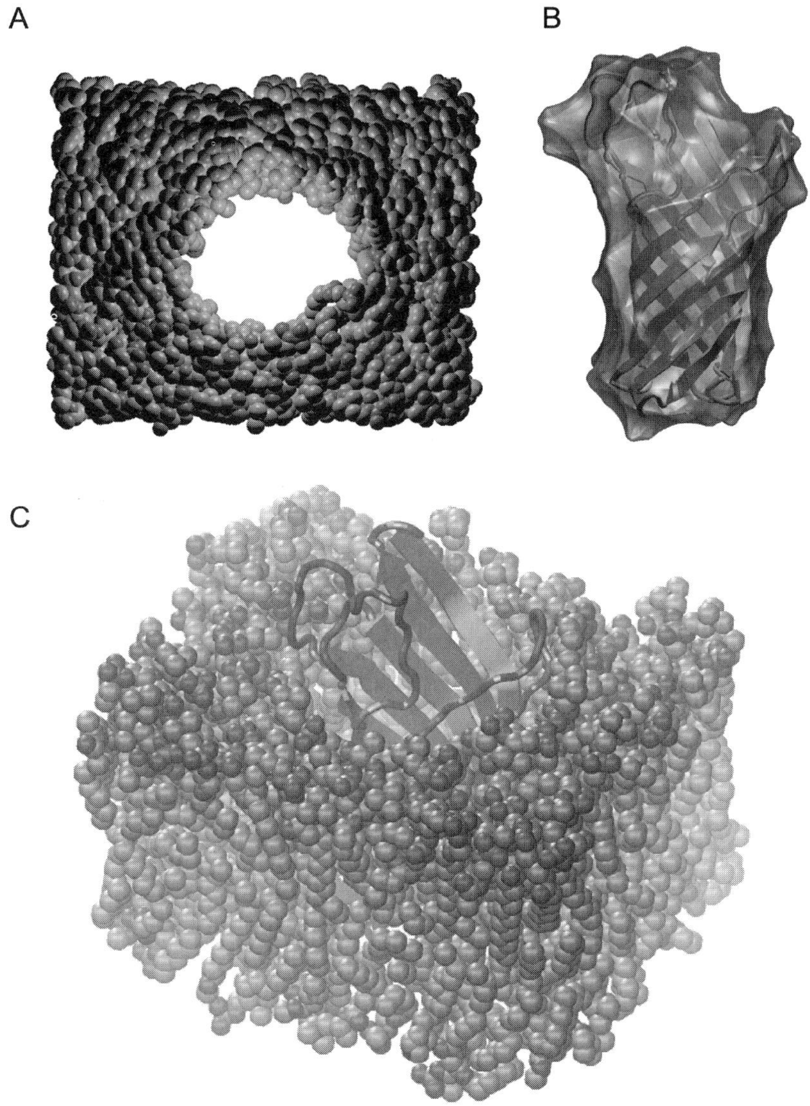

FIG. 3. Set-up of a bilayer simulation. **A** Bilayer of DMPC with a cavity created by the procedure of (Faraldo-Gómez *et al.*, 2002); **B** membrane protein (OmpA) showing the surface accessible area in blue; **C** OmpA embedded in a DMPC bilayer. (See Color Insert.)

any protein geometry. The perturbation of the lipid bilayer is minimized by conducting the process in several stages in which the force applied to lipid atoms is gradually increased until a satisfactory cavity is generated (Faraldo-Gómez *et al.*, 2002).

The majority of membrane protein simulations to date have employed bilayers containing a single lipid species, often DMPC or POPC, for example. However, simulation studies of mixed lipid bilayers have made significant progress (Chiu *et al.*, 2001a; Chiu *et al.*, 2001b; Scott, 2002; Smondyrev and Berkowitz, 1999; Smondyrev and Berkowitz, 2001). Given that a number of membrane proteins exhibit specific, functionally important interactions with lipid molecules (e.g., KcsA and phosphatidyl glycerol; Valiyaveetil *et al.*, 2002) it will be important to extend simulation studies to more complex and biologically realistic lipid mixtures in the near future.

In addition to more complex membrane simulations, there is a need to develop continuum solvent models for the complex bilayer/water environment. By reducing the computational cost of simulations, such models enable much longer timescales to be addressed. For example, generalized Born models have been used with some success in the simulation of folding of small water-soluble proteins (Chowdhury *et al.*, 2003; Simmerling *et al.*, 2002). It is therefore of some interest that progress has been made in extending such continuum models to membranes and membrane proteins (Spassov *et al.*, 2002).

III. Ion Channels

Ion channels are a large and biomedically important family of integral membrane proteins (Hille, 2001). Ion channels play a key role in the electrical activity of cells of the nervous system. However, channels are also found in the membranes of non-excitable cells and a wide range of organisms from viruses to plants. Ion channels form pores in membranes through which ions may pass at high (near diffusion limited) rates (ca. 10^7 ions sec^{-1}). Despite this high rate of transport, channels are selective as to which ions pass; for example, K^+ channels are about 10,000 times more selective for K^+ than for Na^+. Such selectivity arises from structural properties of the channel, which dictates the energetic cost of moving the permeant species from the bulk solution at either sides of the lipid bilayer to the inside of the protein pore. The dehydration energy then will be compensated by the interaction energy of the permeant species and the channel pore-lining surface.

Channels do not stay open all of the time. Instead, they are 'gated' by either the binding of small molecules (e.g., neurotransmitters or intracellular regulators) to the channel protein or by changes in voltage across the membrane. Channel gating is believed to operate via conformational changes that induce structural changes in the pore that alter the energetic profile of the ion permeation pathway.

Traditionally, channels are classified according to the type of ion for which the channel is selective (e.g., K channels). They are also classified according to the factors that control the gating of the channel (e.g., ligand-gated ion channels, voltage-gated ion channels). The K channels are perhaps the most intensively studied family of ion channel (Yellen, 1999, 2002). The major families are: (1) Kv channels, which are activated by a change in transmembrane voltage; (2) Kir, inward rectifier channels, which have a higher conductance for K^+ ions moving into the cell than outwards; and (3) TWIK and related channels, which contain two copies of the selectivity filter motif (see below) in one polypeptide chain (Biggin *et al.*, 2000). X-ray crystallographic methods have been used to obtain structures of bacterial K channels from a number of these families (Doyle, 2003; Doyle *et al.*, 1998; Jiang *et al.*, 2002a). For this reason, K channels have been the subject of more simulation studies than most other membrane proteins. In the following sections, we will focus on simulation studies of a bacterial K channel, KcsA, and of related K channel species.

A. *Ion Permeability and Selectivity in KcsA*

The first crystal structure of a bacterial potassium channel, KcsA was solved at 3.2 Å resolution (Doyle *et al.*, 1998). This structure revealed the central pore-forming motif of a potassium channel. Since then, the structure has been solved at higher resolution, 2.0 and 2.3 Å respectively, for KcsA in the presence of high or low concentrations of K^+ ions (Zhou *et al.*, 2001). These higher resolution structures provided the basis for new theoretical structural and dynamical studies providing insight into ion permeation and gating.

KcsA represents the simplest topology for a K channel. It contains a basic pore-forming unit comprising two transmembrane helices, M1 and M2, linked by a shorter re-entrant pore (P) helix and a loop (Fig. 4). The functional channel is formed from a tetrameric assembly of the 2TM + 1P unit. The selectivity filter motif which forms the loop region is characteristic of K channels and has a highly conserved sequence (TVGYG). Regions of the protein other than the selectivity filter differ between different K channels and confer particular properties of gating and pharmacology.

The selectivity filter is characterized by four K^+ binding sites, S1 to S4. Two rings of four backbone carbonyl oxygen atoms (carbonyl oxygens plus

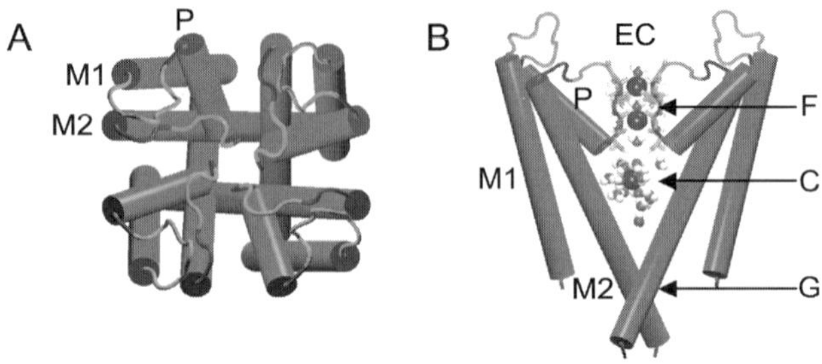

FIG. 4. KcsA: **A** is a view down the pore axis from the extracellular mouth, showing the M1, P, and M2 helices; **B** is a view (of just two subunits) down a perpendicular to the pore axis, showing the extracellular mouth (*EC*), the filter (*F*), the central cavity (*C*), and the gate (*G*) regions. Two K^+ ions (plus waters) are shown in the filter (at sites S1 and S3) along with a third K^+ ion in the central cavity. (See Color Insert.)

sidechain hydroxyl oxygens at site S4) constitute each of these sites, with the oxygens arranged in a square anti-prism. A fifth site, S0 is formed at the extracellular mouth of the channel by four carbonyl oxygens from the protein plus four water molecules. S0 was revealed in the higher resolution crystal structure (Zhou *et al.*, 2001) and had also been seen in simulation studies based on the low resolution structure (Bernèche and Roux, 2001; Sansom *et al.*, 2002).

A rigid model for the selectivity filter was first proposed (Doyle *et al.*, 1998) based on the original structure of KcsA. The dimensions of the filter are such that the exposed carbonyl oxygen atoms can effectively solvate a K^+ ion with a radius of 1.33 Å (Zhou *et al.*, 2001). Conversely, it cannot solvate in the same efficient way a Na^+ ion with a radius of 0.95 Å. However, a variety of simulation studies (Roux, 2002; Sansom *et al.*, 2002) have suggested that the filter has some degree of flexibility. Therefore, the nature of the selectivity filter might be more complex than originally thought, involving a subtle balance of a variety of energetic contributions.

Following the first publication of the KcsA structure, a large number of computational and simulation studies were reported in the literature (Allen *et al.*, 1999, 2000; Chung *et al.*, 2002; Guidoni *et al.*, 1999, 2000; Bernèche and Roux, 2000, 2001; Åqvist and Luzhkov, 2000; Luzhkov and Åqvist, 2000, 2001; Biggin *et al.*, 2001; Ranatunga *et al.*, 2001; Sansom *et al.*, 2000, 2002; Shrivastava and Sansom, 2000, 2002; Shrivastava *et al.*, 2002). Its structural dynamics and the events accompanying movement of ions through the filter region have been described in great detail. In particular, it was shown that ions and water molecules move along the selectivity filter in a concerted fashion on a 1 to 10 ns timescale (Shrivastava and Sansom,

2000). That this is consistent with free energy calculations (Åqvist and Luzhkov, 2000; Bernèche and Roux, 2001) suggests a relatively smooth free energy landscape for ion translocation within the filter such that two configurations, with ions either at S1 and S3 or at S2 and S4, are of approximately equal stability. These simulation results, based on the low resolution crystal structure, were subsequently supported by combined crystallographic and functional studies (Morais-Cabral *et al.*, 2001).

The behaviour and interactions of different species of monovalent cations within the filter of KcsA inserted in a lipid bilayer have also been studied by molecular dynamics (Domene and Sansom, submitted; Shrivastava *et al.*, 2002). In these studies, based on the low resolution and high resolution structures respectively, concerted motions of ions and water within the selectivity filter of the protein, as well as a degree of flexibility of the filter were both observed. Furthermore, pronounced distortions of the filter were observed when it contained water molecules without ions, which is in agreement with the altered conformation of the filter when KcsA was crystallized in the presence of a low concentration of K^+ (Zhou *et al.*, 2001). The two most permeant ions, K^+ and Rb^+, were similar in their interactions with the selectivity filter. The largest ion, Cs^+, surprisingly seemed to exhibit similar behavior to K^+ and Rb^+. In contrast, Na^+ ions (which have a very low rate of permeation (LeMasurier *et al.*, 2001)) tended to distort the filter inwards by binding to a ring of four rather than eight oxygen atoms. The behavior of Na^+ ions was also different in terms of an apparent preference for site S2, rather than either S1/S3 or S2/S4. Na^+ ions were coordinated by six oxygen atoms (four carbonyl oxygen atoms and two the water molecules) which seemed to induce local deformations of the selectivity filter (Domene and Sansom, submitted).

The movements of four crystallographic water molecules, which occupied sites at the "back" of the selectivity filter mediating its interactions with the surrounding P helix ring, were also analyzed during the simulation. No significant changes in the positions of these waters were observed over the course of 15 ns of simulation and no appreciable correlation of their limited motions with the concerted translocations of the ions and water molecules in the selectivity filter was observed. Thus, we may conclude that these waters act as 'structural elements' and are essential to the functional integrity of the filter in a high conductance conformation. This may have important implications for attempts to homology model mammalian K channels.

Immediately beneath the selectivity filter is a central cavity containing ca. 30 water molecules, which seems to stabilize the ion once it has exited the selectivity filter. The dipoles of the four pore helices are oriented towards this cavity, providing additional stability to the K ions (Roux and MacKinnon, 1999). Beneath the cavity is the intracellular channel gate, which is *closed* in the crystallographic structures providing an electrostatic barrier that prevents ions from exiting/entering the channel (Bernèche and Roux, 2001; Biggin *et al.*, 2001; Chung *et al.*, 1999; Shrivastava and Sansom, 2002).

It is important to note that simulations of KcsA have been performed between the different research groups, with a variety of simulation codes and force-fields. Some features of the results (e.g., concerted ion/water translocations and filter flexibility) appear to be robust to the changes in the simulation protocol. Other aspects are more sensitive to the conditions of the simulation. Amongst the simulation variables that may influence the behavior of ions in the filter are: the methods of treatment of long electrostatics, the thermostat and temperature coupling, and the presence or absence of a 150 mM KCl solution at either side of the membrane (in order to more fully simulate physiological conditions; Domene and Sansom, ms. in preparation). In principle, it should be possible to extract information on the sensitivity of simulations to methodology from the literature. However, the absence of a suitable database of simulations (comparable to, for example, the protein data bank (Berman *et al.*, 2000; Bernstein *et al.*, 1977) makes this somewhat difficult in practice.

There have also been attempts to move beyond classical calculations for KcsA. For example, Guidoni and Carloni (2002) extracted snapshots from a classical simulation for use in density functional theory (DFT) calculations. These single point DFT calculations were performed on models of the selectivity filter in isolation as at present it remains unfeasible to carry out electronic structure calculations of the entire channel protein. The cluster models used included the backbone of a group forming two consecutive binding sites, the K^+ ion in a binding site and two water molecules. The results of these calculations indicated some degree of change in electronic polarization of the carbonyl oxygens when a K^+ ion was present at a binding site.

B. *Gating Models: Combining KcsA and MthK*

A central challenge remaining in ion channel biophysics is to understand the mechanism(s) of channel gating. This provides a challenge for direct ''brute force'' simulation approaches because channel gating is thought to occur via slow (> 1 μs) conformational changes in the channel protein. However, by combining structural models of the endpoints of channel gating transitions with simulation studies, we can gain some insight into possible mechanisms. Thus the recent determination of the open pore conformation of a bacterial K channel, a Ca^{2+} activated-K^{+} channel MthK from *Methanobacterium thermoautotrophicum* (Jiang *et al.*, 2002b), provides structural insights into a gating mechanism and enables further biomolecular simulation studies.

The MthK channel structure, when compared with that of KcsA, suggests that gating may occur via hinge-bending at a conserved glycine residue in the M2 helix in order to widen the hydrophobic mouth at the intracellular end of the channel, thus removing the energetic barrier to ion entry/exit. Although this model is persuasive (Jiang *et al.*, 2002b), and is broadly consistent with earlier spectroscopic (Liu *et al.*, 2001; Perozo *et al.*, 2000) and computational (Biggin and Sansom, 2002) studies, it remains at the level of a hypothesis that requires further testing (e.g., to see whether an open channel model of KcsA based on the MthK structure is functionally open and structurally stable).

Molecular modeling and dynamics simulations have been used (Holyoake *et al.*, 2003) to build a model of KcsA in an open state based on the MthK structure (Fig. 5). This assumes that the nature of the K channel open state is conserved between the two channel species. The radius at the intracellular mouth of KcsA changes from 1.4 Å to 6.0 Å on going from the closed state to the (modeled) open state. This change is enough to remove the electrostatic barrier in the energy to the ion permeation (Chung *et al.*, 2002); in other words, ions will be able to pass through.

Simulations of the open state model of KcsA embedded in a membrane-mimetic octane slab (Holyoake *et al.*, 2003) suggest that the model is just as stable in the simulation as is the closed state structure of the channel, in that both show a $C\alpha$ root mean square deviation (RMSD) from the

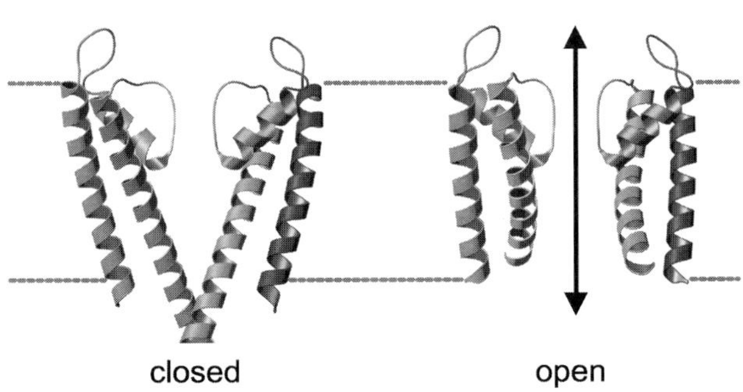

FIG. 5. Comparison of the closed (x-ray) vs. open (modelled on MthK) structures of KcsA. Two of the four subunits are shown. (See Color Insert.)

respective starting model of ca. 3 Å after > 5 ns of simulation. If one tracks the behavior of K^+ ions in the open state simulation (Fig. 6), two key aspects of channel behavior are seen. First, the ions and water molecules within the filter translocate in a concerted fashion (and the filter distorts once all ions have exited); and second, there is rapid exit of ions from the cavity through the open intracellular mouth (i.e., the open gate) of the channel. Thus, the behavior of the filter, to a first approximation, seems to be the same in the open and closed states of KcsA, and the primary gate is located at the intracellular mouth of the channel. A gating mechanism based on change in radius of a narrow hydrophobic region of a pore has been invoked for a number of channels and is consistent with recent simulation studies on a simplified model of a transmembrane pore (Beckstein *et al.*, 2001).

C. *Modeling Mammalian K Channels: Kir*

As discussed previously, the majority of structures of membrane proteins have been determined for bacterial proteins. Homology modeling combined with simulations may be used to extrapolate to the behavior of mammalian channels (Capener *et al.*, 2002). Inward rectifier (Kir) channels are a family of K channels that conduct K^+ ions more readily from outside to inside of the cell than in the opposite direction. Because the transmembrane topology of Kir channels resembles that of KcsA, ion permeation, selectivity, and blocking mechanisms in this family have to be explored via modeling and simulation. However, despite Kir channels sharing a common transmembrane topology with KcsA, they are somewhat

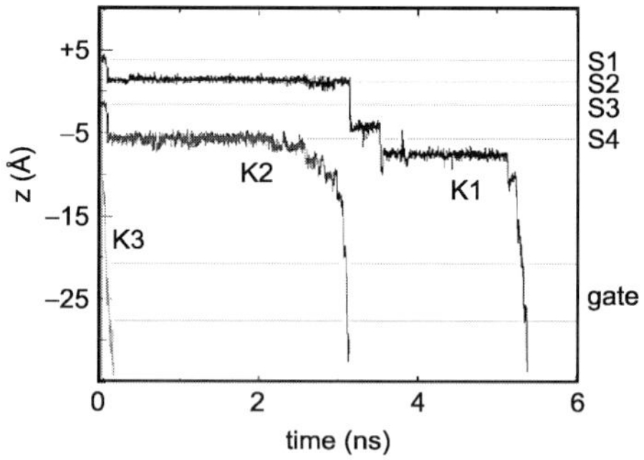

FIG. 6. Potassium ion trajectories through the open state model of KcsA derived from a 10-ns simulation (Holyoake *et al.*, 2003) of the open state model in a membrane mimetic octane slab. The locations of four K^+ binding sites in the filter (S1 to S4) and of the gate region are indicated by the horizontal broken lines.

distant homologues with an overall sequence identity of just ca. 15%. Consequently, difficulties in obtaining an optimal alignment have been encountered, even though extensive mutagenesis studies and related studies (e.g., Minor *et al.*, 1999) have been used to provide additional structural restraints.

Detailed simulation studies of Kir6.2 based upon a homology model (Capener *et al.*, 2000) suggest that a stable model of the transmembrane domain of this channel may be constructed based upon KcsA. This model resembles KcsA in that concerted ion/water translocations occur within the filter on a nanosecond timescale and the filter exhibits a degree of flexibility (Capener and Sansom, 2002). The model and simulations have also been used successfully to rationalize the effects of mutants that alter the single channel conductance (Capener *et al.*, 2003). Because a bacterial homologue of Kir channels has been discovered (Durell and Guy, 2001), it will be of some interest to compare the model structure with an x-ray structure of a bacterial Kir channel once the latter emerges (Doyle, 2003).

D. *Other Channels and Related Transporters*

The determination of the crystal structure of a bacterial mechanosensitive channel MscL (Chang *et al.*, 1998) has enabled MD simulations to investigate the reaction of the protein to the application of surface tension. A potential mechanism of gating was postulated using MD simulations with MscL in isolation, and with the protein embedded in an explicit membrane (Gullingsrud *et al.*, 2001a, 2001b; Gullingsrud and Schulten, 2002) under a range of conditions. It is suggested that changes in the hydrophobic residues in the constricted region of the protein play a key role in the gating process. Further studies on experimentally characterized mutants involving only the C-terminal region (Elmore and Dougherty, 2001) were performed and compared with simulations of the wild type structure. Aspects of protein-lipid interactions were also described.

Simulation studies have also been applied extensively to another family of pore-like proteins, the aquaporins (De Groot and Grubmuller, 2001; de Groot *et al.*, 2000; Jensen *et al.*, 2001; Tajkhorshid *et al.*, 2002; Zhu *et al.*, 2001). In particular, the studies of Schulten and colleagues and of the Grubmuller group have provided a detailed insight into the mechanisms of permeation and selectivity of these pores.

In addition to gating of ion channels and pore proteins, conformational transitions are of great importance in many other membrane transport proteins. These include transport proteins from a number of different families for which structures are beginning to emerge, including ABC transporters (e.g., MsbA [Chang and Roth, 2001] and BtuCD [Locher *et al.*, 2002]) and 12 TM helix transporters (e.g., AcrB [Murakami *et al.*, 2002]). Modeling and simulation studies have been used, for example, to explore the stability and interactions of MsbA models based on a low resolution crystal structure (Campbell *et al.*, 2003). Simulation studies will be required to help elucidate the mechanisms of these and other transporters, although less direct approaches will be required as the slow conformational changes (which will elude direct simulation) are thought to be coupled to solute transport.

IV. Outer Membrane Proteins

A. *Outer Membranes*

Gram-negative bacterial cells are enclosed by two membranes. The outer membrane is highly asymmetric; whereas its inner leaflet has a similar phospholipid composition to that of the inner membrane, its outer leaflet contains lipopolysaccharide molecules (Tamm *et al.*, 2001). These large, sugar-containing lipids are thought to be fairly rigid and are peppered with negative charges and cross-linking cations. The outer membrane is selectively permeable to solutes smaller than ~600 Da because of the presence of pore-forming proteins called porins, rendering it somewhat "leaky" in comparison with the inner membrane (Koebnik *et al.*, 2000). The outer membrane therefore serves as a protective envelope that nevertheless allows the cell to obtain essential nutrients.

B. *Outer Membrane Proteins*

Outer membrane proteins (OMPs) are composed of transmembrane antiparallel β-barrels (Schulz, 2000). The β-strands of the barrels are connected by short turns on the inner (periplasmic) side of the membrane and by long, mobile, sequence-variable loops on the extracellular side. The architecture of the OMP barrel itself varies primarily in the number and stagger of the β-strands, the cross-section of the central pore, and the oligomerization state of the protein. OMPs have a variety of functional roles, including specific and non-specific porins, active transporters, enzymes, structural proteins, and OMPs implicated in virulence and pathogenesis (Buchanan, 1999; Koebnik *et al.*, 2000; Koronakis *et al.*, 2000; Prince *et al.*, 2002; Schulz, 2000; Vandeputte-Rutten *et al.*, 2001; Wiener *et al.*, 2003). Because of the evident importance of OMPs in pathogenesis (Achouak *et al.*, 2001; Lin *et al.*, 2002), along with their structural simplicity and the consequent potential for understanding ion and small molecule transport, they have been investigated by a number of computational techniques (Table I).

C. Porins: Models for Transport, Selectivity, and Gating

Since the elucidation of the first atomic structure of a porin over 10 years ago (Weiss *et al.*, 1991), the crystal structures of several other specific and non-specific porins have been solved (Schulz, 1996). In all cases, they consist of large, homotrimeric β-barrels, but vary in the number of strands and the conformations of the extracellular loops. Thus, the barrels of the sugar-specific porins, typified by the malto-oligosaccharide-specific maltoporin LamB (Schirmer *et al.*, 1995) and the sucrose-specific porin ScrY (Forst *et al.*, 1998), are composed of 18 β-strands, and three loops fold into the barrel to constrict the diameter of the pore. The specificity for sugar molecules is provided by a row of aromatic residues (the "greasy slide"), which make contact with the hydrophobic sugar ring, and a complementary line of polar residues, which complex the sugar hydroxyl groups. There is high sequence homology between ScrY and LamB, and just a few strategically positioned residues in the pore lumen determine specificity for particular sugars via size exclusion (Koebnik *et al.*, 2000). In comparison, the non-specific or general porins allow the passive diffusion of a wide variety of polar molecules and ions. The barrels consist of 16 β-strands, and only one loop (L3) folds into the lumen. In each case, this loop forms the so-called "eyelet" that determines the size exclusion limit of the pore. Across this eyelet there is a variable segregation of positive and negative charges. The resulting electrical field determines the permeability properties of each porin, such as the cation-selective OmpF and the anion-selective PhoE (Cowan, 1993; Koebnik *et al.*, 2000; Schirmer, 1998).

To gain insight into their functional properties, a number of computational methods have been applied to porins. The structure of a porin from *Rhodobacter capsulatus* was analyzed via calculation of the electrostatic potential of the inner molecular surface of the pore (Weiss *et al.*, 1991). The observed negative potential toward the extracellular end of the trimer was used to explain the slight cation selectivity of the porin. However, only standard charges were used to calculate the electrostatic potential, and the possible effects of the low dielectric membrane environment were omitted. This has been circumvented by treating the protein, solvent, and membrane components as continuum regions, each with a particular

TABLE I
Summary of Bacterial Outer Membrane Protein Simulations

Protein	Function	Number of β-strands	Reference
OmpA	Channel and/or anchor	8β	(Bond *et al.*, 2002; Bond and Sansom, 2003)
OmpX	Recognition	8β	*
PagP	Fatty acyl transferase	8β + *N*-terminal helix	*
OpcA	Recognition/adhesion	10β	*
OmpT	Protease	10β	*
OMPLA	Phospholipase	12β (dimeric)	*
R. capsulatus porin	Porins	16β (trimeric)	(Björkstén *et al.*, 1994; Soares *et al.*, 1995).
OmpF porin	Porins	16β (trimeric)	(Watanabe *et al.*, 1997; Suenaga *et al.*, 1998; Tieleman and Berendsen, 1998; Im and Roux, 2002; Robertson and Tieleman, 2002)
FhuA	Ferrichrome transporter	22β + internal "cork" domain	(Faraldo-Gómez *et al.*, 2003)
FecA	Siderophore transporter	22β + internal "cork" domain	*

(* = ms. in preparation)

dielectric constant, and then solving the Poisson-Boltzmann equation, which relates the variation in electrostatic potential to the charge density while considering the solvent ion distribution. This treatment allowed the calculation of the pK_A values of titratable groups and the electrostatic field in the channels of various porins (Dutzler *et al.*, 1999; Karshikoff *et al.*, 1994; Zachariae *et al.*, 2002). From such studies, the electrostatic potentials in both the extracellular mouth region and the constriction zone were suggested to be important in the ion selectivity of each porin, whereas the strong transverse electrostatic field was shown to be strongest around the eyelet, thus facilitating dehydration of polar molecules during passage through the pore. Additionally, the clusters of charged residues in the constriction zone displayed unusual titration behavior, which may stabilize the strong electrostatic field inside the channel, while possibly modulating pore size or activity under changing conditions of pH or ionic concentration, respectively.

Simple studies of electrostatic potential effects on ion transport have been carried out using Brownian dynamics simulations. The electrostatic potential generated by the protein is calculated using the Poisson-Boltzmann equation, whereas the trajectories of ions are simulated explicitly, their motion being governed by the electrostatic interaction with protein charges along with forces representing frictional drag and stochastic collisions caused by solvent molecules. These simulations are much faster than classical all-atom molecular dynamics, and such a representation of ion flux is thought to be suitable for large pores. In an initial study, Brownian dynamics simulations were used to simulate several thousand cation and anion trajectories through OmpF, PhoE, OmpK36, and two OmpF mutants, allowing statistically significant analysis (Schirmer and Phale, 1999). Cations and anions were shown to move along different paths, consistent with the charge segregation around the eyelet region. This highly charged constriction selectively enhanced ion permeabilities for each porin. Encouragingly, electrophysiological properties were accurately reproduced, including ion selectivity and relative conductance. A subsequent Brownian dynamics study carried out on 11 different OmpF mutants with widely varying ion selectivities again resulted in good agreement with experimental data (Phale *et al.*, 2001). Thus, despite the simplifications inherent in the Brownian dynamics technique, the primary characteristics of pore activity in porins were successfully reproduced. Nevertheless, extensions to the method have been made in an attempt to better model the non-equilibrium state of a multi-ion system. This was implemented by combining Brownian dynamics with a Grand Canonical Monte Carlo algorithm to maintain a local ion concentration at the channel entrance, along with a modified Poisson-Boltzmann equation to account for a microscopic transmembrane potential (Im *et al.*, 2000). With the inclusion of an electrostatic reaction field, this method resulted in excellent agreement between calculated and experimental conductance for OmpF (Im and Roux, 2001).

Although Brownian dynamics can be used to study ion trajectories through static protein channels, molecular dynamics techniques are required to investigate the role of conformational change in channel function. For example, early MD studies were carried out in an attempt to explain the experimentally observed voltage gating of porins, which some have attributed to changes at the constriction zone caused by movement of loop L3. Because of the lack of computing power, these simulations were of the porin in a vacuum. It was therefore necessary to constrain parts of the protein to prevent gross deformation from the crystal structure resulting from the absence of the bilayer and solvent. An initial study of *R. capsulatus* porin used simulated annealing, with all mainchain atoms except those of L3 constrained (Björkstén *et al.*, 1994). While the tip of the loop remained stably anchored to the barrel wall, several "mobile zones" in L3 were more flexible than the rest of the loop and were capable of significantly diminishing the size of the pore, although some of these conformational changes may not have been physically realistic. The same group attempted to simulate the voltage gating process in the constrained monomer by gradually unscreening charges in the protein over the course of a simulation (Soares *et al.*, 1995). This was proposed to reflect the perturbation of counterion distribution and water dipole orientation around the highly charged eyelet region, inducing conformational changes in L3. They observed a reversible conformational change, which closed the pore in a part of the loop that overlapped with one of the "mobile zones" from their previous study. Nevertheless, cross-linking experiments have shown that large-scale movement of L3 are not required for voltage gating, although this does not rule out locally restricted movements (Phale *et al.*, 1997). A systematic MD study of OmpF (Watanabe *et al.*, 1997) showed that the strength and extent of constraints could significantly affect the dynamics of the protein, and emphasized the importance of coupling between the L3 loop motions and barrel flexibility, therefore questioning the methodology of the previous simulations. To explore ion conduction, a simulation of solvated OmpF was carried out in the presence of a constant force across the pore representing a transmembrane potential (Suenaga *et al.*, 1998). Loop L3 was shown to be reasonably flexible, and it was suggested that an aspartate residue in L3 may be important in the permeation process. However, the statistics of this study were poor, and again, the absence of an explicit lipid bilayer meant that the protein had to be constrained.

Tieleman and Berendsen (1998) simulated a fully solvated system containing the whole OmpF trimer in an explicit phospholipid bilayer. This groundbreaking 1-ns simulation provided a large amount of information on protein dynamics and behavior of water in the pore, as well as allowing a detailed analysis of lipid-protein interactions (Tieleman *et al.*, 1999). The presence of explicit lipid and solvent resulted in a structurally stable trimer, and a good correlation was observed between protein fluctuations and secondary structure. The pore geometry of all three monomers stayed fairly close to the crystal structure, with just a slight reduction in pore size, and no large motions were observed in the L3 loops, suggesting that conformational changes in this loop are unlikely to be the source of voltage gating. As expected, the translational diffusion and reorientational correlation times of water within each pore were significantly reduced in comparison with their bulk values. Moreover, the large variation in water diffusion and ordering across the pore has consequences for continuum electrostatics models of porins and other ion channels. Although only a few counterions were included in this simulation, Im and Roux (2002) recently performed a 5-ns MD study of a bilayer-embedded OmpF trimer in a 1-M KCl solution. The protein dynamics were similar to the simulation of Tieleman and Berendsen (1998), and no large movements in loop L3 were observed. The high salt concentration allowed for a statistically meaningful interpretation of the ion permeation mechanism. From the average ion density along the pore, it was observed that anions and cations were distinctly separated at the constriction zone, because of the strong transverse electric field. Moreover, because of the positions of charged residues above and below the constriction, the ions took screw-like twisting pathways along the channel. The contributions from water and protein to solvation of the ions varied along the pore in a complementary fashion to maintain full hydration of both ions, as required of an ion channel with high throughput. Finally, more K^+ than Cl^- ions were found in the pore, consistent with the cation selectivity of OmpF. This may be partly explained by the fact that while K^+ ions could permeate the channel when isolated, Cl^- ions were only able to move past the constriction zone, which is overall positively charged.

Recently, computational analysis of ion permeation in porins has come full circle, with a detailed study of OmpF using MD, Brownian dynamics, and Poisson-Nernst-Planck theory (Im and Roux, 2002). At different levels of detail, all three techniques now appear to be able to successfully describe the overall ion distributions along the pore and the protein-ion interactions important for selectivity and gating. It should be noted that computational studies on porins have not only been used to investigate the passage of ions. Thus, the "conjugate peak refinement" method was recently used to analyze the translocation mechanism of maltohexose through the pore of maltoporin (Dutzler *et al.*, 2002). The method is used to find a series of intermediate conformations along the lowest energy pathway, enabling the identification of transition state energy barriers. The study revealed that both the sugar chain and protein undergo few changes in conformation during translocation, and that fast relay of sugar hydroxyl hydrogen-bonds with other sugar hydroxyls and with the protein's line of polar residues (the "ionic track") helps to avoid large energy barriers. This diffusive process thus explained the experimentally observed combination of tight binding of sugar to the pore and efficient sugar translocation. Finally, non-equilibrium or steered MD has been used to study the transport of the dipolar molecules alanine and methylglucose through the OmpF trimer when embedded in fully solvated lipid bilayer (Robertson and Tieleman, 2002). A "virtual spring" was attached to each transported molecule, which was then retracted at constant velocity along the axis of the pore. Contrary to previously observed ion behavior, the molecules did not follow a screw-like trajectory, although they did align strongly with the transverse electric field at the constriction zone. The number of hydrogen bonds made to each transported molecule was maintained during the transport, so that as the pore narrows, hydrogen-bonds were broken with the water and made with the β-barrel, as seen in the ion trajectories of Im and Roux (2002). It appears that the charges near the eyelet have a particularly important role in this mechanism, and ensure that the permeating molecules all show similar behavior at the constriction; this may have important consequences for selectivity.

D. OmpA: An Elusive Ion Channel

OmpA is a small, monomeric OMP expressed constitutively in *E. coli*. It consists of an eight-stranded β-barrel domain that inserts into the outer membrane, along with a globular, periplasmic C-terminal domain (Buchanan, 1999). Its main function is thought to be structural, maintaining the integrity of the cell, although it has also been implicated in recognition events such as bacterial conjugation and pathogenesis. A crystal structure of the OmpA *N*-terminal domain ($OmpA^{NT}$) revealed the β-barrel to be a "solid inverse micelle" consisting of several aqueous cavities formed by an extensive hydrogen-bonding network but with no continuous channel (Pautsch and Schulz, 1998). Nevertheless, a plethora of liposome swelling and planar bilayer electrophysiological experiments have shown that OmpA forms channels *in vitro*. Most recently, Arora *et al.* (2000) showed that OmpA has two distinct open states of small and large flickering conductances; interestingly, only the smaller channels were observed for the *N*-terminal domain. A comparison of known conductance levels and associated exclusion limits would approximately predict a pore diameter lying somewhere between that of gramicidin A (3 Å) and OmpF (10 Å) (Arora *et al.*, 2000).

In an attempt to resolve the apparent paradox between structural and functional data, MD simulations and modeling studies were undertaken for OmpA^{NT}, with a particular emphasis on investigating protein conformational dynamics (i.e., 'gating') that may lead to transient pore formation (Bond *et al.*, 2002). Significant water diffusion was observed along the barrel axis, due to the dynamic nature of the pore sidechains that form the 'rigid' hydrogen-bonded network observed in the crystal structure. Only one region, an Arg-Glu salt bridge around the centre of OmpA^{NT}, proved to be prohibitive to water crossing events (Fig. 7A). We were able to fully open the pore during 5 ns MD simulations by modeling the Arg into a non-native but physically realistic rotamer in order to coordinate it with an alternative, unpaired Glu residue further along the barrel axis (Fig. 7B).

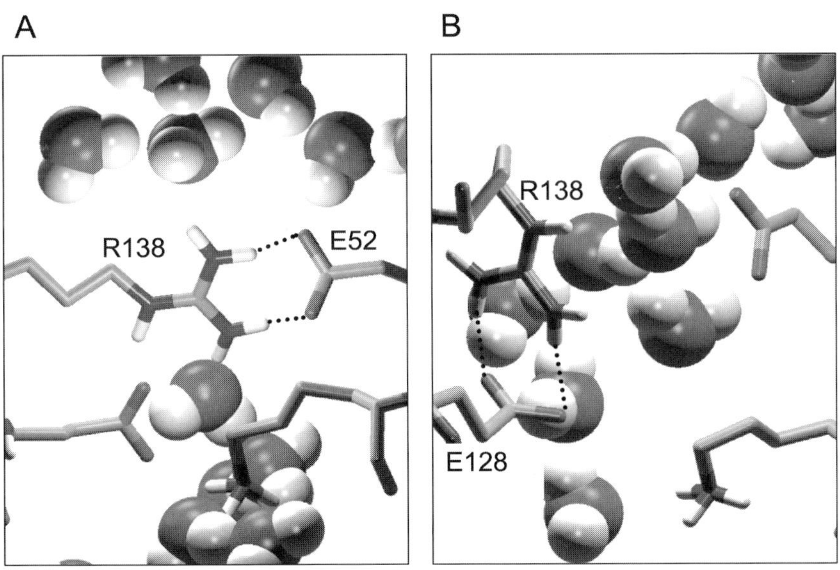

FIG. 7. The gate region in the outer membrane protein OmpA (Bond *et al.*, 2002). Structures are shown from (**A**) a closed state simulation in which water is unable to pass the R138-E52 salt bridge; and (**B**) an open state simulation, in which R138 interacts with E128 (broken black lines); note that an uninterrupted column of water molecules is present. (See Color Insert.)

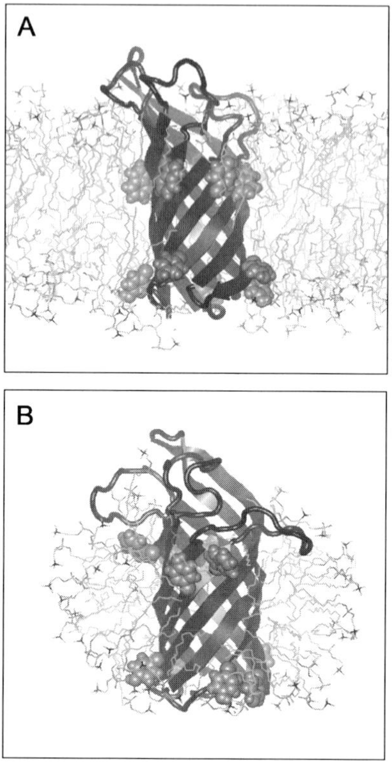

FIG. 8. Simulations of OmpA in (**A**) a lipid bilayer and (**B**) a detergent micelle (Bond and Sansom, 2003). Both structures are taken at 10 ns of simulation. Note how the external loops of OmpA have altered conformation in **B** to interact with the surface of the micelle. (See Color Insert.)

Using the simulation-averaged pore profile, an empirical conductance estimate based on treating the pore as a cylinder of electrolyte with reduced ionic mobility was in excellent agreement with the experiment. We therefore proposed that a perturbation of the Arg-Glu salt bridge may have a critical role in gating, perhaps in response to changes in the transmembrane potential, and may be related to the flickering conductances observed in the electrophysiological experiments.

In a subsequent study, Arora *et al.* obtained the solution NMR structure of OmpA^{NT} in detergent micelles (Arora *et al.*, 2001). Whilst the structure was in close agreement with its x-ray counterpart, the study provided additional information relating to the dynamics of the protein. In particular, a gradient of flexibility was shown to exist along the axis of the β-barrel on the ps-ns timescale. Most recently, simulations of $OmpA^{NT}$ in a detergent micelle (the first such simulations of an integral membrane protein in micelle) have been performed (Bond and Sansom, 2003). Paired 10 ns MD simulations were carried out to compare the behavior of $OmpA^{NT}$ in a detergent micelle and in a phospholipid bilayer (Fig. 8). Good agreement in protein dynamics was found with the NMR study; in particular, the gradient of flexibility along the barrel was reproduced. Differences in protein-lipid and protein-detergent interactions were observed to have significant effects on protein behavior, with the $OmpA^{NT}$ backbone dynamics exhibiting fluctuations ca. 1.5 times greater in the micelle environment than in the lipid bilayer. This increase in overall flexibility resulted in significantly increased pore dynamics, and in particular, led to conformational changes in the Arg-Glu salt bridge, which originally blocked permeation. This in turn resulted in the formation of a continuous pore through the protein. It is therefore clear that changes in the lipid environment can have a drastic effect on ion channel behavior. This should particularly be kept in mind when interpreting essentially static x-ray structural data, and highlights the strength of the MD technique in relating membrane protein structure to function.

V. Future Prospects

The past five years have seen considerable advances in simulations of membrane proteins. These have paralleled progress in experimental structure determination. The future prospects for such studies would seem to lie in two main directions. From a biological perspective, there is a need to extend such simulations to a wide range of membrane proteins, enabling comparisons between different classes of ion channels and transporters. However, while extending the range of simulations, it is also important to attempt improvements in their accuracy, particularly with respect to reproducing ion selectivity.

A. *A Virtual Outer Membrane?*

As discussed above, the structures of a significant number of outer membrane proteins have been determined by X-ray crystallography and, more recently, by NMR of mixed protein/detergent micelles. This makes it possible to perform simulations of a whole family of membrane proteins (Table I). To simulate all OMPs for which structures are known will require development of high throughput simulations. Such methods will be transferable to a wide range of other membrane proteins. To date, most biomolecular simulations have been in-depth studies of a single protein. However, in order to extract the biological signal from the simulation noise, it may be necessary to perform multiple simulations on a range of related proteins. In this way, one addresses the sampling issue from a biological perspective.

This approach will enable us to exploit the increased sampling power of comparative analyses to examine OMP dynamics *vs.* environment (bilayer, micelle, crystal), OMP interactions with lipids, and details of protein mobility *vs.* secondary structure elements and their location relative to the membrane. Such results are an essential first stage in moving towards a virtual outer membrane.

In this context, it is important to note that most simulations of OMPs to date have employed either PC or PE bilayers. However, progress in modeling of lipopolysaccharide membranes (Lins and Straatsma, 2001) will enable us to move towards more complex and more biologically realistic outer membrane simulations.

B. Ab Initio *Molecular Dynamics*

There are some problems for which conventional molecular mechanics forcefields are not sufficient. In particular, these may include the interactions of proteins with ions during transport phenomena, where the exact balance between protein/ion, protein/water, and ion/water interaction energies is essential for biological function. Clearly, as an inorganic ion passes through a channel or through a channel-like region in a transport protein, its environment changes dynamically. Therefore, a first-principle description of the system is fundamental where polarization effects are considered.

It is also important to bear in mind that most biological molecules are surrounded by water and empirical potential functions for liquid water have been under development for the last two or three decades. The parameters for all the water models are empirically adjusted to reproduce water density and the enthalpy of vaporization but they cannot systematically reproduce the freezing point of water or its dipole moment. In order to adequately describe the interaction of water and proteins and of water and ions, a quantum description is once again desirable.

In order to address biophysical processes which involve, for example, charge transfer, polarization effects or chemical reactions, a new molecular dynamics scheme was formulated in 1985, the so-called *ab-initio* or first-principle molecular dynamics (Car and Parrinello, 1985). Here, in contrast to classical molecular dynamics, the interactions between the atoms are determined from first principle electronic structure calculations in the framework of Density Functional Theory (DFT). DFT provides the ground state properties of a many-electron system once the electronic density is available; knowledge of the electronic density is sufficient. Nonetheless, the exact form of the energy functional is not available and it has to be approximated.

Ab-initio molecular dynamics (AIMD) is very demanding computationally. At present, it is generally only feasible to apply AIMD to relatively small systems (ca. 100–200 atoms) for small periods (ca. 10 ps). However, recent improvements leading to the use of the so-called order-N algorithms in codes (Soler *et al.*, 2002) provide an opportunity for applications to a wide range of biological systems, including membrane proteins. In the last two decades, the number of applications of AIMD simulations has already started to grow (Car, 2002; Carloni *et al.*, 2002). Some attempts have already been made on ion channels and related membrane proteins (Guidoni and Carloni, 2002; Guidoni, *et al.*, 1999; Guidoni *et al.*, 2000; Sagnella *et al.*, 1996), and clearly further studies will follow as advances in computation make such methods more widely available.

Given the large scale of membrane protein simulations, it may be more realistic to adopt a hybrid approach, with only certain key regions treated quantum-mechanically. In QM/MM methods, a relatively small region where quantum chemical effects are important is identified; this region is treated quantum-mechanically while the remaining system is modeled using a molecular mechanics force field (see Sherwood, 2000).

References

Achouak, W., Heulin, T., and Pages, J. M. (2001). Multiple facets of bacterial porins. *FEMS Microbiol. Lett.* **199,** 1–7.

Allen, T. W., Bliznyuk, A., Rendell, A. P., Kuyucak, S., and Chung, S. H. (2000). The potassium channel: Structure, selectivity and diffusion. *J. Chem. Phys.* **112,** 8191–8204.

Allen, T. W., Kuyucak, S., and Chung, S. H. (1999). Molecular dynamics study of the KcsA potassium channel. *Biophys. J.* **77,** 2502–2516.

Åqvist, J., and Luzhkov, V. (2000). Ion permeation mechanism of the potassium channel. *Nature* **404,** 881–884.

Arora, A., Abildgaard, F., Bushweller, J. H., and Tamm, L. K. (2001). Structure of outer membrane protein A transmembrane domain by NMR spectroscopy. *Nature Struct. Biol.* **8,** 334–338.

Arora, A., Rinehart, D., Szabo, G., and Tamm, L. K. (2000). Refolded outer membrane protein A of *Escherichia coli* forms ion channels with two conductance states in planar lipid bilayers. *J. Biol. Chem.* **275,** 1594–1600.

Ashcroft, F. M. (2000). "Ion Channels and Disease." Academic Press, San Diego.

Beckstein, O., Biggin, P. C., and Sansom, M. S. P. (2001). A hydrophobic gating mechanism for nanopores. *J. Phys. Chem. B.* **105,** 12902–12905.

Berman, H. M., Westbrook, J., Feng, Z., Gilliland, G., Bhat, T. N., Weissig, H., Shindyalov, I. N., and Bourne, P. E. (2000). The Protein Data Bank. *Nucl. Acids Res.* **28,** 235–242.

Bernèche, S., and Roux, B. (2000). Molecular dynamics of the KcsA K^+ channel in a bilayer membrane. *Biophys. J.* **78,** 2900–2917.

Bernèche, S., and Roux, B. (2001). Energetics of ion conduction through the K^+ channel. *Nature* **414,** 73–77.

Bernstein, F., Koetzle, T., Williams, G., Meyer, E., Brice, M., Rodgers, J., Kennard, O., Shimanouchi, T., and Tasumi, M. (1977). The Protein Data Bank: A computer-based archival file for macromolecular structures. *J. Mol. Biol.* **112,** 535–542.

Biggin, P. C., Roosild, T., and Choe, S. (2000). Potassium channel structure: Domain by domain. *Curr. Opin. Struct. Biol.* **10,** 456–461.

Biggin, P. C., and Sansom, M. S. P. (2002). Open-state models of a potassium channel. *Biophys. J.* **83,** 1867–1876.

Biggin, P. C., Smith, G. R., Shrivastava, I. H., Choe, S., and Sansom, M. S. P. (2001). Potassium and sodium ions in a potassium channel studied by molecular dynamics simulations. *Biochim. Biophys. Acta* **1510,** 1–9.

Björkstén, J., Soares, C. M., Nilsson, O., and Tapia, O. (1994). On the stability and plastic properties of the interior L3 loop in *R. capsulatus* porin. A molecular dynamics study. *Prot. Eng.* **7,** 487–493.

Bocker, J., Brickmann, J., and Bopp, P. (1994). Molecular-dynamics simulation study of an N-decyltrimethylammonium chloride micelle in water. *J. Phys. Chem.* **98,** 712–717.

Bogusz, S., Venable, R. M., and Pastor, R. W. (2000). Molecular dynamics simulations of octyl glucoside micelles: Structural properties. *J. Phy. Chem. B.* **104,** 5462–5470.

Bogusz, S., Venable, R. M., and Pastor, R. W. (2001). Molecular dynamics simulations of octyl glucoside micelles: Dynamic properties. *J. Phys. Chem. B.* **105,** 8312–8321.

Bond, P., Faraldo-Goméz, J., and Sansom, M. S. P. (2002). OmpA—A pore or not a pore? Simulation and modelling studies. *Biophys. J.* **83,** 763–775.

Bond, P., and Sansom, M. S. (2003). Membrane protein dynamics versus environment: simulations of OmpA in a micelle and in a bilayer. *J. Mol. Biol.* **329,** 1035–1053.

Brooks, B. R., Bruccoleri, R. E., Olafson, B. D., States, D. J., Swaminathan, S., and Karplus, M. (1983). CHARMM: A program for macromolecular energy, minimisation, and dynamics calculations. *J. Comp. Chem.* **4,** 187–217.

Buchanan, S. K. (1999). ß-Barrel proteins from bacterial outer membranes: structure, function and refolding. *Curr. Opin. Struc. Biol.* **9,** 455–461.

Campbell, J. D., Biggin, P. C., Baaden, M., and Sansom, M. S. (2003). Extending the structure of an ABC transporter to atomic resolution: modelling and simulation studies of MsbA. *Biochemistry* **42,** 3666–3673.

Capener, C. E., Kim, H. J., Arinaminpathy, Y., and Sansom, M. S. P. (2002). Ion channels: structural bioinformatics and modelling. *Human Molec. Genet.* **11,** 2425–2433.

Capener, C. E., Proks, P., Ashcroft, F. M., and Sansom, M. S. (2003). Filter flexibility in a mammalian channel: models and simulations of ir6.2 mutants. *Biophys. J.* **84,** 2345–2356.

Capener, C. E., and Sansom, M. S. P. (2002). MD Simulations of a K channel model—sensitivity to changes in ions, waters and membrane environment. *J. Phys. Chem. B.* **106,** 4543–4551.

Capener, C. E., Shrivastava, I. H., Ranatunga, K. M., Forrest, L. R., Smith, G. R., and Sansom, M. S. P. (2000). Homology modelling and molecular dynamics simulation studies of an inward rectifier potassium channel. *Biophys. J.* **78,** 2929–2942.

Car, R. (2002). Introduction to density-functional theory and ab-initio molecular dynamics. *Quant. Struct.-Act. Rel.* **21,** 97–104.

Car, R., and Parrinello, M. (1985). Unified approach for molecular-dynamics and density-functional theory. *Phys. Rev. Lett.* **55,** 2471–2474.

Carloni, P., Rothlisberger, U., and Parrinello, M. (2002). The role and perspective of ab initio molecular dynamics in the study of biological systems. *Acc. Chem. Res.* **35,** 455–464.

Chang, G., and Roth, C. B. (2001). Structure of MsbA from *E. coli*: a homolog of the multidrug resistance ATP binding cassette (ABC) transporters. *Science* **293,** 1793–1800.

Chang, G., Spencer, R. H., Lee, A. T., Barclay, M. T., and Rees, D. C. (1998). Structure of the MscL homolog from *Mycobacterium tuberculosis*: a gated mechanosensitive ion channel. *Science* **282,** 2220–2226.

Chiu, S. W., Jakobsson, E., and Scott, H. L. (2001a). Combined Monte Carlo and molecular dynamics simulation of hydrated dipalmitoyl-phosphatidylcholine-cholesterol lipid bilayers. *J. Chem. Phys.* **114,** 5435–5443.

Chiu, S. W., Jakobsson, E., and Scott, H. L. (2001b). Combined Monte Carlo and molecular dynamics simulation of hydrated lipid-cholesterol lipid bilayers at low cholesterol concentration. *Biophys. J.* **80,** 1104–1114.

Chowdhury, S., Lee, M. C., Xiong, G., and Duan, Y. (2003). *Ab initio* folding simulation of the trp-cage mini-protein approaches NMR resolution. *J. Mol. Biol.* **327,** 711–717.

Chung, S. H., Allen, T. W., Hoyles, M., and Kuyucak, S. (1999). Permeation of ions across the potassium channel: Brownian dynamics studies. *Biophys. J.* **77,** 2517–2533.

Chung, S. H., Allen, T. W., and Kuyucak, S. (2002). Conducting-state properties of the KcsA potassium channel from molecular and Brownian dynamics simulations. *Biophys. J.* **82,** 628–645.

Chung, S. H., and Kuyucak, S. (2002). Recent advances in ion channel research. *Biochim. Biophys. Acta* **1565,** 267–286.

Cowan, S. W. (1993). Bacterial porins—lessons from 3 high-resolution structures. *Curr. Opin. Struct. Biol.* **3,** 501–507.

De Groot, B. L., and Grubmuller, H. (2001). Water permeation across biological membranes: Mechanism and dynamics of aquaporin-1 and GlpF. *Science* **294,** 2353–2357.

de Groot, B. L., Heymann, J. B., Engel, A., Mitsuoka, K., Fujiyoshi, Y., and Grubmüller, H. (2000). The fold of human aquaporin 1. *J. Mol. Biol.* **300,** 987–994.

Domene, C., and Sansom, M. (submitted). A potassium channel, ions and water: Simulations studies based on the higher resolution X-ray structure of KcsA. *Biophys. J.*

Doyle, D. A., Cabral, J. M., Pfuetzner, R. A., Kuo, A., Culbis, J. M., Cohen, S. L., Cahit, B. T., and MacKinnon, R. (1998). The structure of the potassium channel: molecular basis of K^+ conduction and selectivity. *Science* **280,** 69–77.

Durell, S. R., and Guy, H. R. (2001). A putative prokaryote member of the Kir family of potassium channels. *Bio. Med. Central Evol. Biol.* **1,** 14.

Dutzler, R., Rummel, G., Albertí, S., Hernández-Allés, S., Phale, P. S., Rosenbusch, J. P., Benedí, V. J., and Schirmer, T. (1999). Crystal structure and functional characterization of OmpK36, the osmoporin of Klebsiella pneumoniae. *Struct.* **7,** 425–434.

Dutzler, R., Schirmer, T., Karplus, M., and Fischer, S. (2002). Translocation mechanism of long sugar chains across the maltoporin membrane channel. *Struct.* **10,** 1273–1284.

Elmore, D. E., and Dougherty, D. A. (2001). Molecular dynamics simulations of wild-type and mutant forms of the *Mycobacterium tuberculosis* MscL channel. *Biophys. J.* **81,** 1345–1359.

Faraldo-Gómez, J., Smith, G. R., and Sansom, M. S. P. (2002). Setup and optimisation of membrane protein simulations. *Eur. Biophys. J.* **31,** 217–227.

Faraldo-Gómez, J., Smith, G. R., and Sansom, M. S. P. (2003). Molecular dynamics simulations of the bacterial outer membrane protein FhuA: a comparative study of the ferrichrome-free and bound states. *Biophys. J* (*in press*).

Feller, S. E. (2000). Molecular dynamics simulations of lipid bilayers. *Curr. Opin. Coll. Interface Sci.* **5,** 217–223.

Forrest, L. R., and Sansom, M. S. P. (2000). Membrane simulations: bigger and better? *Curr. Opin. Struct. Biol.* **10,** 174–181.

Forst, D., Welte, W., Wacker, T., and Diederichs, K. (1998). Structure of the sucrose-specific porin ScrY from *Salmonella typhimurium* and its complex with sucrose. *Nature Struct. Biol.* **5,** 37–46.

Gennis, R. B. (1989). "Biomembranes: Molecular Structure and Function." Springer-Verlag, New York.

Guidoni, L., and Carloni, P. (2002). Potassium permeation through the KcsA channel: a density functional study. *Biochim. Biophys. Acta.* **1563,** 1–6.

Guidoni, L., Torre, V., and Carloni, P. (1999). Potassium and sodium binding in the outer mouth of the K^+ channel. *Biochem.* **38,** 8599–8604.

Guidoni, L., Torre, V., and Carloni, P. (2000). Water and potassium dynamics in the KcsA K^+ channel. *FEBS Lett.* **477,** 37–42.

Gullingsrud, J., Kosztin, D., and Schulten, K. (2001a). Structural determinants of MscL gating studied by molecular dynamics simulations. *Biophys. J.* **80,** 2074–2081.

Gullingsrud, J. R., Kosztin, D., and Schulten, K. (2001b). MscL gating studied by molecular dynamics simulations. *Biophys. J.* **80,** 497.

Gullingsrud, J. R., and Schulten, K. (2002). Gating mechanisms of MscL studied by molecular dynamics simulations using applied surface tension. *Biophys. J.* **82,** 3066.

Hansson, T., Oostenbrink, C., and van Gunsteren, W. F. (2002). Molecular dynamics simulations. *Current Opinion in Structural Biology* **12,** 190–196.

Hille, B. (2001). "Ionic Channels of Excitable Membranes." Sinauer Associates Inc., Sunderland, Mass.

Holyoake, J., Damene, C., Bright, J. N., and Sansom, M. S. P. (2003). KcsA closed and open: modeling and simulation studies. *Eur. Biophys. J.* (in press).

Husslein, T., Moore, P. B., Zhong, Q. F., Newns, D. M., Pattnaik, P. C., and Klein, M. L. (1998). Molecular dynamics simulation of a hydrated diphytanol phosphatidylcholine lipid bilayer containing an alpha-helical bundle of four transmembrane domains of the Influenza A virus M2 protein. *Faraday Disc.* **111,** 201–208.

Im, W., and Roux, B. (2001). Brownian dynamics simulations of ion channels: A general treatment of electrostatic reaction fields for molecular pores of arbitrary geometry. *J. Chem. Phys.* **115,** 4850–4861.

Im, W., and Roux, B. (2002). Ions and counterions in a biological channel: molecular dynamics simulation of OmpF porin from *Escherichia coli* in an explicit membrane with 1 M KC1 aqueous salt solution. *J. Mol. Biol.* **319,** 1177–1197.

Im, W., Seefeld, S., and Roux, B. (2000). A Grand Canonical Monte Carlo-Brownian dynamics algorithm for simulating ion channels. *Biophys. J.* **79,** 788–801.

Isralewitz, B., Gao, M., and Schulten, K. (2001). Steered molecular dynamics and mechanical functions of proteins. *Curr. Opin. Struct. Biol.* **11,** 224–230.

Jensen, M. O., Tajkhorshid, E., and Schulten, K. (2001). The mechanism of glycerol conduction in aquaglyceroporins. *Structure* **9,** 1083–1093.

Jiang, Y., Lee, A., Chen, J., Cadene, M., Chait, B. T., and MacKinnon, R. (2002a). Crystal structure and mechanism of a calcium-gated potassium channel. *Nature* **417,** 515–522.

Jiang, Y., Lee, A., Chen, J., Cadene, M., Chait, B. T., and MacKinnon, R. (2002b). The open pore conformation of potassium channels. *Nature* **417,** 523–526.

Jorgensen, W. L., and Tirado-Rives, J. (1998). Development of the OPLS–AA force field for organic and biomolecular systems. *Abstracts of Papers of the American Chemical Society* **216,** 043–COMP.

Kaminski, G. A., Friesner, R. A., Tirado-Rives, J., and Jorgensen, W. L. (2001). Evaluation and reparametrization of the OPLS-AA force field for proteins via comparasion with accurate quantum chemical calculations on peptides. *J. Phys. Chem. B.* **105,** 6474–6487.

Karplus, M. J., and McCammon, J. A. (2002). Molecular dynamics simulations of biomolecules. *Nature Struct. Biol.* **9,** 646–652.

Karshikoff, A., Spassov, V., Cowan, S. W., Ladenstein, R., and Schirmer, T. (1994). Electrostatic properties of two porin channels from *Escherichia coli. J. Mol. Biol.* **240,** 372–384.

Koebnik, R., Locher, K. P., and Van Gelder, P. (2000). Structure and function of bacterial outer membrane proteins: barrels in a nutshell. *Mol. Microbiol.* **37,** 239–253.

Koronakis, V. A. S., Koronakis, E., Luisi, B., and Hughes, C. (2000). Crystal structure of the bacterial membrane protein TolC central to multidrug efflux and protein export. *Nature* **405,** 914–919.

Koubi, L., Tarek, M., Bandyopadhyay, S., Klein, M. L., and Scharf, D. (2001). Membrane structural perturbations caused by anesthetics and nonimmobilizers: A molecular dynamics investigation. *Biophys. J.* **81,** 3339–3345.

Kuo, A., Gulbis, J. M., Antcliff, J. F., Rahman, T., Lowe, E. D., Zimmer, J., Cuthbertson, J., Ashcroft, F. M., Ezaki, T., and Doyle, D. A. (2003). Crystal Structure of the potassium channel KirBacl.1 in the closed state. *Science* **300,** 1922–1926.

LeMasurier, M., Heginbotham, L., and Miller, C. (2001). KcsA: It's a potassium channel. *J. Gen. Physiol.* **118,** 303–313.

Lin, J., Huang, S. X., and Zhang, Q. J. (2002). Outer membrane proteins: key players for bacterial adaptation in host niches. *Microbes Infect.* **4,** 325–331.

Lins, R. D., and Straatsma, T. P. (2001). Computer simulation of the rough lipopolysaccharide membrane of *Pseudomonas aeruginosa. Biophys. J.* **81,** 1037–1046.

Liu, Y., Sompornpisut, P., and Perozo, E. (2001). Structure of the KcsA channel intracellular gate in the open state. *Nature Struct. Biol.* **8,** 883–887.

Locher, K. P., Lee, A. T., and Rees, D. C. (2002). The *E. coli* BtuCD structure: a framework for ABC transporter architecture and mechanism. *Science* **296,** 1091–1098.

Loussouarn, G., Phillips, L. R., Masia, R., Rose, T., and Nichols, C. G. (2001). Flexibility of the Kir6.2 inward rectifier K^+ channel pore. *Proc. Nat. Acad. Sci. USA* **98,** 4227–4232.

Luzhkov, V. B., and Åqvist, J. (2000). A computational study of ion binding and protonation states in the KcsA potassium channel. *Biochim. Biophys. Acta* **1481,** 360–370.

Luzhkov, V. B., and Åqvist, J. (2001). Mechanisms of tetraethylammonium ion block in the KcsA potassium channel. *FEBS Lett.* **495,** 191–196.

Marrink, S. J., and Mark, A. E. (2002). Molecular dynamics simulations of mixed micelles modeling human bile. *Biochem.* **41,** 5375–5382.

Minor, D. L., Masseling, S. J., Jan, Y. N., and Jan, L. Y. (1999). Transmembrane structure of an inwardly rectifying potassium channel. *Cell* **96,** 879–891.

Moore, P. B., Lopez, C. F., and Klein, M. L. (2001). Dynamical properties of a hydrated lipid bilayer from a multinanosecond molecular dynamics simulation. *Biophys. J.* **81,** 2484–2494.

Morais-Cabral, J. H., Zhou, Y., and MacKinnon, R. (2001). Energetic optimization of ion conduction by the K^+ selectvity filter. *Nature* **414,** 37–42.

Murakami, S., Nakashima, R., Yamashita, E., and Yamaguchi, A. (2002). Crystal structure of bacterial multidrug efflux transporter AcrB. *Nature* **419,** 587–593.

Pautsch, A., and Schulz, G. E. (1998). Structure of the outer membrane protein A transmembrane domain. *Nature Struct. Biol.* **5,** 1013–1017.

Pearlman, D. A., Case, D. A., Caldwell, J. W., Ross, W. S., Cheatham, T. E., Debolt, S., Ferguson, D., Seibel, G., and Kollman, P. (1995). Amber, a package of computer-programs for applying molecular mechanics, normal-mode analysis, molecular-dynamics and free-energy calculations to simulate the structural and energetic properties of molecules. *Comp. Phys. Comm.* **91,** 1–41.

Perozo, E., Liu, Y. S., Smopornpisut, P., Cortes, D. M., and Cuello, L. G. (2000). A structural perspective of activation gating in K^+ channels. *J. Gen. Physiol.* **116,** 5a.

Petrache, H. I., Grossfield, A., MacKenzie, K. R., Engelman, D. M., and Woolf, T. B. (2000). Modulation of glycophorin A transmembrane helix interactions by lipid bilayers: Molecular dynamics calculations. *J. Mol. Biol.* **302,** 727–746.

Phale, P. S., Philippsen, A., Widmer, C., Phale, V. P., Rosenbusch, J. P., and Schirmer, T. (2001). Role of charged residues at the OmpF porin channel constriction probed by mutagenesis and simulation. *Biochem.* **40,** 6319–6325.

Phale, P. S., Schirmer, T., Prilipov, A., Lou, K. L., Hardmeyer, A., and Rosenbusch, J. P. (1997). Voltage gating of *Escherichia coli* porin channels: role of the constriction loop. *Proc. Natl. Acad. Sci. USA* **94,** 6741–6745.

Popot, J. L., and Engelman, D. M. (2000). Helical membrane protein folding, stability, and evolution. *Ann. Rev. Biochem.* **69,** 881–922.

Prince, S. M., Achtman, M., and Derrick, J. P. (2002). Crystal structure of the OpcA integral membrane adhesin from *Neisseria meningitides. Proc. Nat. Acad. Sci. USA* **99,** 3417–3421.

Ranatunga, K. M., Shrivastava, I. H., Smith, G. R., and Sanson, M. S. P. (2001). Side-chain ionisation states in a potassium channel. *Biophys. J.* **80,** 1210–1219.

Robertson, K. M., and Tieleman, D. P. (2002). Orientation and interactions of dipolar molecules during transport through OmpF porin. *FEBS. Lett.* **528,** 53–57.

Roux, B. (2002). Theoretical and computational models of ion channels. *Curr. Opin. Struct. Biol.* **12,** 182–189.

Roux, B., and Berneche, S. (2002). On the potential functions used in molecular dynamics simulations of ion channels. *Biophys. J.* **82,** 1681–1684.

Roux, B., Bernèche, S., and Im, W. (2000). Ion channels, permeation and electrostatics: Insight into the function of KcsA. *Biochem.* **39,** 13295–13306.

Roux, B., and MacKinnon, R. (1999). The cavity and pore helices in the KcsA K^+ channel: electrostatic stabilization of monovalent cations. *Science* **285,** 100–102.

Sagnella, D. E., Laason, K., and Klein, M. L. (1996). Ab initio molecular dynamics study of proton transfer in a polyglycine analog of the ion channel gramicidin. *Biophys. J.* **71,** 1172–1178.

Saiz, L., and Klein, M. L. (2002). Computer simulation studies of model biological membranes. *Acc. Chem. Res.* **35,** 482–489.

Sansom, M. S. P., Shrivastava, I. H., Bright, J. N., Tate, J., Capener, C. E., and Biggin, P. C. (2002). Potassium channels: structures, models, simulations. *Biochim. Biophys. Acta* **1565,** 294–307.

Sansom, M. S. P., Shrivastava, I. H., Ranatunga, K. M., and Smith, G. R. (2000). Simulations of ion channels—watching ions and water move. *Trends Biochem. Sci.* **25,** 368–374.

Schirmer, T. (1998). General and specific porins from bacterial outer membranes. *J. Struct. Biol.* **121,** 101–109.

Schirmer, T., Keller, T. A., Wang, Y. F., and Rosenbusch, J. P. (1995). Structural basis for sugar translocation through maltoporin channels at 3.1A resolution. *Science* **267,** 512–514.

Schirmer, T., and Phale, P. S. (1999). Brownian dynamics simulation of ion flow through porin channels. *J. Mol. Biol.* **294,** 1159–1167.

Schulz, G. E. (1996). Porins: general to specific, native to engineered passive pores. *Curr. Opin. Struct. Biol.* **6,** 485–490.

Schulz, G. E. (2000). β-Barrel membrane proteins. *Curr. Opin. Struct. Biol.* **10,** 443–447.

Scott, H. L. (2002). Modeling the lipid component of membranes. *Curr. Opin. Struct. Biol.* **12,** 495–502.

Shen, L., Bassolino, D., and Stouch, T. (1997). Transmembrane helix structure, dynamics, and interactions: Multi-nanosecond molecular dynamics simulations. *Biophys. J.* **73,** 3–20.

Sherwood, P. (2000). "Hybrid quantum mechanics/molecular mechanics approaches."

Shrivastava, I. H., and Sansom, M. S. P. (2000). Simulations of ion permeation through a potassium channel: molecular dynamics of KcsA in a phospholipid bilayer. *Biophys. J.* **78,** 557–570.

Shrivastava, I. H., and Sansom, M. S. P. (2002). Molecular dynamics simulations and KcsA channel gating. *Eur. Biophys. J.* **31,** 207–216.

Shrivastava, I. H., Tieleman, D. P., Biggin, P. C., and Sansom, M. S. P. (2002). K^+ vs. Na^+ ions in a K channel selectivity filter: a simulation study. *Biophys. J.* **83,** 633–645.

Simmerling, C., Strockbine, B., and Roitberg, A. E. (2002). All-atom structure prediction and folding simulations of a stable protein. *J. Am. Chem. Soc.* **124,** 11258–11259.

Smondyrev, A. M., and Berkowitz, M. L. (1999). Structure of dipalmitotylphosphatidylcholine/cholesterol bilayer at low and high cholesterol concentrations: Molecular dynamics simulation. *Biophys. J.* **77,** 2075–2089.

Smondyrev, A. M., and Berkowitz, M. L. (2001). Molecular dynamics simulation of the structure of dimyristoylphosphatidylcholine bilayers with cholesterol, ergosterol, and lanosterol. *Biophys. J.* **80,** 1649–1658.

Soares, C. M., Björkstén, J., and Tapia, O. (1995). L3 loop-mediated mechanisms of pore closing in porin: a molecular dynamics perturbation approach. *Prot. Eng.* **8,** 5–12.

Soler, J. M., Artacho, E., Gale, J. D., Garcia, A., Junquera, J., Ordejon, P., and Sanchez-Portal, D. (2002). The SIESTA method for ab initio order-N materials simulation. *J. Phys. Condensed Matter* **14,** 2745–2779.

Spassov, V. Z., Yan, L., and Szalma, S. (2002). Introducing an implicit membrane in generalized Born/solvent accessibility continuum solvent models. *J. Phys. Chem. B.* **106,** 8726–2738.

Suenaga, A., Komeiji, Y., Uebayasi, M., Meguro, T., Saito, M., and Yamato, I. (1998). Computation observation of an ion permeation through a channel protein. *Biosci. Reports* **18,** 39–48.

Tajkhorshid, E., Nollert, P., Jensen, M. O., Miercke, L. J. W., O'Connell, J., Stroud, R. M., and Schulten, K. (2002). Control of the selectivity of the aquaporin water channel family by global orientational tuning. *Science* **296,** 525–530.

Tamm, L. K., Arora, A., and Kleinschmidt, J. H. (2001). Structure and assembly of beta-barrel membrane proteins. *J. Biol. Chem.* **276,** 32399–32402.

Tarek, M., Tobias, D. J., Chen, S. H., and Klein, M. L. (2001). Short wavelength collective dynamics in phospholipid bilayers: A molecular dynamics study. *Phys. Rev. Lett.* **8723,** art. no.-238101.

Terstappen, G. C., and Reggiani, A. (2001). In silico research in drug discovery. *Trends Pharmacol. Sci.* **22,** 23–26.

Tieleman, D. P., and Berendsen, H. J. C. (1998). A molecular dynamics study of the pores formed by *Escherichia coli* OmpF porin in a fully hydrated palmitoyloleoylphosphatidylcholine bilayer. *Biophys. J.* **74,** 2786–2801.

Tieleman, D. P., Forrest, L. R., Berendsen, H. J. C., and Sansom, M. S. P. (1999). Lipid properties and the orientation of aromatic residues in OmpF, influenza M2 and alamethicin systems: molecular dynamics simulations. *Biochem.* **37,** 17554–17561.

Tieleman, D. P., Marrink, S. J., and Berendsen, H. J. C. (1997). A computer perspective of membranes: Molecular dynamics studies of lipid bilayer systems. *Biochim. Biophys. Acta* **1331,** 235–270.

Tobias, D. J. (2001). Electrostatics calculations: Recent methodological advances and applications to membranes. *Curr. Opin. Struct. Biol.* **11,** 253–261.

Tobias, D. J., Tu, K. C., and Klein, M. L. (1997). Atomic-scale molecular dynamics simulations of lipid membranes. *Curr. Opin. Coll. Interface Sci.* **2,** 15–26.

Tu, K., Klein, M. L., and Tobias, D. J. (1998a). Constant-pressure molecular dynamics investigation of cholesterol effects in a dipalmitoylphosphatidylcholine bilayer. *Biophys. J.* **75,** 2147–2156.

Tu, K., Tarek, M., Klein, M. L., and Scharf, D. (1998b). Effects of anesthetics on the structure of a phospholipid bilayer: Molecular dynamics investigation of halothane in the hydrated liquid crystal phase of dipalmitoylphosphatidylcholine. *Biophys. J.* **75,** 2123–2134.

Valiyaveetil, F. I., Zhou, Y., and MacKinnon, R. (2002). Lipids in the structure, folding and function of the KcsA channel. *Biochem.* **41,** 10771–10777.

van Gunsteren, W. F., and Berendsen, H. J. C. (1987). "Gromos-87 manual." Biomos BV, Groningen.

Vandeputte-Rutten, L., Kramer, R. A., Kroon, J., Dekker, N., Egmond, M. R., and Gros, P. (2001). Crystal structure of the outer membrane protease OmpT from *Escherichia coli* suggests a novel catalytic site. *EMBO J.* **20,** 5033–5039.

Wallin, E., and von Heijne, G. (1998). Genome-wide analysis of integral membrane proteins from eubacterial, archean, and eukaryotic organisms. *Prot. Sci.* **7,** 1029–1038.

Watanabe, M., Rosenbusch, J., Schirmer, T., and Karplus, M. (1997). Computer simulations of the OmpF porin from the outer membrane of *Escherichia coli. Biophys. J.* **72,** 2094–2102.

Weiss, M. S., Abele, U., Weckesser, J., Welte, W., Schiltz, E., and Schulz, G. E. (1991). Molecular architecture and electrostatic properties of a bacterial porin. *Science* **254,** 1627–1630.

Wendoloski, J. J., Kimatian, S. J., Schutt, C. E., and Salemme, F. R. (1989). Molecular-dynamics simulation of a phospholipid micelle. *Science* **243,** 636–638.

Werten, P. J. L., Remigy, H. W., de Groot, B. L., Fotiadis, D., Philippsen, A., Stahlberg, H., Grubmuller, H., and Engel, A. (2002). Progress in the analysis of membrane protein structure and function. *FEBS Lett.* **529,** 65–72.

Wiener, M. C., Chimento, D. P., Mohanty, A. K., and Kadner, R. J. (2003). The crystal structure of the *E. coli* outer membrane cyanocobalamin transporter BtuB. *Nature Struct. Biol.* In press.

Wiener, M. C., and White, S. H. (1992). Structure of a fluid dioleoylphosphatidylcholine bilayer determined by joint refinement of X-ray and neutron diffraction data. III. Complete structure. *Biophys. J.* **61,** 434–447.

Woolf, T., and Roux, B. (1994). Molecular-dynamics simulation of the gramicidin channel in a phospholipid-bilayer. *Proc. Natl. Acad. Sci. USA* **91,** 11631–11635.

Woolf, T. B. (1996). Molecular dynamics simulations of individual bacteriorhodopsin helices. *Biophys. J.* **70,** A377.

Yellen, G. (1999). The bacterial K^+ channel structure and its implications for neuronal channels. *Curr. Opin. Neurobiol.* **9,** 267–273.

Yellen, G. (2002). The voltage-gated potassium channels and their relatives. *Nature* **419,** 35–42.

Zachariae, U., Koumanov, A., Engelhardt, H., and Karshikoff, A. (2002). Electrostatic properties of the anion selective porin Omp32 from *Delftia acidovorans* and of the arginine cluster of bacterial porins. *Prot. Sci.* **11,** 1309–1319.

Zhou, Y., Morais-Cabral, J. H., Kaufman, A., and MacKinnon, R. (2001). Chemistry of ion coordination and hydration revealed by a K^+ channel-Fab complex at 2.0 Å resolution. *Nature* **414,** 43–48.

Zhu, F. Q., Tajkhorshid, E., and Schulten, K. (2001). Molecular dynamics study of aquaporin-1 water channel in a lipid bilayer. *FEBS Lett.* **504,** 212–218.

LARGE SCALE SIMULATION OF PROTEIN MECHANICS AND FUNCTION

By EMAD TAJKHORSHID, ALEKSIJ AKSIMENTIEV, ILYA BALABIN, MU GAO, BARRY ISRALEWITZ, JAMES C. PHILLIPS, FANGQIANG ZHU, AND KLAUS SCHULTEN*

Theoretical and Computational Biophysics Group, Beckman Institute for Advanced Science and Technology, University of Illinois

I.	Introduction	195
II.	Technology for Simulation and Visualization of Large Biomolecular Systems	198
III.	Aquaporins–Membrane Water Channels	201
IV.	Energy Conversion in ATP Synthase	211
	A. Torque Generation in F_0	213
	B. Torque-Driven Catalysis in F_1	223
V.	Mechanical Signaling in Fibronectin	232
VI.	Outlook	237
	References	239

I. Introduction

During the past five years, life science research has been almost completely transformed. With the sequencing of the genomes of many organisms, in particular that of man, there has emerged a new view of organisms as a networked system of metabolic and signaling pathways. Today, organisms and cells are approached as a healthy functioning whole; studies of disease and therapy focus on failure and repair of pathways. The computer has become an indispensable instrument of all researchers confronted with the vast new data and their integration.

However, the same data remind biomedical scientists that for most proteins structures, functions, mechanisms, and pathways are still elusive, and that science needs to make a strenuous, concerted effort to gain this missing knowledge. Biomolecular modeling is a key methodology that can bridge the knowledge gap and provide atomic level pictures of biomolecular processes. It is adopted today increasingly by computational as well as experimental biologists.

Proteins, the products of genomes, can only be completely understood in the context of their cellular environment. One must not lose sight of the fact that the genome only partly defines a cell; Virchow's famous 19th century

*Corresponding author

ADVANCES IN PROTEIN CHEMISTRY, Vol. 66

0065-3233/03 $35.00

statement, "It takes a cell to make a cell," is still true today. This forces modelers to embed proteins in their proper cellular environment (e.g., membranes and water) to study function and resolve mechanisms. Including the cellular environment is always computationally costly, but the dramatic increase of computing power makes this feasible today. In fact, the authors have already completed numerous molecular dynamics studies of proteins in membrane and aqueous environments, the most recent one including 327,000 atoms and spanning over 10 nanoseconds (ns). A list of large-scale simulations performed in the authors' group during the past decade is presented in Table I. The list illustrates the progress made: Ten years ago, to realize a then-heroic 27,000 atom 200 picosecond (ps). simulation required building a special-purpose computer and over a year of run time; today, simulation for the same period of a much larger system of 270,000 atoms can be accomplished on 128 processors of a widely available machine within a day.

The increase in computing power stems from commodity workstation clusters with 10 to 100 processors and from very large parallel machines with thousands of processors. Harnessing these resources effectively required great effort, to which purpose the authors' group has developed the molecular dynamics program NAMD. For the graphical analysis of the ensuing gigabytes and terabytes of data the group has developed the program VMD. These programs are widely used today since the increasing availability of protein structures has led to most biomedical researchers using structural information for the design and analysis of their experiments. The next section will briefly describe NAMD and VMD.

The subsequent sections will illustrate three exemplary simulations of large scale systems that are presently the subject of intense research. The wide interest in these systems and the exciting research challenges are the main reason why scientists feel compelled to model such large, cumbersome systems. The first example describes simulation of a membrane channel, aquaglyceroporin, that conducts water and small carbohydrates across the cell membrane. Placement of the protein into a membrane environment leads in this case to a 106,000 atom simulation. The second example is a study of ATP synthase, the renowned large protein complex that links an electro-mechanical motor, the F_0 subunit, with a chemo-mechanical motor, the F_1 subunit, through a rotatory elastic axle, the stalk. In this case the two subunits embedded in their appropriate membrane and aqueous environments comprise 112,000 and 327,000 atoms, respectively. The third example investigates a module of the extracellular matrix protein fibronectin, $FN\text{-}III_{10}$ that is involved in cell adhesion and motion. The protein has optimal mechanical elasticity and signals to cell surface receptors, integrins, the tension exerted on it. Atomic force microscopy studies suggest that the

TABLE I
LARGE-SCALE SIMULATIONS PERFORMED BY THE AUTHORS

System	Simulation	Size	Platform (# proc.)	Year, Ref.
bilayer of 200 lipids	NVE, NC	27K	60 node transputer	1993, [50]
membrane-water interface	FMA	32K	SGI Crimson	1995, [148]
binding of estrogen receptor to DNA	FMA, NVT	36K	HP cluster (8)	1997, [78]
apolipoprotein A-I	NVE	46K	HP cluster (4)	1997, [105]
calmodulin	NVE	33K	T3D (64)	1998, [141]
Rieske subunit motion in cytochrome bc_1 complex	SMD	91K	T3E (64)	1999, [60]
bacteriorhodopsin and purple membrane	PME, NVE, NPT	24K	Alpha cluster (8)	2001, [7]
MscL	PME, NPT	55K	T3E (64)	2001, [49]
fibronectin type III_{10}	PME, SMD	126K	Linux cluster (32)	2002, [40], 2003, here
aquaporin-1	PME, NPT	60K	T3E (64)	2001, [149]
BamH1 endonuclease binding to DNA	PME, NPT	65K	T3E (64)	2001, [87]
photosynthetic light harvesting system	PME, NPT	87K	SGI Origin 2000 (4)	2001, [23]
aquaglyceroporin	PME, NPT	106K	TCS (128)	2002, [128] 2003, here
rhodopsin	PME, NPT	40K	T3E (128)	2002, [118]
CD2-CD58 Complex	PME, SMD	91K, 104K	Linux Cluster (32)	2003, [8]
F_0-ATPase	PME, NPT, SMD	112K	T3E (128), TCS (256), Linux cluster (32)	2003, here
F_1-ATPase	PME, NPT, SMD	327K	Platinum (448), TCS (512)	2003, here

The size refers to the number of atoms simulated. The following abbreviations are used: SMD, Steered Molecular Dynamics; FMA, Fast Multipole Algorithm; NC, no cut-off; PME, Particle–Mesh Ewald. TCS and Platinum are computer clusters at the Pittsburgh (PSC) and Urbana (NCSA) supercomputing centers, respectively.

FN-III_{10} module, under the influence of mechanical stretching, unfold completely to a length of about 300 angstrom (Å). Simulation of such an unfolding event in which water plays a key role requires an appropriately long (370 Å) water box leading to a system of 130,000 atoms. The examples illustrate that the simulation size is mainly determined by the choice of environment, not by the actual protein simulated.

The functions of all three proteins investigated are mainly mechanical. To relate the observed mechanics to the architecture of proteins is the domain of so-called steered molecular dynamics simulations, contributing to the founding of the new field of mechanobiology, which studies the role of forces in cellular processes. Here forces appear as key ingredients of processes in cells: as "substrates" that drive reactions, as "products" of molecular motors, and as "signals." All modeling examples presented below employ steered molecular dynamics to accelerate the underlying processes, still requiring simulation times of 1 to 10 ns and longer.

II. Technology for Simulation and Visualization of Large Biomolecular Systems

With continuing increases in high-performance computing technology, the domain of biomolecular simulation has rapidly expanded from isolated proteins in solvent to biomolecular complexes in their native environments. Figure 1 compares simulation sizes of well-known biomolecular systems that have been studied by molecular dynamics (MD): bovine pancreatic trypsin inhibitor, the first protein simulated [91], a pair of DNA binding domains of the estrogen receptor complexed with DNA [10], and the F_1 fraction of ATP synthase (this paper). Today, simulations of biomolecular systems with over 100,000 atoms lasting 10 ns are becoming routine [128] and discoveries of the structures of large biomolecular complexes like the RNA polymerase II [45] or the 70S ribosome complex made from the 30S [21] and 50S [5] subunits forecast simulations involving a million atoms and more. Such simulations become feasible only

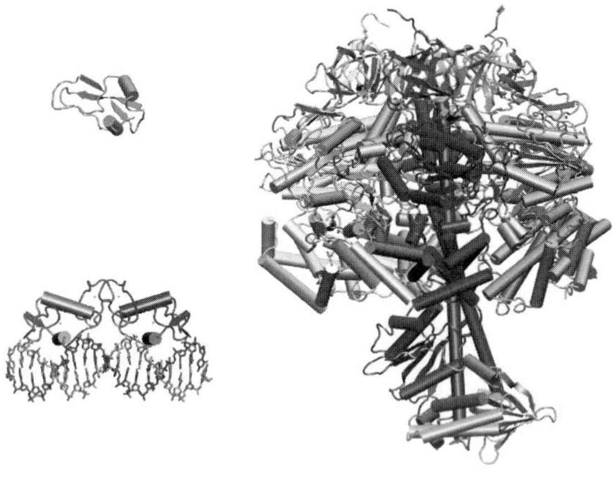

Fig. 1. Simulations have increased in size, from bovine pancreatic trypsin inhibitor (upper left, about 3K atoms), through the estrogen receptor (lower left, 36K atoms), to P_1-ATPase (right, 327K atoms). (Atom counts include solvent.) (See Color Insert.)

through the use of large parallel computers and the development of molecular dynamics programs that run efficiently on such machines. 100,000 atom systems can be simulated with such programs cost effectively on clusters of commodity workstations; for example, a cluster of 24 PCs running Linux and connected by a commodity-switched gigabit ethernet can be constructed for under $1000 per processor and utilized by the new generation of molecular dynamics programs at over 70% parallel efficiency, achieving a nanosecond or more of simulation time per week.

One such simulation program is NAMD [74], which has been developed specifically to simulate large biomolecular aggregates on both commodity clusters and massively parallel supercomputers. NAMD is primarily designed to work with CHARMM force field parameters [89]. Input files for NAMD can be generated using CHARMM [17], X-PLOR [18], or VMD [56]. NAMD has also been extended to read AMBER [139] and GROMACS [83] input file formats. NAMD running on 768 processors has enabled a record-breaking two million atom 5 ns simulation of the ribosome using an AMBER force field (Sanbonmatsu, private communication). The program NAMD has been used for the simulations described below.

NAMD provides the complete functionality needed to carry out standard molecular dynamics simulations. Efficient conjugate gradient minimization, fixed atoms, and harmonic restraints are among the methods available for the initial assembly of aggregates. The resulting system may be equilibrated and simulated in constant temperature and pressure ensembles, with periodic boundary conditions and full long-range electrostatics. NAMD supports steered molecular dynamics (SMD) [58, 61], a computational technique that applies external forces and accelerates reaction events to the nanosecond timescale, as illustrated in the examples below.

NAMD employs several methods to reduce the amount of calculations required by the MD algorithm. First, the complexity of the long-range electrostatic force evaluation is reduced from $O(N^2)$ to $O(N \log N)$ via the particle mesh Ewald (PME) [25] algorithm, which combines a cutoff direct calculation with an FFT-based mesh calculation. Second, nonbonded interactions are split into a smooth long-range component and a complementary short-range component; the period of long-range (electrostatic) evaluation is then increased from every 1 femtosecond (fs) to every 4 fs via an impulse-based symplectic multiple timestepping method. Finally, either the period of short-range nonbonded force evaluation may be increased to 2 fs via multiple timestepping or the timestep itself increased to 2 fs by constraining the lengths of the highest frequency bonds. Timestep limits are imposed by the requirement of energy conservation (larger values result in energy drift); the use of a symplectic integrator allows us to interpret energy conservation as an indicator of accuracy.

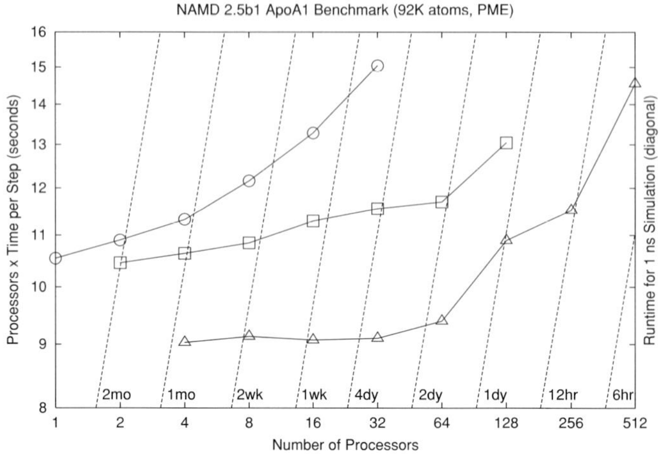

FIG. 2. Computer resources consumed per timestep by NAMD for a 92,000 atom benchmark simulation using 1 fs timesteps and a 12 Å cutoff with PME, long range Coulomb forces being updated every 4 steps. Smaller numbers indicate better performance; perfect linear scaling with processor number is a horizontal line. Dotted diagonal lines indicate the runtime required for 1 ns of simulation. Machines tested are: circles: our group's single-processor 1333 MHz Athlon Linux cluster with switched 100 Mbit Ethernet network; squares: National Center for Supercomputing Applications "Titan" dual-processor 800 MHz Itanium Linux cluster with Myrinet network; triangles: Pittsburgh Supercomputing Center "Lemieux" quad-processor 1000 MHz Alpha cluster with Quadrics network. One can recognize that NAMD running on 32 processors of our Linux cluster requires less than 1 week run time for 1 ns.

The efficiency of NAMD on today's parallel computers is illustrated in Fig. 2. This efficiency is achieved through use of the Charm ++/Converse parallel runtime system [75]. NAMD employs three basic techniques for this purpose: spatial domain decomposition, measurement-based load balancing, and message-driven execution.

In NAMD, the region of space occupied by the atoms in a simulation is divided into regularly sized regions, and these regions are assigned to processors to distribute the data evenly. Processors may be assigned multiple regions, or there may be more processors than regions to distribute. All required force calculations in the simulation, other than the FFT-based PME reciprocal sum, are between atoms in neighboring regions. This serves to minimize communication, which often limits scalability in other programs as the number of atoms in the simulation grows.

In order to distribute the time-consuming force calculations evenly among the available processors, the calculations are divided into sets, such as the set of interactions between atoms in a particular pair of regions.

These sets are initially assigned to processors so as to minimize the number of copies of each region's positions and forces that must be transmitted. During the initial stages of the simulation, the runtime of each set is measured and used to redistribute the calculations more evenly. This process is repeated at regular intervals to maintain performance during the entire simulation.

The order of force calculation in NAMD is not fixed, but rather each set is added to a queue as the required data from the regions it depends on becomes available. The data may arrive from another processor, or result from the integration of atomic coordinates and velocities locally. Calculations that generate results for atoms on other processors are assigned a higher priority. This mechanism allows NAMD to keep all processors doing useful work despite network latency or other minor delays.

Molecular modeling relies heavily on visualization technology. The simulations described below relied on the molecular visualization and analysis program VMD [56]. VMD was employed as a tool for building complete 3D structures of biomolecular aggregates. For example, using VMD, researchers embedded the structure of GlpF (see later) into the membrane and covered it with water. VMD permitted researchers to visualize the trajectories resulting from the simulations and to analyze the data in terms of root-mean square deviation (RMSD), the thickness of the membrane, and the flux of water molecules through channels. VMD permits the visualization of very large structures, for example, of a two million atom ribosome simulation (Sanbonmatsu, private communication). Systems of 100,000 atoms can be readily viewed with VMD on a modern desktop or laptop computer.

III. Aquaporins–Membrane Water Channels

In mammalian cells, driven by osmotic forces, water can pass indiscriminately through the cell lipid membrane, but the conduction is too slow for an active regulation of water homeostasis. Membrane water channels that permit a rapid flux had been previously postulated on these grounds and were identified by Agre and coworkers [110], who termed them *aquaporins* (AQPs). This discovery promised insight into the fundamental physiology of water balance and the pathophysiology of water balance disorders. Soon after, it became evident that AQPs were abundantly present in all kingdoms of life, including mammals, amphibia, insects, plants, and bacteria [2, 12, 53]. In the human genome, eleven different genes have been identified which encode AQPs in organs that are either directly involved in handling large volumes of water, such as the kidneys, or need to continuously maintain a precise level of water, such as the brain. Impaired function of AQPs has been related to many pathophysiological situations such as nephrogenic diabetes insipidus and congenital cataract [2, 12, 29, 82].

AQPs are open channels by design; no gating mechanism has been reported so far for water. However, the water pore in most AQPs can be blocked by mercurial compounds [132]. AQPs are highly selective channels. Despite their high permeability to water, they completely exclude protons, a puzzling peculiarity since protons can be readily transported through a hydrogen bonded chain of water molecules according to the Grotthuss mechanism [48]. Under physiological conditions, the water pores of AQPs are impermeable to ions and other charged species [3, 119, 145]. The exclusion of ions, in particular protons, is central to the biological function of AQPs, enabling them to transport high volumes of water without dissipating the electrochemical potential of cell membranes.

A subfamily of AQPs, known as *aquaglyceroporins*, permit a highly stereoselective passage of small sugar molecules, such as glycerol [13, 51]. The *E. coli* glycerol uptake facilitator, GlpF, is a prominent member of the subfamily, partly because it is the first AQP for which a high resolution structure was solved by x-ray crystallography [38]. GlpF provides the cell with efficient access to extracellular carbon sources and is particularly important under low sugar concentration conditions [116].

The architecture of AQPs was elucidated first by electron microscopy, revealing that AQPs form tetramers in the membrane [27, 135]. Further studies showed that the protein is composed of four functionally independent aqueous pores [70, 124]. The folded monomer is a right-handed helix bundle of six transmembrane α-helices, and includes two reentrant loops that meet each other at the center of the channel [132, 134]. A schematic presentation of an AQP monomer is provided in Fig. 3. The core of the channel is formed by the reentrant loops comprised of two α-helices (cylinders in Fig. 3) that protrude into the protein and return in the form of an inverted helix. The latter exposes the backbone atoms to the interior of the channel that guide single file water and glycerol through the channel [65]. The two half helices are held together by multiple hydrogen bonds between the two highly conserved NPA motifs (Asn-Pro-Ala; see Fig. 3) [115, 140].

At the present time, crystallographically solved high resolution atomic structures are available only for two members of the AQP family, aquaporin-1 (AQP1) [126] and GlpF [38]. As described later, MD simulations of these structures have significantly contributed to our understanding of the molecular basis of function and selectivity in this important family of membrane channels.

FIG. 3. Architecture of an AQP channel. Six transmembrane α-helices form a bundle inside the membrane. Two reentrant loops (shown in darker gray) then fold into the channel and complete the interior of the channel. The loops enter the channel in helical forms, meet each other at the center of the channel, and turn back in an energetically unfavorable extended form. The two conserved NPA motifs, which are located in the middle of the loops, provide multiple hydrogen bonds that keep the loops together. The pathway of substrate transport is along the nonhelical halves of the reentrant loops (light area).

For the simulations described in this section, the system contains a membrane-embedded AQP tetramer solvated by two layers of water molecules, representing the functional form of the channel as present in biological membrane. This resulted in system sizes of more than 100,000 atoms (Fig. 4). Due to the biological function of the channel (water transport), water molecules needed to be explicitly treated in the model. The time step of the simulations was 1 fs. All calculations described in this section lasted a few nanoseconds. This was necessary due to the slow relaxation of the membrane-embedded system. Moreover, since water permeation occurs on a nanosecond time scale in AQPs, multinanosecond simulations were also needed for sufficient sampling of this event. Using the program NAMD described above, it took about 10 days to complete 1 ns on a 32 (1.1 GHz Athlon) processor cluster. The same calculation

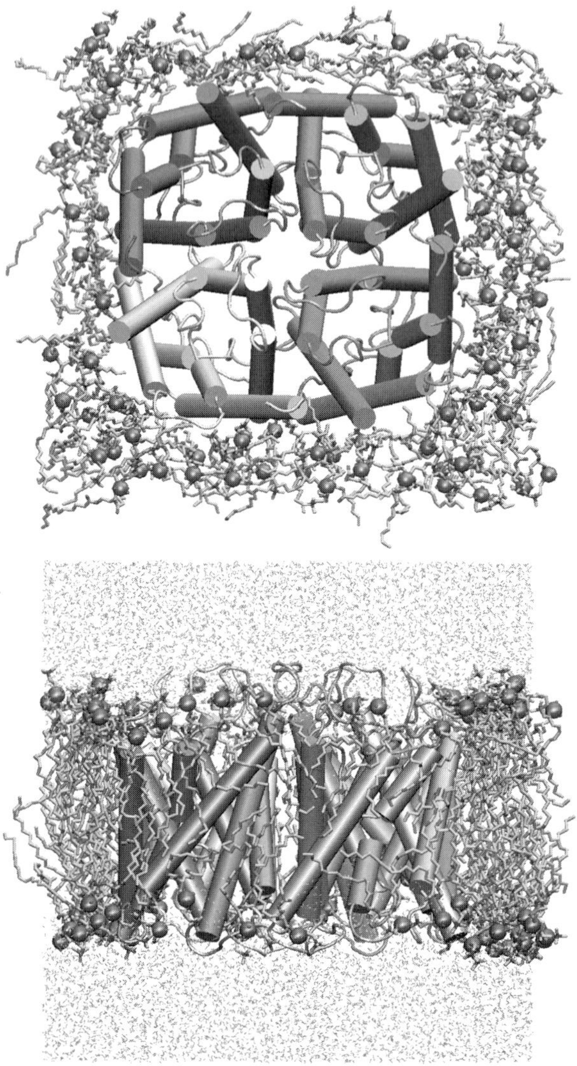

FIG. 4. Top and side views of the tetrameric model of an AQP modeled in a fully hydrated patch of lipid bilayer (water molecules are not shown in the top view) [151]. The monomers are shown using cartoon representations and in different colors. In the side view (bottom), the positions of phosphate groups of lipids are specified using vdW representation. (See Color Insert.)

took 1.2 days on 128 (1 GHz Alpha) processors of the Lemieux cluster at the Pittsburgh Supercomputing Center.

In order to best describe aquaporins under physiological conditions, the following methodologies were applied. First of all, although the system contains over 100,000 atoms, it is still much smaller than macroscopic systems in the cell and, hence, may suffer from boundary effects, such as artificial surface tension. The solution to this problem is to adopt periodic boundary conditions (PBC), where the basic system is a unit cell, but the actual system is infinite and periodic, obtained by copying and translating the unit cell in three dimensions. Applying PBC is especially crucial for systems containing a membrane, in which the membrane is essentially an infinite plane, separating water molecules on its two sides. Langevin Dynamics was used to keep each part of the system at a desired constant temperature (310K). The Langevin piston method was used to maintain the pressure of the system at 1 atmosphere (atm). Such conditions (often referred to as NPT) represent a physiological environment. An efficient algorithm, PME, was used to calculate electrostatic forces without cut-off.

The first attempts of simulating AQPs [26, 149] were based on the medium-resolution electron microscopy structures of AQP1 [95, 113]. AQP1 was modeled by us in its tetrameric form embedded in a solvated palmitoyl-oleoyl-phosphatidyl-choline (POPC) lipid bilayer [149]. Nanosecond simulation [149] of the system for the first time revealed the spontaneous formation of a single file of water inside the channel, a very important feature for the selectivity of AQPs, as will be described later [128]. Examination of critical regions of the channel, particularly the highly conserved NPA motifs at the center, however, showed that the structures were not stable enough for full atomic simulations. The instability was quite evident from the disruption of the hydrogen bonds between the two NPA motifs (Fig. 5) during the simulation [149]. Similar conclusions about the structural details inside the channel were made by other researchers [26] who reported a refined model of AQP1 on the basis of the high-resolution structure of GlpF [38], and were confirmed by comparison with a high-resolution structure of AQP1 [126], which was solved later.

Due to the above-mentioned problems in the simulation of AQP1 models, the focus of MD simulations was shifted from AQP1 to GlpF, as soon as the high-resolution (2.2 Å) crystallographic structure of GlpF [38] became available. GlpF, an aquaglyceroporin, was crystallized in the presence of a high concentration of glycerol, and therefore included three glycerol molecules in the pore region of each monomer [38]. We simulated [65] the tetrameric model of GlpF embedded in a palmitoyloleoyl-phosphatidyl-ethanolamine (POPE) bilayer, which is proper for an *E. coli*

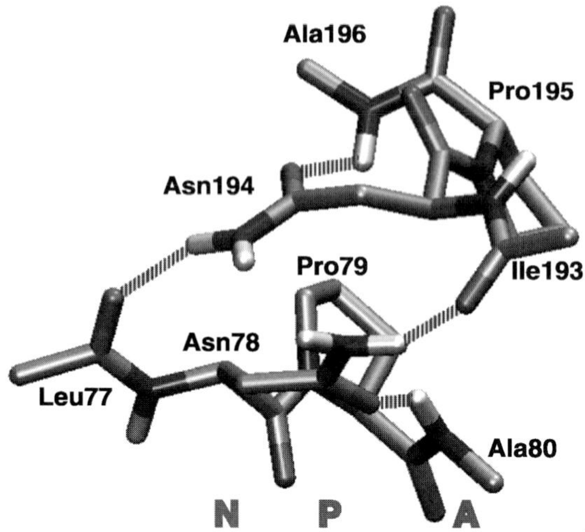

FIG. 5. Hydrogen bonds between the two NPA motifs are essential for the stability of the structure and the function of AQPs. Due to two stable hydrogen bonds between the amido group of the asparagine side chain of each NPA motif with neighboring side chains, one of the amido hydrogens is restrained to be fully exposed toward the interior of the channel, where it forms hydrogen bonds with the permeating substrate. Hydrogen bonds between the two NPA motifs are also important for the stability of the two reentrant loops. (See Color Insert.)

cell membrane. The simulation included more than 106,000 atoms and extended over several nanoseconds [65].

During the simulations, thermal fluctuations promoted significant movements of individual glycerol molecules inside the four channels and allowed us to describe the complete conduction pathway through the channel, a finding that for the first time explained the presence of peculiar secondary structure elements in AQPs [65]. The discovered conduction pathway is mainly formed by the backbone carbonyl groups of the two non-helical parts of the reentrant loops. Together with a highly conserved pair of asparagines at the center of the channel (the NPA motifs) and an arginine located at the extracellular half channel, the carbonyl groups provide hydrogen binding sites that guide the substrate through an otherwise hydrophobic channel in a stepwise fashion [65]. In a helical conformation, these carbonyl groups are mainly involved in intrahelical hydrogen bonds stabilizing the helical arrangement and would not be fully accessible by the substrate permeating the channel. Therefore, the protein has to adopt a non-helical form, although energetically

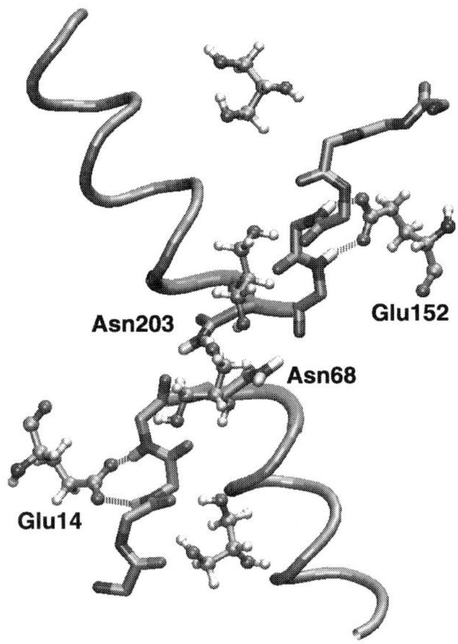

FIG. 6. The curve-linear pathway of substrate conduction in AQPs formed by backbone oxygen atoms of two reentrant loops. Carbonyl oxygens are shown in red. The peculiar secondary structure of the non-helical halves of the loops is stabilized by hydrogen bonds with Glu14 and Glu152. Also shown are the location of the NPA motifs (green), the α-helical parts of the reentrant loops (blue), and the side-chains of Asn68 and Asn203 at the center of the channel. Snapshots of the four glycerol positions (shown in ball and stick representation) from the simulations are included to illustrate the conduction pathway. (See Color Insert.)

unfavorable, to optimize its functionality. It is interesting to note that the non-helical arrangement of these loops is stabilized through hydrogen bonds between the backbone amino groups of each loop and one conserved glutamate (Fig. 6). The glutamate residues are the only burried glutamates negatively charged residues in the transmembrane region of the protein, a fact that clearly reflects their important role in the structure and function of the channel. Water molecules accompanying glycerol in the channel proved to be very important for the conduction process, lubricating the movement of glycerol through competition for the hydrogen binding sites [65].

In order to describe the energetics of glycerol conduction and to study the stereoselectivity of the channel, we performed a series of steered MD (SMD) simulations on the tetrameric GlpF model described [64]. In each run, a steering force was applied to individual glycerol molecules (one per monomer) to accelerate the trans-channel conduction in either direction and for two orientations of glycerol (i.e., head first and tail first). Conduction of glycerol occurs on time scales which are not accessible to equilibrium MD simulations, and therefore one needs to accelerate the event via external forces. By employing a simulation scheme that kept the system close to equilibrium and utilizing Jarzynski's identity [62, 63] linking free energy and irreversible work, we reconstructed the potential of mean force (PMF) from an ensemble of trajectories [64], as shown in Fig. 7. The resulting free energy profile, one of the first ones determined for a biological system, captures major features of the glycerol-channel interaction. The positions of crystallographically observed glycerol molecules [38] closely correspond to wells in the energy profile. Several additional minima, due to multiple hydrogen bonds between glycerol and the channel, were also discovered. The largest barrier against the substrate conduction [64] was found to be located in a region that was suggested to function as the selectivity filter [38].

The constructed PMF (see Fig. 7) exhibits a clear asymmetry. Comparing the cytoplasmic and periplasmic vestibules of the channel, the periplasmic mouth of the channel is characterized by a deep potential well. Although G1pF is a passive channel and in principle, nutrient molecules can be transported either way across the membrane, the asymmetry reflected by PMF might be of importance for the dominant physiological role of the channel, namely uptake of glycerol from the periplasmic space. After entering the cytoplasmic region, glycerol becomes phosphorylated

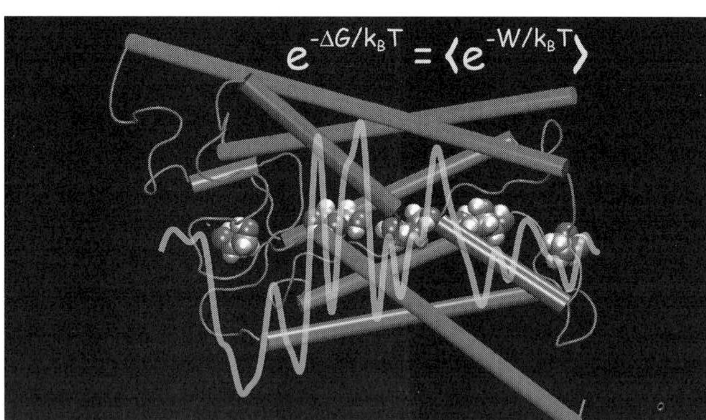

FIG. 7. Potential of mean force (PMF) along the pathway of glycerol conduction in GlpF, constructed from SMD simulations [65]. The PMF is superimposed on a GlpF channel, in which several positions adopted by glycerol along its transport are highlighted. (See Color Insert.)

and cannot exit the cell. Interestingly, the asymmetric PMF corresponds to the asymmetric shape of the channel, when one compares the periplasmic and cytosolic segments of the channel. The significantly bigger protrusion of GlpF to the periplasmic region can provide an attractive site for nutrient molecules, which are usually not abundant. In agreement with this suggestion, it is noteworthy that the asymmetry of the structure is much less pronounced in AQP1, a pure water channel, where substrate (water) is always present in the environment at high concentration.

In another set of simulations [128], we removed all glycerol molecules from GlpF and investigated water permeation through the channel. An extensive set of MD simulations (a total of 15 ns) was performed on the wild-type GlpF and on a designed mutant (see later). During the simulations, single files of water were formed (in 100–200 ps) and maintained in the pore regions of the four channels. Several full permeation events through the 20 Å pore region of the channels were observed in a 4 ns time frame [128]. The calculated conduction rate of native GlpF (1.1 H_2O/ns/channel), which is very close to the value reported by other researchers [27], compares satisfactorily with the experimentally deduced flux for GlpF, namely, 0.5×10^9 H_2O/s [12, 14, 53].

We also designed and prepared a GlpF mutant in which both the size and polarity of the narrowest part of the channel (the selectivity filter) were increased by mutation of two amino acids. The resultant mutant (W48F/F200T) was found to have an increased permeation rate both in measurements (25%) and in simulations (38%) [128], reflecting a surprisingly close agreement between experiment and theory. In both native and mutant species, water occupancy along the channel axis inferred from the simulations was almost exactly matching the electron density of the crystal structures [128].

Detailed analysis of the dynamics of water in the channel lumen in the simulations answered a long-standing puzzle in the function of AQPs. The simulations showed how AQPs employ a global tuning mechanism which permits the fast transport of water, but prevents water-mediated proton transfer. The electric field of the protein dictates a peculiar configuration of water molecules inside the channel that was not reported in any other system before [128]. Starting from the NPA center, water molecules are oriented in opposite directions in the two halves of the channel, with their hydrogen atoms pointing toward the exits as shown in Fig. 8. This arrangement of water molecules in the channel, the wing configuration, is in contrast to the proton wires [16, 108] formed in other water channels, such as gramicidin A [107, 109], synthetic channels [111], and carbon nanotubes [55], and completely blocks proton conduction [128].

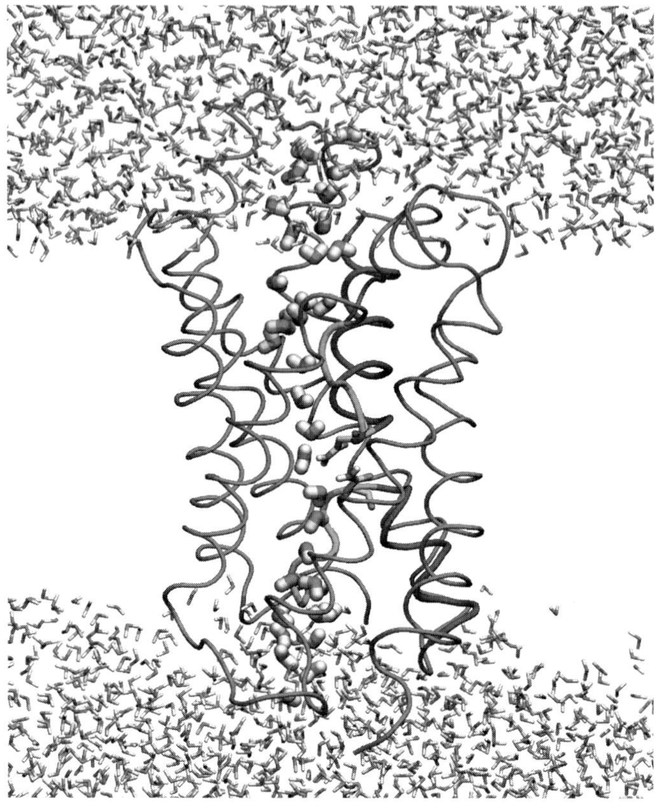

FIG. 8. A snapshot from MD simulation of GlpF. Only one monomer is shown. A single file of water forms in the channel during the simulation. The orientation of water molecules in the single file is reversed in the two halves of the channel due to the electric field of the protein. This bipolar configuration of water (the wing configuration) prevents proton conduction [129]. (See Color Insert.)

As described above, conventional MD simulations can be used to describe water molecules passing through the channel via diffusion, where the net transmembrane water flux through the channel is zero. While the random diffusion of water is always happening, a physiologically more relevant event—measured experimentally—is the net flow of water induced by an osmotic or hydrostatic pressure gradient across the membrane. In fact, the osmotic permeability, p_f, an important quantity for a water channel (defined as the ratio of net water flux to the osmotic pressure difference), cannot be directly obtained by observing water diffusion in equilibrium MD simulations. In order to study such events,

MD simulations are needed in which bulk water located on the two sides of the membrane experience different osmotic or hydrostatic pressures. This is a method that seems to be difficult to implement since under PBC conditions, the bulk water on the two sides of the membrane is actually connected.

We have developed a novel method to overcome this difficulty, which can be used to quantify the osmotic permeability p_f of water channels by MD simulations [150]. In this technique, through application of external forces to every bulk water molecule, one generates a hydrostatic pressure gradient across the membrane that promotes water transport in one direction. The pressure difference can be easily controlled through adjustment of the force applied on individual water molecules. Since the membrane is experiencing different hydrostatic pressures, there is a net force on it from water, which will cause it to move. Constraints can be used to keep the membrane in position, or alternatively, counter forces can be applied to the membrane to make the net force of the whole system zero. This method can be used to generate any magnitude of pressure difference across the membrane, which enables us to observe a net water flux through the channel and measure the channel's osmotic permeability from the simulation. The water flux is measured by counting the water molecules passing through the channel during a certain time. From the ratio of water flux and pressure difference we could determine the osmotic permeability p_f. Simulations performed on GlpF revealed a linear relationship between net flux and applied pressure gradient [150].

IV. Energy Conversion in ATP Synthase

Living cells depend on an efficient transformation of the energy derived from light and foodstuff into the chemical energy of adenosine triphosphate (ATP), the universal energy carrier. The enzyme that facilitates this type of energy conversion in bacteria, chloroplast, and mitochondria is ATP synthase, a complex of two molecular motors mechanically coupled by a common central stalk, as shown in Fig. 9. The membrane unit F_0 converts the transmembrane electrochemical potential into mechanical energy that drives the central stalk rotation inside the solvent exposed F_1 unit. The rotation causes cyclic conformational changes in the F_1 unit catalytic sites, which drive ATP synthesis (binding change mechanism [15]). ATP synthase can also operate in the reverse direction, utilizing the energy released in the process of ATP hydrolysis to pump protons across the membrane.

The F_0 unit consists of three types of subunits: subunit *a*, believed to mediate proton translocation; a dimer of *b* subunits, which extends from

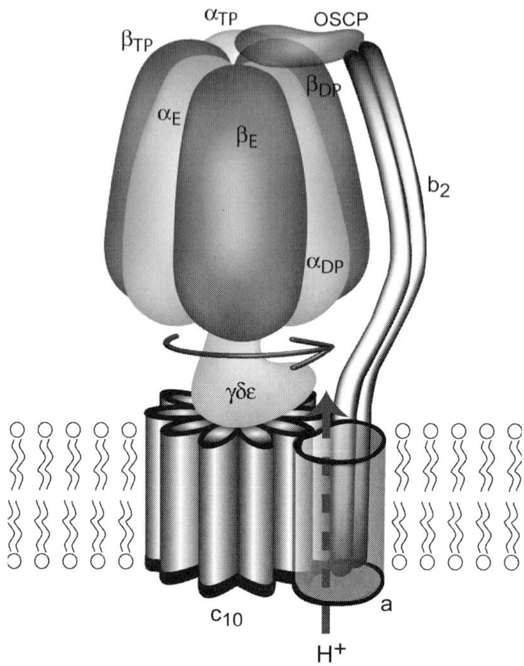

FIG. 9. Schematic view of F_1F_0 ATP synthase. During synthesis, the central stalk ($\gamma\delta\varepsilon$) and the F_0 ring (c_{10}) rotate relative to the rest of the system (direction indicated by the curved arrow). The rotation is driven by a flow of protons (H^+) across the membrane, as indicated by the vectorial arrow, and causes synthesis of ATP in the catalytic sites located at the interface between neighboring α and β subunits. The bovine mitochondrial ATP synthase naming scheme is used to label the subunits.

the membrane to the solvent, mechanically connecting F_0 and F_1, and a ring-like oligomer of *c* subunits [9, 44, 68, 130, 131]. The number of *c* subunits in the oligomer is believed to vary between 9 and 14 for different species; for *E. coli,* which has the simplest F_0 structure, the number was found to be 10 [66]. A single *c* subunit folds in the membrane in a hairpin-like structure of two transmembrane α-helices, the inner one and the outer one, connected by a short polar loop [147]; the loops form a mechanical contact with the central stalk (the γ, ε, and, in the bovine mitochondrial species, δ subunits of the F_1 unit). Residue Asp61 of the *c* subunits, located in the outer transmembrane helix, is believed to be a proton binding site. Its protonation and deprotonation has been proposed to drive rotation of the c_{10} oligomer relative to the ab_2 subunit complex, the motion being coupled to the central stalk rotation in the F_1 motor. A single 360° rotation of the c_{10} complex and the stalk involves ten protons translocating across the membrane, releasing free energy sufficient to produce three ATP molecules. This process is considered further below.

Crystallographic structures of F_1 from bovine mitochondria [1, 43, 92] show three catalytic sites pseudo-symmetrically arranged around the assymetric central stalk, each site in a different configuration. The central stalk of F_1 comprises an α-helical coiled-coil running along most of its length, with a wide base at one end where it contacts F_0 (Fig. 9). The central stalk is surrounded by three α/β subunit pairs in a pseudo-symmetric configuration, the pairs forming the three catalytic sites. F_0 driven rotation of the F_1 central stalk relative to the α/β subunits is thought to drive the subunits, and thus the catalytic sites, through at least three conformational states per rotation, with each site synthesizing one ATP molecule per rotation [15] at a position 100 Å away from where the torque is applied by F_0. The central stalk can spin like a rotor within the $(\alpha\beta)_3$ stator since a secondary, non-rotating stalk fixes $(\alpha\beta)_3$ by connecting it to the F_0 stator. F_1 has been directly visualized performing the reverse of synthesis; in other words, when ATP is hydrolyzed by F_1, the central stalk is observed to spin [90, 97, 144], with changes in $(\alpha\beta)_3$ connected with cyclic ATP hydrolysis presumably producing the required torque. Conformation changes in the α/β pairs appear to be cooperative, the coupling transmitted both between neighboring pairs and through the central stalk [15, 39, 73, 114]. Mutation experiments have identified residues important to α/β/stalk interactions [39, 73] and several schemes for F_1 synthase subunit interactions have been proposed [20, 101, 114, 138]. The rotation of the F_1 central stalk and its coupling to changes in the $(\alpha\beta)_3$ binding sites is considered further below.

A. *Torque Generation in F_0*

While the rotary catalysis mechanism of the F_1 unit operation has been recently demonstrated in a series of spectacular single molecule experiments [54, 76, 90, 96, 97, 144], much less is known about atomic scale events involved in the F_0 motor function. How is proton translocation across the membrane coupled with the mechanical rotation of the *c* ring against the load imposed by the F_1 unit? What protein groups, and possibly water atoms, mediate the proton path to and from the essential Asp61 residues? What domain motions occur in the protein, and how are they coupled to the protonation and deprotonation of the key protein residues? How many protons are needed for a synthesis of one ATP molecule, and what makes the F_0 operation almost 100% efficient? How does the F_0 unit operation depend on the environment, for example, pH? These questions have to be answered before any quantitative understanding of the protein motor operation can be achieved.

The principal problem that limits understanding of F_0 operation on the atomic scale is lack of a sufficiently complete protein structure. The only available crystallographic structure obtained for the mitochondrial F_{1}-c_{10} complex at 3.9Å resolution [125] does not include subunits *a* and *b*, which are critical for the protein function. Several structural models have been developed on the basis of NMR experiments with individual subunits of F_0 in polar solvents and detergents, disulfide cross-linking, scanning mutagenesis, and analysis of suppressor mutations [32, 47, 68, 72, 112]. It is not clear, however, how far the protein structure in detergent is similar to that in membranes [47]. In this regard, all-atom MD simulations provide a unique opportunity to computationally investigate the proposed models on the atomic scale, estimate their stability, identify functionally relevant domain motions, and provide parameters necessary for modeling the protein motor operation on the physiological (millisecond) time scale.

To perform an MD study, a minimal structural model of the F_0 motor was built using the available experimental data on subunits from *E. coli* ATP synthase, which is known to have the simplest structure [34, 67]. The model included subunit *a* and a 10-mer of subunits *c*, about 14,300 atoms total. Although most functionally relevant events are assumed to occur at the interface between subunit *a* and two nearest subunits *c*, all 10 subunits *c* were included to assure the oligomer stability in the membrane/solvent environment. Since it is still unclear whether subunits *b* are involved in proton translocation[1] [31], and their exact position in F_0 is not known, they were not included in the model. The structure of the c_{10} complex was identical to the one provided by Fillingame [30]. The only available structure of subunit *a*, which had been obtained from an ac_{12} complex [112], was docked to the c_{10} structure and then modeled. The outer transmembrane helices (TMH) of the four subunits *c* of the ac_{12} complex closest to subunit *a* were aligned to the outer TMH's of four sequential subunits *c* in the c_{10} complex; then subunit *a* was merged to the latter, forming an ac_{10} structure. Since the structure of subunit *a* was determined in a polar environment, whereas the *c* oligomer structure was obtained in a detergent mimicking the hydrophobic environment inside biological

[1] Reported experimental data remain controversial: Fillingame and co-workers [52] found that proton translocation (but not active transport) is possible if subunits *b* (then named Psi) are missing, whereas Takeyama [129] and Monticello [94] concluded that they are required for any proton transport in F_0. Since all these studies had been done before the first atomistic structure [1] was available, the role of subunits *b* in proton conduction is not clear yet.

membranes, obtaining the correct structure of the interface between subunits *a* and *c* required additional modeling. The system underwent 10,000 minimization steps followed by 130 ps equilibration *in vacuo* at 4K with all subunits *c* constrained. While the temperature was too low for dihedral angle rotation, preventing substantial re-orientation of any relevant structural group, the equilibration allowed the helices in subunit *a* to assume the conformation that fits the structure of the *c* subunits complex.

To correctly model the operation of F_o, the protein was embedded in a membrane/solvent environment, which stabilized the protein structure. The membrane was modeled as a 130 × 130 Å patch of a pre-equilibrated POPE bilayer. After the protein had been embedded in the bilayer, lipid molecules that overlapped with the protein were removed. The protein-lipid complex was then solvated with TIP3 water using the *solvate* plugin of program VMD. Although an implicit solvent model could provide a better performance of the MD simulations, using TIP3 water was necessary to observe events in which individual water molecules may participate, for example, interacting with solvent exposed protein residues or possible spontaneous formation of the proton inlet or outlet half-channels. The water volume was large enough to provide the minimum distance between the protein and its periodic images of over 30Å, which was sufficient to screen out the electrostatic interactions. To electrically neutralize the system, which was required for simulations with periodic boundary conditions, chlorine and sodium ions were added using VMD extended through custom Tcl scripts. The resulting system, which included about 112,000 atoms, underwent minimization and equilibration for about 3 nanoseconds in the Charmm27 forcefield [69].

The principal goal of the work was to understand microscopic events that couple proton translocation across the membrane with the mechanical rotation of the *c* subunit oligomer. A single cycle of the F_0 operation involves processes that take place on dramatically different time scales: protonation/deprotonation of the key protein residues occurs about a billion times as fast as a revolution of the *c* oligomer. Neither the first nor the second process can be addressed directly by state of the art MD simulations: it is not feasible to simulate proton transfer over distances as large as about 15Å (a half of the membrane hydrophobic layer thickness), and the millisecond time scale of the mechanical rotation is far beyond reach of conventional MD methods. Given the limitation, can simulations help us learn anything about the microscopic mechanisms of the F_0 operation? We believe that this is possible, even though it is difficult. To bridge the time scale gap, we employed a multiscale approach, assuming that F_0 operates as a molecular ratchet in which the rotor shows rotational diffusion in a potential biased by the proton motive force [71]. Given that ratchets are known to be well-described by stochastic models [93], we developed such a model to extend the investigation of F_0 to the physiological (millisecond) time scale. In addition to the model, a series of MD simulations was performed to obtain quantitative parameters and test different scenarios of key events in F_0. Unlike the earlier-idealized one-dimensional ratchet model suggested in [32], which essentially describes a generic protein motor, our approach directly relates the atomistic structure of F_0 to the stochastic model. As we show below, using the atomistic structure not only leads to a more realistic dynamical description, but it also provides insight into the relationship between the protein structural features and its function.

The established overall view of the F_0 operation assumes that the c_{10} complex, being in contact with the central stalk of the F_1 motor, rotates relative to the ab_2 complex. Each subunit c has a proton binding site, Asp61, which is located in the middle of the membrane hydrophobic layer. Given that the membrane-spanning part of the c subunits is formed almost entirely by hydrophobic residues, which cannot mediate proton pathways, the binding site can only change its protonation when it is located against TMH4 of subunit a. The latter includes several polar groups, which are thought to mediate two proton half-channels. The inlet channel leads from the proton-rich periplasm to the center of the membrane, and the outlet channel connects the membrane center to the cytoplasm. The binding site in subunits c can assume the protonated (neutral) or deprotonated (charged) form depending on the environment, which may affect the effective pK_a value. The subunit c located at the interface with subunit a, which is formed by polar protein residues that include positively charged Arg210, assumes the deprotonated state. However, for the c_{10} complex to rotate, the binding site needs to be protonated before it leaves the interface: The energy penalty for exposing an electrically charged group to the hydrophobic membrane environment is too high. Therefore, a proton from the periplasm travels via the inlet half-channel and protonates the binding site. By a mechanism not yet understood, the rotation of the c_{10} complex brings, after an almost 360° turn, that binding site close to the a subunit again, causing it to release a proton, which travels via the outlet half-channel to the cytoplasm. Each step of the proposed scenario needs to be investigated in detail; here, we focus on two events: rotation of the entire c_{10} oligomer relative to the ab_2 complex, and rotation of individual membrane-spanning TMH's in the protein subunits. A more complete description of our simulations and model can be found elsewhere [4].

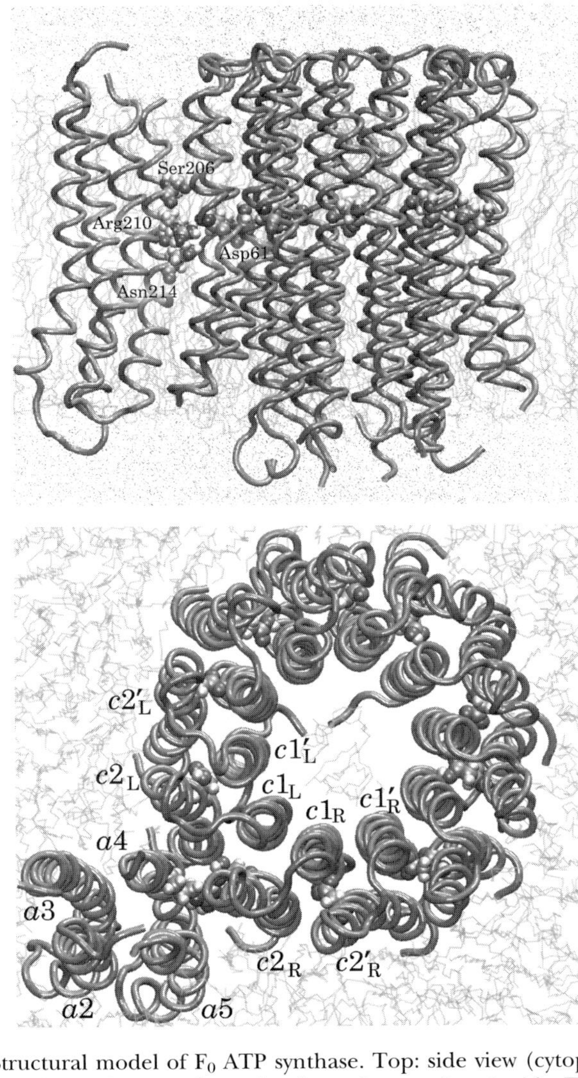

FIG. 10. Structural model of F_0 ATP synthase. Top: side view (cytoplasmic side up, periplasmic side down); bottom: front view from the cytoplasmic side. The F_1 unit is not shown for clarity. (See Color Insert.)

A series of steered MD simulations was performed to examine the feasibility of rotation of the c_{10} complex relative to subunit a in the lipid bilayer. To avoid distortion of the protein structure in the process of rotation on the nanosecond time scale (about 10^5 times faster than the physiological rotation rate), forces were applied not only to the polar loops of subunits *c*, which form contact with the central stalk, but to all 3170 backbone atoms of all *c* subunits. The backbone atoms of TMH's 2, 3, and 5 of subunit *a* were restrained to prevent subunit *a* from being dragged along with lipid molecules surrounding the protein. TMH4 of subunit *a*, which forms the interface with subunits *c*, was not restrained. The rotation axis was normal to the membrane plane, and the center of rotation was at the center of mass of the *c* subunit oligomer. The magnitude of the force acting on each atom was proportional to the distance between the atom and the rotation axis. The simulations were performed in the NVT ensemble. To control the system temperature, dissipate the heat produced by the applied forces, and additionally reduce distortion of the protein structure, NAMD Langevin forces were applied to all heavy atoms in the system. To allow a full revolution of the c_{10} complex, Asp61 residues in all subunits *c* were protonated.

The angle of the c_{10} oligomer rotation as a function of simulation time *t* is shown in Fig. 11 for several values of applied torque. The angle was calculated by averaging the rotation angles for each *c* subunit, which were computed using the subunit center of mass positions at time *t* and in the beginning of the MD simulations. For all torque values, the protein structure remained stable, and the angular velocity increased with the torque. Interestingly, the rotation substantially slowed down for torques of less

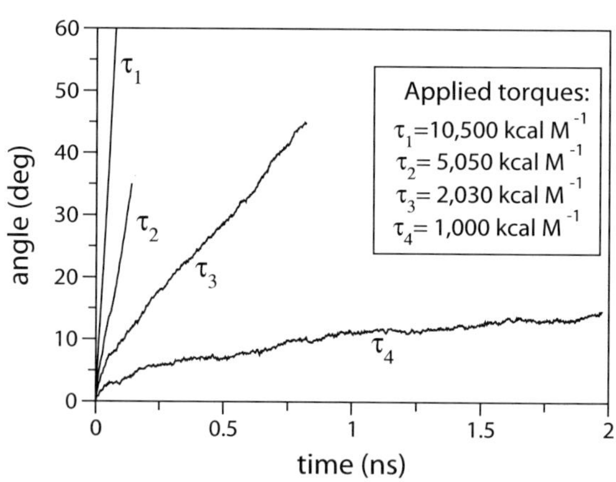

FIG. 11. Forced rotation of the c_{10} oligomer in the ac_{10} protein complex in membrane/solvent. The system included about 112,000 atoms. The simulations were performed using the program NAMD2 with full electrostatics and Langevin thermostat damping of 5 ps^{-1}.

than 2,000 kcal/M, indicating a transition to a high friction regime at lower angular velocities.

Another series of MD simulations addressed the feasibility of rotation of individual TMH's in subunit *c* that had been proposed by Fillingame and co-workers [35–37]. NMR and cross-linking data showed that the position of Asp61 in the outer TMH of the *c* subunits depends on its protonation: In the protonated state, Asp61 is hidden in the hydrophobic core of subunit *c*, whereas in the deprotonated state, Asp61 extends from the outer TMH towards Arg210 of subunit *a* [68, 72, 112]. In order to release (or bind) a proton, Asp61 needs to approach the terminus of one of the proton half-channels, which are believed to be residues Ser206 and Asn214 in subunit *a*; both residues are located at the interface with subunits *c* next to the positive Arg210 residue of subunit *a* [35]. It was therefore suggested that a rotation of the outer TMH of subunit *c* ($c2_L$) around its axis by about 120° (or 240° in the other direction) is necessary to bring Asp61 to the interface before it can release a proton (note that, at the interface, the proximity of Arg210 also facilitates deprotonation of Asp61) [35]. The scenario proposed in [35] also included rotations of TMH4 of subunit *a* and the outer TMH of the other nearest subunit *c* ($c2_R$), which were thought to bring Asp61 close to the terminus of the other half-channel, allowing its re-protonation necessary for the c_{10} complex to rotate further. While the idea of individual TMH rotation successfully explains the observed conformational changes in subunit *c* upon protonation or deprotonation of Asp61, the particular scenario raises a number of questions. Most importantly, deprotonated Asp61 is likely to form a salt bridge with Arg210, which could effectively prevent rotation of either TMH4 of subunit *a* or the outer TMH of the *c* subunit.

To find out if the rotation of the c_{10} complex can occur, a 1 ns steered rotation of the c_{10} oligomer with one or two deprotonated Asp61 residues was performed. We observed formation of the salt bridge between Arg210 and Asp61, which dragged TMH4 of subunit *a* along with the c_{10} complex, breaking apart the four-helix bundle structure of the subunit. To preserve the structure in another 1 ns simulation, all TMH's of subunit *a* were restrained; that, however, resulted in unwinding the outer TMH of the *c* subunit. Therefore, simple rotation of the c_{10} complex appears unlikely to break the salt bridge; instead, the rotation leads to significant distortion of the subunit's structure. These results suggest a concerted rotation of the c_{10} complex and the $c2_R$ TMH. To investigate this possibility, a special algorithm for rotating the TMH was used. The $c2_R$ TMH has a kink caused by the Pro64 residue located right below Asp61; the helix symmetry axis changes its direction at the kink. Therefore, a rotation of the entire helix around a single axis requires application of an artificially high torque because of the steric hindrance. To minimize the resistance, the rotation axes were assigned individually to every residue of the helix. As in the earlier simulations, the forces were applied to every heavy atom, and the force magnitude was proportional to the distance from the atom to the corresponding rotation axis. Driven by these forces, the helix rotated entirely within its reptation tube formed by the surrounding proteins and lipids. To the best of our knowledge, this type of protein domain motion has not been investigated before.

In the simulation of the combined rotation of the c_{10} oligomer and the $c2_R$ helix, the two rotations were performed in several steps for technical reasons. First, the c_{10} oligomer was rotated clockwise by about 37 degrees in a 1 ns simulation. After that, the $c2_R$ helix was rotated counterclockwise by 220°. The salt bridge stayed intact, and no significant distortions of the structure were observed. Then, the nearest binding site clockwise from $c2_R$ (which belongs to the $c2'_R$ helix, see Figs 10 and 12) was deprotonated. The simulation of the clockwise rotation of the c_{10} complex was continued. After rotating c_{10} by about 10° more, a complex of three charged residues was formed, as shown in Fig. 12. The dissociation energy of the salt bridge was dramatically reduced by the presence of another charged residue, making it possible to transfer the salt bridge from one subunit *c* to the other. Indeed, when we continued the simulation of the clockwise

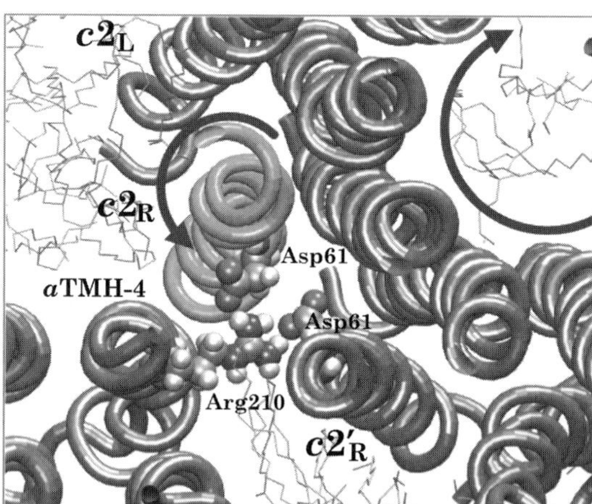

FIG. 12. Concerted rotation of the *c* subunit outer helix and the c_{10} complex in a lipid bilayer. The highlighted helix has been forced to rotate counterclockwise by 220°. The moment of the salt bridge transfer between the two neighboring *c* subunits is shown. The initial structure is shown in Fig. 10. (See Color Insert.)

rotation, the salt bridge transfer was observed. In this simulation, we demonstrated a mechanism by which the rotation of the c_{10} oligomer can proceed when the deprotonation of the binding sites take place.

While MD proved to be very successful in describing separate microscopic events that are involved in the process of the torque generation by F_0, it cannot provide an overall account of the F_0 function on the physiological time scale. Instead, we constructed a simplified, mathematical model of the F_0 motor, based on the stochastic equations of motion.

The model was derived from the microscopic description by reducing the number of degrees of freedom in the system. It was assumed that all torque generating events take place at the interface of the *a* and *c* subunits (see Fig. 13). The only motions allowed in the model are the rotations of the transmembrane helices and the c_{10} complex, which were already investigated by MD. Only one residue from each helix is considered explicitly in the model. Those residues are the binding sites (Asp61s) of the ten *c* subunits and the Arg210 residue of subunit *a*. The conformations

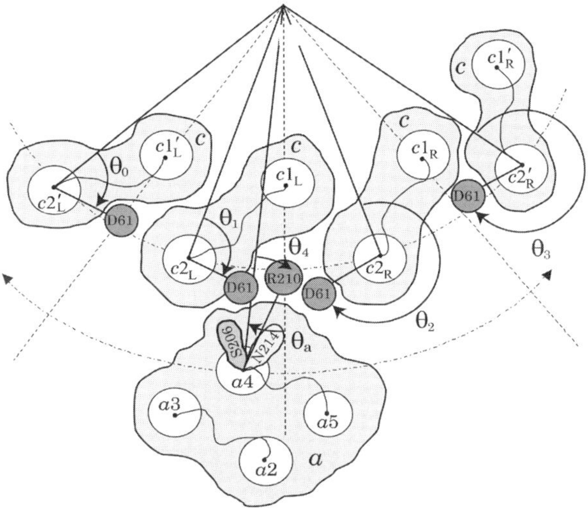

FIG. 13. Mechanical model for F_0 (view from cytoplasm). The c_{10} complex is fixed (only four *c*-subunits are shown) while the *a* subunit can move in either direction. The second trans-membrane helix ("*c*2") of each *c* subunit can rotate independently, moving the key D61 residues, which are the proton binding sites. Similarly, the fourth helix of the *a* subunit ("*a*4") can turn, moving the Arg210 residue. The proton transfer is assumed to occur between the terminal residue of the periplasm channel (N214) and the binding site on the $c2_R$ helix, and between the terminal residue of the cytoplasm channel (S206) and the binding site on the $c2_L$ helix. All mechanical motions are confined to the plane of the figure. The system is fully described by the six angles and the protonation state of the two aspartates on helices $c2_L$ and $c2_R$.

of the residues are restricted geometrically, mimicking their covalent attachment to the backbone of the protein. It is assumed that the rotation of the residues is equivalent to the rotation of their parent helices and vice versa.

Intuitively, the motor operates as follows. When a proton binds or unbinds from the binding site, it alters the potential that governs the motion of all helices. A repeated, asymmetric alternation of the potential transforms the random rotary fluctuations of the helices and c_{10} oligomer into a unidirectional motion. The direction of motion is dictated by the counteraction of the proton electro-chemical gradient and the load torque from F_1.

A typical result of the mathematical model is reproduced in Fig. 14, where the angular coordinates of the key residues are plotted versus time. The conformations of these residues are constrained by the Potential of Mean Force (PMF) acting on their parent helices.

In this particular simulation, the PMF acting on the parent helices of the binding sites has an asymmetric double-well shape. Therefore, the binding sites spend different amounts of time at the minima of the potential (angles θ, and θ_2). The barrier between the minima is low enough to permit the binding sites to change their orientations frequently. The Arg210 residue (θ_4) is directed mostly towards the center of the c_{10} oligomer, as it is dictated by the PMF acting on the *a*4 helix. The latter potential has only one minimum at zero, such as when Arg210 is oriented towards to the center of c_{10}. Angle θ_a in Fig. 14 indicates the relative position of the *a* subunit with respect to the c_{10} oligomer. In this simulation, the oligomer rotates against the physiological load of 41pN-nm imposed by the F_1 unit.

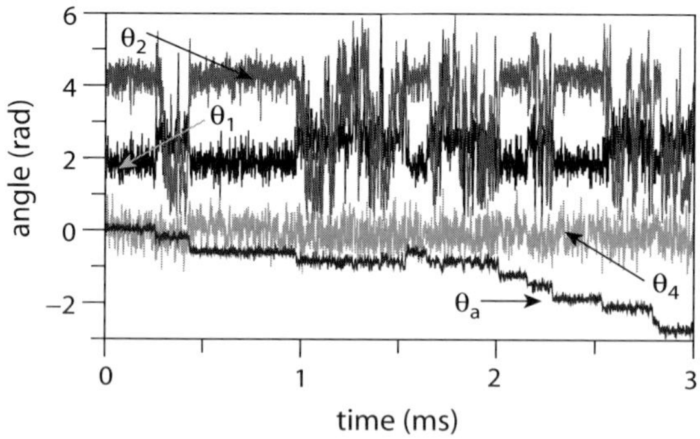

FIG. 14. Stochastic simulations of the events involved in the F_o function: time evolution of rotation angles θ_1, θ_2, θ_4, and θ_a. The angles are defined in Fig. 13. The rotation of *a* subunit takes place in steps (angle θ_a).

The rotation proceeds in steps, with the average velocity of 0.1 revolution per ms.

Thus, we have demonstrated how a microscopic event, discovered in a large-scale MD simulation, can be linked to the physiological time scale through mathematical modeling. The present study is only the beginning of our computational effort directed to solve the great puzzle of the operating mechanism of the F_0 ATP synthase. Our mathematical model is open to accommodate new structural information, specifically anticipating the atom resolution crystal structure of F_0.

B. *Torque-Driven Catalysis in F_1*

ATP synthase is a double motor. In the previous section we have focused on the F_0 component of this motor turning a proton motive force into a torque. This torque is coupled through the stalk, mainly its γ subunit which includes the long coiled-coil α-helix structure, to catalysis in the F_1 ATP binding sites. The mechanical properties of the stalk are obviously an important part of nature's design of ATPase. A key question arises: Is the stalk rigid? Does turning of its base move all parts as a rigid body at the same angular velocity all way up to the tip, or is it highly elastic such that it develops first a twist of its bottom segment before its main segment embedded in $(\alpha\beta)_3$ begins rotation? Possibly, if the latter scenario applies, the partial twist of the stalk induces a change of shape in the stalk, altering the coil-coil helix arrangement. It is even possible that the action of the rotating stalk on F_1 catalysis can only be understood on the basis of the correct shape change of the stalk. Indeed, this change and, hence, the mechanism of coupling between stalk torque and catalysis might be different for the forward (synthesis) and backward (hydrolysis) reaction of ATP synthase.

The rotation of the stalk had been investigated already in the groundbreaking simulation of Böckmann and Grubmüller [11]. However, these authors did not rotate the base of the stalk–which in fact was not even included in the simulation–but only rotated the stalk center, closer to the catalytic binding sites, enforcing the rotation equally along the entire stalk section. As a result, geometric change of the stalk induced by its base segment, elastic energy storage in this part, and results of the interaction of the base with surrounding structures might have been missed in this simulation.

Once rotation of the stalk is properly understood, one can focus on the torque-catalysis coupling in F_1. Masaike *et al.* [90] have carried out so-called targeted MD simulations (TMD) [120] that address this coupling. The TMD simulations apply external forces to the system that pulled F_1

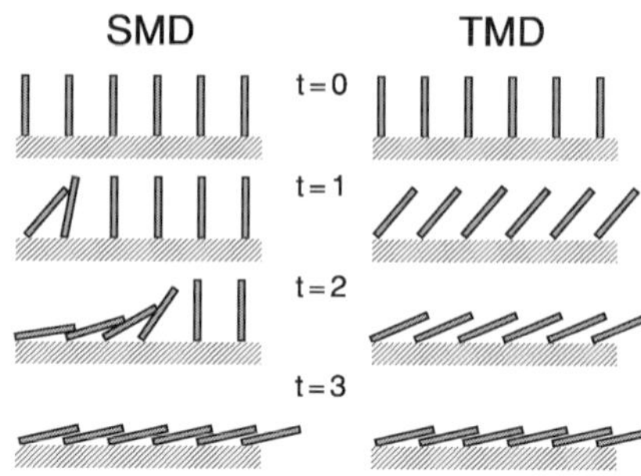

FIG. 15. Loss of causal relationships in TMD simulations. A row of dominoes moves from the initial upright state to the final collapsed state. SMD applies a force to the leftmost domino and predicts the correct series of events. TMD predicts instead a uniform angular motion of all dominoes, failing to capture the "domino effect." Notice that the TMD dominoes do not even contact each other until the final state is reached.

ATPase from its observed structure S0 to another structure S1 that results from S0 through 120° rotation. The forces pull all S0 atoms to their new positions in S1 along the shortest paths. Such an interpolation-like method makes it difficult to distinguish behavior resulting from properties of the studied system from behavior imposed by the simulation procedure. To explain the shortcoming of TMD, we consider in Fig. 15 its application to observing the toppling of a line of dominoes leading to the well-known domino effect. One can discern that TMD misses entirely the causal chain of the domino effect, instead predicting falsely that all dominoes without collision rotate in concert from an upright to a horizontal position. One wishes that a simulation applies forces as they actually occur in the normal function of ATPase, namely, forces that are consistent with a torque at the base of the stalk. Only such simulations can reveal the actual mechanism of ATPase.

ATP synthase is an extremely complex machine that functions on a millisecond time scale. Many aspects of the mechanism of F_1 torque–catalysis coupling are not explained by existing experimental data [11, 20, 38, 88, 101]. A description of how F_1 couples applied torque to catalytic site changes requires answers to these questions: (1) What are the changes to central stalk structure and torque transmission along the stalk during rotation? (2) What is the sequence of deformations and interactions between/within F_1 subunits that lead to catalytic site conformation changes? (3) How do the conformations of the catalytic sites change throughout a rotation?

Experimental studies of F_1 torque production [90, 103], the interactions among F_1 subunits, and how these relate to catalytic function [73, 122] provide partial answers to these questions. However, mechanical details, properties, and intermediate states of the rotating system, required for an understanding of torque–catalysis coupling, cannot be obtained experimentally. There is also the danger that model simulations on a nanosecond time scale do not capture all of the action of this machine and, if one accelerates motion to induce transformations over nanoseconds, that the machine may break, or at the minimum, distort and become artificially caught in a distortion. One must expect many such problems; simulation studies of ATP synthase likely have a long way to go in revealing the underlying physical mechanisms, yet on the other hand the currently feasible simulations may already provide a first glimpse into the mechanics of ATP synthase.

To simulate torqued rotation in the F_1 unit, we performed periodic boundary MD simulations on a 327,000 atom system including ATP synthase F_1 nucleotides, water, and ions, and applied a torque to the central stalk. The F_1 structure was built from coordinates of DCCD-inhibited bovine mitochondrial ATP synthase [43] (RCSB code 1e79), which has the most complete structure of the central stalk among available structures. To construct our model system, the inhibitor-modified β_{DP}-Glu-99 was changed back to its native structure, and two small γ chain gaps were modeled in place. We placed three ATP molecules in the α subunits, and an ATP and ADP molecule in the β_{TP} and β_{DP} respectively, even though the 1e79 structure shows ADP bound at both the β_{DP} and the β_{TP} catalytic sites. An Mg^{2+} was placed at each of the five occupied nucleotide binding sites. To fill in cavities with water molecules, water positions suggested by DOWSER [146] and additional proximity criteria were used. A system was then produced by solvating the assembly with a water shell four molecules thick, producing a system size of 100,000 atoms; the protein displayed large backbone conformation changes when equilibrated in this system. A larger system was then created by placing the complex in a rectangular prism of water and ions large enough to allow periodic boundary simulations, for a total of 92,000 water molecules and 400 ions, producing a total system size of 327,000 atoms. This system was equilibrated for 0.2 ns as an NPT ensemble using (as for all F_1 MD trajectories described here) a 1 fs timestep with PME electrostatics, followed by 0.2 ns equilibration as an NVE ensemble.

We use a modified form of our SMD torque method [60] to simulate the torque that the F_0 subunit normally applies to F_1 in fully assembled ATP synthase, applying torque to restrain a subset of atoms to constant angular velocity (see Fig. 16). Restraining only angular velocity makes a minimum

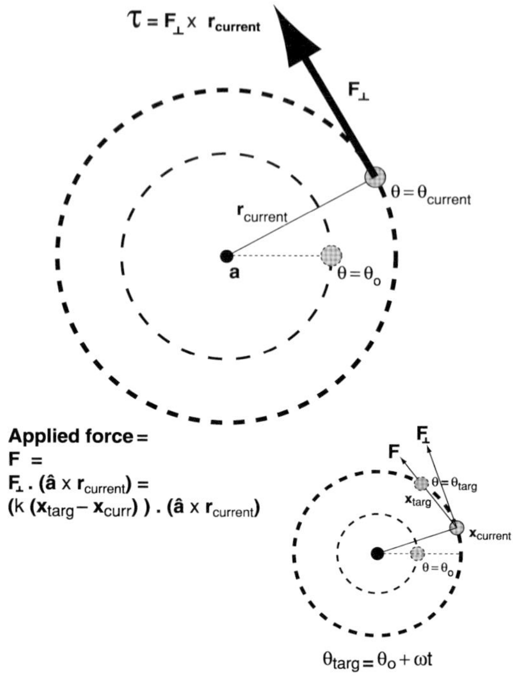

FIG. 16. Torque is applied to rotate 16 C_α atoms at constant angular velocity $\omega = \sim 24^\circ/\mathrm{ns}$ without restraining motion in the directions of **a** and **r**. Force is applied to the component tangent to $\mathbf{r}_{\mathrm{current}}$ of a harmonic restraint to the fictitious "target" atom.

of assumptions about the details of the as yet unknown F_1-F_0 interface. Although the method can be shown to be unstable in any system if applied with a sufficiently high target angular velocity, testing showed this not to be a problem with this system at the angular velocities we employed. The axis of rotation for torque application was chosen as the pseudo-symmetry axis of F_1, determined as a fit through the averaged positions of backbone atoms in the N-terminal region of the three α subunits, a portion of F_1 with near-C3 symmetry. We applied torque to 16 atoms within 5 Å of the presumptive F_0 interface to reflect the physiological torque transmission (see Fig. 17a).

In our first round of simulations we applied torque to enforce 240° of 24°/ns stalk rotation, requiring 10 ns of simulation time, then subjected the system at different stages of rotation to a total of 8.0 ns of equilibration. The 18 ns of dynamics required 652,000 hours of processing time, utilizing at different times between 32 and 512 processors of NCSA and PSC supercomputer clusters (see Fig. 2).

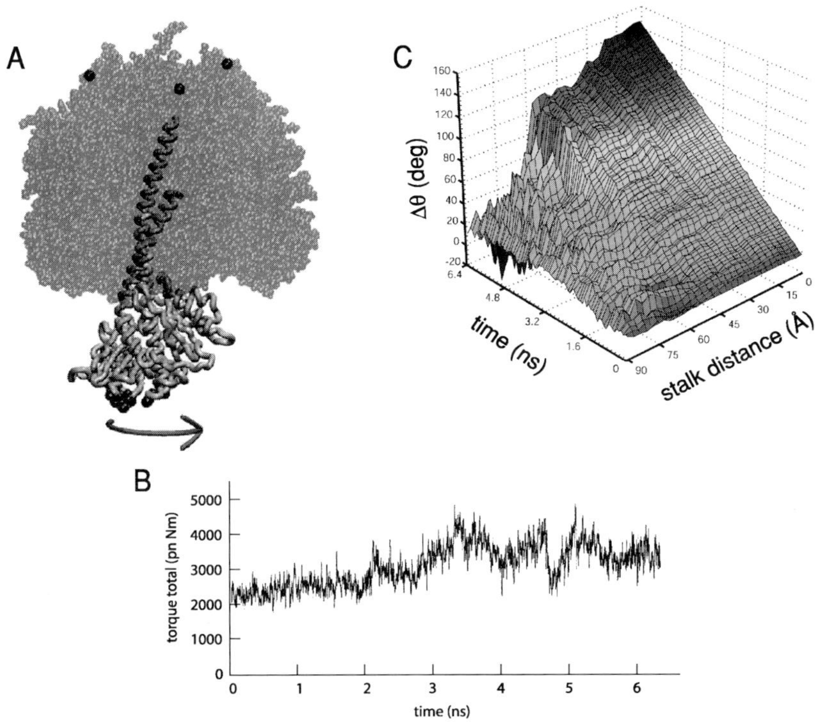

FIG. 17. Events during F_1 central stalk rotation. (a) Torque application to F_1. The central stalk is shown as thick tubes, $(\alpha\beta)_3$ is shown as translucent spheres. Torque is applied to 16 C_α atoms (black spheres on stalk), the three C_α atoms at the N-termini of the three α chains are held fixed (black spheres in $(\alpha\beta)_3$). (b) Total applied torque during the first 6.5 ns of rotation. At each timestep the torque applied to each of the angular-velocity constrained atoms is summed; a 100 fs windowed average is shown here. (c) Twist propagation up the central stalk. $\Delta\theta$ is computed for C_α atoms, stalk height is measured along the rotation axis, parallel to the stalk major axis.

The total external torque required to maintain 24°/ns constant angular velocity of the 16 SMD constrained atoms is shown in Fig. 17b, and ranges between roughly 50 to 100 times the experimentally observed total torque of 40 pN·nm [77]. The total torque, averaged with a 0.1 ps window, fluctuates around 2300 pN·nm for the first 2 ns of torque application, then displays three cycles of increase/decrease between 2.0 and 6.5 ns, with local maxima at 2.2, 3.3, and 5.1 ns, reaching a maximum of 4820 pN·nm. There is a 2000 pN·nm torque increase between 2.8 and 3.5 ns, a 1500 pN·nm increase from 4.0 to 4.7 ns, a sharp drop, then an increase of 2000 pN·nm from 4.8 to 5.1 ns. To measure the propagation of twist up the stalk, we graph the best rotation fit of slices of the stalk structure, proceeding along the rotation axis, in Fig. 17c. The enforced rotation propagates 60 Å up the stalk from the foot along the solvent-exposed region of the stalk, until about 6.4 ns (153°), with the region 0–15 Å along the stalk acting nearly like a rigid body. In the region 60–90 Å along the stalk, where the stalk is buried in the $(\alpha\beta)_3$ complex, the overall propagation of rotation in the stalk sections along the rotation axis appears to halt at about 3.0 ns (72°). The discontinuity in average rotation at 60 Å and 3.0 ns seems like an unphysical stretching of the bonds connecting sections of the stalk, but actually represents the sum of forward rotation of one of the two helices making up the γ-subunit coiled-coil and local backwards rotation of the other. The behavior is not captured by the averaged rotation graph in Fig. 17c since in this region the stalk is acting neither like a rigid body nor a single twisted rod. As shown in Fig. 18, the longer C-terminal helix (γ-192 to 272) continues to rotate around the symmetry axis, but the shorter N-terminal helix (γ-1 to 50) tilts across the symmetry axis, since it winds around the N-terminal helix, as shown in Fig. 18, making a backwards rotation contribution to the total average-fit rotation shown in Fig. 17b.

To characterize the interaction between the central stalk and the three γ subunits during the rotation, we examine the vdW interaction between the

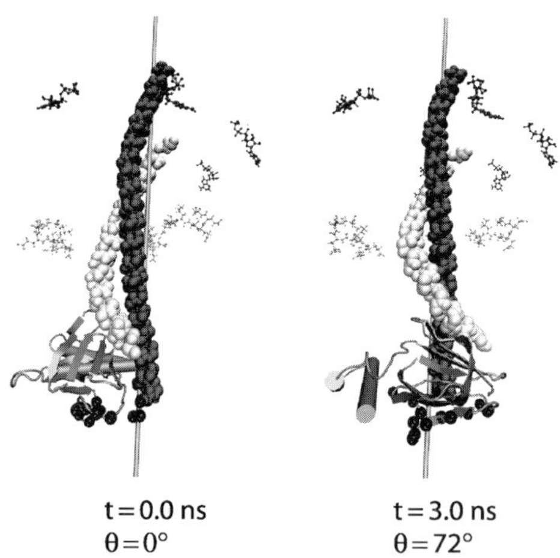

FIG. 18. Winding of γ subunit coiled-coil. The longer helix (γ:197 to 272) of the coiled-coil is shown as dark gray vdW spheres, the shorter helix (γ:1 to 50) as light gray vdW spheres. The δ subunit is shown in cartoon representation (medium gray), with secondary structure recalculated for each configuration. For spatial reference, the rotation axis, DELSEED sequences (light gray CPK representation), bound nucleotides (dark gray CPK), and torqued atoms (black spheres) are shown.

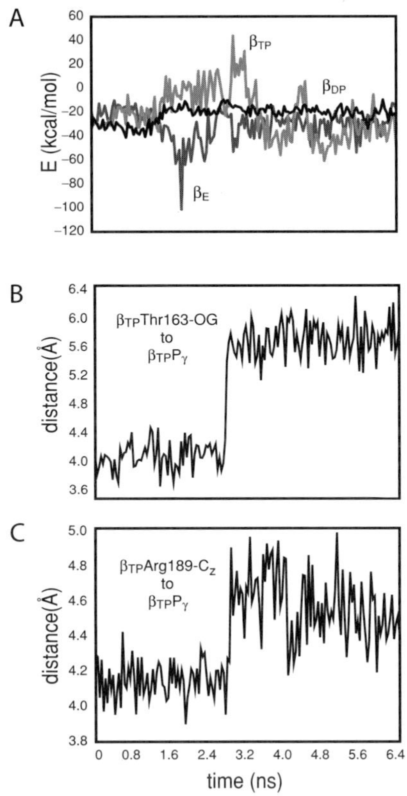

FIG. 19. Subunit interaction and ATP unbinding during central stalk rotation. (a) vdW interaction energy of a subset (see text) of central stalk groups with the three β subunits. Light gray line, β_{TP}; black line, β_{DP}, thick dark gray line, β_E. (b),(c) distances between two β_{TP} phosphate pocket-forming residues and β_{TP}-ATP-P_γ.

β subunits and the portion of the γ stalk (γ 12 to 19, 25 to 30, 234 to 242) which was observed to have close interactions with the C-terminal domains of β_{DP} and β_E in the AlF_4^- hydrolysis intermediate structure of F_1 [92]. As shown in Fig. 19a, the favorable decrease in vdW interaction energy with the γ section is observed for β_E around 1.8 ns, an unfavorable increase is observed for β_{TP} at 3.0 ns, and relatively little change is observed for β_{DP}.

If the system is obeying the binding change model, we expect the β_{TP} subunit to move to an open configuration as the central stalk rotates; the two parts of the ATP catalytic site in β_{TP}, the nucleotide binding pocket and the phosphate binding pocket (which binds P_i before synthesis and ATP-P_γ after synthesis), should unbind from ATP. In Fig. 19b,c, the

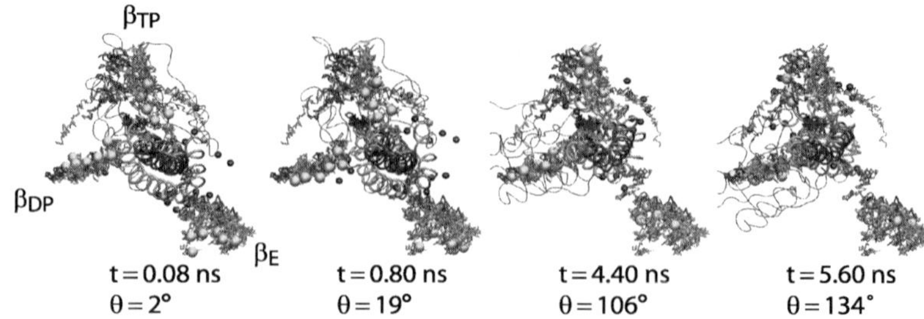

FIG. 20. Motion of DELSEED regions (β:394 to 400) during torque application. The β_{TP} movement is a non-physiological distortion of the F_1 structure. Movements of all C_α atoms of the three DELSEED regions are traced as thin line segments, progressing in color from red to green to blue over the 5.6 ns of motion depicted, with the current positions shown as yellow spheres. Torqued atoms are shown as black spheres, the γ subunit as a purple line, the γ coiled-coil as green and brown tubes. The distortion of the β_{TP} DELSEED region is seen in the movement from the green trace area at 4.40 ns to the blue trace area at 5.60 ns. (See Color Insert.)

charged groups of two phosphate binding pocket residues, β_{TP} Thr-163 and β_{TP} Arg-189, move away from β_{TP} ATP-P_γ at roughly 2.8 ns (67°) of rotation.

The SMD-enforced stalk rotation was carried out to 250°, but with nucleotide bound in a configuration compatible with at most 120° of stalk rotation, the extra rotation was performed in order to re-wind any untwisting of the gamma stalk that the DCCD-inhibition may have caused [19, 43]. A likely point to mark the end of the useful trajectory, after which the enforced stalk rotation is damaging the F_1 structure, is the movement of the β_{TP} DELSEED sequence at 134° (5.6 ns) of stalk rotation, as shown in Fig. 20. The β_{TP} DELSEED is rotating towards α_{DP}, severely distorting F_1.

As the enforced central stalk rotation proceeds, two events consistent with a synthesis binding change mechanism take place in F_1 as far away as the 80 to 100 Å distance between the stalk foot and the catalytic sites: central stalk-β subunit cooperative interactions (Fig. 19a), and ATP unbinding at the phospate pocket of the catalytic β_{TP} site (Fig. 19b). These two energy-requiring events take place during the first large increase in total torque application, between 2.8 and 3.5 ns (c.f. Fig. 17).

The helix winding shown in Fig. 18 may play an important role in energy storage and torque transmission in F_1 function, although the behavior may be largely an artifact of subjecting a system to unphysically high forces. The γ-subunit C-terminal helix, which rotates along with the stalk base, has a direct connection to the putative F_0 c_{10} ring interface, very near the pseudo-symmetry rotation axis. The γ-subunit N-terminal helix, which is observed to tilt as it wraps around the C-terminal helix, is connected to the F_0

interface less directly, via the β-sandwich of the δ subunit, and at a greater distance from the rotation axis than the C-terminal helix. This would provide a c_{10} ring rotating around the symmetry axis with a longer moment arm with which to apply force to the N-terminal helix, allowing the helices to store and release elastic energy independently with additional possible elastic energy storage in deformation of the δ subunit beta sheet, as in the deformation of β-subunit β-sheets modeled by Sun *et al.* [127].

The torque increase and coincident increase in stalk-β_{TP} interaction energy depicted in Fig. 17b and Fig. 19a is consistent with what is required for a step forward in synthesis: an energy input to unbind ATP from β_{TP} to transform its catalytic site to an open state. This energy input is connected with an overall hinge-opening motion of the β_{TP} and in the ATP unbinding from β_{TP} P_γ. The attractive stalk-β_E interactions are consistent with the synthesis-direction transition to a tightly closed binding site which β_E must undergo. The relatively small motions required to transform β_{DP} to a β_{TP}-state are reflected in the much smaller changes observed in the stalk-β_{DP} interaction energy observed.

The unbinding of two residues from β_{TP}-ATP-P_γ appears to be an opening of the β_{TP} site, a movement to an open state to release newly synthesized ATP, a clear step of synthesis-direction binding site change, nearly 100 Å away from the region of external torque application. Since diffusion does not take place during the simulation time, and since the nucleotides remain bound throughout stalk rotation, the continued presence of ATP in the β_{TP} will eventually cause the rotating stalk to distort the structure into non-physiological configurations. Later simulations should remove ATP from the β_{TP} site after a geometric or energetic unbinding criterion is met.

Considering the millisecond time scales of F_1 function, along with its large size and multiple subunit interactions, a complimentary ''top-down'' approach to modeling is necessary; the needed kinetic model of F_1 might contain a small number of springs and rods, along with mechanical catches and levers linked to the states of the catalytic sites. However, as with all biomolecular systems, such a modeling approach requires deciding which degrees of freedom are important for function and which can be discarded; the problem is made especially difficult by the complexity and large size of F_1. There is still too little experimental data on mechanical changes within F_1 during catalysis to clearly guide construction of a top-down model, although several models have been proposed [72, 103, 104, 127, 136, 137] based on available structure and functional data.

The ''bottom-up'' approach of whole-system SMD simulations eliminates the problem of deciding which elements of the system to simulate, it seeks to identify important details and emergent properties of F_1 by preserving as much structural detail as possible while making a minimum of assumptions beyond what is experimentally known. We plan to combine both of these approaches in our further studies of F_1.

An ultimate goal of our work is to integrate our descriptions of torque generation in F_0 and torque utilization in F_1 into a single model. An understanding of the elastic properties of both the central and secondary stalks will allow us to connect independent stochastic models of F_0 and F_1.

V. Mechanical Signaling in Fibronectin

Biological cells are surrounded by the extracellular matrix (ECM), a flexible network containing several classes of proteins secreted by cells themselves. Fibronectin (FN), found in all vertebrates, is the first structurally well-characterized ECM protein, yet its functional properties are not fully understood. FN is an important mechanical component of the ECM and acts as a specific adhesive, forming elastic FN fibrils that connect cells via transmembrane protein integrins and guide cell movement [57]. Cross-linked through a disulfide bond at the C-terminus, two identical FN subunits form a large dimeric FN molecule (450–500 kD) consisting of ~20 different modules in each subunit. About 15 modules are type III (FN-III) modules, structurally exhibiting a so-called β-sandwich motif that has been also identified in many force-bearing proteins, such as immunoglobulin-like (Ig) domains from the muscle protein titin [59].

Mechanical stress has been found to regulate the assembly of FN fibrils, a process termed fibrillogenesis, through integrin receptors that mechanically couple intracellular actin filaments to extracellular FN molecules (see reviews [42, 121]). It has been hypothesized that the stretching of FN fibrils unfolds individual FN-III modules [33], providing the necessary extension to the FN fibril that can be stretched four times as long as its relaxed length [100]. In addition to providing elasticity, the unfolding of FN-III modules is functionally relevant to mediating fibrillogenesis by exposing otherwise-buried cryptic sites [42, 133]. Supporting this unfolding hypothesis, recent fluorescence studies found that cells not only integrate fibronectin in an extended conformation into fibrils, but also overstretch many fibrils so that FN-III modules may become unfolded [6].

The mechanical stability of individual FN-III modules has been probed in force-unfolding experiments using atomic force microscopy (AFM) [98, 99, 117]. The force-extension profiles obtained from AFM experiments display characteristic saw-tooth patterns, implying that these protein modules unfold in a one-by-one fashion. The unfolding is reversible because fully stretched modules refold in seconds after releasing the external force, a crucial property for elastic components. With genetically engineered identical repeats of FN-III modules, the mechanical stability of individual modules can be quantified in terms of ruptured force peaks. Suprisingly, although FN-III modules possess quite similar tertiary structures, they demonstrate distinctly different mechanical stabilities with peak rupture force ranging from 75 to 200 pN at a pulling velocity of 0.6 μm/s [98]. More interestingly, FN-III modules such as FN-III$_1$ may have multiple distinct stable conformations, indicating the existence of unfolding intermediates. The mechanical variablity and the intermediate states of FN-III modules may be closely related to the function of FN.

The structural reseason why FN-III modules exhibit such mechanical properties is not clear. Although about half FN-III modules have been resolved at high resolution [81, 123], it is difficult to connect the static structures to the dynamic unfolding scenarios that one sees in AFM experiments. To link these static structures to their dynamically changing non-equilibrium states induced under external forces, steered molecular dynamics (SMD) simulations have been introduced to provide an atomic-level description of the unfolding processes for FN-III modules [22, 40, 79, 80]. The SMD method has also been successfully applied to study the mechanical unfolding of other mechanical proteins such as titin domains (reviewed in [41, 59]).

In early approaches the simulated FN-III modules were either solvated in a small water sphere (typically ~70 Å in diameter) [22, 80] or treated with implicit solvent models [102]. While the spherical solvent models are sufficiently accurate for short stretching (i.e., within 60 Å), they produce both surface tension and protein de-solvation artifacts, both of which become increasingly pronounced as a protein is stretched beyond the surface of the water sphere. On the other hand, the implicit models overlooked the interactions between the protein and water molecules, which play a key role in disrupting the inter-strand hydrogen bonds that stabilize the whole protein structure [86]. To address the problem of insufficient solvation and obtain a more faithful description of the unfolding dynamics, FN-III modules have been solvated in long water boxes, resulting in large systems of more than 100,000 atoms [40]. Figure 21 displays such a model with the water box long enough to allow a single FN-III$_{10}$ module to be completely extended over 300 Å without encountering a boundary, and accounting for full electrostatics [40]. The system contains 130,000 atoms in total, requiring 170 hours running on our group's cluster of 32 1.33GHz Athlon processors (see Fig. 2) for one nanosecond of simulation with the program NAMD [74].

Two SMD protocols, constant velocity and constant force stretching, have been applied to our systems. SMD-using restraints moving with constant velocity simulates the stretching of protein domains by an AFM cantilever.

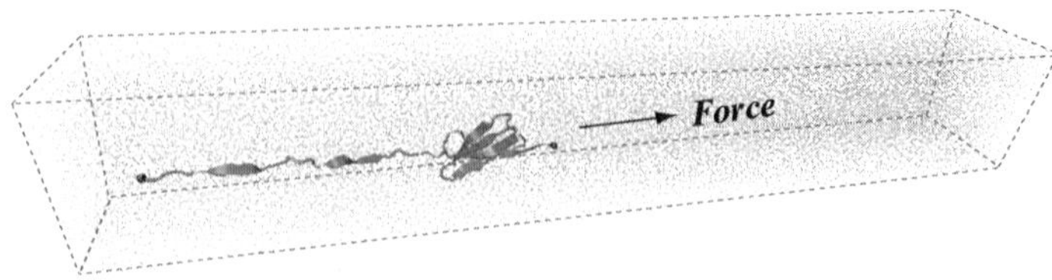

FIG. 21. An FN-III$_{10}$ module was solvated and stretched in a 370 Å long water box (130,000 atoms). During the simulations the N-terminus was fixed, and the force was applied along the vector pointing from the N-terminus to the C-terminus.

One terminus is fixed during the simulation, which corresponds to attaching the protein to a fixed substrate in the AFM experiment. The other terminus is restrained to a point in space (restraint point) by an external (e.g., harmonic) potential. The restraint point is then shifted in a chosen direction [85], typically along the vector pointing from the C-terminus to the N-terminus.

SMD simulations revealed that the main force-bearing elements of FN-III modules arise primarily from the β-strands close to the termini, specifically, from hydrogen bonding between A- and B-strands and between F- and G-strands, rather than the whole β-sheets [40]. This is similar to titin domain I27, the mechanical stability of which is attributed to the β-strands interweaving two terminal portions. Despite the structure similarity with titin Ig domains, however, FN-III modules are typically weaker than Ig domains, and some modules exhibit stable intermediates with extension much longer than its fully folded length [98]. Unlike the single unfolding pathway that titin modules exhibit, FN-III modules may have multiple unfolding pathways and intermediate states as observed in previous simulations [22, 40, 102]. As shown in Fig. 22 during the unfolding of FN-III$_{10}$, the termini of the module were straightened first upon mechanical stress, entering an intermediate termed ''twisted state.'' As the unfolding proceeds, the hydrophobic core was disrupted and a pair of backbone hydrogen bonds between Arg6 and Asp23 near the C-termini broke, permitting extension up to 25 Å. The protein subsequently entered an intermediate named ''aligned state,'' in which the two β sheets of FN-III$_{10}$ aligned along the direction of the external force. The two intermediates, twisted and aligned states, are universal to other FN-III modules [22], extending modules up to ~70% of their original length without unraveling the tertiary structure. The aligned intermediate thus may expose partially buried cryptic binding sites. From the aligned state, FN-III$_{10}$ progresses along three alternative unfolding pathways: separating first the A-strand from the module, separating the A- and G-strands simultaneously, or separating first the G-strand. When unraveling of the β-strands

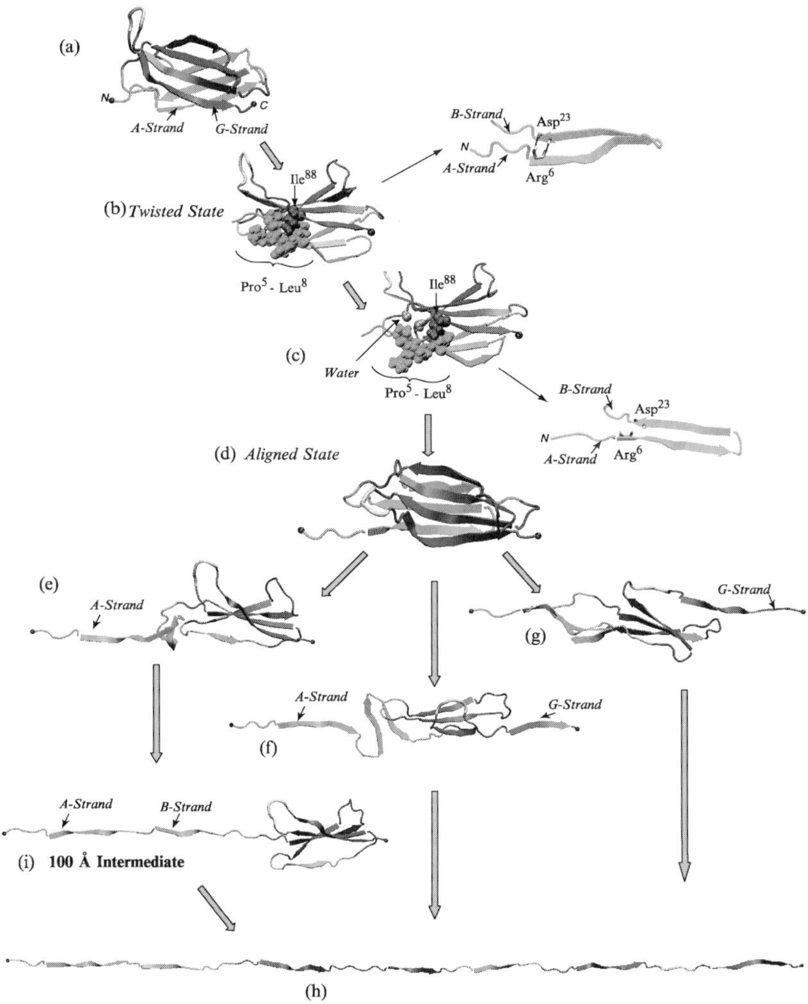

FIG. 22. Unfolding pathways for FN-III_{10} revealed by SMD simulations. (a) Equilibrated FN-III_{10}. (b) The randomly coiled termini are straightened slightly upon applying a constant force. (c) Rupture of two key inter-strand hydrogen bonds between A- and B-strands permits solvation of the hydrophobic core periphery. (d) Alignment of the two β-sheets before unraveling any β-strands. Separation of the β-strands begins by either separation of (e) A-strand first, (f) A- and G-strands simultaneously, or (g) G-strand first. (h) Fully unraveled structure. (i) In case that the A-strand separates first, a stable intermediate at 100 Å extension arises.

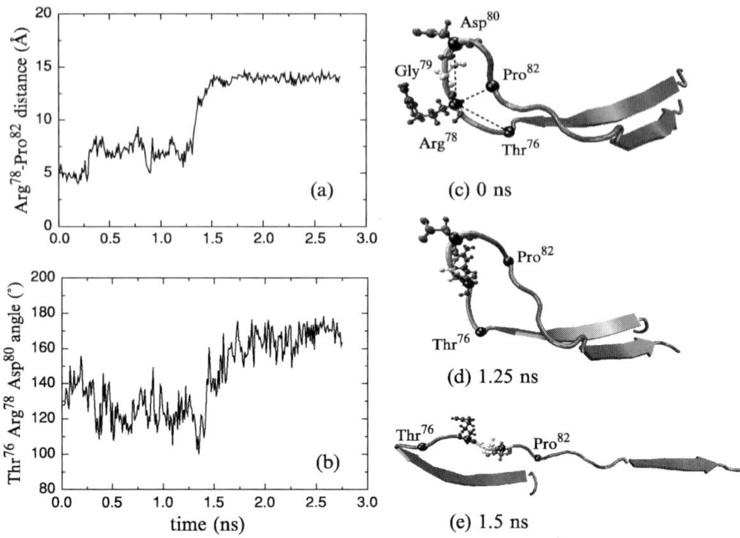

FIG. 23. Conformational changes of the RGD-containing loop upon a constant external force. (a) The width of the RGD loop is measured by the distance between the C_α atoms of Arg78 and Pro82. (b) Angle formed by the C_α atoms of residues Thr76, Arg78, and Asp80 illustrates a change from a bent ($\sim 110^\circ$) to a more planar ($\sim 170^\circ$) conformation. (c) Snapshot of F- and G-strands and the connecting loop in the native structure. (d) The RGD loop remains intact in the intermediate state at 100 Å. (e) The RGD loop is straightened out after the protein passes the intermediate state.

began with the A-strand, a stable intermediate at 100 Å was observed. With four-fold extension compared to folded FN-III$_{10}$, the intermediate may serve as a basis for fibronectin self-assembly through a proposed β-strand swapping mechanism [84].

FN-III$_{10}$ mediates cell adhesion to the ECM via its integrin binding motif, Arg-Gly-Asp (RGD), that is located at the apex of the loop connecting β-strands F and G (Fig. 23). The binding of the RGD loop to certain integrins, such as $\alpha_5\beta_1$, is assisted by a synergy site on the neighboring domain FN-III$_9$. SMD simulations found that this interaction can be regulated either by directly detaching the RGD motif [40, 80] or through increasing the distance between the synergy site and the RGD loop [79]. By separating the G-strand from the remaining fold, the distance between the apex of the RGD-containing loop and the module surface was shortened. This conformational change reduces the affinity to integrin. The RGD loop thus constitutes a mechanosensitive switch for recognition by integrin receptors [80]. SMD simulations of the FN-III$_{9\text{-}10}$ dimer [79] identified an intermediate in which the length of the linker chain between the two modules is increased by 17 Å, a change that could also switch off the signaling between FN modules and transmembrane protein integrins.

Recently published crystal structures of integrin $\alpha_V\beta_3$ [142, 143] provide a further opportunity to study the signaling pathways between integrin and fibronectin modules. Simulations of integrin $\alpha_V\beta_3$ containing RGD peptide in the binding pocket with 136,000 atoms are in progress and may reveal how integrins mediate the adhesion with FN ligands. However, the ultimate goal of simulations is not only to reveal the dynamic processes of FN-III unfolding or dissociation from integrins, but also to understand in combination with experiments how the individual FN modules contribute to the overall mechanical properties of FN fibrils and how cell receptors and the ECM proteins respond and exchange mechanical signals.

VI. Outlook

We have illustrated above the state of art in large scale biomolecular modeling. Such modeling has become feasible only through parallel computing utilizing hundreds and soon thousands of processors. Fortunately, the necessary computers are available to many researchers and the needed new generation of molecular dynamics and molecular visualization programs have been developed and are widely shared. During the next years the new programs (e.g., NAMD) will mature and most likely merge with the familiar, feature-rich first generation of modeling programs (e.g., CHARMM and AMBER).

Interestingly, most large scale simulation projects in our and other groups have been carried out in close collaboration with experimentalists. The projects have not been accomplished for the record books, but were motivated by a desire to extend the resolution of observation to atomic level dynamics and to explain thereby the mechanisms underlying important cellular processes. As stated already above, biomolecular modeling is being adopted today by experimentalists to guide and explain observations; often these researchers simulate large systems.

Indeed, size is not really a key obstacle today and will be less so in the near future. Nevertheless, it seems frivolous to describe extremely large systems uniformly at the atomic level when one is only interested in a small part, for example, an active site. Unfortunately, the old dream of computational approaches that permit one to dial an effective resolution for different parts of a system is apparently hard to realize; legions of researchers have attempted a solution of such multi-scale modeling, but successes have been rare. Computational biologists seek to replace an explicit simulation of water molecules in bulk water by a so-called implicit solvent model; much effort has been invested in this approach. Other approaches replace an atomic-level simulation of DNA strands by elastic rod models to simplify the description of protein-DNA complexes. Attempts have also been made to systematically reduce the number of simulated atoms by placing a majority of atoms into a class of ''slaved'' atoms that move along with the other atoms according to a fixed mathematical formula [107]. None of the approaches described can yet be employed in an automated way with controlled accuracy. These approaches may also be ill-fated due to computer power permitting more and more easily full atom descriptions and outrunning the development of the necessarily complex multi-scale algorithms.

But universally accepted as a bottleneck is the presently short 10 to 100 ns time scale of simulations that is even for the fastest cellular processes five orders of magnitude too short. Parallelization is not a choice in this regard since long time dynamics is inherently sequential; a general parallel solution to the time scale problem has not been found and very well may not exist. The only remedy is an application of external forces in so-called steered molecular dynamics simulations that accelerate the process. The problem with this approach is not so much that the forces applied are artificial, but that one does not know *a priori* in which direction in the multi-dimensional space of biopolymer conformations forces should be applied. Such knowledge implies that the reaction path and hence the mechanism for the cellular process studied is known, but often one neither knows this path nor does one even have a suitable hypothesis. The problem may be overcome with a method called interactive molecular dynamics [46] that applies forces like steered molecular dynamics, but does so interactively in a running simulation, permitting one to tinker with a system and develop the needed hypotheses.

Simulations of 100,000 atoms and more accomplished so far, for the purpose of explaining the physical mechanism of cellular processes, focus as much as smaller-scale simulations on key moieties of biopolymers such as channel lining, force bearing secondary structure elements, active sites, or even electronic degrees of freedom[23, 25]. The main difference between small-scale and large-scale operations is that mechanisms are captured by the latter in the context of integral functions of biopolymeric machineries. Also, large-scale simulations are necessary when one does not know from the outset the essential components of a cell's machinery. For example, presently the mechanistic essentials of ATP synthase that couple reversibly proton currents to stalk rotation, and stalk rotation to catalysis, are all but unknown forcing large scale simulations to avoid missing key features.

A further argument for large simulations are the recently discovered structures of the most basic machineries of living cells, like redox chains in bioenergetic membranes, electrical channels of neural membranes, or the genome storage and expression machinery, which involve large scale systems that cannot be readily subdivided.

The ultimate argument for large-scale simulations focuses on the most fundamental question of life sciences: Where in the hierarchies of assemblies found in cells does the step from inorganic matter to living systems occur? Living systems are all made of many molecular components that self-assemble, control and repair each other, and self-replicate. In the past, molecular biologists have taken the molecular machines in living cells apart and learned of their composition. Now they seek to learn how the molecular constituents of cells assemble into machines, cellular structures, and complete living cells. Eventually, large-scale simulations investigating assemblies of biopolymers will provide life scientists with a unique tool to answer the question, "What is life?"

References

1. Abrahams, J., Leslie, A., Lutter, R., and Walker, J. (1994). Structure at 2.8-Å resolution of F_1-ATPase from bovine heart mitochondria. *Nature* **370,** 621–628.
2. Agre, P., Bonhivers, M., and Borgnia, M. J. (1998). The aquaporins, blueprints for cellular plumbing systems. *J. Biol. Chem.* **273,** 14659–14662.
3. Agre, P., Lee, M. D., Devidas, S., and Guggino, W. B. (1997). Aquaporins and ion conductance. *Science* **275,** 1490.
4. Aksimentiev, A., Balabin, I., Fillingame, R. H., and Schulten, K. (2003). Exploring protein motors on multiple time scales: F_0 ATP synthase. Submitted.
5. Ban, N., Nissen, P., Hansen, J., Capel, M., Moore, P. B., and Steitz, T. A. (1999). Placement of protein and rna structures into a 5 Å-resolution map of the 50S ribosomal subunit. *Nature* **400,** 841–847.
6. Baneyx, G., Baugh, L., and Vogel, V. (2002). Supramolecular chemistry and self-assembly special feature: fibronectin extension and unfolding within cell matrix fibrils controlled by cytoskeletal tension. *Proc. Nat. Acad. Sci. USA* **99,** 5139–5143.
7. Baudry, J., Tajkhorshid, E., Molnar, F., Phillips, J., and Schulten, K. (2001). Molecular dynamics study of bacteriorhodopsin and the purple membrane. *J. Phys. Chem. B,* **105,** 905–918.
8. Bayas, M. V., Schulten, K., and Leckband, D. (2003). Simulations of the forced detachment of the CD2-CD58 complex. *Biophys. J.* **84,** 2223–2233.
9. Birkenhager, R., Hoppert, M., Deckers-Hebestreit, G., Mayer, F., and Altendorf, K. (1995). The F_0 complex of the *Escherichia coli* ATP synthase. Investigation by electron spectroscopic imaging and immunoelectron microscopy. *Eur. J. Biochem.* **230,** 58–67.
10. Bishop, T. C., Kosztin, D., and Schulten, K. (1997). How hormone receptor–DNA binding affects nucleosomal DNA: The role of symmetry. *Biophys J.* **72,** 2056–2067.
11. Böckmann, R. A., and Grubmüller, H. (2002). Nanoseconds molecular dynamics simulation of primary mechanical energy transfer steps in F_1-ATP synthase. *Nat. Struct. Biol.* **9,** 198–202.
12. Borgnia, M., Nielsen, S., Engel, A., and Agre, P. (1999). Cellular and molecular biology of the aquaporin water channels. *Annu. Rev. Biochem.* **68,** 425–458.

13. Borgnia, M. J., and Agre, P. (2001). Reconstitution and functional comparison of purified GlpF and AqpZ, the glycerol and water channels from *Escherichia coli*. *Proc. Nat. Acad. Sci. USA* **98,** 2888–2893.
14. Borgnia, M. J., Kozono, D., Calamita, G., Maloney, P. C., and Agre, P. (1999). Functional reconstitution and characterization of AqpZ, the E-coli water channel protein. *J. Mol. Biol.* **291**(5), 1169–1179.
15. Boyer, P. D. (2000). Catalytic site forms and controls in ATP synthase catalysis. *Biochim. Biophys. Acta Bioenerg.* **1458(2–3),** 252–262.
16. Brewer, M. L., Schmitt, U. W., and Voth, G. A. (1999). The formation and dynamics of proton wires in channel environments. *Biophys. J.* **80(4),** 1691–1702.
17. Brooks, B. R., Bruccoleri, R. E., Olafson, B. D., States, D. J., Swaminathan, S., and Karplus, M. (1983). CHARMM: A program for macromolecular energy, minimization, and dynamics calculations. *J. Comput. Chem.* **4,** 187–217.
18. Brünger, A. T. (1992). X-PLOR, Version 3.1: *A System for X-ray Crystallography and NMR.* The Howard Hughes Med. Inst. and Dept. Mol. Biophys. Biochem. Yale University.
19. Capaldi, R. A., and Aggeler, R. (2002). Mechanism of F_1F_0-type ATP synthase, a biological rotary motor. *Trends Biochem. Sci.* **27,** 154–160.
20. Cherepanov, D. A., Mulkidjaninan, A. Y., and Junge, W. (1999). Transient accumulation of elastic energy in proton translocating ATP synthase. *FEBS Letters.* **449,** 1–6.
21. Clemons, Jr, W. M., May, J. L. C., Wimberly, B. T., McCutcheon, J. P., Capel, M. S., and Ramakrishnan, V. (1999). Structure of a bacterial 30S ribosomal subunit at 5.5 Å resolution. *Nature* **400,** 833–840.
22. Craig, D., Krammer, A., Schulten, K., and Vogel, V. (2001). Comparison of the early stages of forced unfolding of fibronectin type III modules. *Proc. Nat. Acad. Sci. USA* **98,** 5590–5595.
23. Damjanović, A., Kosztin, I., Kleinekathoefer, U., and Schulten, K. (2002). Excitons in a photosynthetic light-harvesting system: A combined molecular dynamics, quantum chemistry and polaron model study. *Phys. Rev. E.* **65,** 031919.
24. Damjanović, T., Ritz, and Schulten, K. (2002). Excitation transfer in the peridinin-chlorophyll-protein of *Amphidinium carterae. Biophys. J.* **79,** 1695–1705.
25. Darden, T., York, D., and Pedersen, L. (1993). Particle mesh Ewald. An N·log(N) method for Ewald sums in large systems. *J. Chem. Phys.* **98,** 10089–10092.
26. de Groot, B. L., Engel, A., and Grubmüller, H. (2001). A refined structure of human aquaporin-1. *FEBS Letters* **504,** 206–211.
27. de Groot, B. L., and Grubmüller, H. (2001). Water permeation across biological membranes: Mechanism and dynamics of aquaporin-1 and GlpF. *Science* **294,** 2353–2357.
28. de Groot, B. L., Heymann, J. B., Engel, A., Mitsuoka, K., Fujiyoshi, Y., and Grubmüller, H. (2000). The fold of human aquaporin 1. *J. Mol. Biol.* **300,** 987–994.
29. Deen, P. M. T., and van Os, C. H. (1998). Epithelial aquaporins. *Curr. Opin. Cell Biol.* **10(4),** 435–442.
30. Dmitriev, O., Jones, P. C., and Fillingame, R. H. (1999). Structure of the subunit *c* oligomer in the F_1F_0 ATP synthase: Model derived from solution structure of the monomer and cross-linking in the native enzyme. *Proc. Nat. Acad. Sci. USA* **96,** 7785–7790.
31. Dunn, S. D., Revington, M., Cipriano, D. J., and Shilton, B. H. (2000). The *b* subunit of *Escherichia coli* ATP synthase. *J. Bioenerg. Biomembr* **32**(4), 347–355. Alignment of the b-subunit to the a subunit.

32. Elston, T., Wang, H., and Oster, G. (1998). Energy transduction in ATP synthase. *Nature* **391,** 510–513.
33. Erickson, H. (1994). Reversible unfolding of fibronectin type III and immunoglobulin domains provides the structural basis for stretch and elasticity of titin and fibronectin. *Proc. Nat. Acad. Sci. USA* **91,** 10114–10118.
34. Fillingame, R. H. (1997). Coupling H^+ transport and ATP synthesis in F_1F_0-atp synthases: Glimpses of interacting parts in a dynamic molecular machine. *J. Exp. Biol.* **200,** 217–224.
35. Fillingame, R. H., Angevine, C. M., and Dmitriev, O. Y. (2002). Coupling proton movements to *c*-ring rotation in F_1F_0 ATP synthase: aqueous access channels and helix rotations at the *a-c* interface. *Biochim. Biophys. Acta Bioenerg.* **1555**(1–3), 29–36. Helix rotation scenario; proton half-channel termini S206 and N210 in the a subunit.
36. Fillingame, R. H., Jiang, W., and Dmitriev, O. Y. (2000). Coupling H^+ transport to rotary catalysis in F-type ATP synthases: structure and organization of the transmembrane rotary motor. *J. Exp. Biol.* **203,** 9–17.
37. Fillingame, R. H., Jiang, W., Dmitriev, O. Y., and Jones, P. C. (2000). Structural interpretation of F_0 rotary function in the *Escherichia coli* F_1F_0 ATP synthase. *Biochim. Biophys. Acta* **1458,** 387–403.
38. Fu, D., Libson, A., Miercke, L. J. W., Weitzman, C., Nollert, P., Krucinski, J., and Stroud, R. M. (2000). Structure of a glycerol conducting channel and the basis for its selectivity. *Science* **290,** 481–486.
39. Futai, M., Omote, H., Sambongi, Y., and Wada, Y. (2000). Synthase (H^+ ATPase): Coupling between catalysis, mechanical work, and proton translocation. *Biochim. Biophys. Acta Bioenerg.* **1458**(2–3), 276–288.
40. Gao, M., Craig, D., Vogel, V., and Schulten, K. (2002). Identifying unfolding intermediates of FN-III_{10} by steered molecular dynamics. *J. Mol. Biol.* **323,** 939–950.
41. Gao, M., Lu, H., and Schulten, K. (2002). Unfolding of titin domains studied by molecular dynamics simulations. *J. Muscle Res. Cell Motil* **23,** 513–521.
42. Geiger, B., Bershadsky, A., Pankov, R., and Yamada, K. Transmembrane crosstalk between the extracellular matrix and the cytoskeleton. *Nat. Rev. Mol. Cell. Biol.* **2,** 793–805.
43. Gibbons, C., Montgomery, M. G., Leslie, A. G. W., and Walker, J. E. (2000). The structure of the central stalk in bovine F_1-ATPase at 2.4 Å resolution. *Nat. Struct. Biol.* **7,** 1055–1061.
44. Girvin, M. E., Rastogi, V. K., Abildgaard, F., Markley, J. L., and Fillingame, R. H. (1998). Solution structure of the transmembrane H^+-transporting subunit *c* of the F_1F_0 ATP synthase. *Biochem.* **37,** 8817–8824.
45. Gnatt, A. L., Cramer, P., Fu, J. H., Bushnell, D. A., and Kornberg, R. D. (2001). Structural basis of transcription: an RNA pplymerase II elongation complex at 3.3 Å resolution. *Science* **5523,** 1876–1882.
46. Grayson, P., Tajkhorshid, E., and Schulten, K. (2003). Mechanisms of selectivity in channels and enzymes studied with interactive molecular dynamics. *Biophys. J.*
47. Groth, G. (2000). Molecular models of structural arrangement of subunits and the mechanism of proton translocation in the membrane domain. *Biochim. Biophys. Acta* **1458,** 417–427. Arrangements of F0 subunits using scanning mutagenesis combined with c subunit structure.
48. d. Grotthuss, C. J. T. (1806). *Ann. Chim.* **LVIII,** 54.

49. Gullingsrud, J., Kosztin, D., and Schulten, K. (2001). Structural determinants of MscL gating studied by molecular dynamics simulations. *Biophys. J.* **80,** 2074–2081.
50. Heller, H., Schaefer, M., and Schulten, K. (1993). Molecular dynamics simulation of a bilayer of 200 lipids in the gel and in the liquid crystal-phases. *J. Phys. Chem.* **97,** 8343–8360.
51. Heller, K. B., Lin, E. C., and Wilson, T. H. (1980). Substrate specificity and transport properties of the glycerol facilitator of *Escherichia coli. J. Bacteriol* **144,** 274–278.
52. Hermolin, J., Gallant, J., and Fillingame, R. H. (1983). Topology, organization, and function of the psi subunit in the f_0 sector of the H^+-ATPase of *escherichia coli. J. Biol. Chem.* **258**(23), 14550–14555. G9 in the b subunit located close to P240 in the a subunit.
53. Heynmann, J. B., and Engel, A. (1999). Aquaporins: Phylogeny, structure, and physiology of water channels. *News in Physiol. Sci.* **14,** 187–193.
54. Hirono-Hara, Y., Noji, H., Nishiura, M., Muneyuki, E., Hara, K. Y., Yasuda, R., Kinosita, K. Jr., and Yoshida, M. (2001). Pause and rotation of F_1-ATPase during catalysis. *Proc. Nat. Acad. Sci. USA* **98,** 13649–13654.
55. Hummer, G., Rasaiah, J. C., and Noworyta, J. P. (2001). Water conduction through the hydrophobic channel of a carbon nanotube. *Nature* **414,** 188–190.
56. Humphrey, W., Dalke, A., and Schulten, K. (1996). VMD – Visual Molecular Dynamics. *J. Mol. Graph.* **14,** 33–38.
57. Hynes, R. O. (1990). Fibronectins. Springer-Verlag, New York.
58. Isralewitz, B., Baudry, J., Gullingsrud, J., Kosztin, D., and Schulten, K. (2001). Steered molecular dynamics investigations of protein function. *J. Mol. Graph. Model.* **19,** 13–25. Also in *Protein Flexibility and Folding,* (L. A. Kuhn and M. F. Thorpe, Ed.), *Biol. Mod. Series,* Elsevier.
59. Isralewitz, B., Gao, M., and Schulten, K. (2001). Steered molecular dynamics and mechanical functions of proteins. *Curr. Opin. Struct. Biol.* **11,** 224–230.
60. Izrailev, S., Crofts, A. R., Berry, E. A., and Schulten, K. (1999). Steered molecular dynamics simulation of the Rieske subunit motion in the cytochrome bc_1 complex. *Biophys. J.* **77,** 1753–1768.
61. Izrailev, S., Stepaniants, S., Isralewitz, B., Kosztin, D., Lu, H., Molnar, F., Wriggers, W., and Schulten, K. (1998). Steered molecular dynamics. *In* Computational Molecular Dynamics: Challenges, Methods, Ideas, (P. Deuflhard, J. Hermans, B. Leimkuhler, A. E. Mark, S. Reich, and R. D. Skeel, Eds.), *Lecture Notes in Comput. Sci. Eng.* Vol. 4, Springer-Verlag, Berlin, 39–65.
62. Jarzynski, C. (1997). Equilibrium free-energy differences from nonequilibrium measurements: a master equation approach. *Phys. Rev. E.* **56,** 5018–5035.
63. Jarzynski, C. (1997). Nonequilibrium equality for free energy differences. *Phys. Rev. Lett.* **78,** 2690–2693.
64. Jensen, M. Ø., Park, S., Tajkhorshid, E., and Schulten, K. (2002). Energetics of glycerol conduction through aquaglyceroporin GlpF. *Proc. Nat. Acad. Sci. USA* **99,** 6731–6736.
65. Jensen, M. Ø., Tajkhorshid, E., and Schulten, K. (2001). The mechanism of glycerol conduction in aquaglyceroporins. *Structure* **9,** 1083–1093.
66. Jiang, W., Hermolin, J., and Fillingame, R. H. (2001). The preferred stoichiometry of *c* subunits in the rotary motor sector of *Escherichia coli* ATP synthase is 10. *Proc. Nat. Acad. Sci. USA* **98,** 4966–4971.
67. Jones, P. C., and Fillingame, R. H. (1998). Genetic fusion of subunit *c* in the F_0 sector of H^+-transporting ATP synthase. *J. Biol. Chem.* **273**(45), 29701–29705.

68. Jones, P. C., Jiang, W., and Fillingame, R. H. (1998). Arrangement of the multi-copy H^+-translocating subunit *c* in the membrane sector of the *Escherichia coli* F1F0 ATP synthase. *J. Biol. Chem.* **273,** 17178–17185.
69. Mackerell, Jr., A. D., Brooks, B., Brooks III, C. L., Nilsson, L., Roux, B., Won, Y., and Karplus, M. (1998). Charmm: The energy function and its parameterization with an overview of the program. John Wiley & Sons, Chichester, 271–277.
70. Jung, J. S., Preston, G. M., Smith, B. L., Guggino, W. B., and Agre, P. (1994). Molecular structure of the water channel through aquaporin CHIP – the hourglass model. *J. Biol. Chem.* **269,** 14648–14654.
71. Junge, W., Lill, H., and Engelbrecht, S. (1997). Atp synthase: An electrochemical transducer with rotary mechanics. *Trends Biochem. Sci.* **22,** 420–423.
72. Junge, W., Pänke, O., Cherepanov, D. A., Gumbiowski, K., Müller, M., and Englebrecht, S. (2001). Inter-subunit rotation and elastic power transmission in F_0F_1-ATPase. *FEBS Letters* **504**(3), 152–160.
73. Kagawa, Y., Hamamoto, T., and Endo, H. (2000). The α/β interfaces of $\alpha_1\beta_1$, $\alpha_3\beta_3$, and F_1: domain motions and elastic energy stored during γ rotation. *J. Bioenerg. Biomembr.* **32,** 471–484.
74. Kalé, L., Skeel, R., Bhandarkar, M., Brunner, R., Gursoy, A., Krawetz, N., Phillips, J., Shinozaki, A., Varadarajan, K., and Schulten, K. (1999). NAMD2: Greater scalability for parallel molecular dynamics. *J. Comput. Phys.* **151,** 283–312.
75. Kalé, L. V., and Krishnan, S. (1996). Charm++: Parallel programming with message-driven objects. *In* Parallel Programming Using C++, (G. V. Wilson and P. Lu, Eds.), MIT Press, 175–213.
76. Kato-Yamada, Y., Noji, H., Ryohei Yasuda, K. K. J., and Yoshida, M. (1998). Direct observation of the rotation of ε subunit in f_1-atpase. *J. Biol. Chem.* **273,** 19375–19377.
77. Kinosita, K. Jr., Yasuda, R., and Adachi, K. (2000). A rotary molecular motor that can work at near 100% efficiency. *Phil. Trans. Roy. Soc. B* **355,** 473–489.
78. Kosztin, D., Bishop, T. C., and Schulten, K. (1997). Binding of the estrogen receptor to DNA: The role of waters. *Biophys. J.* **73,** 557–570.
79. Krammer, A., Craig, D., Thomas, W. E., Schulten, K., and Vogel, V. (2002). A structural model for force regulated integrin binding to fibronectin's RGD-synergy site. *Matrix Biol.* **21,** 139–147.
80. Krammer, A., Lu, H., Isralewitz, B., Schulten, K., and Vogel, V. (1999). Forced unfolding of the fibronectin type III module reveals a tensile molecular recognition switch. *Proc. Nat. Acad. Sci. USA* **96,** 1351–1356.
81. Leahy, D. J., Aukhil, I., and Erickson, H. P. (1996). 2.0 Å crystal structure of a four-domain segment of human fibronectin encompassing the RGD loop and synergy region. *Cell* **84,** 155–164.
82. Li, J., and Verkman, A. S. (2001). Impaired hearing in mice lacking aquaporin-4 water channels. *J. Biol. Chem.* **276,** 31233–31237.
83. Lindahl, E., Hess, B., and van der Spoel, D. (2001). GROMACS 3.0: a package for molecular simulation and trajectory analysis. *J. Mol. Mod.*
84. Litvinovich, S. V., Brew, S., Aota, S., Akiyama, S., Haudenschild, C., and Ingham, K. (1998). Formation of amyloid-like fibrils by self-association of a partially unfolded fibronectin type III module. *J. Mol. Biol.* **280,** 245–258.
85. Lu, H., Isralewitz, B., Krammer, A., Vogel, V., and Schulten, K. (1998). Unfolding of titin immunoglobulin domains by steered molecular dynamics simulation. *Biophys. J.* **75,** 662–671.

86. Lu, H., and Schulten, K. (2000). The key event in force-induced unfolding of titin's immunoglobulin domains. *Biophys. J.* **79,** 51–65.
87. Lynch, T. W., Kosztin, D., McLean, M. A., Schulten, K., and Sligar, S. G. (2002). Dissecting the molecular origins of protein-nucleic acid recognition: Hydrostatic pressure and molecular dynamics. *Biophys. J.* **82,** 93–98.
88. Ma, J., Flynn, T. C., Cui, Q., Leslie, A. G. W., Walker, J. E., and Karplus, M. (2002). A dynamics analysis of the rotation mechanism for conformational change in F_1-ATPase. *Structure* **10,** 921–931.
89. MacKerell, A. D. Jr., Bashford, D., and Bellott, M. (1992). Self-consistent parameterization of biomolecules for molecular modeling and condensed phase simulations. *FASEB J.* **6**(1), A143–A143.
90. Masaike, T., Mitome, N., Noji, H., Muneyuki, E., Yasuda, R., Kinosita, Jr., K., and Yoshida, M. (2000). Rotation of F_1-ATPase and the hinge structures of the β subunit. *J. Exp. Biol.* **203,** 1–8.
91. McCammon, J. A., Gelin, B. R., and Karplus, M. (1977). Dynamics of folded proteins. *Nature* **267,** 585–590.
92. Menz, R. I., Walker, J. E., and Leslie, A. G. W. (2001). Structure of bovine mitochondrial F_1-ATPase with nucleotide bound to all three catalytic sites: Implications for the mechanism of rotary catalysis. *Cell* **106,** 331–341.
93. Mogilner, A., Wang, H., Elston, T., and Oster, G. (2002). Molecular motors: theory & experiment. *In* Comput. Cell Biol., (C. Fall, E. Marland, J. Wagner and J. Tyson, Eds.), Springer, New York, 321–380.
94. Monticello, R. A., and Brusilow, W. S. (1994). Role of the delta subunit in enhancing proton conduction through the F0 of the *Escherichia coli* F1F0 ATPase. *J. Bacteriol.* **176**(5), 1383–1389.
95. Murata, K., Mitsuoka, K., Hirai, T., Walz, T., Agre, P., Heymann, J. B., Engel, A., and Fujiyoshi, Y. (2000). Structural determinants of water permeation through aquaporin-1. *Nature* **407,** 599–605.
96. Noji, H., Häsler, K., Junge, W., Kinosita, Jr., K., Yoshida, M., and Engelbrecht, S. (1999). Rotation of *escheria coli* F_1-ATPase. *Biochem. Biophys. Res. Comm.* **260,** 597–599.
97. Noji, H., Yasuda, T., Yoshida, M., and Kinosita, Jr., K. (1997). Direct observation of the rotation of F1-ATPase. *Nature* **386,** 299–302.
98. Oberhauser, A., Badilla-Fernandez, C., Carrion-Vazquez, M., and Fernandez, J. (2002). The mechanical hierarchies of fibronectin observed with single molecule AFM. *J. Mol. Biol.* **319,** 433–447.
99. Oberhauser, A. F., Marszalek, P. E., Erickson, H., and Fernandez, J. (1998). The molecular elasticity of tenascin, an extracellular matrix protein. *Nature* **393,** 181–185.
100. Ohashi, T., Kiehart, D. P., and Erickson, H. P. (1999). Dynamics and elasticity of the fibronectin matrix in living cell culture visualized by fibronectin-green fluorescent protein. *Proc. Nat. Acad. Sci. USA* **96,** 2153–2158.
101. Oster, G., and Wang, H. (2000). Reverse engineering a protein: the mechanochemistry of ATP synthase. *Biochimi. Biophys. Acta* **1458,** 482–510.
102. Paci, E., and Karplus, M. (1999). Forced unfolding of fibronectin type 3 modules: an analysis by biased molecular dynamics simulations. *J. Mol. Biol.* **288,** 441–459.
103. Pänke, O., Cherepanov, D. A., Gumbowski, K., Engelbrecht, S., and Junge, W. (2001). Viscoelastic dyanamics of actin filaments coupled to rotary F-ATPase: angular torque profile of the enzyme. *Biophys. J.* **81,** 1220–1233.

104. Pänke, O., and Rumberg, B. (1999). Kinetic modeling of rotary CF_0F_1-ATP synthase: storage of elastic energy during energy transduction. *Biochim. Biophys. Acta* **1412,** 118–128.

105. Phillips, J. C., Wriggers, W., Li, Z., Jonas, A., and Schulten, K. (1997). Predicting the structure of apolipoprotein A-I in reconstituted high density lipoprotein disks. *Biophys. J.* **73,** 2337–2346.

106. Phillips, R., Dittrich, M., and Schulten, K. (2002). Quasicontinuum representations of atomic-scale mechanics: From proteins to dislocations. *Ann. Rev. Mat. Res.* **32,** 219–233.

107. Pomès, R., and Roux, B. (1996). Structure and dynamics of a proton wire: a theoretical study of H^+ translocation along the single-file water chain in the gramicidin A channel. *Biophys. J.* **71**(1), 19–39.

108. Pomès, R., and Roux, B. (1996). Theoretical study of H^+ translocation along a model proton wire. *J. Phys. Chem.* **100**(7), 2519–2527.

109. Pomès, R., and Roux, B. (2002). Molecular mechanism of H^+ conduction in the single-file water chain of the gramicidin channel. *Biophys. J.* **82,** 2304–2316.

110. Preston, G. M., and Agre, P. (1991). Isolation of the cDNA for erythrocyte integral membrane-protein of 28-kD – member of an ancient channel family. *Proc. Nat. Acad. Sci. USA* **88,** 11110–11114.

111. Randa, H. S., Forrest, L. R., Voth, G. A., and Sansom, M. S. P. (1999). Molecular dynamics of synthetic leucine-serine ion channels in a phospholipid membrane. *Biophys. J.* **77**(5), 2400–2410.

112. Rastogi, V. K., and Girvin, M. E. (1999). Structural changes linked to proton translocation by subunit *c* of the ATP synthase. *Nature* **402,** 263–268.

113. Ren, G., Reddy, V. S., Cheng, A., Melnyk, P., and Mitra, A. K. (2001). Visualization of a water-selective pore by electron crystallography in vitreous ice. *Proc. Nat. Acad. Sci. USA* **98,** 1398–1403.

114. Ren, H., and Allison, W. S. (2000). On what makes the γ subunit spin during ATP hydrolysis by F_1. *Biochim. Biophys. Acta Bioenerg.* **1458,** 221–233.

115. Reuss, L. (1998). Focus on "effect of expressing the water channel aquaporin-1 on the CO2 permeability of *Xenopus* oocytes." *Am. J. Physiol.* **274,** C297–C298.

116. Rickey, D. P., and Lin, E. E. C. (1972). Importance of facilitated diffusion for effective utilization of glycerol by *Escherichia coli*. *J. Bacteriol.* **112**(2), 784–790.

117. Rief, M., Gautel, M., Schemmel, A., and Gaub, H. (1998). The mechanical stability of immunoglobulin and fibronectin III domains in the muscle protein titin measured by AFM. *Biophys. J.* **75,** 3008–3014.

118. Saam, J., Tajkhorshid, E., Hayashi, S., and Schulten, K. (2002). Molecular dynamics investigation of primary photoinduced events in the activation of rhodopsin. *Biophys. J.* **83,** 3097–3112.

119. Sansom, M. S. P., and Law, R. J. (2001). Membrane proteins: Aquaporins—channels without ions. *Curr. Biol.* **11,** R71–R73.

120. Schlitter, J., Engels, M., Krüger, P., Jacoby, E., and Wollmer, A. (1993). Targeted molecular dynamics simulation of conformational change—application to the $T \leftrightarrow R$ transition in insulin. *Mol. Simul.* **10**(2–6), 291–308.

121. Schwarzbauer, J. E., and Sechler, J. L. (1999). Fibronectin fibrillogenesis: a paradigm for extracellular matrix assembly. *Curr. Opin. Cell Biol.* **11,** 622–627.

122. Senior, A. E., Nadanaciva, S., and Weber, J. (2002). The molecular mechanism of ATP synthesis by F_1F_o-ATP synthase. *Biochim. Biophys. Acta* **1553,** 188–211.

123. Sharma, A., Askari, J. A., Humphries, M. J., Jones, E. Y., and Stuart, D. I. (1999). Crystal structure of a heparin- and integrin-binding segment of human fibronectin. *EMBO J.* **18,** 1468–1479.
124. Shi, L. B., Skach, W. R., and Verkman, A. S. (1994). Functional independence of monomeric CHIP28 water channels revealed by expression of wild-type-mutant heterodimers. *J. Biol. Chem.* **269,** 10417–10422.
125. Stock, D., Leslie, A. G. W., and Walker, J. E. (1999). Molecular architecture of the rotary motor in ATP synthase. *Science* **286,** 1700–1705.
126. Sui, H., Han, B. G., Lee, J. K., Walian, P., and Jap, B. K. (2001). Structural basis of water-specific transport through the AQP1 water channel. *Nature* **414,** 872–878.
127. Sun, S., Chandler, D., Dinner, A. R., and Oster, G. (2003). Elastic energy storage in F_1-ATPase. *Eur. Biophys. J.* In press.
128. Tajkhorshid, E., Nollert, P., Jensen, M. Ø., Miercke, L. J. W., O'Connell, J., Stroud, R. M., and Schulten, K. (2002). Control of the selectivity of the aquaporin water channel family by global orientational tuning. *Science* **296,** 525–530.
129. Takeyama, M., Noumi, T., Maeda, M., and Futai, M. (1988). F_0 portion of *Escherichia coli* h^+-ATPase. *J. Biol. Chem.* **31**(5), 16106–16112.
130. Takeyasu, K., Omote, H., Nettikadan, S., Tokumasu, F., Iwamoto-Kihara, A., and Futai, M. (1996). Molecular imaging of *Escherichia coli* F0F1-ATPase in reconstituted membranes using atomic force microscopy. *FEBS Lett.* **392,** 110–113.
131. Valiyaveetil, F. I., and Fillingame, R. H. (1997). On the role of Arg-210 and Glu-219 of subunit *a* in proton translocation by the *Escherichia coli* F_1F_0 ATP synthase. *J. Biol. Chem.* **272,** 32635–32641.
132. Verkman, A. S., and Mitra, A. K. (2000). Structure and function of aquaporin water channels. *Am. J. Phys. – Renal Phys.* **278,** F13–F28.
133. Vogel, V., Thomas, W. E., Craig, D. W., Krammer, A., and Baneyx, G. (2001). Structural insights into the mechanical regulation of molecular recognition sites. *Trends Biotechnol.* **19,** 416–423.
134. Walz, T., and Ghosh, R. (1997). Two-dimensional crystallization of the light-harvesting I reaction centre photounit from *Rhodospirillum rubrum*. *J. Mol. Biol.* **265,** 107–111.
135. Walz, T., Hirai, T., Murata, K., Heynmann, J. B., Mitsuoka, K., Fujiyoshi, Y., Smith, B. L., Agre, P., and Engel, A. (1997). The three-dimensional structure of aquaporin-1. *Nature* **387,** 624–627.
136. Wang, H., and Oster, G. (2002). Ratchets, power strokes, and molecular motors. *Appl. Phys. A* **75,** 315–323.
137. Wang, H. Y., and Oster, G. (1998). Energy transduction in the F_1 motor of ATP synthase. *Nature* **396,** 279–282.
138. Weber, J., and Senior, A. E. (2000). ATP synthase: what we know about ATP hydrolysis and what we do not know about ATP synthesis. *Biochim. Biophys. Acta Bioenerg.* **1458,** 300–309.
139. Weiner, P. K., and Kollman, P. A. (1981). AMBER: Assisted model building with energy refinement. A general program for modeling molecules and their interactions. *J. Comput. Chem.* **2**(3), 287–303.
140. Wistow, G. J., Pisano, M. M., and Chepelinsky, A. B. (1991). Tandem sequence repeats in transmembrane channel proteins. *Trends Biochem. Sci.* **16,** 170–171.
141. Wriggers, W., and Schulten, K. (1999). Investigating a back door mechanism of actin phosphate release by steered molecular dynamics. *PROTEINS: Struct., Funct., Genet.* **35,** 262–273.

142. Xiong, J. P., Stehle, T., Diefenbach, B., Zhang, R., Dunker, R., Scott, D. L., Joachimiak, A., Goodman, S. L., and Arnaout, M. A. (2001). Crystal structure of the extracellular segment of integrin $\alpha v\beta 3$. *Science* **294,** 339–345.
143. Xiong, J. P., Stehle, T., Zhang, R., Joachimiak, A., Frech, M., Goodman, S. L., and Arnaout, M. A. (2002). Crystal structure of the extracellular segment of integrin $\alpha v\beta 3$ in complex with an arg-gly-asp ligand. *Science* **296,** 151–155.
144. Yasuda, R., Noji, H., Yoshida, M., Kinosita Jr., K., and Itoh, H. (2001). Resolution of distinct rotational substeps by submillisecond kinetic analysis of F_1-ATP-ase. *Nature* **410,** 898–904.
145. Zeidel, M. L., Ambudkar, S. V., Smith, B. L., and Agre, P. (1992). Reconstitution of functional water channels in liposomes containing purified red cell CHIP28 protein. *Biochem.* **31,** 7436–7440.
146. Zhang, L., and Hermans, J. (1996). Hydrophilicity of cavities in proteins. *PROTEINS: Struct., Funct., Genet.* **24,** 433–438.
147. Zhang, Y., and Fillingame, R. H. (1994). Essential aspartates in subunit *c* of F_1F_0 ATP synthase. *J. Biol. Chem.* **269,** 5473–5479.
148. Zhou, F., and Schulten, K. (1995). Molecular dynamics study of a membrane–water interface. *J. Phys. Chem.* **99,** 2194–2208.
149. Zhu, F., Tajkhorshid, E., and Schulten, K. (2001). Molecular dynamics study of aquaporin-1 water channel in a lipid bilayer. *FEBS Lett.* **504,** 212–218.
150. Zhu, F., Tajkhorshid, E., and Schulten, K. (2002). Pressure-induced water transport in membrane channels studied by molecular dynamics. *Biophys. J.* **83,** 154–160.

STRUCTURE/FUNCTION CORRELATIONS OF PROTEINS USING MM, QM/MM, AND RELATED APPROACHES: METHODS, CONCEPTS, PITFALLS, AND CURRENT PROGRESS

By A. SHURKI AND A. WARSHEL

Department of Chemistry, University of Southern California, Los Angeles, California 90089-1062

I. Introduction 249
II. Classical Force Fields and Their Use 250
A. Potential Functions and Molecular Mechanics 250
B. Molecular Dynamics and Direct Simulations 252
C. Proper and Improper Treatments of Long-Range Effects 254
D. FEP and LRA Calculations by All-atom Models 255
III. Evaluation of Electrostatic Free Energies of Macromolecules 258
A. Computational Models 258
B. Calculating Binding Free Energies 261
C. Calculations of Protonation States 262
IV. Methods For Simulation of Chemical Processes in Enzymes 263
A. Quantum Mechanical Studies of the Reacting Fragments in the Gas Phase 263
B. QM/MM Molecular Orbital Methods 265
C. The EVB as a Reliable QM/MM Method 269
D. QM Treatments of the Entire Protein 271
V. Studying Enzyme Catalysis 274
A. General Considerations 274
B. Reactant State Destabilization or TS Stabilization: General Considerations and Definition 278
C. Modeling the Overall Reaction and Reproducing the Corresponding Catalytic Effect 279
D. Examining How Enzymes Really Work 282
VI. Concluding Remarks 301
References 302

I. Introduction

One of the primary goals of modern molecular biology is to provide a reliable correlation between the structure and function of biomolecules. The importance of this challenging task has become even more evident with the advance of the human genome project and the emergence of proteomics. Clearly, the "translation" of the vast, accumulating genetic information to functional information requires moving from sequence to structure and from structure to function. Fortunately, the enormous progress in

ADVANCES IN PROTEIN CHEMISTRY, Vol. 66

0065-3233/03 $35.00

structural studies makes the second step somewhat independent from the first step, since we already have an extensive set of structural benchmarks.

This review will consider the progress and the current state of structure-function correlation of biomolecules. Enzyme activity will be emphasized but other aspects of protein functions will also be addressed.

The review will start with a review of classical force field approaches and their use in macromolecular modeling. That section will briefly cover molecular mechanics, molecular dynamics, and evaluation of free energy using the free energy perturbation (FEP) and linear response approximation (LRA) approaches. It will also emphasize the need for proper long-range treatments and the problem with using a single very long trajectory to explore biological functions. The next section will consider models for studies of electrostatic energies in proteins and their use in evaluation of pK_a's, redox potentials, and binding free energies. The main part of the review deals with the modeling of enzyme catalysis. This part will consider several modeling approaches ranging from the empirical valence bond (EVB) to different molecular hybrid quantum mechanics/molecular mechanics (QM/MM) orbital approaches as well as approaches that attempt to treat the entire enzyme quantum mechanically. After reviewing the methods, a careful account of current progress in studies of enzymatic reactions emphasizing the use of modeling approaches will be used to elucidate different mechanistic proposals.

II. Classical Force Fields and Their Use

A. *Potential Functions and Molecular Mechanics*

Functional properties of a protein can be evaluated by knowing the exact structure/energy relationship of the given protein. This relationship is given explicitly by the corresponding Born Oppenhimer potential surface. Such potential surfaces can be evaluated in principle by using quantum mechanics; however, such approaches are presently too expensive to effectively model large molecules. Alternatively, one can utilize the knowledge that macromolecules are assembled by the same type of bonds that connect the atoms in small molecules. Thus, one can describe large molecules as a collection of small molecular fragments where the overall potential surface is expressed as a sum of contributions from bonded atoms and interactions between nonbonded atoms. Such a presentation is usually obtained by using sets of analytical functions that approximate the true potential surface and are called potential functions or force fields. The functional forms and parameters of molecular force fields are taken from studies of small molecules with the implicit assumption that these functions are transferable from small to large molecules. Molecular potential functions are usually given in the form

$$U(\boldsymbol{s}) = U_{b,\theta}(\boldsymbol{b}, \boldsymbol{\theta}) + U_{\phi}(\phi) + U_{nb}(\boldsymbol{r}) \tag{1}$$

where the internal coordinates vector, $\boldsymbol{s}$, includes $\boldsymbol{b}, \boldsymbol{\theta}, \boldsymbol{\phi}$, and $\boldsymbol{r}$, which are, respectively, the vectors of bond lengths, bond angles, torsion angles, and the vector of the Cartesian coordinates that are used to evaluate the nonbonded distances. The right hand side terms of Eq. 1 are usually given as simple functions of the corresponding internal coordinates (see e.g., [1]). The first term, for example, is represented by

$$U_{b,\theta}(\boldsymbol{b}, \boldsymbol{\theta}) = \frac{1}{2}\sum_i K_{b,i}\left(b_i - b_{0,i}\right)^2 + \frac{1}{2}\sum_i K_{\theta,i}\left(\theta_i - \theta_{0,i}\right)^2 + \textit{cross terms} \tag{2}$$

The introduction of potential functions enabled conformational analysis via computers (e.g., [2, 3]). However, the evaluation of first and second derivatives with respect to internal coordinates appeared to be almost impossible for general topologies (e.g., ring molecules). Thus, it is the formulation and treatment of potential functions in terms of the corresponding Cartesian coordinates [4] that paved the way for conformational analysis of general molecular systems including proteins [5] and for molecular dynamics simulations of macromolecules (see below).

The parameters of the different potential functions (e.g., K_b and K_θ) can be optimized. This is done by calculating different independent molecular properties for an initial set of parameters and then minimizing the difference between the calculated and observed properties by adjusting the parameters. Performing such a fitting while considering energies, structures, and vibrations and when using a least-squares procedure generates the so-called consistent force field (CFF) (see [4, 6]).

Modeling protein functions by force field approaches may lead to irrelevant conclusions if one ignores the solvent environment around the given protein. The corresponding solvent effect is not a small perturbation but a major contribution to the overall energy and forces. Ignoring the effect of the solvent around a protein active site, for example, can lead to entirely incorrect ionization states of the catalytic residues.

The effect of the solvent around proteins can be considered by either all-atom solvent models [7–9] or simplified solvent models (e.g., [10]) as a part of the overall potential function. One can also use implicit solvent models (e.g., [11]). Models of solvent effects will be discussed in section III.

Calculations of properties of molecules using the corresponding potential functions are called molecular mechanics (MM). This term reflects the fact that the representation of a molecule by a force field is similar to the use of a mechanical model of balls and springs, where the balls represent the atoms and the springs determine the stretching and bending motions.

MM approaches involve several techniques aimed at determining different molecular properties. For example, with a given set of analytical potential functions one can evaluate the molecular equilibrium geometries and the vibrations around these configurations. The simplest way to accomplish this task is by using Cartesian representation as described in [6]. In addition, the use of MM approaches can be extended to complex systems such as molecular crystals. In these systems one exploits the periodicity of the given crystal and drastically reduces the complexity of the problem. Here again the use of Cartesian representation is particularly effective [6].

The availability of first and second derivative sets allows one to use effective methods in conformational studies of medium size molecules [1, 6]. However, exploring the conformational space of macromolecules by MM approaches in not so effective. In such systems one deals with enormous dimensionality and thus colossal numbers of local minima. Here one may try to generate a large number of initial configurations at different regions of the conformational space and then find the corresponding local minima. Unfortunately, such an approach requires an extremely large amount of computer time. Exploring the configurational space of macromolecules can be done more effectively by the Monte Carlo and molecular dynamics approaches.

B. Molecular Dynamics and Direct Simulations

Molecular dynamics (MD) simulations evaluate the motion of atoms in a given system and provide the positions (trajectory) of these atoms as a function of time. The trajectories are calculated by solving the classical equations of motion for the considered molecule. Such calculations require the knowledge of all the forces that act on the system. Fortunately, these forces are trivially obtained from the same Cartesian first derivatives of the potential functions developed originally in [4] for energy minimization studies ($F_i = -dU/dx_i$). The actual evaluation of classical trajectories is done numerically using the analytical forces (see [8]).

The propagation of atoms of a given system in classical trajectories corresponds to a fixed total energy (determined by the specified initial conditions). However, the evaluation of statistical mechanical averages implies that the system included in the simulation is a part of a much larger system (ensemble) whose atoms are not considered in an explicit way. Thus, in order to simulate a given macroscopic property at a specified temperature, one should introduce into the system some kind of a "thermostat" that will keep it at the desired temperature. This can be easily accomplished by assuming, for example, equal partition of kinetic energy among all degrees of freedom. The different approaches for constant temperature simulations ranging from very simple (e.g., [1]) to sophisticated considerations are described elsewhere [8].

The appeal of MD approaches is associated with their ability to simulate, at least in principle, true microscopic behaviors of macromolecules. The weakness is associated with the fact that some properties reflect relatively slow processes that occur during an extremely long time and cannot be simulated by any current computer.

The emergence of MD simulations in studies of biological systems can be traced back to a simulation of the dynamics of the primary event in the visual process [12] that correctly predicted a photoisomerization process of around 100 fs (femto seconds). A subsequent study [13] attempted to examine the heat capacity of bovine pancreatic trypsin inhibitor (BPTI) by a very short simulation of this protein in vacuum. However, at the early stages of the fields development, it was very difficult to obtain meaningful results or to reach a reasonable convergence (e.g., the heat capacity was drastically underestimated [14], reflecting artificial relaxation motions). Nevertheless, ultra-fast photobiological processes could be simulated even at this early stage [12, 15].

The gradual increase of computer power made it eventually possible to obtain meaningful average properties of macromolecules by MD simulations (see section II.D). Nevertheless, the perception that direct-MD simulations can provide a simple prescription for studying biological systems can be highly deceptive. In particular, there is a tendency to believe that slow biological processes can be understood by a single long-time simulation. For example, in a discussion of MD simulations of ion current in [16], it has been stated that, "one might expect to need about 1 μs of time in order to see an ion fully traverse the pore; this huge calculation might give an enormous payoff in our understanding of what forces control the ion at various stages of permeation." Unfortunately, the assumption that a single long trajectory is sufficient to teach us about a biological process tends to encourage superficial and frequently irrelevant studies. More specifically, simulations of biological systems require proper treatment of many complex aspects (e.g., long-range effects) which might not be properly represented in the given simulation program. Obviously, a single simulation run is not capable of ensuring the validity of the simulated result. Moreover, it is frequently essential to obtain an average over many trajectories in order to obtain the relevant macroscopic observable. Here, it is essential to consider the simulation as an "experimental" approach, and to examine whether key features of the given system can be reproduced by changing the simulation conditions. In particular, it is crucial to examine the reproduction of properties, which are relevant to the process under study (e.g., ion binding energies in studies of ion current, or pK_a of catalytic groups in studies of enzymatic reactions). Even after these verification studies, it is essential to explore the process of interest by many simulation runs and determine the sensitivity of the results to the simulation conditions. Of course, a single very-long simulation does not allow one to perform any validation study. Here, it is essential to develop indirect approaches that would allow one to examine the relevant process with relatively short simulation times. This includes effective free energy calculations, using subsystems with proper boundary conditions, etc. Such approaches will be considered in later sections.

C. *Proper and Improper Treatments of Long-Range Effects*

Explicit all-atom simulations of macromolecules embedded in their actual solvent environment are extremely challenging, even with the current computer power. The key problem is associated with the long-range nature of electrostatic interactions. The number of pair-wise interactions increases like the square of the number of atoms in the system (N^2) while the number of atoms in a spherical shell increases like the square of the radius of the system. This leads to a vast number of interactions if all are taken into account in simulations of macromolecules. The customary approach is to include only the interactions within a relatively small cutoff distance. This approach, however, leads to major problems in treatments of energies of charged groups in proteins (for a demonstration see [17]).

Early realization of the need for complete long-range treatments led to a gradual development of a series of models and ideas. One of the first advances was the demonstration that solvated systems can be represented by spherical models surrounded by a bulk region [18]. The main conceptual advance of these models was the introduction of buffer (surface) constraints that forced the surface region to behave as if it is a part of an infinite system [9, 18]. The surface constraints models and in particular the surface constraint all-atom solvent (SCAAS) model [9] include special polarization constraints that forced the surface region to have the angular polarization of the corresponding infinite system. These constraints were found to be crucial for proper treatments of long-range electrostatic effects. The importance of proper electrostatic constraint has not been widely appreciated and many popular treatments emphasize the proper treatments of the temperature in the surface region, which can be easily satisfied by any "thermostat" in the so-called "stochastic boundary conditions'' [19], rather than considering the polarization of this region and emphasizing proper treatment of electrostatic energies.

Although the use of the SCAAS model provides a proper way of overcoming the long-range problem, it still requires the expensive evaluation of many interactions when one uses no cutoff and spherical models of significant size. Probably the first attempt to reduce the cutoff problem, and thus to evaluate a smaller number of interactions in spherical models was introduced by Kowajima and Warshel [20], who developed a special spherical Ewald-type model. A subsequent development [21] led to the more convenient local reaction field (LRF) model. The stability and reliability of this model has been validated repeatedly [17, 21, 22]. Alternative long-range treatments for spherical models are also available (e.g., [23]), but these models do not yet involve proper surface constraints.

Another way of addressing the long-range problem is using periodic boundary conditions and different versions of the Ewald method [24–26]. Periodic Ewald treatments greatly improve the simulated structural properties but at present we are not aware of the validation of such approaches in the evaluation of electrostatic energies in proteins (validation means obtaining the same results as those without cutoff and with a very large unit cell). In fact, it is not clear that the present implementations of the periodic Ewald treatments can give correct electrostatic free energies in proteins (see discussion in [22]). The very slow realization that proper treatments of long-range effects is crucial for proper modeling of protein functions is, in fact, still diagnostic of the field. Many workers emphasize the length of the simulations and the number of atoms in the simulation system instead of validating the calculated electrostatic free energies.

D. FEP and LRA Calculations by All-Atom Models

Thermodynamic cycles of the same type used in chemistry (e.g., the Born-Haber cycle) can describe many biochemical processes. Such cycles allow one to replace the very challenging direct calculations of physical processes with calculations of much simpler physical or non-physical processes. For example, the calculation of the difference in the binding energy of the substrate to the protein in reactant state (RS) and in the transition state (TS) can be replaced by the difference between the free energy required to bring RS to TS, both in water and in protein, which involves calculation of the corresponding free energy barriers [27]. Similarly, the difference between the pK_a of an acid in the protein and in water can be calculated by the non-physical process of ''mutating'' the protonated acid to its ionized form in the protein and in water instead of calculating the binding energies of the protonated and ionized forms of the acid [28, 29]. The use of thermodynamic cycles in microscopic calculations of free energies of biological processes had been introduced in the early 1980s [28, 29] and it is now a major part of many approaches in the free-energy calculations of biological processes (e.g., [7]).

Evaluation of free energies by statistical mechanical approaches is extremely time-consuming, due to sampling problems. Fortunately, it is possible in some cases to obtain meaningful results using perturbation approaches. Such calculations are usually done by the so-called free-energy perturbation (FEP) method [30, 31] and the related umbrella sampling (US) method [31]. The FEP method evaluates the free energy associated with the change of the potential surface from U_1 to U_2 by gradually changing the potential surface using the relationship

$$U_m(\lambda_m) = U_1(1 - \lambda_m) + U_2\lambda_m \tag{3}$$

where λ_m is a parameter that changes between $(0 = \lambda_m = 1)$. The free-energy increment, associated with the change of U_m, can be obtained by [31]

$$\exp\{-\delta G(\lambda_m \rightarrow \lambda_{m+1},)\beta\} = \langle \exp\{-(U_{m+1}, -U_m)\beta\}\rangle_m \tag{4}$$

where $\langle\ \rangle_m$ indicates that the given average is evaluated by propagating trajectories over U_m. The overall free-energy change is now obtained by changing λ_m in n equal increments and evaluating the sum of the corresponding ΔG:

$$\Delta G(U_1 \rightarrow U_2) = \sum_{m=0}^{n-1} \delta G(\lambda_m \rightarrow \lambda_{m+1}) \tag{5}$$

The FEP approach has been used extensively in studies of free energies of biological systems (e.g., [7, 32]).

In many cases, such as binding large ligands, it is very hard to perform converging FEP calculations. In such cases it is extremely useful to estimate the free energy of biological processes by an equation derived by Lee *et al.* [33] and used in studies of ligand binding to proteins. This equation expresses the free energy associated with changing the potential of the system from U_1 to U_2 by

$$\Delta G(U_1 \rightarrow U_2) = \frac{1}{2}\left(\langle U_2 - U_1\rangle_1 + \langle U_2 - U_1\rangle_2\right) \tag{6}$$

The derivation of this equation was based on the assumption that the linear response approximation (LRA) [34] is valid. Namely, the protein and solvent environments respond linearly to the force associated with the given process. This assumption is, in fact, the basis of the macroscopic electrostatic theory where the free energy of charging a positive ion is given by the well-known result $\Delta G = U(Q = 1)/2$, where U is the electrostatic potential of the given charge [10]. Although it is hard to accept that the LRA can provide a reliable way of describing the energetics of macromolecules or of realistic molecular systems, it was found by simulation studies that it is a reasonable approximation, in particular for processes that depend on electrostatic effects [35–37].

Many biological processes can be formulated in terms of the probability of being on a given potential surface along a specified reaction coordinate x. The corresponding free energy $\Delta g(x)$ is obtained by combining the FEP procedure with the US approach [31]. The combined FEP/US approach gives the free-energy function $\Delta g(x)$ as [35, 38, 39]

$$\Delta g(x') = \Delta G_m - \beta^{-1}\ln\langle\delta(x - x')\exp\{-\beta(E - U_m)\}\rangle_{U_m} \tag{7}$$

where E is the potential surface of interest and U_m is the mapping potential that keeps the system near x'.

Another quantity of interest is the so-called potential of mean force (PMF). This quantity represents the probability of being at a given x while averaging on all other degrees of freedom, and it can be written as

$$\Delta G(x_m) = \Delta G(x_{m-1}) + \delta G(x_{m-1} \rightarrow x_m) \tag{8}$$

$$\exp\{\delta G(x_{m-1} \rightarrow x_m)\beta\} = \langle\exp\{-[E(x_m) - E(x_{m-1})]\beta\}\rangle_{E(x_{m-1})} \tag{9}$$

Interestingly, the rather simple alternative evaluation of the PMF by the FEP/US method using Eq. (7) has involved some conceptual difficulties. It seems that the idea of "umbrella sampling" was sometimes taken too literally, where it was assumed that one should run different independent runs with different U_m and then try to "overlap" the corresponding results from the second term of Eq. (7) with an arbitrary constant. The obvious idea (e.g., [35, 38, 39]) that ΔG_m of Eq. (7) should give this constant has been considered as a significant advance (e.g., see discussion in [40]).

The above approaches and in particular the FEP/US and PMF methods are in principle rigorous. However, it must be emphasized that the convergence of such approaches is extremely slow and that even proper convergence does not guarantee proper results. For example, FEP calculations with improper long-range treatments cannot provide proper electrostatic energy. Furthermore, studies that do not use surface constraint spherical models require very large simulation systems and thus involve slow convergence. The best advice in using such method is to perform validation studies and to see, for example, if the given method can reproduce observed electrostatic energies with different sizes of the explicit simulation system (e.g. [22, 41]). Interestingly, the PMF approach, which appears the most rigorous, may make one ''blind'' to problems with the simulation model. For example, PMF studies of ion penetration in ion channels do not tell us about the absolute energy of moving the ion from water to the given channel site. Thus, one might be led to believe that the electrostatic model is correct even if the absolute energies are entirely incorrect (see discussion in [42]).

III. Evaluation of Electrostatic Free Energies of Macromolecules

A. *Computational Models*

One of the most effective correlations between the structure and function of proteins is provided by the corresponding electrostatic energies (e.g., [10]). In order to obtain reliable information from such a correlation, it is essential to have a reliable way for computing electrostatic free energies in macromolecules. The treatments of electrostatic energies in proteins have evolved through different models with different degrees of sophistication and sometimes with major oversimplifications. Early models that treated the protein as a uniform low-dielectric sphere (e.g., the influential Tanford-Kirkwood (TK) model [43]) overlooked the crucial contribution of self-energy of the charged groups and the important role of the protein permanent dipoles (see [44]). The first consistent treatment of electrostatic energies in proteins [45] represented all the relevant electrostatic interactions microscopically but used a simplified model of Langevin-type dipoles on a grid to represent the solvent. This protein dipoles Langevin dipoles (PDLD) model facilitated consistent studies of the role of electrostatic energies in protein function with the limited computer power of the mid-1970's (see [10]). Such studies considered microscopically the shape of the protein, its local polarity as well as the effect of the solvent around the protein, and to a limited extent the effect of the protein reorganization (which was modeled by energy minimization). Subsequent advance of discretized continuum (DC) models, which are frequently referred to as Poisson-Boltzmann (PB) models, have allowed macroscopic continuum approaches to take into account the shape of the protein [46, 47]. Early DC models (until the late 80's) involved some of the major inconsistencies of the TK model. More specifically, these models ignored the microscopic nature of the protein permanent dipoles (see discussion in ref. [48]). However, recent PB models have represented these dipoles explicitly and actually became semimacroscopic models [49], which give more stable (though not necessarily accurate) results than the more microscopic PDLD models. This interesting feature is due to the fact that the microscopic models involve compensation of large numbers while the macroscopic models assume implicitly that the compensation exist by using a protein dielectric constant ε_p. Thus, PB models give excellent results when dealing with surface groups (where the solvation energy in the protein is similar to that in water) but become more problematic when one deals with groups in the protein interiors. The problem is that the results depend strongly on the assumed ε_p, which apparently has little to do with the actual protein dielectric but merely with the contributions that are treated implicitly (see [50]). A promising way to exploit the stability of the semimacroscopic models, and yet to keep a clear physical picture, is the semimacroscopic version of the PDLD model (the PDLD/S model [51]). The PDLD/S model takes the PDLD model and scales the corresponding contributions by assuming that the protein has a dielectric constant ε_p. In order to reduce the unknown factors in ε_p, it is useful to move to the PDLD/S-LRA model, where the PDLD/S energy is evaluated within the LRA approximation of Eq. (6) with an average on the configurations generated by MD simulation of the charged and the uncharged states (see [52–54]). In this way one uses MD simulations to automatically generate protein configurations for the charged and uncharged forms of the given "solute" and then average these contributions according to Eq. (6). Because the protein reorganization is considered explicitly, we have less uncertainties with ε_p. It is also important to mention that the recently introduced molecular mechanics Poisson-Boltzmann surface area (MM-PBSA) model [55] is basically an adaptation of the PDLD/S-LRA idea of MD generation of configurations for implicit solvent calculations that only calculates the average of the configurations generated with the charged solute (the first term of Eq. (6)).

The tendency to focus on electrostatic benchmarks of surface groups in proteins (whose energy is similar to the corresponding energy in water) has slowed down the realization of the nature of ε_p and the validity of different electrostatic models (see [54]). Apparently, only studies of groups in protein interiors can provide proper validation of macroscopic electrostatic models. With the proper benchmark one finds that the main problem of the PB models is the proper selection of ε_p and that different ε_p should be used for different properties and interactions (see [54]).

The conceptual problems associated with the use of macroscopic models do not exist in principle when one uses microscopic simulations. However, the use of such models in evaluation of electrostatic models is still extremely challenging. First, many workers are still unfamiliar with the need for proper long-range treatments. Second, despite the emergence of FEP calculations of electrostatic energies in proteins (e.g., ref. [51]), there are still major convergence problems in both FEP- and LRA-type studies (e.g., [53,56]).

Electrostatic energies in proteins can be formulated in terms of a two-step thermodynamic cycle. The first step involves the transfer of each ionized group from water to its specific protein site where all other groups are neutral, and a second step where the interaction between the ionized groups is turned on [54]. Thus, we can write the energy of moving ionized groups from water to their protein site as

$$\Delta G = \sum_{i=i}^{N} \Delta G_i + \sum_{i>j} \Delta G_{ij} \tag{10}$$

where the first and second terms correspond respectively to the first and second steps. ΔG_i, which represent the change in self-energies of the i-th ionized group upon transfer from water to the protein site, can be evaluated by the PDLD/S-LRA approach with a relatively small ε_{p}. However, the charge-charge interaction terms (the ΔG_{ij}) are best reproduced by using the simple Coulomb's-type law

$$\Delta G_{ij} = 332 Q_i Q_j / r_{ij} \varepsilon_{ij} \tag{11}$$

where ΔG_{ij} is given in kcal/mol, r_{ij} in Å and ε_{ij} is a relatively high distance dependent dielectric constant [44]. Although the fact that ε_{ij} is large has been supported by mutation experiments (e.g., [41, 57]) and conceptual considerations [10, 58], the use of Eq. (11) is still considered by many as a poor approximation for PB treatments (without realizing that the PB approach depends entirely on an unknown ε_{p}). Interestingly, the so-called generalized Born (GB) model, whose usefulness is now widely appreciated [59], is basically a combination of Eq. (11) and the Born energy of the individual charges. Thus, the GB model is merely a version of an earlier treatment [10]. More specifically, as pointed out originally by Warshel and coworkers [10, 60], the energy of an ion pair in a uniform dielectric medium can be written as [10]

$$\Delta G^{+-} = 332/r + \Delta G_{sol} = \Delta G_{sol}^{\infty} + 332/r\varepsilon \tag{12}$$

where the free energies are given in kcal/mol and the distances in Å. ΔG_{sol} is the solvation energy of the ions and ΔG_{sol}^{∞} is the solvation of the ions at infinite separation. Equation (12) gives (see also [10, 60])

$$\begin{aligned} \Delta G_{sol} = \Delta G_{sol}^{\infty} - (332/r)(1 - 1/\varepsilon) = -166[(1/a^{+} + 1/a^{-})(1 - 1/\varepsilon) \\ + (2/r)(1 - 1/\varepsilon)] \end{aligned} \tag{13}$$

where the a′s are the Born′s radii of the indicated ions. Equation (13) with some empirical modifications leads to the GB treatment [61]. Thus, the widely accepted GB treatment is in fact a glorified Coulomb's law treatment [49]. Of course, nothing is wrong with such a treatment once we realize that the use of large ε for the second term in Eq. (12) is justified even for protein interiors. Furthermore, it is also important to realize that the GB "derivation" corresponds to media with a constant ε and that the use of this equation should involve the replacement of ΔG_{sol}^{∞} by the ΔG^{+} and ΔG^{-} of the ions in their specific protein site. Here again, the main issue is the validity of Eq. (11), not its legitimization by the seemingly "rigorous" GB formulation.

B. *Calculating Binding Free Energies*

Studies of biocatalysis require in some cases the evaluation of binding energies and protonation states. Here we briefly consider several approaches that were found to be effective in such studies.

The binding free energy of ligands to proteins can be evaluated in principle by the FEP approach of Eqs. (3)–(5). This approach gives reasonable results in studies of difference in binding due to small "mutations" of the given ligand [7] (e.g., replacement of NH_2 by OH), but it involves major convergence problems when absolute binding free energy of medium size ligands is required. In such cases, one has to "mutate" the ligand to "nothing," and at present, the reported results are quite disappointing, except in cases of very small ligands. A much more useful alternative is offered by the LRA approach of Eq. (6). The LRA approach is particularly effective in calculating the electrostatic contribution to the binding energy [33, 62]. With this approximation one can express the binding energy as

$$\Delta G_{bind} = \frac{1}{2}\left[\left\langle U^{p}_{elec,l}\right\rangle_{l} + \left\langle U^{p}_{elec,l}\right\rangle_{l'} - \left\langle U^{w}_{elec,l}\right\rangle_{l} - \left\langle U^{w}_{elec,l}\right\rangle_{l'}\right] + \Delta G^{nonelec}_{bind} \quad (14)$$

where $U^{p}_{elec,\ell}$ is the electrostatic contribution for the interaction between the ligand and its surrounding. p and w designate protein and water, respectively; l and l′ designate the ligand in its actual charge form and the "nonpolar" ligand where all the residual charges are set to zero. In this expression the terms $\langle U_{elec,l} - U_{elec,l'}\rangle$, which are required by Eq. (6), are replaced by $\langle U_{elec,l}\rangle$ since $U_{elec,l'} = 0$. Now, the evaluation of the nonelectrostatic contribution $\Delta G^{nonelac}_{bind}$ is still very challenging because these contributions might not follow the LRA. A useful option, which was used in references [33] and [62], is to evaluate the contribution to the binding free energy from hydrophobic effects, van der Waals, and water penetration using the PDLD/S approach. Another powerful option is the so-called linear interaction energy (LIE) approach [37]. This approach adopts the LRA approximation for the electrostatic contribution but neglects the $\langle U_{elec,l} \rangle_{l'}$ terms. The binding energy is then expressed as

$$\Delta G_{bind} \approx \alpha \left[\left\langle U^{p}_{elec,l} \right\rangle_{l} - \left\langle U^{w}_{elec,l} \right\rangle_{l} \right] + \beta \left[\left\langle U^{p}_{vdW,l} \right\rangle_{l} - \left\langle U^{w}_{vdW,l} \right\rangle_{l} \right] \tag{15}$$

where α is a constant that is approximately 1/2 in many cases and ß is an empirical parameter that scales the van der Waals (vdW) component of the protein-ligand interaction. A careful analysis of the relationship between the LRA and LIE approaches and the origin of the α and β parameters is given in reference [62].

Of course, more implicit models can estimate the binding energy. This includes the use of the PDLD/S-LRA model for both the electrostatic and nonelectrostatic components (see [33] and [62]) and the PB approach augmented by estimates of the hydrophobic contributions using the calculated surface area of the ligand.

The idea that conformational averaging is useful for free energy calculations seems quite obvious, generally understood, and widely used [55, 63]. What is much less understood and somehow still ignored by a large part of the computational community is the realization that a proper LRA treatment requires averaging on both the charged (or polar) and nonpolar states of the ligand. The frequently neglected average on the nonpolar state (the second term of Eq. (6)) plays a crucial role in proteins as it reflects the effect of the protein preorganization (see also V.D.1). The role of this term in binding calculations has been discussed in [62] and its importance has been illustrated in an impressive way in studies of the fidelity of DNA polymerase [64].

C. Calculations of Protonation States

The results obtained from calculations of binding energies and activation energies depend on the ionization states of the protein residues. The evaluation of an ionization state of a given residue requires one to determine the corresponding intrinsic pK_a (the pK_a when all other ionized groups are uncharged) and the shift of this pK_a due to charge-charge interactions. The evaluation of the intrinsic pK_a by fully microscopic FEP or LRA calculations is quite challenging [53, 56, 65] and may involve errors of several pK_a units. In many cases and particularly in cases of surface groups one can obtain more stable results by using semimacroscopic models such as the PDLD/S-LRA and PB (see the previous sub-section). The problem is, however, obtaining reliable results for groups in the protein interiors (see discussion in [54]).

IV. Methods for Simulation of Chemical Processes in Enzymes

A major part of this review is dedicated to studies of enzyme catalysis. Before addressing this issue we will consider general approaches for modeling chemical processes in enzymes and in solutions. In doing so we will focus on strategies capable of providing activation barriers for enzymatic reactions.

The study of chemical reactions is more challenging than studies of simple conformational changes that can be explored by MM methods. Here the potential surfaces cannot be described by simple force fields since chemical bonds are being broken and formed during such processes. The common prescription of obtaining potential surfaces for chemical reactions involves the use of quantum mechanical computational approaches; such approaches have become quite effective in treating small molecules in the gas phase (e.g., [66]). However, here we are interested in chemical reactions in very large systems and the selection of a reasonable approach is far from obvious. In principle there are several ways to evaluate activation energies in enzymatic reactions. However, at present only a few approaches give reasonable results. Below we will consider the main options and discuss their effectiveness.

A. *Quantum Mechanical Studies of the Reacting Fragments in the Gas Phase*

One seemingly obvious option of modeling enzymatic reactions is to study the energetics of the given reaction in the gas phase. With such a model one can use a relatively high level and rigorous quantum mechanical approaches. While such studies (e.g., [67]) have been instrumental in providing insight about the reacting system, they are problematic. That is, the entire issue of catalysis is related to the effect of the environment. Thus, omitting the environment from the calculations prevents one from exploring key catalytic effects. Furthermore, the effect of the environment is not a small perturbation, but frequently involves enormous energy contributions. This point can be realized by considering the fact that the solvation energies of a simple ion can be around −100 kcal/mol and those of an infinitely separated ion pair around −200 kcal/mol.

Sometimes the reacting system is considered together with a few protein residues. Although this can help in providing some insight, it cannot be considered as a reasonable model of the enzyme active site. Such approaches frequently assume that some residues that are ionized in the complete system are also ionized but in the model used. However, the ionized form of these residues can be extremely unstable without the solvation effect of the rest of the protein and thus should not be ionized in that particular subsystem. Some specific examples of the problem associated with gas phase models will be given in section V.C.1.

Despite our warning about the risk of using gas phase models, there are clear exceptions. Most notable are the cases of large metal clusters, where the effect of the environment might be relatively unimportant. In such cases one can obtain instructive mechanistic information from *ab initio* studies that do not include the effect of the environment around the reacting cluster [68, 69].

A seemingly reasonable approximation to the full QM/MM treatment is the use of gas phase *ab initio* calculations to obtain the charges and force field of the substrate (solute), and the use of these charges and force field in free energy calculations in solution and in the enzyme active site [70]. This uncoupled approach is sometimes referred to as the QM-FE method [70], but this name gives the incorrect impression that it reflects a proper free energy perturbation (FEP) study of a QM surface and can be confused with true QM/MM-FEP approaches. Thus, we will refer to such approaches as the solvated gas phase (SGP) approaches. The SGP treatment, first introduced by Jorgensen and coworkers [71] in studies of solution reactions, has some useful aspects (see below) but also involves several major problems. First, the solute charges in the enzyme active site may be very different from those in the gas phase. This can lead to significant inaccuracies in studies of reactions where the difference between the solvation energies of the correct charges (those obtained in solution) and the solvation of the gas phase charges can be more than 100 kcal/mol (see [38]). Second, the SGP force field constrains the reacting system along the gas phase reaction coordinate [70, 71]. This prevents one from capturing the solute entropic contributions. Nevertheless, the SGP method does allow one to perform proper FEP calculations of the solvation energies of the gas phase charges in enzymes and solutions. In this respect it has an advantage over QM/MM approaches that do not perform a proper configurational average. In fact, all the deficiency of the SGP approach could be easily removed by moving to the EVB approach (section IV.C). In both approaches one calibrates the potential surface and charges using gas phase information (e.g., quantum mechanical results), but the EVB provides a consistent way of transferring this information to the solvating environment (whether enzyme or solution).

B. QM/MM Molecular Orbital Methods

The realization that the effect of the environment on the reacting fragments must be included in studies of enzymatic reactions led to the development of the hybrid Quantum Mechanics/Molecular Mechanics (QM/MM) approach [45]. This approach divides the simulation system (e.g., the enzyme/substrate complex) into two regions. The inner region, region I, contains the reacting fragments that are represented quantum mechanically. The surrounding protein/solvent region, region II, is represented by a molecular mechanics force field. The Hamiltonian of the complete system can then be written as

$$\mathrm{H} = \mathrm{H}_{\mathrm{QM}} + \mathrm{H}_{\mathrm{QM/MM}} + \mathrm{H}_{\mathrm{MM}} \tag{16}$$

Where the H_{QM} is the QM Hamiltonian, $\mathrm{H}_{\mathrm{QM/MM}}$ is the Hamiltonian that couples regions I and II, and H_{MM} is the Hamiltonian of region II. H_{QM} is evaluated by any standard QM approach, which can be either *ab initio* or semiempirical. The total potential can then be expressed as

$$\mathrm{V}_{\mathrm{total}} = \langle\Psi|H_{\mathrm{QM}} + \mathrm{H}_{\mathrm{QM/MM}} + \mathrm{H}_{\mathrm{MM}}|\Psi\rangle = \mathrm{E}_{\mathrm{QM}} + \langle\Psi|\mathrm{H}_{\mathrm{QM/MM}}|\Psi\rangle + \mathrm{E}_{\mathrm{MM}} \tag{17}$$

The nature of the first two terms in the above equation can be best realized by thinking on a simple molecular orbital representation of a solute/solvent system. If we consider this system as a supermolecule we can express its molecular orbital wave function as

$$\phi_i = \sum_{\mu} v^S_{\mu i}\chi^S_{\mu} + \sum_{\mu} v^s_{\lambda i}\chi^s_{\lambda} \tag{18}$$

where S and s designate the solute and solvent, respectively, v are the MO coefficients and χ are AO wave functions. We also assume that the atomic orbitals χ^S and χ^s are orthogonal. The coefficients $\mathrm{v}_{\mu i}{}^S$ and $\mathrm{v}_{\lambda i}{}^s$ can be obtained by solving the LCAO SCF equation

$$\mathbf{F}\mathbf{v}_i = \varepsilon_i\mathbf{v}_i \tag{19}$$

for the supersystem, where the matrix **F** can be separated to blocks, describing the solute-solute, solvent-solvent, and the solute-solvent interactions

$$\begin{vmatrix} F^S & F^{Ss} \\ F^{Ss} & F^s \end{vmatrix} \tag{20}$$

the matrix elements of **F** are given in [60, 72]. The assumption that the orbitals are orthogonal to each other implies that $F^{Ss} = 0$ (see [60]).

Now with some simple manipulations and with the assumption that the electron-electron repulsion integral between atoms A and B in the solute and solvent regions, respectively, is given by e^2/r_{AB}, we obtain for any solvent configuration

$$F^S_{\mu\mu} \cong (F^S_{\mu\mu})_0 - \sum_\sigma \frac{e^2 q_B}{r_{AB}} = (F^S_{\mu\mu})_0 - U_A \tag{21}$$

where $\mu \in$ A and U_A designates the total electrostatic potential from the solvent atoms at the site of atom A. This equation can be used in the more general case where the solvent charge distribution is not fixed and can be polarized by the field of the solute [60]. Thus, the leading term in the solute-solvent coupling Hamiltonian is obtained by adding the potential from the solvent atoms to the solute Hamiltonian. Now, the total potential energy is given by

$$V_{total} = E_S(\mathbf{F}^S) + E'_{Ss} + E_{ss} \tag{22}$$

where $E_S(\mathbf{F^S})$ is the energy obtained quantum mechanically with the **F** matrix that includes the given electrostatic potential from the solvent (the vector of all the U_A's). E'_{Ss} is the non-electrostatic solute-solvent interaction term and E_{ss} is the solvent-solvent classical force field. At this level of approximation the nonelectrostatic term is evaluated by the standard classical van der Waals potential function. In studies of very large solute molecules, we sometimes divide the solute region into quantum and classical parts. The ''connection'' between the quantum and the classical regions is treated by a classical force field (which is included in E_{Ss}'), where the quantum atoms at the boundaries are connected to dummy hydrogen-like atoms in order to balance the electrons in the quantum system (see later).

Interestingly, the seemingly simple implementation of the QM/MM method was not obvious at the early 70's, where most attempts to introduce the effect of the solvent were done in a formal way at the configurations interaction level. In fact, the acceptance of the QM/MM approach took a long time, perhaps due to the difficulty of realizing that medium-range electrostatic effects can be incorporated in quantum treatments using classical concepts. At any rate, the QM/MM method is now widely used and we can only mention several works (e.g., [73–84]). Several key issues of the implementation and the use of QM/MM approaches are considered below.

The treatment of the connectivity between regions I and II poses a basic problem. Obviously the division between these regions is artificial and one would like to make it as physical as possible. For example, when dealing with the boundary between two bonded atoms j and k, where j is in region I and k is in region II, the following strategy can be used: a phantom atom k′ can be introduced along the k → j vector. This phantom (linked) atom can then be included in the QM region.

Alternatively, using hybrid orbitals (as was done in the original work of Warshel and Levitt [45]) allows one to avoid the use of a linked atom (k′) by using a single hybrid orbital, located on atom k and pointing towards atom j. In general the properties of the linked atom or the hybrid orbital (and the corresponding semiempirical integrals) can be parameterized in a way that the quantum atom j will behave as if it were actually bonded to atom k. This hybrid orbital idea has been elegantly extended in recent works that made use of the related localized orbitals approach [73, 74]. Even though the use of localized orbitals is not complicated, most linked atom treatments involve the use of standard Cartesian MO's. Here one usually represents the linked atom by an hydrogen atom (sometimes with a modified core potential). However, the linked atom interacts with several orbitals rather than with one hybrid bonding orbital; this makes the definition of the boundaries and the corresponding parameterization somewhat problematic.

The linked atom problem is frequently presented as the key problem in QM/MM approaches (e.g., [76]). Fortunately, as much as enzyme catalysis is concerned, this problem is much less serious than commonly assumed. Since such studies involve a comparison between the enzyme and solution reactions, the effect of the linked atoms is canceled to a significant extent. Moreover, the validity of this cancellation assumption can be examined by simply increasing the size of the QM region.

A popular version of the QM/MM approach is the so-called "our own n-layered molecular orbital and molecular mechanics" (ONIOM) model [85]. This approach treats the system with several layers of quantum mechanical methods. Usually the inner layer is treated on the most rigorous level, the middle region is treated by an approximated QM method, and the outer region by a MM force field. In a typical case of two QM methods the energy of the combined system is expressed by

$$E_{total} = (E_2)_{region\ I+region\ II} + (E_1 - E_2)_{region\ I} \tag{23}$$

where E_1 and E_2 are the energies calculated by the rigorous and approximated QM methods, respectively. Unfortunately, this simple approach does not correspond to a consistent multi-layer QM/MM treatment. That is, a more consistent treatment would involve the representation of regions I and II by methods 1 and 2, respectively, with a consistent coupling between the two regions, where the Hamiltonian of method 1 includes the effect of region II. Instead, E_1 is obtained basically by a gas phase treatment. In fact, the implicit idea of the ONIOM method (using several levels of QM methods) was implemented consistently some time ago in a pseudopotential method [86] and in the FDFT method [87] which represents region I by full DFT and region II by frozen density DFT while using a consistent coupling between these regions (see section IV.D)

The reliability of the QM/MM approaches is far from obvious. For reliable free energy calculations, it is essential to use accurate QM methods and to perform an extensive configurational sampling of the multidimensional reacting systems. Unfortunately, regular semiempirical approaches are not sufficiently accurate. However, it should be noted that their accuracy can be increased by forcing them to reproduce the energetics of the reference solution reactions [88–90]. This recalibration has been inspired by the calibration concepts of the EVB approach (see next subsection).

In principle the most reliable potential surfaces can be obtained using *ab initio* (*ai*)–*QM/MM* approaches. At present, however, the enormous computer time needed for obtaining proper sampling by the *QM*(*ai*)/*MM* approaches makes such studies close to impossible. A novel way to reduce this problem is provided by using the *EVB* potential as a reference for the *QM*(*ai*)/*MM* calculations [91]. This approach allows one to evaluate the free energy profile of the *EVB* surface using FEP calculations and then calculates the free energy $\Delta G(EVB \rightarrow QM\ (ai)/MM)$ of moving from the *EVB* to the *ab initio* surface, both at the RS and TS regions. Such calculations can sometimes be done by a single-step FEP calculation that involves MD runs on the *EVB* surface. However, when the *EVB* surface is not sufficiently similar to the *ab initio* surface it is hard to obtain converging results. This problem can be drastically reduced by using the LRA approach [33], and by evaluating $\Delta G(EVB \rightarrow QM\ (ai)/MM)$ by [92]:

$$\Delta G(EVB \rightarrow QM(ai)/MM) = 0.5\Big[\big\langle E_{QM(ai)/MM} - EVB\big\rangle_{EVB} + \big\langle E_{QM(ai)/MM} - E_{EVB}\big\rangle_{QM(ai)/MM}\Big] \tag{24}$$

where $\langle\ \rangle$ designates an MD average of the designated potential.

An interesting approach was used by Zhang *et al.* [74], who consistently took into account the effect of the solvent environment in their *ab initio* free energy calculations. This approach, however, constrains the reacting fragments to move along a predefined reaction coordinate obtained from *ab initio QM/MM* calculations and thus neglects the effect of the solute fluctuations.

In summary, a significant advantage of the *QM*(MO)/*MM* approach is the fact that it can be easily integrated with standard *QM* program packages. However, the problems mentioned above still slow down the progress in realistic studies of enzymatic reactions by such approaches. Nevertheless, advances have been and will continue to be made in studies of different enzymatic reactions (see [45, 74, 76, 79, 80, 82, 93, 94]).

C. *The EVB as a Reliable QM/MM Method*

As stated above, reliable studies of enzyme catalysis require accurate results for the difference between the activation barriers in enzymes and solutions. The early realization of this point led to a search for a method that can be calibrated using experimental and theoretical information of reactions in solution. It also became apparent that in studies of chemical reactions it is more physical to calibrate surfaces that reflect bond properties (i.e., valence bond-based [VB] surfaces) than to calibrate surfaces that reflect atomic properties (e.g., MO-based surfaces). Furthermore, it appeared to be very advantageous to force the potential surfaces to reproduce the experimental results of the broken fragments at infinite separation in solution. This could be easily accomplished with the VB picture. The resulting empirical valence bond (*EVB*) method has been discussed extensively elsewhere [1, 95] but its main features will be outlined below.

The EVB is a *QM/MM* method that describes reactions by mixing resonance states (or more precisely diabatic states) that correspond to classical valence-bond (VB) structures, which describe the reactant intermediate (or intermediates) and product states. The potential energies of these diabatic states are represented by classical MM force fields of the form

$$\varepsilon_i = \alpha^i_{gas} + U^i_{intra}(\mathbf{R}, \mathbf{Q}) + U^i_{Ss}(\mathbf{R}, \mathbf{Q}, \mathbf{r}, \mathbf{q}) + U_{ss}(\mathbf{r}, \mathbf{q}) \qquad (25)$$

Here **R** and **Q** represent the atomic coordinates and charges of the reacting atoms in the diabatic states, while **r** and **q** are those of the surrounding protein and solvent. α'^{i}_{gas} is the gas-phase energy of the $\boldsymbol{i}^{\text{th}}$ diabatic state (where all the fragments are taken to be at infinity) U_{intra} (**R**, **Q**) is the intramolecular potential of the solute system (relative to its minimum), U_{Ss} (**R, Q, r, q**) represents the interaction between the solute (S) atoms and the surrounding (s) solvent and protein atoms. U_{ss} (**r, q**) represents the potential energy of the protein/solvent system (''ss'' designates surrounding-surrounding). ε_i given by Eq. (25) form the diagonal elements of the EVB Hamiltonian (H_{ii}). The off-diagonal elements of the Hamiltonian H_{ij}, are represented by simple exponential function of the distances between the reacting atoms. The H_{ij} elements are assumed to be the same in the gas phase, in solutions, and in the proteins. The ground state energy E_g is obtained by solving

$$\mathbf{H}_{EVB}\mathbf{C}_g = E_g\mathbf{C}_g \tag{26}$$

here, C_g is the ground state eigenvector and E_g provides the *EVB* potential surface.

The EVB treatment provides a natural picture of intersecting electronic states, which is useful for exploring environmental effects on chemical reactions in condensed phases [1, 95]. The ground-state charge distribution of the reacting species (''solute'') polarizes the surroundings (''solvent'') and the charges of each resonance structure of the solute, then interact with the polarized solvent [1]. This coupling enables the EVB model to capture the effect of the solvent on the quantum mechanical mixing of the different states of the solute. For example, in cases where ionic and covalent states are describing the solute, when the solvent stabilizes the ionic state to a greater extent the resulting ground state has more ionic character and more solvation energy.

MD trajectories on the *EVB* surface of the reactant state can provide the free energy function Δg that is needed to calculate the activation energy $\Delta g^{\ddagger}$. However, since trajectories on the reactant surface will rarely reach the transition state, it is usually necessary to run a series of trajectories on potential surfaces that gradually drive the system from the reactant to the product state [38]. The *EVB* approach accomplishes this by changing the system adiabatically from one diabatic state to another. In the simple case of two diabatic states, this ''mapping'' potential can be written

$$\varepsilon_{\mathrm{m}} = (1-\lambda_{\mathrm{m}})\varepsilon_1 + \lambda_{\mathrm{m}}\varepsilon_{\mathrm{z}} \qquad (0 \leq \lambda_{\mathrm{m}} \leq 1) \tag{27}$$

as a linear combination of the reactant and product potentials (ε_1 and ε_2). Where λ is changed from 0 to 1 in fixed increments (m = 0, 1, 2, . . . , M). Potentials with one or more of the intermediate values of λ will force the system to fluctuate near the TS.

The free energy ΔG_{m} associated with changing λ from 0 to λ_{m} is evaluated by the FEP procedure described elsewhere (e.g., [1]). The free energy functional is then obtained by the FEP-umbrella sampling (FEP/US) method of Eq. (7), which is now written as

$$\Delta g(x') = \Delta G_m - \beta^{-1}\ln\left\langle \delta(x-x')\exp\left\{-\beta\left(E_g - \varepsilon_m\right)\right\}\right\rangle_{\varepsilon_m} \tag{28}$$

where ε_m is the mapping potential that keeps x in the region of x'. If the changes in λ are sufficiently gradual, the free energy functionals $\Delta g(x')$ obtained with several values of m overlap over a range of x', and patching together the full set of $\Delta g(x')$ gives the complete free energy curve for the reaction.

The *EVB* method satisfies some of the main requirements for reliable studies of enzymatic reactions. Among the obvious advantages of the *EVB* approach is the facilitation of proper configurational sampling and converging free energy calculations. This includes the inherent ability to evaluate non-equilibrium solvation effects [88]. Another important feature of the *EVB* method is the ability to correctly capture the linear relationship between activation free energies and reaction energies (LFER) observed in many important reactions (e.g., [1, 96]). Furthermore, the *EVB* benefits from the above-mentioned ability to treat consistently the solute-solvent coupling. This feature is essential not only in allowing one to model properly charge-separation reactions but also in allowing for a reliable and convenient calibration. Calibrating *EVB* surfaces using *ab initio* calculations was found to provide quite reliable potential surfaces. For example, it has been found that the *EVB* approach reproduces accurately the structure and energetics of water clusters around H_3O^+ [97, 98].

The seemingly simple appearance of the *EVB* method may have led to the initial impression that this is an oversimplified qualitative model rather than a powerful quantitative approach. However, eventually the model has been widely adopted as a general model for studies of reactions in large molecules and in condensed phases [97, 98]. Several very closely related versions have been put forward with basically the same ingredients as in the *EVB* method (see discussion in [99]). Nevertheless, it has been argued [100] that various adaptations of the *EVB* method are useful since they involved calibration on *ab initio* surfaces rather than on experimental information. However, as previously mentioned, *EVB* potential surfaces have been calibrated quite early by *ab initio* surfaces [38].

The *EVB* approach has been used extensively in studies of different enzymatic reactions. Some of those studies will be considered in subsequent sections.

D. *QM Treatments of the Entire Protein*

A *QM* treatment of the entire protein/substrate/solvent system is possible, at least in principle. A promising progress in this direction has been offered by the divide and conquer (D&C) approach. This approach, which was originally developed for *ab initio* DFT studies [101], divides a large system into many subsystems, determines the electron density of each subsystem separately, and collects the corresponding contributions from all subsystems to obtain the total electron density and the energy of the system. Unfortunately, the treatment of the entire protein by an *ab initio* approach is extremely expensive and cannot be used in free energy calculations of enzymatic reactions. Thus, the most current effort in treating the entire protein by *QM* approaches has been invested in semi-empirical treatments with different tricks of accelerating the solution of the large SCF problem. Recent efforts (which have been inspired in part from the FDFT idea mentioned below) resulted in the so-called frozen density matrix approach [102, 103], which fixes the molecular orbital coefficients of the region around the reacting fragments. Such a treatment freezes the density matrix in the regions that surround the reaction region. Unfortunately, this approach does not offer a practical tool for *ab initio* studies of enzymes, since the cost of evaluating the relevant integrals is larger than the cost of evaluating the density matrix.

Overall, we feel that the development of D&C approaches is very important as a methodological advance and as a continuing challenge for computational chemistry. However, the potential of the method in terms of studies of enzymatic reactions is somewhat problematic. That is, trying to include the entire enzyme in the *QM* region usually results in less emphasis on the relevant reacting fragments or on a proper configurational averaging. Moreover, the effort in terms of computer time and intellectual resources in *QM* description of regions far from the active site must reduce the focus on the relevant issues. Also, the public might tend to believe that an approach that treats the whole system quantum mechanically is more reliable than a *QM/MM* approach that focuses on the *QM* description of the reacting fragments, while considering their surrounding classically. With the currently available computer resources one usually obtains more reliable results by focusing on calibration and validation of a relatively small reacting region. In fact, simulation methods should be judged by their ability to reproduce observables and not by their presumed formal rigor.

Despite the above reservations, there is no doubt that progress in representing the entire protein quantum mechanically will continue in the future. One of the most promising options is provided by the frozen DFT (FDFT) and the constraint DFT (CDFT) approaches (e.g., [104, 105, 232]). The basic idea behind these approaches (which should not be confused with the above-mentioned frozen density matrix approach) is to treat the entire protein solvent system quantum mechanically with the DFT formulation, but to freeze (or to constrain) the electronic densities of the groups in region II. This way the entire system is treated by *ab initio* DFT and the coupling between regions I and II is evaluated by a formally rigorous non-additive kinetic energy functional. This approach is formally more rigorous than the above-mentioned frozen density matrix approach. That is, here we do not have to worry about orthogonality between the wave functions of regions I and II since all that counts in the DFT treatment is the electron density. In principle, due to the properties of the DFT, if we can correctly guess both the densities in region II and the relevant functionals then our energy is correct. Furthermore, the coupling between region I and II is formally rigorous.

The FDFT approach presents a general way of coupling two subsystems by means of an orbital-free and first-principle-effective potential. This makes it possible to cast the concept of "embedding potential" in DFT terms. The FDFT approach is related in a formal way to the work of Cortona [106], but his study did not deal with the issue of embedding the subsystem in a larger system, which is the most important advance made by the development of the FDFT method. Wesolowski and Warshel [87] were the first to notice that the coupling term in any hybrid method can be obtained by partial minimization of the total energy functionals.

A rather subtle computational advantage of the FDFT/CDFT approach is the fact that even though it evaluates the interaction between region I and II quantum mechanically, it evaluates the interaction within region II classically. Thus, the FDFT/CDFT approach focuses on the energy of region I, and its correct quantum mechanical interaction with region II. Using the *EVB* as a reference potential for the CDFT allows one to evaluate the free energy of reactions in condensed phases and proteins while treating the solute-solvent interaction by an *ab initio* quantum mechanical approach [104]. This approach can be used, for example, to properly represent metal–to–ligands charge transfer in metalloenzymes (for a preliminary study see [105, 232]). The CDFT embedding idea has been adopted in related fields such as studies of molecules on metal surfaces (e.g., [107] and see discussion in [108]).

In concluding this section it is important to comment on the Car-Parrinello Molecular Dynamics (CPMD) approach [109], which emerged in recent years as a powerful way to study complex molecular systems. One element of this approach is an elegant way of performing *ab initio* MD by treating the expansion coefficients of the wave function as dynamical variables. However, this feature is not as unique as frequently assumed, since any approach that gives the first derivative vector of the *QM* energy can be used in a straightforward way for *QM/MD* simulations (e.g., [12]). The CP approach was applied, for example, to studies of proton transfer in solution [110]. However, these studies used periodic boundary conditions with too few water molecules in each unit cell to truly represent the reorganizational aspects of the solvent motion. The use of periodic boundary conditions makes it hard to apply this method to enzymatic reactions and gas phase calculations of the reacting fragments and a few protein residues, and results in extremely problematic studies (see discussion in section IV.A). Nevertheless, embedding the CP method in an MM surrounding and considering it a *QM/MM* method may provide a promising option [111–113].

V. Studying Enzyme Catalysis

A. General Considerations

Although this review describes approaches for general studies of structure function correlation of proteins, we will focus on enzymatic reactions. Simulations of other protein functions are reviewed in [15].

Understanding enzymatic reactions and realizing what makes them so efficient is one of the challenges of modern biophysics. Although many elements of this puzzle were elucidated by biochemical and structural studies, the source of the catalytic power of enzymes is not entirely understood. Statements such as, "the enzyme binds the transition state stronger than the ground state," or that "the catalytic groups are perfectly oriented" are not sufficient to explain this catalytic power since the real questions are how this differential binding is accomplished and what the actual catalytic groups are.

In order to set up the problem we will consider a generic enzymatic reaction using the equation

$$E + S \overset{K}{\leftrightarrow} ES \overset{k_{cat}}{\rightarrow} ES^{\ddagger} \rightarrow EP \rightarrow E + P \tag{29}$$

where E, S, and P are the enzyme substrate and products respectively, *ES*, *EP*, and $ES^{\ddagger}$ are the enzyme-substrate complex, enzyme product complex, and transition state, respectively, and $K = k_1/k_{-1}$

As was shown so eloquently by Wolfenden and coworkers (e.g., [114]) many enzymes evolved by optimizing k_{cat}/K_M, where $K_M = (k_{-1} + k_{cat})/k_1$ and can be approximated as $K_M \sim k_{-1}/k_1$. However, this and related findings did not identify the factors responsible for the catalytic effect. As will be shown in the next section, the key question is related to the reduction of the activation barrier in the chemical step. Unfortunately, even mutation experiments, which were found to be extremely useful in identifying catalytic factors [115], cannot tell us in a unique way the origin of the catalytic effect (see [116]). What is needed here is a quantitative tool for structure-function correlation and the ability to determine which are the individual contributions to the overall catalytic effect. It is gradually becoming clear that this requirement can be best accomplished using computer simulation; the rest of this review will consider both the current state of this approach and the conclusions that have emerged from simulation studies.

The analysis of enzyme catalysis is virtually futile without a proper definition of the relevant questions. First, when talking about catalysis it is essential to ask, "Catalysis relative to what?" Here we must define a reference reaction and the most obvious reference is the uncatalyzed reaction in water (see Fig. 1). Since the mechanism in water can be different than that in the enzyme, different mechanisms should be considered along with the effect of having different environments. Fortunately, a difference in mechanism can be classified as a "chemical effect" (e.g., having a general base instead of a water base), and such effects are well understood (see step a→b in Fig. 2). Thus we can focus on the difference in the effect of the environment, comparing the rate constant of a reaction that involves the *same* mechanism and binds the same chemical groups but

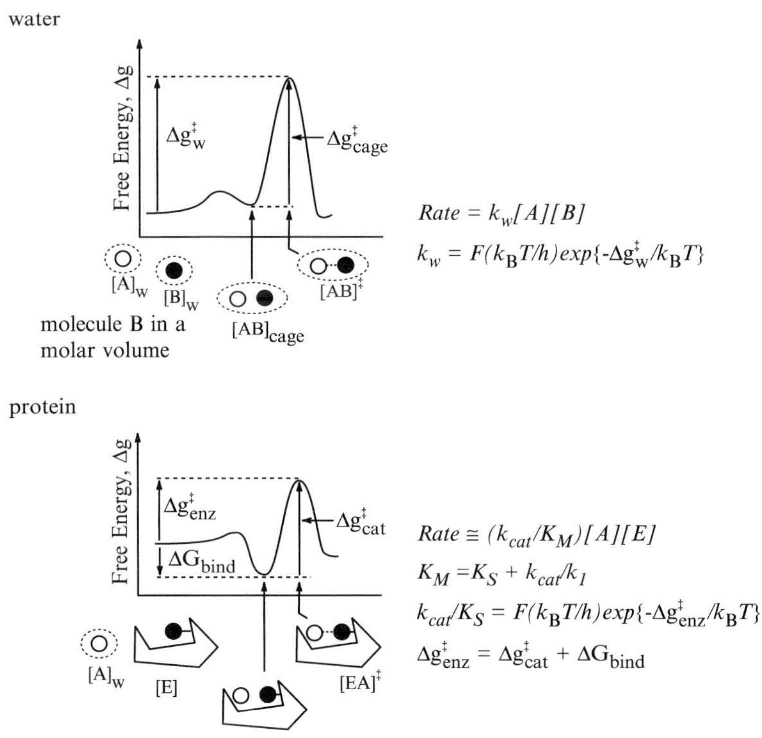

FIG. 1. Comparing the activation-free energy profile for a reference reaction in water and the corresponding reaction in an enzyme active site. The figure illustrates the relationship between $\Delta g^{\ddagger}_{cat}$, $\Delta g^{\ddagger}_{cage}$, and $\Delta g^{\ddagger}_{w}$.

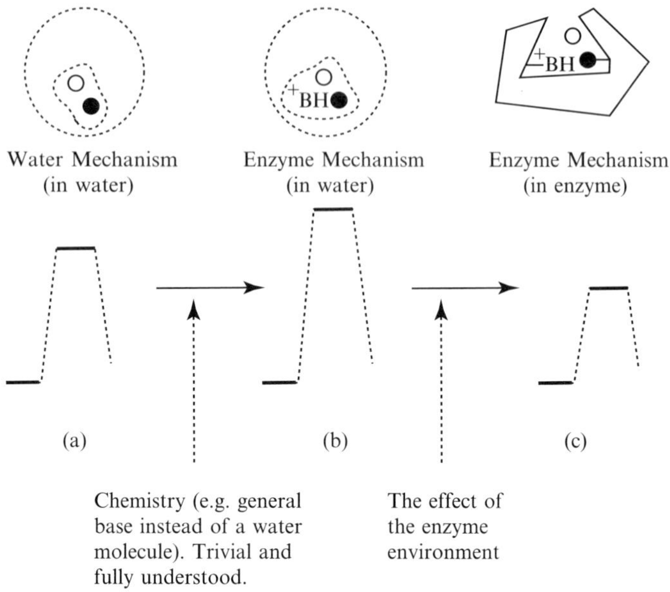

FIG. 2. Illustrating the concept of reference state for enzymatic reactions. The idea is to separate the effect of the enzyme environment ($\Delta g_c^{\ddagger} - \Delta g_b^{\ddagger}$) from the possible effect of having a different mechanism for the reaction in the enzyme and in water ($\Delta g_b^{\ddagger} - \Delta g_a^{\ddagger}$). This separation is best accomplished by comparing the enzymatic reaction to the *same* reaction in water.

is conducted in water (step b→c in Fig. 2). The idea that the reaction in water must serve as a reference reaction is sometimes criticized by those who say that it is better to use the gas phase reaction. However, as seen from Fig. 3, a proper thermodynamic analysis of the difference between the reaction in the gas phase and the enzyme should also tell us about the difference between the gas phase and solution reactions; thus it addresses the origin of the difference between the reaction in enzyme and in solution. Regardless of the cycle that we choose to use, the elucidation of the origin of enzyme catalysis boils down to understanding the origin of the difference between the activation barrier in water ($\Delta g_w^{\ddagger}$) and the activation barrier in the protein ($\Delta g_p^{\ddagger}$). However, the enzyme can reduce $\Delta g_p^{\ddagger}$ by binding the substrate with both an equal strength in the reactant state (RS) and in the transition state (TS) (in which case $\Delta g_p^{\ddagger} - \Delta g_w^{\ddagger} = \Delta G_{bind}$) and/or by reducing the activation barrier $\Delta g_{cat}^{\ddagger}$ for the chemical step. Since the factors that control the binding step are well understood (although may be difficult to evaluate quantitatively), the real puzzle is related to the

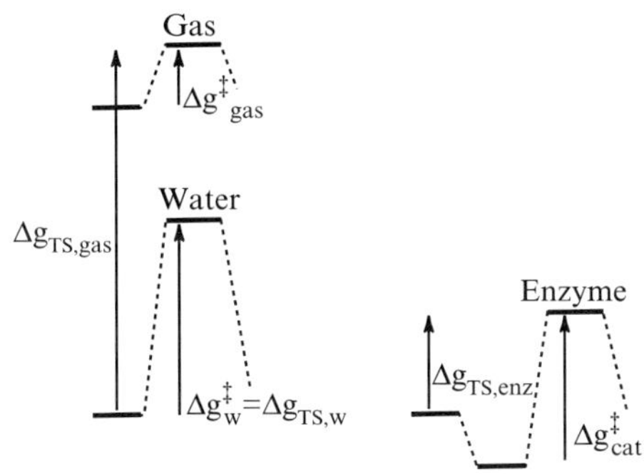

FIG. 3. Showing that the selection of the reference reaction in water provides an absolute framework for discussing enzyme catalysis in choosing the proper reference (i.e., the reactants at infinite separation in water). The figure shows that other reference states (e.g., the reactants in the gas phase) are possible, but their use must involve the proper thermodynamic cycle, which will relate them to the reaction in water.

factors that govern the reduction of the activation barrier of the chemical step ($\Delta g^{\ddagger}_{cat}$). The fact that $\Delta g^{\ddagger}_{cat}$ is reduced by the enzyme was recognized by Polanyi [117] and Pauling [118] long ago, but their insight has not elucidated the origin of this reduction.

Experimentally, even the question of whether the enzyme works by stabilizing the RS or TS is not easily resolved, although the type of mutational analysis introduced in [116] can help in this respect. This difficulty persists despite a recent attempt by Bruice and Benkovic [119] to address this issue in a general way. These authors pointed out that log k_w is correlated with log K_{TS} (where $K_{TS} = k_w\ K_M/k_{cat}$) but not with log k_{cat}. They then argued that this fact supports the RS destabilization proposal. However, this is a simple reflection of the fact that while k_{cat}/K_M does not change in a drastic way between different enzymes, the corresponding k_w changes over many orders of magnitudes [114]. Thus, k_{cat}/K_M is relatively constant and k_w is correlated with K_{TS}. As Wolfenden and coworkers [120] already noted, the above observation merely reflects the fact that different reactions in water might have very different activation barriers due to their different chemistry, while reactions in enzymes must have activation barriers in a similar range to provide a reasonable activity. This tells us that enzymes must provide more catalysis to those reactions, which are slow in water. However, this fact does not tell us how this catalysis is being accomplished. In view of the above discussion we should focus on two questions: (i) What contributions create difference between $\Delta g^{\ddagger}_{cat}$ and $\Delta g^{\ddagger}_{w}$ and (ii) how these contributions operate (i.e., do they destabilize the RS or stabilize the TS).

Many proposals (e.g., see [88] for a partial list) have attempted to address the above questions by either direct or indirect ways. However, most of the proposals have not been defined on a completely logical level or properly analyzed by their proponents. Some proposals have not even considered a proper thermodynamic cycle (see later). In fact, a major problem in the field is the use of soft definitions that cannot be examined conceptually or computationally. As will be shown below, with a clear definition of the problem and with a combination of experimental and computational studies it is possible to explore and quantify the origin of the catalytic power of enzymes.

In our discussion and analysis it will be convenient in some cases to take $\Delta g^{\ddagger}_{cage}$ as a reference reaction to the reaction in a water cage, where the reactants are already at an interaction distance (see Fig. 1). This analysis does not reflect any arbitrary assumption since the relationship between $\Delta g^{\ddagger}_{cage}$ and $\Delta g^{\ddagger}_{w}$ can be established in a rigorous way (see [88]).

B. *Reactant State Destabilization or TS Stabilization: General Considerations and Definition*

While analyzing the origin of the reduction of $\Delta g^{\ddagger}$, we consider two general options. One is reactant state destabilization (RSD) and the other is TS stabilization (TSS). Interestingly, although it is common to suggest that enzymes work by TSS, most early proposals and almost all explicit proposals involved RSD. An interesting case is Pauling's insightful proposal [118] (considered as one of the first TSS proposals), where it is suggested that ''the substrate molecule is attracted to the enzyme, and caused by the forces of the attraction to assume the strained state which favors the chemical reaction.'' Because this important proposal was formulated before the emergence of clear ideas about the nature of enzyme active sites, it is hard to know which forces were thought to lead to the above effect. However, the concept of strained state is reserved to the RSD proposal.

In defining RSD it is important to avoid confusion and to choose a proper reference state. In doing so, we must remember that we are focusing on catalysis, so that the reference must be some form of the uncatalyzed reaction. This can be either the reaction in water (see previous section) or the reaction of the un-evolved enzyme. Thus we have the following possible reference states that define the problem: (i) RSD is defined relative to the energetics of un-evolved enzyme (the enzyme that was not evolved for catalysis and the enzyme that will be obtained by mutating all the residues that help the catalysis), and (ii) RSD is defined by considering the energy of the reactive part of the substrate relative to its energy in water. Only in this way can one quantify and examine Jencks' concept of using the binding energy of distant parts to destabilize the reactive part [121]. With this definition we accept the fact that the total binding energy (relative to water) is negative, but we allow for destabilization of a part of the molecule at the expense of strong binding of other parts far from the reaction region. Again we must have the water as a references state in order to define stability or instability of the different parts.

Now, all the well-known RSD catalytic proposals (see also section V.D.2, V.D.3, and V.D.6) are easily classified by the above definitions. This includes the steric strain effect, which corresponds to increasing the energy of RS of the reactive part of the molecule by a steric strain (this is relative to the unstrained state in water), the desolvation idea, where one destabilizes the reactive RS relative to water by placing it in a nonpolar environment, and finally the entopic proposal, for which the RS is destabilized relative to water since its motion is restricted so that the entropic stabilizing contribution is eliminated. We will also have RSD by the definition of (i), since a non-evolved entropic enzyme (the one that is not able to restrict the substrate motions) will allow the bound substrate to move in the RS as much as in water.

C. *Modeling the Overall Reaction and Reproducing the Corresponding Catalytic Effect*

Any reliable analysis of enzyme catalysis must start with an attempt to reproduce the observed reduction of the corresponding activation of free energy. It is also essential to be able to determine the actual mechanism in the active site. A method that cannot reproduce the observed catalytic effect might not be so useful in examining individual contributions to this effect. The progress in addressing these problems is reviewed later.

1. High-Level Calculations of Incomplete Models May Lead to Low-Level Results

It is sometimes very tempting to use standard computer packages with high-level *ab initio* treatments but without the protein environment. Although such approaches might be insightful, they frequently lead to incorrect perspectives. Several examples of the use of such models are given below, along with a discussion of the corresponding traps. The most alarming problems occur when one ignores the contribution of solvation effects to the proton affinity (pK_a) of ionizable groups. Studies with this omission and particularly with non-electroneutral systems may result in artificial protonation states of key catalytic residues (e.g., Lysine residues tend to be neutral in this gas phase model) and completely incorrect energies for proton transfer processes. A typical example is provided by the study of Futatsugi *et al.* [122], who presumed to explore the mechanism of p21ras. These workers included in their calculations a GTP molecule; two water molecules, an Mg^{+2} ion and fragments of Thr35, Ser17, and Lys16. Such a model misses the interaction of the charged system with its surrounding and leads to major artifacts. That is, the study starts with a protonated Lys16, for instance, but this residue should not be protonated in the gas phase environment used in the calculations (see discussion in [123]).

Another example is provided by gas phase studies of the ionization state of HIV proteases [124, 125]. Unfortunately, it is impossible to correctly evaluate pK_as in proteins without including the solvation effects of the protein/solvent environment. Furthermore, in the case of the acid pair in HIV proteases, gas phase calculations (e.g., [124]) lead to an artificial low-barrier hydrogen bond (see section V.D.7) and confuse experimentalists (e.g., [126]) who assumed that high-level calculations of subsystems should provide reliable results.

Reasonable gas phase treatments have paid more attention to electroneutrality and to the system chosen. However, when such treatments are applied to mechanistic questions, they may favor the wrong mechanism because they ignore the effect of the protein environment. An interesting example is provided by a recent *ab initio* study [127] of the catalytic reaction of T7 DNA polymerase. This instructive study concluded that the reaction mechanism involves the substrate as a base. However, EVB studies of the same system [128] reproduced the observed catalytic effect with a mechanism that involves Asp654 as a base. Obviously, it is impossible to examine this mechanism without including the protein environment, since this environment determines the pK_a of Asp654.

Another type of problem occurs when one uses models designed to calculate solvation processes in water to examine enzyme catalysis. In this case the energetics may be very reasonable since enzyme and solution reactions are much more similar than the enzyme and the gas phase reactions. However, such an approach precludes any chance to study catalysis, since the enzyme and solution reactions are modeled in the same way (see discussion of [129] in subsection V.D.7).

Finally, we find it useful to illustrate why uncritical use of first principle MD simulations of biological systems should not be confused with a useful analysis of such systems. One example is given by the previously-mentioned studies of HIV proteases [124], while another more instructive example is given by the recent work of Cavalli and Carloni [130]. These workers concluded that Gln61 is the general base in the catalytic reaction of the ras/GAP complex (see [123] for a discussion). To determine whether a residue may serve as a base in a given reaction, it is essential to evaluate the free energy of proton transfer (PT) between the nucleophile and this residue. It is also useful to examine the pK_a of the proposed base (in its protein site). Such a study that used the EVB approach [131] has shown that Gln61 cannot be a general base in ras and that the γ-phosphate is more suitable for this role. Nevertheless, Cavalli and Carloni studied this system by the CPMD approach without being aware of the previous theoretical study and the need to validate their results by pK_a calculations. They started with a subsystem that included the substrate, Gln61, a few other residues, and the attacking water molecule (whose orientation was selected based on problematic force field calculations). Next they performed a very short MD run while constraining the distance between the oxygen of the attacking water and the γ-phosphate to 1.8 Å. The water proton collapsed to Gln61 during this relaxation and this led Cavalli and Carloni to conclude that Gln61 must be the general base. Of course, pushing the attacking oxygen to a bonding distance from the phosphate would force the proton to migrate to the closest base. However, this is not a way to examine reaction mechanism or to identify a general base

2. *Enzymes that were Studied by Reasonable Calculations*

The ability to obtain reliable free energy surfaces for enzymatic reactions has been restricted almost exclusively to the *EVB* method until about the mid-nineties. *QM/MM* studies with semiempirical *QM* Hamiltonians started to give stable results with the work of Bash and coworkers [89], when they adopted the EVB idea of calibration on the energetics of solution reactions. The past few years had witnessed a significant progress in the reliability of semiempirical *QM/MM* calculations with reasonable PMFs and proper sampling, although EVB studies are probably still more reliable. The systems studied by reasonably reliable approaches include lysozyme [1, 28, 45, 132], serine proteases [51, 133–135], ribonuclease [136], acetylcholinesterase [137, 138], staphylococcal nuclease [139], carbonic anhydrase [140, 141], DNA polymerase [142], ras [123], triosephosphate isomerase [143, 144], aldose reductase [145–147], lactate dehydrogenase [148], malate dehydrogenase [89], isocitrate dehydrogenase [149], chorismate mutase [81, 150], tyrosine phosphatase [151], orotidine 5-monophosphate decarboxylase (ODCase) [90, 152], and others. Most of the above studies involved *EVB* and *QM/MM* semiempirical MO approaches but some studies with *ab initio QM/MM* methods (e.g., [74]) also started to give reasonable results.

Interestingly, even reasonable *QM*(MO)/*MM* studies of enzyme catalysis still use different models for the reaction in the enzyme and in solution (e.g., [153] uses *QM/MM* model for the enzyme and the COSMO model for the solution reaction). Using different models makes it impossible to reach unique conclusions about the origin of the catalytic effect.

In almost any case that was analyzed carefully and properly, it was found that the catalytic effect is mainly due to electrostatic contributions. The issue whether this reflects RS destabilization or TS stabilization is still controversial, though consistent analyses (see section V.D.1) seem to point towards TS stabilization by electrostatic effects as the key catalytic factor. Nonetheless, it is not entirely clear how the electrostatic contributions are stored and distributed and what is the upper limit of other catalytic factors (see later).

D. Examining How Enzymes Really Work

Various proposals attempt to explain the origin of enzyme catalysis (a partial list can be found in [88, 116]). Although some of these proposals are reasonable, many are poorly defined and it is essential to reformulate them before examining their validity. In fact, it appears that some proposals violate the law of thermodynamics, and thus cannot be examined by using thermodynamic cycles. It seems that the prerequisite to any analysis of a catalytic proposal is the existence of a well-defined logical argument that can actually be tested.

In the following sections we will present key catalytic proposals in clear, logical terms and then consider simulation studies that explore their validity.

1. Electrostatic Contributions to Catalysis

Consistent computational analysis of enzyme catalysis have found that the difference between $\Delta g^{\ddagger}_{cat}$ and $\Delta g^{\ddagger}_{cage}$ is mainly due to electrostatic effects. The calculations indicated that enzymes ''solvate'' their TS's more than the corresponding *TS*'s in the reference solution reactions [154]. The nature of this ''solvation'' effect appeared to be far from obvious. That is, the calculated interaction energy between the *TS* charges of the reacting atoms and the enzyme was found to be similar to the corresponding interaction energies in solution. This finding indicated that the entire electrostatic free energy associated with the formation of the *TS* must be examined [1] rather than only the interaction energy at the *TS*. This includes the penalty for the reorganization of the environment upon "charging" the *TS*. In fact, most of the catalytic effect appeared to be due to the reorganization energy. To quantify this catalytic effect it is useful to express the electrostatic free energy of the *TS* charges by the LRA formulation of Eq. 6 that can be written in the present case as

$$\begin{aligned}\Delta G(Q^{\ddagger}) = 0.5(\langle U(Q=Q^{\ddagger}) - U(Q=0)\rangle_{Q=Q^{\ddagger}} + \langle U(Q=Q^{\ddagger}) \\ - U(Q=0)\rangle_{Q=0}) = 0.5(\langle \Delta U\rangle_{Q^{\ddagger}} + \langle \Delta U\rangle_0)\end{aligned} \tag{30}$$

where U is the solute-solvent interaction potential, Q designates the residual charges of the solute atoms with $Q^{\ddagger}$ indicating the *TS* charges and $<\Delta U>_Q$ designates an average over configurations obtained from an MD run with the given solute charge distribution (see section II.D.). The first term in Eq. (30) is the previously mentioned interaction energy at the *TS*, where $Q = Q^{\ddagger}$, which is similar in the enzyme and in solution. The second term expresses the effect of the environment preorganization. If the environment is randomly oriented toward the *TS* in the absence of charge (as is the case in water), then the second term is zero and we obtain

$$\Delta G(Q^{\ddagger})_w = \frac{1}{2}\langle \Delta U\rangle_{Q^{\ddagger}} \tag{31}$$

where the electrostatic free energy is half of the average electrostatic potential [10].

However, in the preorganized environment of an enzyme we obtain a significant contribution from the second term and the overall $\Delta G(Q^{\ddagger})$ is more negative than in water. This extra stabilization is the catalytic effect of the enzyme. Another way to see this effect is to realize that in water, where the solvent dipoles are randomly oriented around the uncharged form of the TS, activation-free energy includes the free energy needed to reorganize the solvent dipoles towards the changed *TS*. On the other hand, the reaction in the protein costs less reorganization energy since the active site dipoles (associated with polar groups, charged groups, and water molecules) are already partially preorganized toward the *TS* charges [1]. The reorganization energy is related to the well-known Marcus′ reorganization energy, but is not equal to it. More specifically, the Marcus′ reorganization energy [155] is related to the transfer from the reactant to the product state while here we deal with charging the *TS* (see discussion in ref. [88]). Nevertheless, calculations of the Marcus′ reorganization energy in enzyme and in solution are also consistent with the above idea, and it has been repeatedly found that λ_p is smaller than λ_w (see [88, 116, 143, 148]). At this point it is important to clarify that the reduction of $\Delta g^{\ddagger}$ by the previously mentioned preorganization effect cannot be satisfied by having a non-polar enzyme active site. This is important in view of the proposal that the reduction of λ is due to having non-polar active sites (e.g., [156]). Enzymes active-site reduce λ while being very polar.

2. *The Entropic Proposal*

The idea that entropic effects are a major source of enzyme catalysis has been invoked in many proposals (e.g., [70, 121, 157, 158]) and many textbooks (a partial list is given in [88]). These proposals have assumed (correctly) that the large configurational space of the reacting fragments is drastically reduced by the enzyme, but then concluded that this leads to large increase in k_{cat}. The entropic proposal was frequently presented in somewhat appealing yet ill-defined terms (e.g., "the entropy trap effect"), but was rarely formulated in a correct way. That is, in order to determine the contribution of the substrate entropy to catalysis, one has to find the difference between the contribution in the enzyme $(\Delta S^{\ddagger}_{cat})'$ and in water $(\Delta S^{\ddagger}_{w})'$ (the prime designates the entropic change from the configurational space of the reacting fragments rather than the overall entropic change [159]). Instead of considering free energy diagrams that involve these contributions, (see below) it has been customary to talk in a vague way of the negative value of the binding entropy $\Delta S'_{bind}$ as the entropic contribution to catalysis. For example, as shown clearly in [15, 88, 159], the considerations of Jencks (e.g., [121]) and Kollman and coworkers (e.g., [157]) are equivalent to the incorrect assumption that

$$-T(\Delta\Delta S^{\ddagger})' \cong -T\Delta S'_{bind} \tag{32}$$

Actually, the entropic contribution to catalysis must be defined by a proper thermodynamic cycle and this gives [88]

$$-T(\Delta\Delta S^{\ddagger})' = -T((\Delta S^{\ddagger}_{cat})' - (\Delta S^{\ddagger}_{w})') = -T(\Delta S'_{TS,p} - \Delta S'_{TS,w}) - T\Delta S'_{bind} \tag{33}$$

Thus, the expression used in Eq. (32) is incomplete and overestimates the actual entropic contribution by $-T(\Delta S'_{TS,p} - \Delta S'_{TS,w})$. That is, the assumption that $\Delta S'_{bind}$ gives the entropic contribution to catalysis misses entirely the transition state contributions. Unfortunately, $\Delta S'_{TS,w}$ is quite large and can be similar to $\Delta S'_{RS,w}$ or $-\Delta S'_{bind}$, thus canceling most of the assumed entropic effect (see later). In other words, the assumption that the entropic contribution to catalysis is given by the binding entropy is unjustified. Theoretical evaluation of $(\Delta\Delta S^{\ddagger})'$ is extremely challenging since the convenient use of the gas phase harmonic approximation leads to a major overestimation (see discussion in [160]). At present the only reported attempt to properly estimate the entropic contribution to catalysis involves a restrain-release approach (RRA) [159]. This approach has some elements in common with the approach used by Hermans and Wang in studies of binding entropies [161]. The RRA evaluates $\Delta S'$ for the *RS* and *TS* of the given system by imposing Cartesian restraints on the system and evaluating the free energy associated with the release of these restrains. The resulting free energy $\Delta G'$ includes both entropic and enthalpic contributions. This enthalpic contribution makes it hard to isolate the entropic part. Fortunately, the enthalpic contribution to $\Delta G'$ can be eliminated by performing series of simulations with different restraint coordinates and finding the one that yields the smallest $\Delta G'$. Application of this approach to the catalytic reaction of subtilisin produced much smaller entropic contributions to catalysis than usually assumed [159]. This is due to the fact that many of the motions that are free at the *RS* in the reference reaction are also free at the *TS* (thus, $\Delta S'_{TS,w} \neq 0$). In order to illustrate the above point we may consider the reaction of subtilisin that involves a proton transfer from a serine to a histidine and a nucleophilic attack of the ionized serine on the substrate carbonyl. It is natural to assume that the histidine residue loses its motion at the *TS* in solution. However, the protonated histidine can move as freely at the *TS* as the unprotonated histidine at the *RS* (see [159]).

The idea that entropic effects are the major factor in enzyme catalysis has been supported by the work of Kollman and coworkers [157]. However, the reported studies did not involve the proper thermodynamic cycle (see analysis in [15, 88]). Furthermore, the energy diagrams used by these workers added ΔS_{bind} to the free energy profile of the water reaction. This is problematic since the free energy in water cannot depend on the specific enzyme studied (using this method different mutants will give different $\Delta g^{\ddagger}_{w}$).

3. *Steric Strain, Confinement Effects, and the NAC Proposal*

Several proposals for the origin of enzyme catalysis involve ground state steric strain [162–164]. The original strain hypothesis of Phillips and coworkers (e.g., [162]) and related subsequent works (e.g., [164]) invoked the idea of "molding" the substrate towards the *TS* by strong steric forces. This idea was inconsistent with simulation studies, which demonstrated that enzymes are flexible and do not apply strong steric forces (e.g., [1, 45]). A refined proposal, which is related in some respects to the strain hypothesis, has been put forward by Bruice and coworkers (e.g., [165]). They proposed that a steric confinement of the enzyme active site brings the substrate to a so-called near attack conformation (NAC). They then suggested that this conformation is closer to the corresponding TS than in the reference reaction in water. However, the definition of this proposal was based on defining critical distance and angle where the NAC is supposed to occur, rather than on free energy surfaces that could be related directly to the difference between $\Delta g^{\ddagger}_{w}$ and $\Delta g^{\ddagger}_{cat}$. More specifically, the activation free energy can only be defined by the difference between the free energy at the *TS* and the lowest point at the *RS* minima, or by the free energy in the *TS* and the overall free energy of the *RS* (see [15]). Thus, selecting an arbitrary point along the reaction coordinate as a reference for the evaluation of the activation energy cannot give unique results. Furthermore, the NAC idea has not explained whether the difference between the reaction in enzyme and in water is due to electrostatic or steric effects and thus does not address the origin of the catalytic effect.

Another crucial problem has been the lack of a proper energy diagram that will tell us whether the NAC effect is an RSD proposal (as is implied from the wording in the corresponding papers) or a TSS proposal. In fact, if the NAC reflects a TSS effect, it does not provide any new proposal but merely a non-unique way of looking on transition state stabilization. That is, as stated in section V. A., any catalytic proposal must be defined with a clear reference state. Taking the reactants in water as a reference state leads to two possible definitions of the NAC proposal (see Fig. 4). The first option is a clear RSD steric proposal, which might or might not be true. The second option is a regular TSS proposal where the stabilization of the *TS* by electrostatic effects leads also to shifts in the RS minimum toward the *TS* position. The resulting "NAC effect" has little to do with the implications of the NAC proposal.

The increasing popularity of the NAC hypothesis along with the absence of a lucid definition motivated Shurki *et al.* [166] to reformulate the NAC idea and to try to quantify it. The main reason for the reformulation was the realization that the NAC or any related confinement (or steric) proposal must be defined in terms of the corresponding reduction in the activation-free energy. Since the definition of the steric effect might not be unique, it was essential to use two alternative definitions and two operational computational methods. The first method (which is aimed at examining Bruice's structural concept) defined the NAC effect as the reduction in activation barrier due to the presence of a confined active site. Here the question is: How large is the modification of the surface for the reaction in water due to the presence of the steric effect of the enzyme? With this definition, it is sufficient to ask the extent to which $\Delta g^{\ddagger}_{w}$ would be reduced if the reactive fragment in water would be confined to the same configurational space as in the enzyme. This catalytic

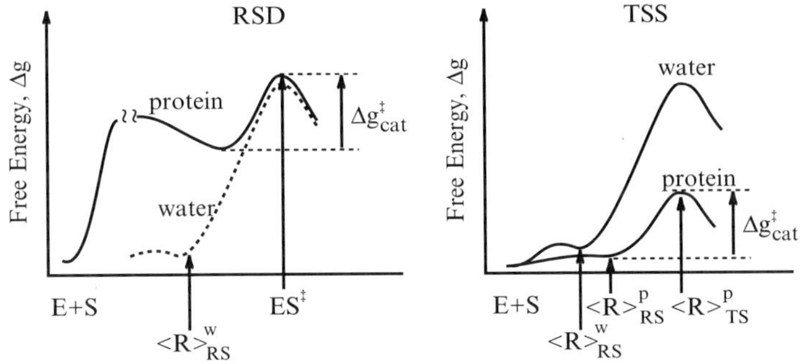

FIG. 4. Describing the two possible definitions of the NAC effect in terms of the free energy profile of the reacting fragments along the reaction coordinate. (a) The NAC effect is associated with ground state steric destabilization. That is, the free-energy surface of the reacting fragments is pushed up in the RS and thus leads to small $\Delta g_{cat}^{\ddagger}$. The modification of the water-free energy surface by the enzyme is presumably due to the binding of the non-reactive parts of the substrate (this energy is not included in the figure). (b) The enzyme stabilizes the TS by electrostatic effects and in doing so pushes the RS minimum closer to the TS position. This option is the regular TSS proposal, which is not related to the implications of the NAC proposal (see [233] for more clarification).

effect was evaluated by adding an external restraint potential, V', to the potential surface of the substrate in water, V_{w}, and forcing the probability distribution, $\rho(r)$, of both the reactants RS and the TS to satisfy the relationship

$$\rho(r)_{V_w+V'}^{RS} \cong \rho(r)_{V_p}^{RS} \tag{34}$$

$$\rho(r)_{V_w+V'}^{TS} \cong \rho(r)_{V_p}^{TS} \tag{35}$$

where V_{p} is the potential surface of the substrate in the protein. Reduction of $\Delta g_w^{\ddagger}$ upon addition of a restraint potential, which generates the protein probability distribution, is a logical definition of the NAC effect that can also be measured. The actual effect of V' was then examined by FEP approaches with several alternative thermodynamic cycles [166].

The second method defined the steric effect as the sum of all the non-electrostatic contributions and then examined the size of the steric contributions. Thus, the evaluation of the steric effect is done by eliminating the environment electrostatic contribution, leaving only the steric contribution to affect the activation free energy, and comparing the reaction in water and in the protein. In this method, all but the environment-substrate electrostatic interactions were considered (namely water–water, protein–protein, water–protein, and substrate–substrate). Thus, in these conditions, the difference in free energy of activation between the two distinct environments ($\Delta g^{\ddagger}_{w}$ and $\Delta g^{\ddagger}_{cat}$) stands for the non-electrostatic, steric, contributions.

Both methods for examining the effect of the protein restraint were used in the specific case of haloalkane dehalogenase. Fig. 5 shows the reactant state (RS) and the TS probability distribution of the substrate both in water and in protein. As can be seen from the figure, the probability distribution in the protein is different than the corresponding distribution in water. However, calculations showed that the free energy needed for changing ρ_w to ρ_p by the constraint V' was rather small. Namely, it was found that V′ reduces $\Delta g^{\ddagger}_{w}$ by less than 3 kcal/mol, whereas the total catalytic effect is ~7 kcal/mol. Moreover, comparing the results from the restraint-release approach to those obtained by eliminating the electrostatic interaction showed that pure steric component of the NAC contribution amounts to less than 1 kcal/mol. Apparently the remaining contribution was found to reflect electrostatic stabilization of the *TS*, which has little to do with the original idea of steric contributions to enzyme catalysis. This point, which was already alluded to in the above discussion of Fig. 4, should be explained now in view of our findings. According to our calculations, the fact that the average *RS* configuration in the protein, $\langle R\rangle^{p}_{RS}$, is closer to $\langle R\rangle^{p}_{TS}$ than $\langle R\rangle^{w}_{RS}$ reflects the flattening of the free energy surface by the protein electrostatic potential (case b in Fig. 4) and not a genuine NAC/steric effect. This point becomes even

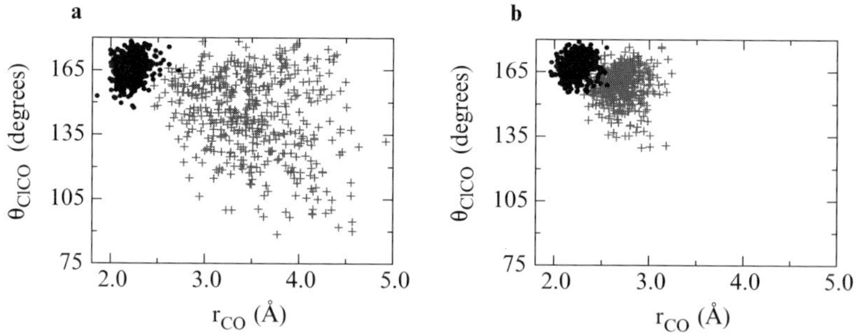

FIG. 5. The distribution of r_{CO} and θ_{ClCO} during MD trajectories in the RS (light plus signs) and TS (black dots). (a) in water and (b) in the protein. It is seen that the difference is especially pronounced in the geometric distribution of the RS structures in water relative to that of the protein, having larger ranges of both r_{CO} distance and θ_{ClCO}. Based on the figure, it appears that the protein system shows greater similarity of the RS to the TS.

clearer in the extreme case of electrostatic stabilization of the *TS* so that the reaction barrier in the protein goes to zero. Based on the NAC model, one can conclude that the NAC effect contributes to the entire catalytic effect. This, however, is an artificial result of having an ill-defined NAC concept, since we considered a model where all the catalysis and the reduction of $\Delta g^{\ddagger}$ is due to electrostatic stabilization (for more recent analysis of this issue see [233]).

If the electrostatic contributions in haloalkane dehalogenase were leading to RSD effects (Fig. 4a), we would have an effect that could be classified as an NAC-type effect (with the proper redefinition of the NAC effect as the effect of configurational restriction). This effect could have been classified as "desolvation-induced NAC." That is, one can assume that in the S_N2 reference reaction of haloalkane dehalogenase there is a large loss of solvation upon moving the negative nucleophile from infinity to the NAC position. This would mean that the enzyme desolvation effects lead to reduction of the difference between $\langle R \rangle_{RS}$ and $\langle R \rangle_{TS}$, and to a NAC effect. However, our careful calculations indicated that the main loss of solvation in the water reaction occurs after passing the NAC point. Moreover, a proper thermodynamical analysis of a desolvation-induced NAC (using "absolute" reference states as in Fig. 4) indicates that this NAC proposal is a regular desolvation proposal, which is inconsistent with proper calculations of the binding energies of the *TS* and *RS* (see also section V.D.6).

To summarize the section, it is important to point out that the NAC proposal does not really explain the origin of enzyme catalysis. That is, this proposal tells us that the enzyme reduces $\langle R \rangle_{RS}$ relative to water. It does not tell us, however, how this reduction is being accomplished (steric or electrostatic effect), and whether it is due to *RS* destabilization or *TS* stabilization effect. If the NAC proposal implies a pure steric effect (i.e., the catalysis is due to van der Waals forces) then its contribution to catalysis is very small [166]. If it implies *RS* destabilization by entropic effects, as was done by Kollman and coworkers [167], these effects are rather small (see previous section). The only way for the NAC effect to be significant is to reflect a part of the overall *TS* stabilization using electrostatic effects. In this case, the apparent steric effect reflects the response of the solvent to the solute charges. Here it is important to realize that the change in solute charges is coupled to both the changes of the solvent and the solute components of the reaction coordinate. As demonstrated in [166], the change in the solvent coordinates leads to the largest contributor of catalysis; this change is not an NAC effect, as it does not depend on the solute coordinate.

4. Dynamical Proposals

The proposal that special "dynamical" effects play a major role in enzyme catalysis (e.g., [168, 169]) has become quite popular in recent years (e.g., [170–178]). However, a significant part of this popularity indicates confusion about the nature of dynamical effects and the requirement from catalytic contributions, which must be related to the reference reaction in solution. Many workers overlooked apparently the difference between the well-known fact that all chemical and biological processes involve atomic motions and the requirement from true dynamical contributions to catalysis. Since this issue has been analyzed in great length in several recent reviews [15, 88], we consider only the key points here.

Several definitions can be used to describe dynamical effects. In order to exhibit dynamical contribution to catalysis, any definition must find a different magnitude of dynamical contributions to the rate constant in the enzyme and in water. All the reasonable definitions result in very small dynamical contributions [15]. The chemical physics community agrees (see refs. in [88]) that all the dynamical effects are contained in the transmission factor, a contribution that corrects the absolute rate theory for re-crossing of the reactive trajectories (see [15] for a clear definition). All reported simulation studies (e.g., [88, 134]) found that the transmission factors are similar in enzyme and in solution and do not differ by much more than a unity in the enzyme (e.g., [88, 175]).

Another definition states that dynamical effects are related to the availability of special coherent motions. In this way the dynamical proposal implies that enzymes "activate" special types of coherent motions that are not available in the solution reaction. The difference between the reaction in enzyme and in solution cannot be accounted for by evaluating the corresponding $\Delta g^{\ddagger}$ using non-dynamical Monte Carlo (MC) methods. In other words, if the results from MC and MD are identical, then we do not have dynamical contributions to catalysis. Careful and systematic studies (e.g., [88, 179]) have shown that the reactions in both enzymes and solutions involved large electrostatic fluctuations. However, these fluctuations follow the Boltzmann distribution and thus do not provide dynamical contributions to catalysis.

Attempts to imply that dynamical effects are associated with the so-called non-equilibrium solvation effects [169] have been very problematic (see [88]). Furthermore, it has been clearly demonstrated that the difference between the non-equilibrium solvation effects in enzyme and solution is an integral part of the difference between the corresponding activation barriers [88].

Apparently, there is no single experimental finding that can be used to consistently support the dynamical hypothesis. Most of the experiments that were used to support this proposal have not compared the catalyzed and uncatalyzed reactions, and thus have not addressed the issue of catalysis (see discussion in [88]). Instructive NMR experiments (e.g., [176]) demonstrated the involvement of different motions in enzymatic reactions. The obvious existence of motions that have components along the reaction coordinate does not constitute a dynamical effect unless these motions are shown to be coherent. Most likely, all the motional effects identified so far are related to entropic factors (i.e., to change in the available configurational space) rather than to real dynamical effects.

In summary, it is useful to recognize that consistent simulation studies found no evidence for dynamical contributions to catalysis [88]. Studies that were used to invoke such effects (e.g., [172]) did not evaluate any actual activation free energy or transmission factors.

5. Vibrationally Enhanced Tunneling Does Not Lead to a Large Catalytic Effect

Recent studies (see [177] and refs. in [88]) have suggested that vibrationally enhanced tunneling (VET) plays a major role in enzyme catalysis. Some workers (e.g., [178]) assumed that here exists an entirely new phenomenon that makes TST inapplicable to enzymatic reactions. However, the VET effect is not new and is common to many chemical reactions in solution [180–182]. Moreover, the VET is strongly related to TST. When the solvent fluctuates and changes the energy gap (see [179, 180]) the light atom sees a fluctuating barrier that allows, in some cases, for a larger rate of tunneling. As shown in [179], these fluctuations are taken into account in the statistical factor of the classical TST; the same is true when quantum effects are considered. Thus, the recent finding that the solvent coordinates should be considered in tunneling studies is not new and does not mean that this effect is important in catalysis.

Hwang *et al.* were the first to evaluate the contribution of tunneling and other nuclear quantum effects to enzyme catalysis [183]. Since then, particularly in the past few years, there has been a significant increase in simulations of quantum mechanical-nuclear effects in enzyme reactions. These types of approaches range from the quantized classical path (QCP) (e.g., [88, 184, 185]) to the centroid path integral approach [186, 187], the vibrational transition state theory [188], and the molecular dynamic with quantum transition (MDQT) surface-hopping method [189]. Most studies did not yet examine the reference water reaction and thus could only evaluate the quantum mechanical contribution to the enzyme rate constant rather than the corresponding catalytic effect. However, studies that explored the actual catalytic contributions (e.g., [88, 184, 185]) concluded that the quantum mechanical contributions are similar for the reaction in the enzyme and in solution, and thus do not contribute to catalysis.

Interestingly, the MDQT approach of Hammes-Schiffer and coworkers [189] allowed them to explore the quantum mechanical transmission factor. It was found that even with quantum mechanical considerations the transmission factor is not so different than a unity, and thus we do not have a large dynamical correction to the TST rate constant.

6. Reactant State Destabilization by Desolvation and Other Electrostatic Effects

Many catalytic proposals involve the idea of reactant state destabilization (RSD). These proposals involve the previously mentioned strain and entropy proposals and the popular concept that enzymes provide a non-polar (sometimes described as gas phase-like) environment that destabilizes highly charged ground states (e.g., [190, 191] and refs. in [152]). The validity of the desolvation idea has been examined carefully (e.g., [1]) and was shown to reflect improper thermodynamic cycles that do not use a proper reference state. This amounts to ignoring the desolvation energy associated with taking the *RS* from water to a hypothetical non-polar enzyme site. With a proper reference state, a polar *TS* is less stable in non-polar sites than in water and the RSD does not help in increasing k_{cat}/K_M [1]. In fact, there is no evolutional pressure for this mechanism. Interestingly, many desolvation models (e.g., [191, 192]) involve ionized residues in non-polar environments. Such residues would be unionized in non-polar sites. Moreover, in any specific case when the structure of the active site is known, one finds by current electrostatic models a very polar (rather than non-polar) active site environment near the chemically active part of the substrate.

In order to illustrate the problems with the desolvation RSD proposal we will consider the case of ODCase [90, 193–195]. This enzyme is considered to be the most proficient enzyme (having the largest $(k_{cat}/K_M)/k_w$ and an extremely large rate acceleration $k_{cat}/k_w \sim 10^{17}$, see [194]). Obviously the origin of this enormous catalytic effect is of great interest. Before the evaluation of the structure of ODCase, Lee and Houk proposed [191] that the enzyme provides a non-polar low-dielectric environment, and works by a desolvation mechanism. However, it was then demonstrated that this proposal is based on an incorrect thermodynamic cycle and proposed that the enzyme must provide a very polar environment [196]. The simultaneous elucidation of the enzyme structure by several research groups (see refs. in [197]) revealed that the active site contains two aspartic residues whose carbonyl groups are located near the carboxyl group of the substrate (Fig. 6).

An initial examination of this system has suggested that the interaction between the two negatively charged carboxylates and the oratate should lead to a major RSD effect (e.g., [90]). This idea, which was almost instantly embraced by many [197], seemed to be supported by the calculations of Wu *et al.* [90], who performed *QM/MM*-PMF calculations that reproduced the catalytic effect and found that their RS is destabilized by the enzyme. However, Warshel and coworkers [152] pointed out that the proposed electrostatic destabilization is questionable since the strong electrostatic interaction of the carboxyl groups would result in protonation of one of these groups. Further work of Warshel *et al.* [152] pointed out that the binding calculations of Wu *et al.* (which were used to support the RSD proposal) produced almost identical destabilizations to the *RS* and the *TS*, and thus could not reproduce the catalytic effect (which was obtained by the PMF calculations). They also noted that the binding calculations used to support the RSD proposal are much more challenging than the PMF calculations and require special attention to the treatment of the corresponding electrostatic free energy. The FEP calculations of Warshel *et al.*[152], which involved the SCAAS boundary conditions and proper treatment of long-range effects, did not reproduce significant RSD. Furthermore, it was pointed out that the reacting system includes the entire pyrimidine ring of OMP, along with a protonated Lys72 amino group that most likely donates a proton to the ring during the reaction. Stabilizing interactions with the Lys72 and other residues largely cancel the repulsive interactions with the two aspartates. When this larger region is considered as the reacting system (Fig. 6), one finds that

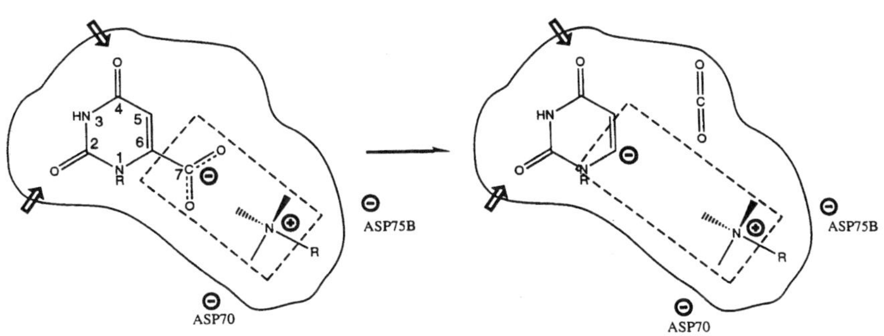

FIG. 6. Description of the RS and TS in the reaction of ODCase. As seen from the figure and as explained in [152], the reaction involves an increase in the dipole moment of the reacting fragments.

the dipole moment of the system increases in the TS. This change in the dipole moment is favored by the Asp residues of the enzyme. These calculations support the view that the catalytic effect of ODCase results mainly from *TS* stabilization, rather than from destabilization of the GS. In fact, considering the reacting system as a dipole moment provides an instructive example of the preorganization effect. That is, the reference reaction in water involves significant solvent reorganization upon the formation of the *TS* with its large dipole moment. In the enzyme, on the other hand, the two ionized Aspartic groups are already placed properly to stabilize the *TS* dipole.

Warshel *et al.* [152] have provided additional reasons why the RSD mechanism cannot be operative in ODCase. A recent study of Kollman and coworkers [198] also questions the RSD mechanism.

Recent studies of Wolfenden and coworkers [199, 200] have provided a strong evidence against the RSD proposal. These studies demonstrated that mutations of Asp96 and other residues that were supposed to destabilize the orotate led to weaker rather than stronger binding. As predicted in [152], this result is inconsistent with the RSD since destabilization of the RS should result in a reduction of the binding energy.

7. *Ionic HBs and LBHB as an Illustration of the Need of Energy-Based Analysis*

In this section we will consider the Low Barrier Hydrogen Bond (LBHB) proposal [201–203] as an example of an ill-defined proposal, and demonstrate how important it is to define and examine such a proposal by clear energy considerations. We will start by showing how the well-known idea that hydrogen bonds (HBs) contribute to catalysis was reformulated as the LBHB proposal, where the only new element (covalent contribution to catalysis) was not properly defined. After giving a proper testable definition we will consider the specific LBHB proposal in serine proteases and show why this proposal does not account for the catalytic effect. We will then consider some of the poor theoretical studies of the LBHB proposal where the apparent difficulty of focusing on the relevant issues and the use of incorrect models (e.g., gas phase models) lead to incorrect conclusions. We will conclude by considering several proper theoretical studies.

i. How to Examine the Only New Element in the LBHB Proposal The idea that hydrogen bonds (HBs) contribute to enzyme catalysis is not new. The first concrete support to this idea can be traced to the structural identification of the oxyanion hole in subtilisin [204]. However, this observation did not involve any estimate of the relevant catalytic energy. Subsequent theoretical studies [133, 134, 154] have established the idea that the overall electrostatic effect of preorganized hydrogen bonds contribute in a major way to enzyme catalysis. These theoretical predictions were confirmed by mutation experiments, clearly showing that a single hydrogen bond can contribute around 5 kcal/mol to an ionic transition state [115, 205]. The results of some specific mutation experiments were subsequently reproduced by FEP/US calculations [134].

After the experimental demonstration of *TS* stabilization by hydrogen bonds, it was proposed by several workers that HBs stabilize *TS*s by a special non-electrostatic method which was termed a low barrier hydrogen bond (LBHB) [201–203]. This LBHB proposal has been poorly defined and rationalized by circular arguments (see discussion in [206]). The continuing confusion about this issue makes it a textbook example for the need of energy-based considerations. Such considerations are presented below.

The LBHB proposal has suggested that catalytic hydrogen bonds involve a single rather than a double minimum[201–203]. This definition does not allow one to distinguish the LBHB proposal from the previous electrostatic proposal of ionic HBs. In order to do so one must define the LBHB proposal in a clear way, which reflects the energetics of the system and can be related to enzyme catalysis. At present the best (and in some respects the only) way to define the LBHB proposal is to use the Valence-Bond representation. This representation can be treated in a simplified 2-state version of the 3-state model of Coulson and Danielsson [207, 208], augmented by the EVB solvent effect [1]. In this representation (Fig. 7), one sees how the pure zero order states ϕ_1 = X-H Y^-, ϕ_2 = X^- H-Y are mixed by an off-diagonal term H_{12} (resonance term). When the energy gap (g_2–g_1) at $r_1{}^0$ is similar in magnitude to H_{12} we have a single minimum with large covalent character, which is qualified to be defined as LBHB. Otherwise, the electrostatic contribution is larger than the covalent contribution. The transition between an ionic HB and LBHB can be quantified by considering the reorganization energy λ, the mixing term H_{12}, and the proton transfer free energy ΔG_{PT} (see discussion in [209]). In other words we are dealing here with a competition between the localized [X^- H-Y], [X-H Y^-] picture and the delocalized [$X^{-1/2}\cdots H\cdots Y^{-1/2}$] picture. In the gas phase, the delocalized picture tends to dominate, while in solution the localized is more important. With these limiting cases in mind we can ask what is new in the LBHB proposal. Obviously the idea that preorganized HBs that stabilize ionic TSs contribute to catalysis is not new (see earlier). Thus, the only new element in the LBHB proposal is the idea that the covalent delocalized character, which leads to the single energy minimum, is the origin of the catalytic effect. In this respect, it should be clear that HBs in solution have a significant covalent character (see early works [210]). Furthermore, for the LBHB proposal to be valid

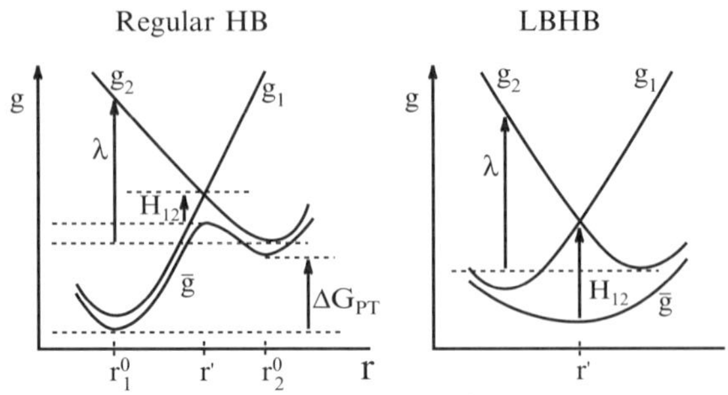

FIG. 7. (a) A two-state VB model for an ionic hydrogen bonded system (see [209]). The free energies g_1 and g_2 correspond to the states (X-H Y^-) and (X^- H-Y). The ground state surface E_g (with a corresponding free energy surface $\bar{g}$) is obtained from the mixing of the two states. The donor and the acceptor are held at a distance R. The equilibrium distances for isolated X-H and H-Y fragments are designated by $r_1{}^0$ and $r_2{}^0$. λ and ΔG_{PT} designate the reorganization energy and proton transfer energies respectively. (b) The LBHB limit occurs when $|H_{12}| > \lambda + \Delta G_{PT}$.

the covalent character must be larger in the enzyme than in solution, and the corresponding difference must be the source of the LBHB catalytic effect. Obviously, these issues cannot be examined without evaluating the relevant energies.

Before we consider energy calculations it is useful to clarify that the LBHB proponents frequently used misleading definitions that should not be used in scientific arguments. These include attempts to equate LBHB with ionic HBs (e.g., [211, 212]), while attributing to Warshel and coworkers the irrelevant proposal of neutral HBs. For example, it was stated that, ''Warshel and coworkers, and Guthrie to a certain extent, maintains that an ionic hydrogen bond, such as that formed in an LBHB, is destabilized by the presence of a polar cavity relative to that of a traditional, weak, neutral hydrogen bond''[211]. Warshel and coworkers are those who pointed out the stability rather than instability of preoriented ionic HBs in polar sites and that the issue is the difference between localized and delocalized ionic HBs rather than their relationship to neutral HB. The LBHB proponents also attempted to confuse the LBHB proposal with the nature of the donor acceptor distance by talking on short, strong HB (SSHB) instead of LBHB. Although the donor acceptor distance changes the magnitude of H_{12}, the LBHB proposal should be analyzed by the potential surface of the proton motion [209]. Other problems may reflect unfamiliarity with logical arguments, where the explanation why the ionic state would be unstable in a (hypothetical) nonpolar site [209], is understood as it meant to say that the ionic state is unstable in polar sites.

Before consulting early papers about the LBHB proposal, it is strongly suggested to take a critical view instead of assuming that it is an unresolved controversial issue. For example, the argument that NMR shifts prove the validity of the LBHB proposal is circular and nonconclusive [206, 209]. The NMR shift might reflect a large covalent mixing but it may as well reflect the inductive electrostatic effect in the X^- H–Y system (see [209, 213]). In fact, solid-state NMR evidence that the proton is bound to the donor in serine protease invalidates the LBHB proposal in serine protease by showing that the proton is bound to the catalytic histidine. At any rate, even if the NMR shifts are due to a delocalization effect it does not mean that this is the source of the catalytic effect.

Another example of an irrelevant and incorrect experimental argument is the claim of Kreevoy and Clealand [214], summarized by [215], "Warshel *et al.* suggested that electrostatic interactions alone can account for the stabilization energy in enzymatic reactions and, therefore, regarded LBHBs as being of no importance in enzymes [216]. Kreevoy, Cleland [214], and others [217], on the other hand, argued that electrostatics alone cannot explain the special physicochemical properties, such as...". First, Warshel *et al.* excluded LBHB based on clear energy considerations rather than by just pointing out the effectiveness of electrostatic interactions. Second, the proposal that electrostatic effects cannot lead to spectral changes displays an apparent misunderstanding of the nature of electrostatic effects in proteins. For example, a negatively charged group near a histidine residue will change the electronic distribution of this group and thus, change its spectra. Similarly, electrostatic solvation effects play a crucial role in controlling spectra in proteins. Third, as stated above, HBs in proteins always have some covalent characters but this does not make this contribution a catalytic factor. Again, only energy-based analysis can establish the magnitude of the LBHB contribution.

ii. Energy-Based Analysis of the LBHB Proposal The entire issue of the validity of the LBHB proposal is related to the interaction between the environment and the VB states of the given ionic HB. The LBHB proponents who originally assigned to LBHBs in enzymes the enormous energy (~20 kcal/mol of gas phase LBHB [201]), argued that the enzyme environment is nonpolar. Obviously, such desolvation arguments are not useful without actual calculations of the relevant polarity and the corresponding solvation effect. Performing such calculations in a reliable way is the best way to examine the LBHB proposal.

An excellent case for the analysis of the LBHB proposal is offered by the catalytic triad of serine proteases. Frey and coworkers [202] put forward this system as a prime example of LBHB catalysis. They basically proposed that the proton of His57 is shared between this residue and Asp102, forming an $O^{-1/2} \cdots H \cdots N^{+1/2}$ system at the TS of the hydrolytic reaction (see [51] for the definition of the relevant system). Warshel and coworkers argued, in turn, (based on early calculations [51, 218]) that the pK_a of Asp102 is lower than that of His57 in the protein, and thus the proton must be on His57. This argument was supported by NMR studies [219], which were ignored by the LBHB proponents. Now, instead of addressing the well-defined pK_a issue, a new argument has been put forward by Cassidy *et al.* [220]. These workers considered the highly relevant system of chymotrypsin with a TS analogue (which will be referred to here as I^-) where the pK_a of His57 is shifted from ~7 to ~11. They then argued that the pK_a shift is due to a binding-induced steric strain between Asp102 and His57, the release of which supposedly leads to LBHB. The logical flaws of this proposal are significant. First, the source of the idea of strain release as the source of the LBHB is alarming. The LBHB proponents [214] brought the "proton sponge" N,N,N′,N′-tetramethyl-1, 8-diamino-naphthalene ($2H^+$) as a proof for the LBHB stabilization, challenging Warshel and coworkers to explain the pK_a shift in this molecule, claiming, "If strong hydrogen bonding is not the source of internal stabilization for $3H^+$, we will have to invent new physicochemical phenomena to explain the basicity of compound 3" ($3H^+$ is the $2H^+$ compound with additional 2,7, methoxyl substituents) [221]. However, the shift was easily explained [209] not by LBHB or by inventing new phenomena, but by the large sterically induced destabilization of the *deprotonated* form rather than any LBHB stabilization of the protonated form. After this clarification we find suddenly that the steric effect in the "proton sponge" is used to explain LBHB in serine protease. Unfortunately, there is no obvious reason (or evidence) for binding-induced steric strain between Asp102 and His57. It is very likely that a negatively charged TS analogue should increase the pK_a of the His57 and thus increase ΔpK_a between Asp102 and His57. This would reduce, rather than increase, the LBHB contribution. Yet many workers (e.g., [222, 223]) accepted the argument of Cassidy *et al.* without wondering what the origin of the new strain proposal is, and why the system does not follow the logically expected trend from its electrostatic interactions.

Although we believe that the trend expected from the $I^- \cdots His^+$ interaction based on simple electrostatic considerations should have excluded the proposal of [220] as a valid illustration of the LBHB catalytic effect, we found it useful to examine the actual energetics of the system. The detailed analysis, which will be reported elsewhere [224], is based on the PDLD/S-LRA method (section III.A), which is one of the most reliable ways of evaluating pK_as in proteins and on EVB calculations. It was found that the ion pair state is more stable than the nonpolar pair state by ~6–8 kcal/mol in the presence of the TS analogue and about ~5–7 kcal/mol in the absence of the TS analogue. It should be noted that the same calculations reproduce all the known facts on the system including the pK_a shifts. Thus, regardless of the reasons for the pK_a shifts, the LBHB does not exist in either the free enzyme or in the enzyme with the bound TS analogue. In addition, it is clear from the enormous contribution of the calculated solvation energy that the environment here is very polar and does not resemble the nonpolar environment envisioned by the LBHB proponents. Note in this respect that we consider our analysis to be valid due to its ability to reproduce the observed pK_as using a realistic model of the protein inhibitor system.

Detailed examination of the LBHB proposal for the HBs in the oxyanion hole of subtilisin was already presented by Warshel and Papazyan [209]. Here again, it was found that the catalysis is due to the preoriented $[t^{-}\cdots H\text{–}N]$ system and not to the $[t^{-1/2}\cdots H\cdots N^{-1/2}]$ of LBHB (t^{-} designates the oxyanion TS). Recent EVB analysis [225] examined the LBHB proposal in ketosteroid isomerase. It was found that the HB between Tyr16, the substrate enolate has a double minimum, and that the charge transfer contribution is similar in the protein and in water. In order to fully determine the magnitude of pure LBHB contribution (if any), a study of the reaction in water with water molecules rather than tyrosine as an HB donor would be needed.

iii. Example of a Poor Use of "High-Level" Calculations The wide availability of easy-to-use *ab initio* packages coupled with a superficial understanding of what can be modeled by different approaches made the LBHB problem a glaring example of poor-level theoretical studies. Many studies illustrate the fact that high-level calculations of incorrect systems and ill-defined questions are not so useful. The most obvious cases involve gas phase calculations as a model for enzymes. For example, McAllister and coworkers (e.g., [211, 212]) presumed that an ionic HB and two water molecules in the gas phase will teach them about the effect of enzyme microsolvation on LBHBs. Unfortunately, the configuration chosen, with one water molecule near each oxygen in a $[O\cdots H\cdots O^{-}]$ system (see [212]), stabilized the LBHB from acting oppositely to typical solvation effects. Another even more prevailing problem is the calculation of irrelevant quantities. Typically, calculations of the energetics of ionic HBs are confused with LBHB calculations (see previous), or calculations of the

O···O distance are presented as LBHB studies. In other cases we find studies of entirely irrelevant systems reported as LBHB studies (e.g., studies of the proposed LBHB in serine proteases, considering a proton transfer between Ser195 and His57 instead of His57 and Asp102 [223]). Another more sophisticated problem is associated with attempts to study LBHB in enzymes while considering the given substrate and a few residues in a water model [129]. Obviously, without studies with an actual enzyme model we cannot examine whether the LBHB is catalytic or not. Interestingly, there is a tendency of some experimentalists (e.g., [203]) to use the results of improper calculations as a support of the LBHB idea.

iv. Proper Computational Studies of Catalytic HBs in Enzymes do not Support the LBHB Proposal As stated repeatedly above, probably the most effective way to examine the LBHB proposal is by proper calculations of relevant quantities. Even mutational studies (e.g., [226]) that can give important hints about the origin of the observed effect by changing the intrinsic pK_a of the HB donor (or acceptor) but cannot provide conclusive information [51, 209, 225]. As was pointed out, above current EVB studies do not support the LBHB idea. Other QM/MM studies have not yet reach the quantitative level of the EVB method. In particularly, such studies have not provided quantitative pK_as and did not involve actual free energy calculations. Nevertheless, the few relevant QM/MM studies do not support the LBHB idea (e.g., [80, 83, 213]). For example, it was found that the QM/MM energy differences for the two extreme points on the LBHB proton transfer profile of citrate synthase are sufficiently large to exclude the previously proposed LBHB [83]. Similarly, the QM/MM study of [213] indicated that there is no LBHB contribution in the catalytic reaction of serine proteases. At any rate, the LBHB proposal is not supported at present by any consistent simulation study.

8. Other Proposals

In addition to the catalytic mechanisms considered above, there are significant number of additional proposals for the origin of enzyme catalysis (e.g., see list in [88]). These include the orbital steering [227], induced fit [228], and entatic state [229] proposals. As argued in the previous sections, the only way to validate these proposals, or any others, is to first try to see if they can be defined in a clear way and then to examine their validity in specific cases by energy considerations. For example, the orbital steering proposal can be eliminated by very simple energy consideration (chapter 9 of [1]). Similarly, the induced fit idea, which has been used extensively in discussing DNA fidelity (e.g., [230]), can be shown to be somewhat irrelevant. The induced fit proposals state that the binding process brings the catalytic groups to their catalytic position, thus helping catalysis. Unfortunately, the real question is what is the origin of the reduction of $\Delta g^{\ddagger}_{cat}$. Telling us that in the ES complex the catalytic groups are properly placed does not help to identify the catalytic groups (e.g., the protein dipoles), their proper orientation, or how they help in stabilizing the *TS*. Finally, Williams′ entatic state proposal [229], which is sometimes brought as a hint of a preorganization idea, is in fact an *RS* strain idea. That is, the only explicit proposal in [229] states that, ''the protein makes the metal sites entatic by forcing upon the metal bond distances and bond angles which are compromise between the demands of its two oxidation states.'' This is a typical strain idea, where in the specific case of a redox reaction the protein deforms the redox center away from its reactant state toward the *TS*. Not only does the strain mechanism does not provide an effective catalytic mechanism does in general, but it also does not operate in the specific case of the cupper proteins considered by Williams (see [231]).

VI. Concluding Remarks

The field of computer modeling offers a powerful way of understanding biochemical and biophysical systems on a molecular level [1]. This review considers some of the key advances in the field from both technical and conceptual points of view. The review starts by considering different strategies for modeling biological processes and emphasizing the evaluation of the relevant free energies and the requirement of reliable simulation approaches. The second part of the review focuses on computer simulations of enzyme catalysis. This part points out the crucial role of the proper definition of catalysis (i.e., the importance of having a clear reference state). It also illustrates that many catalytic proposals are poorly defined, and that in almost all cases it is essential to have energy-based definition in order to explore the merit of a given catalytic proposal. Obviously, the final examination of an energy-based catalytic proposal requires accurate modeling. Thus, we provide a detailed description and a critical evaluation of different methods for simulation of enzymatic reactions. We discuss the problems with models that do not include the enzyme in the study and point out that the apparent rigor of incomplete models should not be confused with the reliability of such models. We consider the potential and actual current performance of *QM/MM* and related approaches. It is pointed out that while the *EVB* provides the most reliable results, alternative *QM*(MO)/*MM* methods are starting to provide reasonable results as well. However, these *QM*(MO)/*MM* approaches do not yet involve calculations of free energy of the combined solute solvent systems with *ab initio QM* surfaces. Nevertheless, the general direction of the field is very promising. There is no doubt that the next few years will include massive progress in the use of reliable *QM/MM* and related approaches. It is also certain that there will be a growing realization that consistent simulation studies provide the ultimate tool for understanding structure function correlation in proteins.

Acknowledgments

This work was supported by NIH grant GM 24492 and NSF grant MCB-0003872.

References

1. Warshel, A. (1991). Computer modeling of chemical reactions in enzymes and solutions. John Wiley & Sons, New York.
2. Hendrickson, J. B. (1961). *J. Am. Chem. Soc.* **83,** 5437.
3. Burkert, U., and Allinger, N. L. (1982). Molecular Mechanics. American Chemical Society, Washington D.C.
4. Lifson, S., and Warshel, A. (1968). A consistent force field for calculation of conformations, vibrational spectra and enthalpies of cycloalkanes and n-alkane molecules. *J. Chem. Phys.* **49,** 5116–5129.
5. Levitt, M., and Lifson, S. (1969). *J. Mol. Biol.* **46,** 269–279.
6. Warshel, A., and Lifson, S. (1970). Consistent force field calculations. II. Crystal Structure, Sublimation Energies, Molecular and Lattice Vibrations, Molecular Conformations and Enthalpies of Alkanes. *J. Chem. Phys.* **53,** 582–594.
7. Kollman, P. (1993). Free energy calculations: applications to chemical and biochemical phenomena. *Chem. Rev.* **93,** 2395–2417.
8. Allen, M. P., and Tildesley, D. J. (1987). Computer Simulation of Liquids. Oxford University Press, Oxford.
9. King, G., and Warshel, A. (1989). A surface constrained all-atom solvent model for effective simulations of polar solutions. *J. Chem. Phys.* **91**(6), 3647–3661.
10. Warshel, A., and Russell, S. T. (1984). Calculations of electrostatic interactions in biological systems and in solutions. *Q. Rev. BioPhys.* **17,** 283–421.
11. Miertus, S., Scrocco, E., and Tomasi, J. (1981). Electrostatic interaction of a solute with a continuum. A direct utilization of ab initio molecular potentials for the prevision of solvent effects. *J. Chem. Phys.* **55,** 117–129.
12. Warshel, A. (1976). Bicycle-Pedal model for the first step in the vision process. *Nature* **260,** 679–683.
13. McCammon, J. A., Gelin, B. R., and Karplus, M. (1977). Dynamics of folded proteins. *Nature* **267,** 585–590.
14. Levitt, M. (1983). Molecular dynamics of native proteins II. Analysis and Nature of Motion. *J. Mol. Biol.* **168,** 621–657.
15. Warshel, A., and Parson, W. W. (2001). Dynamics of biochemical and biophysical reactions: insight from computer simluations. *Quart. Rev. Biophys.* **34,** 563–670.
16. McCleskey, E. W. (2000). Ion channel selectivity using an electric stew. *Biophys. J.* **79,** 1691–1692.

17. Åqvist, J. (1996). Calculation of absolute binding free energies for charged ligands and effects of long-range electrostatic interactions. *J. Comp. Chem.* **17**(14), 1587–1597.
18. Warshel, A. (1979). Calculations of chemical processes in solutions. *J. Phys. Chem.* **83,** 1640–1650.
19. Brooks III, C. L., and Karplus, M. (1983). Deformable stochastic boundaries in molecular dynamics. *J. Chem. Phys.* **79,** 6312–6325.
20. Kuwajima, S., and Warshel, A. (1988). The extended Ewald method: A general Treatment of long-range electrostatic interactions in microscopic simulations. *J. Chem. Phys.* **89,** 3751–3759.
21. Lee, F. S., and Warshel, A. (1992). A local reaction field method for fast evaluation of long-range electrostatic interactions in molecular simulations. *J. Chem. Phys.* **97,** 3100–3107.
22. Sham, Y. Y., and Warshel, A. (1998). The surface constrained all atom model provides size independent results in calculations of hydration free energies. *J. Chem. Phys.* **109,** 7940–7944.
23. Saito, M. (1994). Molecular dynamics simulations of proteins in solution: artifacts caused by the cutoff approximation. *J. Chem. Phys.* **101,** 4055–4061.
24. Figueirido, F., Del Buono, G. S., and Levy, R. M. (1997). On the finite size corrections to the free energy of ionic hydration. *J. Phys. Chem. B* **101**(29), 5622–5623.
25. Bogusz, S., Cheatham III, T. E., and Brooks, B. R. (1998). Removal of pressure and free energy artifacts in charged periodic systems via net charge corrections to the Ewald potential. *J. Chem. Phys.* **108,** 7070–7084.
26. Hummer, G., Pratt, L. R., and Garcia, A. E. (1996). Free energy of ionic hydration. *J. Phys. Chem.* **100,** 1206–1215.
27. Warshel, A. (1992). Computer simulations of enzymatic reactions. *Curr. Opinion in Struct. Biol.* **2,** 230–236.
28. Warshel, A. (1981). Calculations of enzymic reactions: calculations of pKa, proton transfer reactions, and general acid catalysis reactions in enzymes. *Biochemistry* **20,** 3167–3177.
29. Wong, C. F., and McCammon, J. A. (1986). Dynamics and design of enzymes and inhibitors. *J. Am. Chem. Soc.* **108,** 3830–3832.
30. Zwanzig, R. W. (1954). High-temperature equation of state by a perturbation Method. I. Nonpolar Gases. *J. Chem. Phys.* **22,** 1420.
31. Valleau, J. P., and Torrie, G. M. (1977). Modern theoretical chemistry: a guide to Monte Carlo for statistical mechanics. 2. (pp. 169–194). Berne, B. J. (Ed.), Byways, Vol. **5**. New York: Plenum Press.
32. Warshel, A., and Åqvist, J. (1991). Electrostatic energy and macromolecular function. *Ann. Rev. Biophys. Chem.* **20,** 267–298.
33. Lee, F. S. *et al.* (1992). Calculations of antibody-antigen interactions: microscopic and semi-microscopic evaluation of the free energies of binding of phosphorylcholine analogs to McPC603. *Prot. Eng.* **5,** 215–228.
34. Kubo, R., Toda, M., and Hashitsume, N. (1985). Statistical Physics II: Nonequilibrium Statistical Mechanics. Springer-Verlag, Berlin.
35. Hwang, J.-K., and Warshel, A. (1987). Microscopic examination of free energy relationships for electron transfer in polar solvents. *J. Am. Chem. Soc.* **109,** 715–720.
36. Kuharski, R. A. *et al.* (1988). Molecular model for aqueous ferrous ferric electron transfer. *J. Chem. Phys.* **89,** 3248–3257.

37. Åqvist, J., and Hansson, T. (1996). On the validity of electrostatic linear response in polar solvents. *J. Phys. Chem.* **100,** 9512–9521.
38. Hwang, J.-K. *et al.* (1988). Simulation of free energy relationships and dynamics of SN2 reactions in aqueous solution. *J. Am. Chem. Soc.* **110**(16), 5297–5311.
39. King, G., and Warshel, A. (1990). Investigation of the free energy functions for electron transfer reactions. *J. Chem. Phys.* **93,** 8682–8692.
40. Roux, B. (1995). The calculation of the potential of mean force using computer simulations. *Comp. Phys. Comm.* **91,** 275–282.
41. Alden, R. G. *et al.* (1995). Calculations of electrostatic energies in photosynthetic reaction centers. *J. Am. Chem. Soc.* **117,** 12284–12298.
42. Burykin, A. *et al.* (2002). Simulations of ion current in realistic models of ion channels: The KcsA potassium channel. *Proteins: Struc., Func., and Gen.* **47,** 265–280.
43. Tanford, C., and Kirkwood, J. G. (1957). Theory of protein titration curves. I. General equations for impenetrable spheres. *J. Am. Chem. Soc.* **79,** 5333.
44. Warshel, A., Russell, S. T., and Churg, A. K. (1984). Macroscopic models for studies of electrostatic interactions in proteins: limitations and applicability. *Proc. Natl. Acad. Sci. USA* **81,** 4785–4789.
45. Warshel, A., and Levitt, M. (1976). Theoretical studies of enzymic reactions: dielectric, electrostatic and steric stabilization of the carbonium ion in the reaction of lysozyme. *J. Mol. Biol.* **103,** 227–249.
46. Warwicker, J., and Watson, H. C. (1982). Calculation of the electric potential in the active site cleft due to alpha-helix dipoles. *J. Mol. Bio.* **157,** 671–679.
47. Sharp, K. A., and Honig, B. (1990). Electrostatic interactions in macromolecules: theory and applications. *Ann. Rev. Biophys. Biophys. Chem.* **19,** 301–332.
48. Warshel, A. (1987). What about protein polarity? *Nature* **333,** 15–18.
49. Warshel, A., and Papazyan, A. (1998). Electrostatic effects in macromolecules: fundamental concepts and practical modeling. *Curr. Opinion in Struct. Biol.* **8,** 211–217.
50. King, G., Lee, F. S., and Warshel, A. (1991). Microscopic simulations of macroscopic dielectric constants of solvated proteins. *J. Chem. Phys.* **95,** 4366–4377.
51. Warshel, A. *et al.* (1989). How do serine proteases really work? *Biochemistry* **28,** 3629–3673.
52. Lee, F. S., Chu, Z. T., and Warshel, A. (1993). Microscopic and semimicroscopic calculations of electrostatic energies in proteins by the POLARIS and ENZYMIX programs. *J. Comp. Chem.* **14,** 161–185.
53. Sham, Y. Y., Chu, Z. T., and Warshel, A. (1997). Consistent calculations of pKa′s of ionizable residues in proteins: semi-microscopic and macroscopic approaches. *J. Phys. Chem. B* **101,** 4458–4472.
54. Schutz, C. N., and Warshel, A. (2001). What are the dielectric "constants" of proteins and how to validate electrostatic models. *Proteins: Struc., Func., and Gen.* **44,** 400–417.
55. Kollman, P. A. *et al.* (2000). Calculating structures and free energies of complex molecules: combining molecular mechanics and continuum models. *Acc. Chem. Res.* **33,** 889–897.
56. Buono, G. S., Figueirido, F. E., and Levy, R. M. (1994). Intrinsic pKa′s of ionizable residues in proteins: an explicit solvent calculation for lysozyme. *Proteins: Struct., Func., and Gen.* **20,** 85–97.
57. Johnson, E. T., and Parson, W. W. (2002). Electronic interactions in an integral membrane protein. *Biochem.* **41,** 6483–6494.
58. Sham, Y. Y., Muegge, I., and Warshel, A. (1998). The effect of protein relaxation on charge-charge interactions and dielectric constants of proteins. *Biophys. J.* **74,** 1744–1753.

59. Bashford, D., and Case, D. A. (2000). Generalized born models of macromolecular solvation effects. *Ann. Rev. Phys. Chem.* **51,** 129–152.
60. Luzhkov, V., and Warshel, A. (1992). Microscopic models for quantum mechanical calculations of chemical processes in solutions: LD/AMPAC and SCAAS/AMPAC calculations of solvation energies. *J. Comp. Chem.* **13,** 199–213.
61. Still, W. C. *et al.* (1990). Semianalytical treatment of solvation for molecular mechanics and dynamics. *J. Am. Chem. Soc.* **112,** 6127–6129.
62. Sham, Y. Y. *et al.* (2000). Examining methods for calculations of binding free energies: LRA, LIE, and PDLD-LRA, and PDLD/S-LRA calculations of ligands binding to an HIV protease. *Proteins: Struct. Funct. Genet.* **39,** 393–407.
63. You, T. J., and Bashford, D. (1995). Conformation and hydrogen ion titration of proteins: a continuum electrostatic model with conformational flexibility. *Biophys. J.* **69,** 1721–1733.
64. Florian, J., Goodman, M. F., and Warshel, A. (2002). Theoretical investigation of the binding free energies and key substrate-recognition components of the replication fidelity of human DNA polymerase b. *J. Phys. Chem. B* **106,** 5739–5753.
65. Warshel, A., Sussman, F., and King, G. (1986). Free energy of charges in solvated proteins: microscopic calculations using a reversible charging process. *Biochemistry* **25,** 8368–8372.
66. Pople, J. A. (1999). Quantum chemical models (Nobel lecture). *Angew. Chem. Int. Ed. Engl.* **38,** 1894–1902.
67. Wladkowski, B. D., Krauss, M., and Stevens, W. J. (1995). Ribonuclease a catalyzed transphosphorylation: an ab initio theoretical study. *J. Phys. Chem.* **99**(17), 6273–6276.
68. Friesner, R., and Dunietz, B. (2001). Large-scale ab initio quantum chemical calculations on biological systems. *Acc. Chem. Res.* **34,** 351–358.
69. Bloomberg, M. R. A., and Siegbahn, P. E. M. (2001). A quantum mechanical approach to the study of reaction mechanisms of redox active metalloenzymes. *J. Phys. Chem.* **105,** 9376–9386.
70. Stanton, R. V. *et al.* (1998). Combined ab initio and free energy calculations to study reactions in enzymes and solution: amide hydrolysis in trypsin and aqueous solution. *J. Am. Chem. Soc.* **120,** 3448–3457.
71. Chandrasekhar, J., and Jorgensen, W. L. (1984). SN2 reaction profiles in the gas phase and aqueous solution. *J. Am. Chem. Soc.* **106,** 3049–3059.
72. Warshel, A., and Lappicirella, A. (1981). Calculations of ground-and excited-state potential surfaces for conjugated heteroatomic molecules. *J. Am. Chem. Soc.* **103,** 4664–4673.
73. Théry, V. *et al.* (1994). Quantum mechanical computations on very large molecular systems: the local self-consistent field method. *J. Comp. Chem.* **15,** 269–282.
74. Zhang, Y., Liu, H., and Yang, W. (2000). Free energy calculation on enzyme reactions with an efficient iterative procedure to determine minimum energy paths on a combined ab initio QM/MM potential energy surface. *J. Chem. Phys.* **112**(8), 3483–3492.
75. Gao, J. (1996). Hybrid quantum and molecular mechanical simulations: an alternative avenue to solvent effects in organic chemistry. *Acc. Chem. Res.* **29,** 298–305.
76. Bakowies, D., and Thiel, W. (1996). Hybrid models for combined quantum mechanical and molecular approaches. *J. Phys. Chem.* **100,** 10580–10594.
77. Field, M. J., Bash, P. A., and Karplus, M. (1990). A combined quantum mechanical and molecular mechanical potential for molecular dynamics simulations. *J. Comp. Chem.* **11,** 700–733.

78. Friesner, R., and Beachy, M. D. (1998). Quantum mechanical calculations on biological systems. *Curr. Op. Struct. Biol.* **8,** 257–262.
79. Monard, G., and Merz, K. M. (1999). Combined quantum mechanical/molecular mechanical methodologies applied to biomolecular systems. *Acc. Chem. Res.* **32,** 904–911.
80. Garcia-Viloca, M., Gonzalez-Lafont, A., and Lluch, J. M. (2001). A QM/MM study of the racemization of vinylglycolate catalysis by mandelate racemase enzyme. *J. Am. Chem. Soc* **123,** 709–721.
81. Marti, S. *et al.* (2001). Transition structure selectivity in enzyme catalysis: a QM/MM study of chorismate mutase. *Theor. Chem. Acc.* **105,** 207–212.
82. Field, M. *et al.* (2002). Stimulating enzyme reactions: challenges and perspectives. *J. Comp. Chem.* **23,** 48–58.
83. Mulholland, A. J., Lyne, P. D., and Karplus, M. (2000). Ab Initio QM/MM study of the citrate synthase mechanism. A low-barrier hydrogen bond is not involved. *J. Am. Chem. Soc.* **122,** 534–535.
84. Cui, Q. *et al.* (2001). A QM/MM implementation of the self-consistent charge density functional tight binding (SCC-DFTB) Method. *J. Phys. Chem. B* **105,** 569–585.
85. Vreven, T., and Morokuma, K. (2000). The ONIOM (our own N-layered integrated molecular orbital + molecular mechanics) method for the first singlet excited (S1) state photoisomerization path of a retinal protonated Schiff base. *J. Chem. Phys.* **113,** 2969–2975.
86. Vaidehi, N., Wesolowski, T. A., and Warshel, A. (1992). Quantum-mechanical calculations of solvation free energies. A combined ab initio pseudopotential free-energy perturbation approach. *J. Chem. Phys.* **97,** 4264–4271.
87. Wesolowski, T. A., and Warshel, A. (1993). Frozen density functional approach for ab initio calculations of solvated molecules. *J. Phys. Chem.* **97,** 8050–8053.
88. Villa, J., and Warshel, A. (2001). Energetics and dynamics of enzymatic reactions. *J. Phys. Chem. B* **105,** 7887–7907.
89. Cunningham, M. A. *et al.* (1997). Simulation of the enzyme reaction mechanism of malate dehydrogenase. *Biochemistry* **36**(16), 4800–4816.
90. Wu, N. *et al.* (2000). Electrostatic stress in catalysis: structure and mechanism of the enzyme orotidine monophosphate decarboxylase. *Proc. Natl. Acad. Sci. USA* **97**(5), 2017–2022.
91. Bentzien, J. *et al.* (1998). Hybrid ab initio quantum mechanics/molecular mechanics calculations of free energy surfaces for enzymatic reactions: the nucleophilic attack in subtilisin. *J. Phys. Chem. B* **102,** 2293–2301.
92. Štrajbl, M., Hong, G., and Warshel, A. (2002). *Ab initio* QM/MM simulation with proper sampling: "First Principle" calculations of the free energy of the auto-dissociation of water in aqueous solution *J. Phys. Chem. B.* **106,** 13333–13343.
93. Alhambra, C. *et al.* (1999). Quantum mechanical dynamical effects in an enzyme-catalyzed proton transfer reaction. *J. Am. Chem. Soc.* **121,** 2253–2258.
94. Peräkylä, M., and Kollman, P. A. (1997). A simulation of the catalytic mechanism of aspartylglucosaminidase using ab initio quantum mechanics and molecular dynamics. *J. Am. Chem. Soc.* **119,** 1189–1196.
95. Åqvist, J., and Warshel, A. (1993). Simulation of enzyme reactions using valence bond force fields and other hybrid quantum/classical approaches. *Chem. Rev.* **93,** 2523–2544.
96. Warshel, A., Schweins, T., and Fothergill, M. (1994). Linear free energy relationships in enzymes. Theoretical analysis of the reaction of tyrosyl-tRNA synthetase. *J. Am. Chem. Soc.* **116,** 8437–8442.

97. Vuilleumier, R., and Borgis, D. (1998). An extended empirical valence bond model for describing proton transfer in $H^{+}(H_{2}0)n$ clusters and liquid water. *Chem. Phys. Let.* **284,** 71–77.
98. Schmitt, U. W., and Voth, G. A. (1998). Multistate empirical valence bond model for proton transport in water. *J. Phys. Chem. B* **102,** 5547–5551.
99. Florian, J. (2002). Comment on molecular mechanics for chemical reactions. *J. Phys. Chem. A* **106**(19), 5046–5047.
100. Truhlar, D. G. (2002). Reply to comment on molecular mechanics for chemical reactions. *J. Phys. Chem. A* **106**(19), 5048–5050.
101. Yang, W. (1991). Direct calculation of electron density in density-functional theory. *Phys. Rev. Lett.* **66,** 1438–1441.
102. Lee, T. S., and Yang, W. (1998). Frozen density matrix approach for electronic structure calculations. *Int. J. Quant. Chem.* **69,** 397–404.
103. Ermolaeva, M. D., van der Vaart, A., and Merz, K. M. Jr. (1999). Implementation and testing of a frozen density matrix: divide and conquer algorithm. *J. Phys. Chem. A* **103,** 1868–1875.
104. Wesolowski, T., Muller, R. P., and Warshel, A. (1996). Ab initio frozen density functional calculations of proton transfer reactions in solution. *J. Phys. Chem.* **100,** 15444–15449.
105. Hong, G. *et al.* (2000). Constraining the electron densities in DFT method as an effective way for ab initio studies of metal catalyzed reactions. *J. Comp. Chem.* **21,** 1554–1561.
106. Cortona, P. (1991). Self-consistently determined properties of solids without band-structure calculations. *Phys. Rev. B* **44,** 8454–8458.
107. Klüner, T. *et al.* (2001). Prediction of electronic excited states of adsorbates on metal surfaces from first principles. *Phys. Rev. Lett.* **86,** 5954–5957.
108. Wesolowski, T. (2002). Comment on "prediction of electronic excited states of adsorbates on metal surfaces from first principles." *Phys. Rev. Lett.* **88,** 209701.
109. Car, R., and Parrinello, M. (1985). Unified approach for molecular dynamics and density-functional theory. *Phys. Rev. Lett.* **55,** 2471–2474.
110. Tuckerman, M. *et al.* (1997). On the quantum nature of the shared proton in hydrogen bonds. *Science* **275,** 817–820.
111. Woo, T. K. *et al.* (1997). A combined car-parrinello QM/MM implementation for ab initio molecular dynamics simulations of extended systems: application to the transition metal catalysis. *J. Phys. Chem. B* **101,** 7877–7880.
112. Rothlisberger, U. *et al.* (2000). A comparative study of galactose oxidase and active site analogs based on QM/MM Car-Parrinello simulations. *J. Biol. Inorg. Chem.* **5,** 236–250.
113. Eichinger, M. *et al.* (1999). A hybrid method for solutes in complex solvents: Density functional theory combined with empirical force fields. *J. Chem. Phys.* **110,** 10452–10467.
114. Wolfenden, R., and Snider, M. J. (2001). The depth of chemical time and the power of enzymes as catalysts. *Acc. Chem. Res.* **34,** 938–945.
115. Leatherbarrow, R. J., Fersht, A. R., and Winter, G. (1985). Transition-state stabilization in the mechanism of tyrosyl-tRNA synthetase revealed by protein engineering. *Proc. Natl. Acad. Sci. USA* **82,** 7840–7844.
116. Warshel, A. (1998). Electrostatic origin of the catalytic power of enzymes and the role of preorganized Active Sites. *J. Biol. Chem.* **273,** 27035–27038.
117. Polanyi, M. (1921). *Z. ElektroChem.* **27,** 143–152.

118. Pauling, L. (1946). Molecular architecture and biological reactions. *Chem. Eng. News* **24,** 1375–1377.
119. Bruice, T. C., and Benkovic, S. J. (2000). *Chemical basis for enzyme catalysis. BioChem.* **39,** 6267–6274.
120. Snider, M. J. *et al.* (2000). Temperature effects on the catalytic efficiency, rate enhancement, and transition state affinity of cytidine deaminase, and the thermodynamic consequences for catalysis of removing a substrate "anchor" *BioChem.* **39,** 9746–9753.
121. Jencks, W. P. (1986). *Catalysis in Chemistry and Enzymology.* Dover Publication, New York.
122. Futatsugi, N. *et al.* (1999). Ab initio study of the role of lysine 16 for the molecular switching mechanism of ras protein p21. *Biophys J.* **77,** 3287–3292.
123. Glennon, T. M., Villa, J., and Warshel, A. (2000). How does GAP catalyze the GTPase reaction of Ras? A computer simulation study. *Biochemistry* **39,** 9641–9651.
124. Piana, S., and Carloni, P. (2000). Conformational flexibility of the catalytic asp dyad in HIV-1 protease: an ab initio study on the free enzyme. *Proteins: Struct. Funcy. Genet* **39,** 26–36.
125. Beveridge, A. J., and Heywood, G. C. (1993). A quantum mechanical study of the active site of aspartic proteinases. *Biochemistry* **32,** 3325–3333.
126. Northrop, D. B. (2001). Follow the protons: a low-barrier hydrogen bond unifies the mechanism of the aspartic proteases. *Acc. Chem. Res.* **34,** 790–797.
127. Abashkin, Y. G., Erickson, J. W., and Burt, S. K. (2001). Quantum chemical investigation of enzymatic activity in DNA polymerase ß. A mechanistic study. *J. Phys. Chem. B* **105,** 287–292.
128a. Florian, J., Goodman, M. F., and Warshel, A. (2003). Computer simulation studies of the fidelity of DNA polymerases. *Biopolymers* **68,** 286–299.
128b. Florian, J., Goodman, M. F., and Warshel, A. (2003). Computer simulation of the chemical catalysis of DNA polymerases: discriminating between alternative nucleotide insertion mechanisms for T7 DNA polymerase. *J. Am. Chem. Soc.* **125,** 8163–8177.
129. Kim, K., Kim, D., Lee, J. Y., Tarakeshwar, P., and Oh, K. S. (2002). Catalytic mechanism of enzymes: preorganization, short strong hydrogen bond, and charges buffering. *Biochemistry* **41**(16), 5300–5306.
130. Cavalli, A., and Carloni, P. (2001). Enzymatic GTP Hydrolysis: insights from an ab initio molecular dynamics study. *J. Am Chem. Soc.* **124,** 3763–3768.
131. Langen, R., Schweins, T., and Warshel, A. (1992). On the mechanism of guanosine triphosphate hydrolysis in ras p21 proteins. *Biochemistry* **31,** 8691–8696.
132. Sun, D. P., Liao, D. I., and Remington, S. J. (1989). Electrostatic fields in the active sites of lysozymes. *Proc. Natl. Acad. Sci. USA* **86,** 5361–5365.
133. Rao, S. N. *et al.* (1987). Free energy perturbation calculations on binding and catalysis after mutating Asn 155 in subtilisin. *Nature* **328,** 551–554.
134. Warshel, A., Sussman, F., and Hwang, J.-K. (1988). Evaluation of catalytic free energies in genetically modified proteins. *J. Mol. Biol.* **201,** 139–159.
135. Soman, K. *et al.* (1989). Electrical potentials in trypsin isozymes. *BioChem.* **28,** 9918–9926.
136. Glennon, T. M., and Warshel, A. (1998). The energetics of the catalytic reaction of ribonuclease A: a computational study of alternative mechanisms. *J. Am. Chem. Soc.* **120,** 10234–10247.
137. Fuxreiter, M., and Warshel, A. (1998). Origin of the catalytic power of acetylcholinesterase: computer simulation studies. *J. Am. Chem. Soc.* **120,** 183–194.

138. Vagedes, P. *et al.* (2000). The deacylation step of acetylcholinesterase. *J. Am Chem. Soc.* **122,** 12254–12262.
139. Åqvist, J., and Warshel, A. (1989). Calculations of free energy profiles for the staphylococcal nuclease catalyzed reaction. *Biochemistry* **28,** 4680–4689.
140. Åqvist, J., and Warshel, A. (1992). Computer simulation of the initial proton transfer step in human carbonic anhydrase I. *J. Mol. Biol.* **224,** 7–14.
141. Åqvist, J., Fothergill, M., and Warshel, A. (1993). Computer simulation of the CO2/HCO-3 interconversion step in human carbonic anhydrase I. *J. Am. Chem. Soc.* **115,** 631–635.
142. Fothergill, M. *et al.* (1995). Structure-energy analysis of the role of metal ions in phosphodiester bond hydrolysis by DNA polymerase I. *J. Am. Chem. Soc.* **117**(47), 11619–11627.
143. Åqvist, J., and Fothergill, M. (1996). Computer simulation of the triosephosphate isomerase catalyzed reaction. *J. Biol. Chem.* **271,** 10010–10016.
144. Bash, P. A. *et al.* (1991). Computer simulation and analysis of the reaction pathway of triosephosphate isomerase. *Biochemistry* **30,** 5826–5832.
145. Varnai, P., and Warshel, A. (2000). Computer simulation studies of the catalytic mechanism of human aldose reductase. *J. Am. Chem. Soc.* **122,** 3849–3860.
146. Lee, Y. S. *et al.* (1998). Catalytic mechanism of aldose reductase studied by the combined potentials of quantum mechanics and molecular mechanics. *Biophys. Chem.* **70,** 203–216.
147. Várnai, P., Richards, W. G., and Lyne, P. D. (1999). Modelling the catalytic reaction in human aldose reductase. *Proteins: Struct. Funct. Genet.* **37,** 218–227.
148. Yadav, A. *et al.* (1991). Role of solvent reorganization energies in the catalytic activity of enzymes. *J. Am. Chem. Soc.* **113,** 4800–4805.
149. Hurley, J. H., and Remington, S. J. (1992). Contribution of charged side-chains, Mg2+, and solvent exclusion to enzymatic beta-decarboxylation of alpha-keto acids. *J. Am. Chem. Soc.* **114,** 4769–4773.
150. Lyne, P. D., Mulholland, A. J., and Richards, W. G. (1995). Insights into chorismate mutase catalysis from a combined QM/MM simulation of the enzyme reaction. *J. Am. Chem. Soc.* **117,** 11345–11350.
151. Hansson, T., Nordlund, P., and Åqvist, J. (1997). Energetics of nucleophile activation in a protein tyrosine phosphatase. *J. Mol. Biol.* **265,** 118–127.
152. Warshel, A. *et al.* (2000). Remarkable rate enhancement of orotidine 5′-monophosphate decarboxylase is due to transition state stabilization rather than ground state destabilization. *BioChem.* **39,** 14728–14738.
153. Dinner, A. R., Blackburn, G. M., and Karplus, M. (2001). Uracil-DNA glycosylase acts by substrate autocatalysis. *Nature* **413,** 752–754.
154. Warshel, A. (1978). Energetics of enzyme catalysis. *Proc. Natl. Acad. Sci. USA* **75,** 5250–5254.
155. Marcus, R. A. (1956). On the theory of oxidation-reduction reactions involving electron transfer I. *J. Chem. Phys.* **24,** 966–978.
156. Krishtalik, L. I. (1980). Catalytic acceleration of reactions by enzymes. Effect of screening of a polar medium by a protein globule. *J. Theor. Biol.* **86,** 757–771.
157. Kollman, P. A. *et al.* (2001). Elucidating the nature of enzyme catalysis utilizing a new twist on an old methodology: quantum mechanical-free energy calculations on chemical reactions in enzymes and in aqueous solution. *Acc. Chem. Res* **34,** 72–79.
158. Blow, D. (2000). So do we understand how enzymes work? *Structure* **8,** R77–R81.
159. Villà, J. *et al.* (2000). How important are entropy contributions in enzymatic catalysis? *Proc. Natl. Acad. Sci. USA* **97,** 11899–11904.

160. Štrajbl, M., Florian, J., and Warshel, A. (2001). Ab initio evaluation of the free energy surfaces for the general base/acid catalyzed thiolysis of formamide and the hydrolysis of methyl thioformate: a reference solution reaction for studies of cysteine proteases. *J. Phys. Chem. B* **105,** 4471–4484.
161. Hermans, J., and Wang, L. (1997). Inclusion of loss of translational and rotational freedom in theoretical estimates of free energies of binding. Application to a complex of benzene and mutant T4 lysozyme. *J. Am. Chem. Soc.* **119,** 2707–2714.
162. Ford, L. O. J., Machin, P. A., Phillips, D. C., and Tijian, R. J. (1974). Crystal structure of lysozyme-tetrasaccharide lactone complex. *J. Mol. Bio.* **88,** 349–371.
163. Khanjin, N. A., Snyder, J. P., and Menger, F. M. (1999). Mechanism of chorismate Mutase: contribution of conformational restriction to catalysis in the claisen rearrangement. *J. Am Chem. Soc.* **121,** 11831–11846.
164. Tapia, O. A., and Safront, V. S. (1994). Enzyme catalysis and transition structures in vacuo. Transition structures for the enolization, carboxylation and oxygenation reactions in ribulose-1,5-bisphosphate carboxylase/oxygenase enzyme (Rubisco). *J. Chem. Soc. Faraday Trans. 1* **90,** 2365–2374.
165. Bruice, T. C. (2002). A view at the millennium: the efficiency of enzymatic catalysis. *Acc. Chem. Res.* **35**(3), 139–148.
166. Shurki, A. *et al.* (2002). How much do enzymes really gain by restraining their reacting fragments?. *J. Am Chem. Soc.* **124,** 4097–4107.
167. Kollman, P. A., Kuhn, B., and Perokyla, M. (2002). Computational studies of enzyme-catalyzed reactions: where are we in predicting mechanisms and in understanding the nature of enzyme catalysis. *J. Phys. Chem. B* **106,** 1537–1542.
168. Careri, G., Fasella, P., and Gratton, E. (1979). Enzyme dynamics: the statistical physics approach. *Ann. Rev. Biophys. Bioeng.* **8,** 69–97.
169. Karplus, M., and McCammon, J. A. (1983). Dynamics of proteins: elements and function. *Ann. Rev. BioChem.* **53,** 263–300.
170. Kohen, A. *et al.* (1999). Enzyme dynamics and hydrogen tunnelling in a thermophilic alcohol dehydrogenase. *Nature* **399,** 496–499.
171. Basran, J., Sutcliffe, M. J., and Scrutton, N. S. (1999). Enzymatic H-transfer requires vibration-driven extreme tunneling. *Biochemistry* **38**(10), 3218–3222.
172. Radkiewicz, J. L., and Brooks, C. L. III, (2000). Protein dynamics in enzymatic catalysis: exploration of dihydrofolate reductase. *J. Am. Chem. Soc.* **122,** 225–231.
173. Cameron, C. E., and Benkovic, S. J. (1997). Evidence for a functional role of the dynamics of glycine-121 of *Escherichia coli* dihydrofolate reductase obtained from kinetic analysis of a site-directed mutant. *Biochem.* **36,** 15792–15800.
174. Berendsen, H. J. C., and Hayward, S. (2000). Collective protein dynamics in relation to function. *Curr. Opin. Struct. Biol.* **10,** 165–169.
175. Neria, E., and Karplus, M. (1997). Molecular dynamics of an enzyme reaction: proton transfer in TIM. *Chem. Phys. Lett.* **267,** 23–30.
176. Eisenmesser, E. Z. *et al.* (2002). Enzyme dynamics during catalysis. *Science* **295,** 1520–1523.
177. Kohen, A., and Klinman, J. P. (1999). Hydrogen tunneling in biology. *Chem. Biol.* **6,** R191–R198.
178. Sutcliffe, M. J., and Scrutton, N. S. (2000). Enzyme catalysis: over-the-barrier or through-the-barrier? *Trends Biochem. Sci* **25,** 405–408.
179. Warshel, A. (1984). Dynamics of enzymatic reactions. *Proc. Natl. Acad. Sci. USA* **81,** 444–448.

180. Warshel, A. (1982). Dynamics of reactions in polar solvents. Semiclassical trajectory studies of electron-transfer and proton-transfer reactions. *J. Phys. Chem.* **86,** 2218–2224.
181. Borgis, D., and Hynes, J. T. (1991). Molecular-dynamics simulation for a model nonadiabatic proton transfer reaction in solution. *J. Chem. Phys.* **94,** 3619–3628.
182. German, E. K. (1981). *J. Chem. Soc. Faraday Trans. 1,* **77,** 397–412.
183. Hwang, J.-K. *et al.* (1991). Simulations of quantum mechanical corrections for rate constants of hydride-transfer reactions in enzymes and solutions. *J. Phys. Chem.* **95,** 8445–8448.
184. Hwang, J.-K., and Warshel, A. (1996). How important are quantum mechanical nuclear motions in enzyme catalysis? *J. Am. Chem. Soc.* **118,** 11745–11751.
185. Feierberg, I., Luzhkov, V., and Åqvist, J. (2000). Computer simulation of primary kinetic isotope effects in the proposed rate limiting step of the glyoxalase I catalyzed reaction. *J. Biol. Chem.* **275,** 22657–22662.
186. Gillan, M. J. (1987). Quantum-classical crossover of the transition rate in the damped double well. *J. Phys. C. Solid State Phys.* **20,** 3621–3641.
187. Voth, G. A. (1996). Path-integral centroid methods in quantum statistical mechanics and dynamics. *Adv. Chem. Phys.* **93,** 135–218.
188. Alhambra, C. *et al.* (2000). Quantum dynamics of hydride transfer in enzyme catalysis. *J. Am. Chem. Soc.* **122**(34), 8197–8203.
189. Billeter, S. R. *et al.* (2001). Hydride transfer in liver alcohol dehydrogenase: quantum dynamics, kinetic isotope effects, and role of enzyme motion. *J. Am Chem. Soc.* **123,** 11262–11272.
190. Crosby, J., Stone, R., and Lienhard, G. E. (1970). Mechanisms of thiamine-catalyzed reactions. Decarboxylation of 2-(1-carboxy-1-hydroxyethyl)-3,4-dimethylthiazolium chloride. *J. Am. Chem. Soc.* **92,** 2891–2900.
191. Lee, J. K., and Houk, K. N. (1997). A Proficient enzyme revisited: the predicted mechanism for orotidine monophosphate decarboxylase. *Science* **276,** 942–945.
192. Lightstone, F. C. *et al.* (1997). Non-enzymatic and enzymatic hydrolysis of alkyl halides: A haloalkane dehalogenation enzyme evolved to stabilize the gas-phase transition state of an SN2 displacement reaction. *Proc. Natl. Acad. Sci. USA* **94,** 8417–8420.
193. Appleby, T. C. *et al.* (2000). The crystal structure and mechanism of orotidine 5′-monophosphate decarboxylase. *Proc. Natl. Acad. Sci. USA* **97,** 2005–2010.
194. Miller, B. G. *et al.* (2000). Anatomy of a proficient enzyme: the structure of orotidine 5′-monophosphate decarboxylase in the presence and absence of a potential transition state analog. *Proc. Natl. Acad. Sci. USA* **97,** 2011–2016.
195. Harris, P. *et al.* (2000). Structural basis for the catalytic mechanism of a proficient enzyme: Orotidine 5′-monophosphate decarboxylase. *BioChem.* **39,** 4217–4224.
196. Warshel, A., and Florian, J. (1998). Computer simulations of enzyme catalysis: finding out what has been optimized by evolution. *Proc. Natl. Acad. Sci. USA* **95,** 5950–5955.
197. Rouhi, A. M. (2000). The buzz about a remarkable enzyme. *Chem. Eng. News* **78,** 42–46.
198. Lee, T. S. *et al.* (2001). An alternative explanation for the catalytic pProficiency of orotidine 5′-phosphate decarboxylase. *J. Am. Chem. Soc.* **123,** 12837–12848.
199. Miller, B. B., Short, S., and Wolfenden, R. (2001). Role of enzyme-ribofuranosyl contacts in the ground state and transition state for orotidine 5′-phosphate decarboxylase: a role for substrate destabilization. *BioChem.* **40,** 6227–6232.

200. Miller, B., and Wolfenden, R. (2002). Catalytic proficiency: the unusual case of OMP decarboxylase. *Ann. Rev. BioChem.* **71,** 847–885.
201. Cleland, W. W., and Kreevoy, M. M. (1994). Low-barrier hydrogen bonds and enzymic catalysis. *Science* **264,** 1887–1890.
202. Frey, P. A., Whitt, S. A., and Tobin, J. B. (1994). A low-barrier hydrogen bond in the catalytic triad of serine proteases. *Science* **264,** 1927–1930.
203. Cleland, W. W., Frey, P. A., and Gerlt, J. A. (1998). The low barrier hydrogen bond in enzymatic catalysis. *J. Biol. Chem.* **273,** 22529–22532.
204. Alden, R. A. *et al.* (1971). Something about subtilisin BPN structure. *Biochem. Biophys. Res. Comm.* **45,** 337.
205. Carter, P., and Wells, J. A. (1990). Functional interaction among catalytic residues in subtilisin BPN′. *Proteins: Struct. Func. Gen.* **6,** 240–248.
206. Guthrie, J. P. (1996). Short strong hydrogen bonds: can they explain enzymic catalysis? *Chem. & Biol.* **3,** 163–170.
207. Coulson, C. A., and Danielsson (1954). Ionic and covalent contributions to the hydrogen bond. Part I. *Arkiv for Fysik* **8,** 239–244.
208. Coulson, C. A., and Danielsson (1954). Ionic and covalent contributions to the hydrogen bond. Part II. *Arkiv for Fysik* **8,** 245–255.
209. Warshel, A., and Papazyan, A. (1996). Energy considerations show that low-barrier hydrogen bonds do not offer a catalytic advantage over ordinary hydrogen bonds. *Proc. Natl. Acad. Sci.* **93,** 13665–13670.
210. Warshel, A., and Weiss, R. M. (1980). An empirical valence bond approach for comparing reactions in solutions and in enzymes. *J. Am. Chem. Soc.* **102**(20), 6218–6226.
211. Pan, Y., and McAllister, M. A. (1998). Characterization of low-barrier hydrogen bonds 6. Cavity polarity effects on the formic acid-formate anion model system. An ab Initio and DFT Investigation. *J. Am. Chem. Soc.* **120,** 166–169.
212. Pan, Y., and McAllister, M. A. (1997). Characterization of low-barrier hydrogen Bonds. 5. Microsolvation of enol-enolate. An ab initio and DFT investigation. *J. Org. Chem.* **62,** 8171–8176.
213. Molina, P. A., Sikorski, R. S., and Jensen, J. H. (2002). NMR chemical shifts in the low-pH form of a-chymotrypsin. A QM/MM and ONIOM-NMR Study. *Theoretical Chemistry Accounts* **109,** 100–107.
214. Cleland, W. W., and Kreevoy, M. M. (1995). *Science* **269,** 104–104.
215. Schiøtt, B. *et al.* (1998). Characterization of the short strong hydrogen bond in benzoylacetone by ab initio calculations and accurate diffraction experiments. implications for the electronic nature of low-barrier hydrogen bonds in enzymatic reactions. *J. Am Chem. Soc.* **120,** 12117–12124.
216. Warshel, A., Papazyan, A., and Kollman, P. A. (1995). On low-barrier hydrogen bonds and enzyme catalysis. *Science* **269,** 102–104.
217. Gerlt, J. A. *et al.* (1997). Understanding enzymic catalysis: The importance of short, strong hydrogen bonds. *Chem. Biol.* **4,** 259–267.
218. Warshel, A., and Russell, S. (1986). Theoretical correlation of structure and energetics in the catalytic reaction of trypsin. *J. Am. Chem. Soc.* **108,** 6569–6579.
219. Ash, E. L. *et al.* (1997). A low-barrier hydrogen bond in the catalytic triad of serine proteases? Theory versus experiment. *Science* **278,** 1128–1132.
220. Cassidy, C. S., Lin, J., and Frey, P. A. (1997). A new concept for the mechanism of action of chymotrypsin: the role of the low-barrier hydrogen bond. *Biochemistry* **36,** 4576–4584.

221. Frey, P. A. (1995). On low-barrier hydrogen bonds and enzyme catalysis response. *Science* **269,** 104–106.
222. Viragh, C. *et al.* (2000). NMR evidence for a short, strong hydrogen bond at the active site of a cholinesterase. *Biochemistry* **39,** 16200–16205.
223. Kim, Y., and Ahn, K.-H. (2001). Theoretical study of the role of low-barrier hydrogen bonds in enzyme catalysis: a model of proton transfer in serine protease. *Theor. Chem. Acc* **106,** 171–177.
224. Schutz, C. N, and Warshel, A. (2003). The low barrier hydrogen bond (LBHB) proposal revisited: the case of the Asp···His pair in serine proteases.
225. Feierberg, I., and Aqvist, J. (2002). Computational modelling of enzymatic keto-enol isomerization reactions. *Theor. Chem. Acc.* **108,** 71–84.
226. Stratton, J. R., Pelton, J. G., and Kirsch, J. F. (2001). A novel engineered subtilisin BPN′ lacking a low-barrier hydrogen bond in the catalytic triad. *Biochemisty* **40,** 10411–10416.
227. Storm, D. R., and Koshland, D. E. Jr. (1970). A Source for the special catalytic power of enzymes: orbital steering. *Proc. Natl. Acad. Sci.* **66,** 445–452.
228. Koshland, D. E. Jr. (1958). Application of a theory of enzyme specificity to protein synthesis. *Proc. Natl. Acad. Sci. USA* **44,** 98–104.
229. Williams, R. J. P. (1972). The entatic state. *Cold Spring Harb. Symp. Quant. Biol.* **36,** 53–62.
230. Leulliot, N., and Varani, G. (2001). Current topics in RNA-protein recognition: control of specificity and biological function through induced fit and conformational capture. *Biochemistry* **40,** 7947–7956.
231. Ryde, U. *et al.* (1996). The Cupric geometry of blue copper proteins is not strained. *J. Mol. Bio.* **261,** 586–596.
232. Olsson, M. H. M., Hong, G., and Warshel, A (2003). Frozen density functional free energy simulations of redox proteins: computational studies of the reduction potential of plastocyanin and rusticyanin. *I. Am. Chem. Soc.* **125,** 5025–5039.
233. Štrajbl, M., Shurki, A., Kato, M., and Warshel, A. (2003). The apparent NAC effect in chorismate mutase reflects electronic transition state stabilization. *J. Am. Chem. Soc.* **125,** 10228–10237.

CATALYSIS AND SPECIFICITY IN ENZYMES: A STUDY OF TRIOSEPHOSPHATE ISOMERASE AND COMPARISON WITH METHYL GLYOXAL SYNTHASE

By QIANG CUI* AND MARTIN KARPLUS[†]

*Department of Chemistry and Theoretical Chemistry Institute University of Wisconsin, Madison, 1101 University Avenue, Madison, WI 53706 and [†]Department of Chemistry and Chemical Biology Harvard University Cambridge, Massachusetts 02138; and Laboratoire de Chimie Biophysique, ISIS Université Louis Pasteur, 67000 Strasbourg, France

I. Introduction ... 315
II. Computational Methods ... 320
III. Preparation of the Active Site: The TIM "Lid" Transition ... 324
IV. Analysis of Mechanisms of the TIM Catalyzed Reactions: The Effect on the Activation Barrier ... 327
A. Alternative Catalytic Pathways ... 328
B. Comparison with Model Reactions ... 331
C. Perturbation Analysis of Important Residues ... 336
D. Substrate Polarization Analysis by Mutagenesis ... 339
V. Dynamic Effects in the TIM Reaction ... 344
A. Classical Transmission Coefficient ... 345
B. Contribution of Tunneling in TIM Catalysis ... 349
C. The Effects of Protein Motion ... 351
VI. Catalytic Specificity: Comparison of TIM and MGS ... 358
VII. Conclusions ... 366
References ... 368

I. Introduction

Enzymes are molecules developed through evolution to catalyze specific reactions at physiological temperatures. They overshadow most designed catalysts in terms of their efficacy. Rate increases by a factor of 10^{17} relative to the reaction in solution have been observed (Wolfenden, 1995; Wolfenden, 2001). Understanding how enzymes do this is one of the major challenges of structural molecular biology. Although many proposals have been put forward to "explain" enzyme catalysis (Jencks, 1987; Fersht, 1999), it seems to us that the best approach is to focus on a specific enzyme and study its mechanism in detail. An analogy is provided by the in-depth experimental and theoretical analyses of the $H+H_2$ exchange reaction, which provided fundamental insights into how simple chemical reactions occur (Karplus *et al.*, 1965; Truhlar, 1976). For this review, we have chosen the enzyme triosephosphate isomerase (TIM).

ADVANCES IN PROTEIN CHEMISTRY, Vol. 66

0065-3233/03 $35.00

It has been referred to as a "perfect enzyme," in that the rate of the chemical transformation itself is increased sufficiently so that it no longer is the rate limiting step, which appears to be product release (Albery and Knowles, 1976a,b; Knowles and Albery, 1977). Our primary focus is on the role of theoretical studies in elucidating the details of the TIM mechanism; comparisons with experiment are made as appropriate.

It is useful to begin by writing an Arrhenius-like expression for k_{cat}, the rate constant for the chemical reaction catalyzed by the enzyme. It has the form

$$k_{cat} = A(T)e^{-\Delta G^{\dagger}/RT} \quad (1)$$

where $\Delta G^{\dagger}$ is the activation free energy and A(T) is the pre-exponential factor; R is the gas constant and T is the absolute temperature. For some enzymatic reactions, the concentrations in living systems are such that k_{cat}/K_m, where K_m is the Michaelis constant, is the appropriate quantity to examine (Fersht, 1999), but our concern is with the chemical step. It is clear from Eq. (1) that contributions to the rate enhancement can arise by lowering $\Delta G^{\dagger}$ or increasing A(T), relative to their values in the uncatalyzed reaction. In what follows, we examine both A(T) and $\Delta G^{\dagger}$, though it is evident that because the latter appears in the exponential it is likely to be more important than the former, which has only a linear effect.

Triosephosphate isomerase (TIM) is a homo-dimeric enzyme that catalyzes the conversion of dihydroxyacetone phosphate (DHAP) and glyceraldehyde 3-phosphate (GAP), which is an important step in the glycolytic pathway (Herlihy *et al.*, 1976; Albery and Knowles, 1976a,b; Knowles and Albery, 1977). The detailed catalytic mechanism of TIM and the origins of its catalytic power have been the subjects of numerous experiments (Lolis *et al.*, 1990; Davenport *et al.*, 1991; Komives *et al.*, 1991; Lodi *et al.*, 1991; Harris *et al.*, 1997; Harris *et al.*, 1998; Belasco *et al.*, 1980; Nickbarg *et al.*, 1988; Pompliano *et al.*, 1990; Zhang *et al.*, 1999) and theoretical studies (Bash *et al.*, 1991; Åqvist *et al.*, 1996; Alagona *et al.*, 1984; Alagona *et al.*, 1986, 1995; Peräkylä, 1997; Peräkylä *et al.*, 1996). Despite these efforts, the precise chemical mechanism of the multi-step reaction catalyzed by TIM is not fully elucidated (see Scheme I). The observed proton exchange between solvent and the enzyme with bound substrate suggested that an intermediate is involved in the catalyzed reaction (Albery and Knowles, 1976a,b) although there may be more than one; the identity of the intermediate(s) is not clear from experiment. There is agreement on the first step (transfer of a proton from DHAP to Glu 165 to form an enediolate) and the last step (transfer of a proton from G1uH 165 to an enediolate to form GAP), but there are three alternative proposals for the

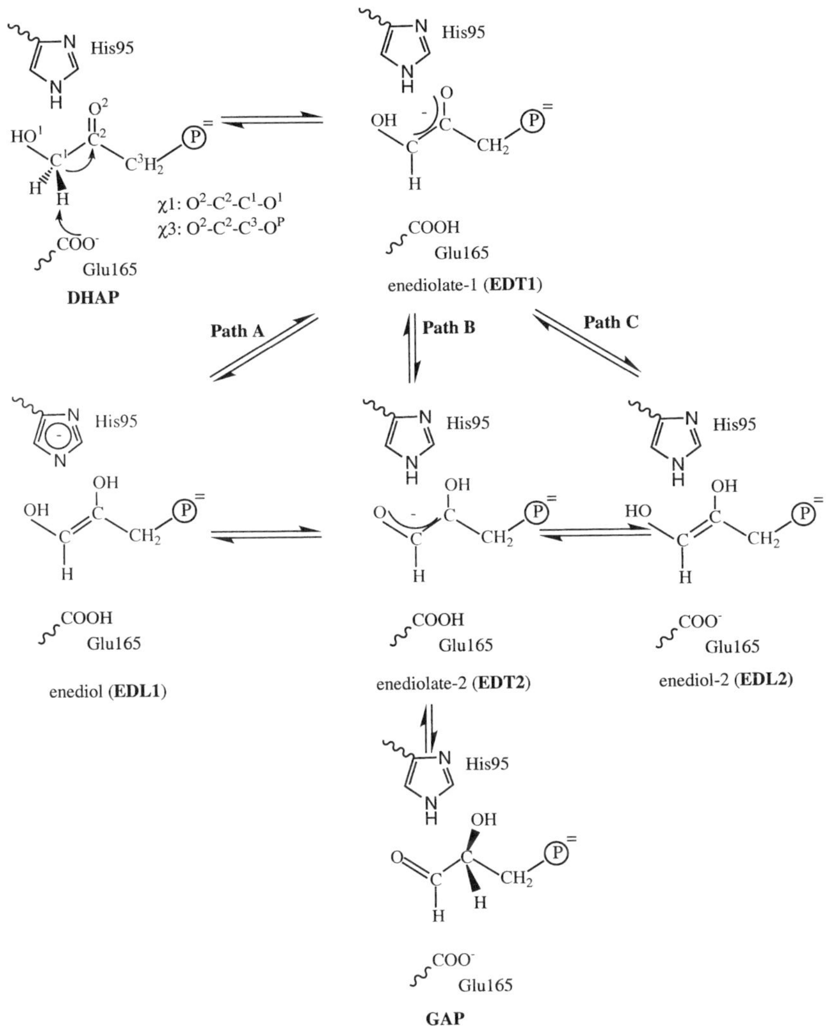

SCHEME I.

intermediate step (transfer of a proton from O^1 of the enediolate EDT1 to O^2 so as to form EDT2), and there is some support for each possibility (see later). As to the origin of catalytic efficiency, which is intimately related to the mechanistic issues, several factors that have been emphasized include reduced reorganization energy in the active site (Åqvist *et al.*, 1996) and possible strong hydrogen bond(s) to the intermediate (Cleland *et al.*, 1998; Harris *et al.*, 1997; Harris *et al.*, 1998). Since the TIM reaction involves proton transfers, it has also been proposed that proton tunneling makes a substantial contribution to catalysis, based on (secondary) kinetic isotope measurements (Alston *et al.*, 1996). Moreover, based on theoretical models for proton transfers, mainly in solution, it has been established that the energetics and tunneling efficiency of proton transfer reactions are sensitive to the separation of donor and acceptor atoms. Because enzymes fluctuate significantly at physiological temperature (Brooks *et al.*, 1988), it is possible that the motion (or "dynamics") of enzymes makes a substantial contribution to catalysis, although most available experimental "evidence" for this is indirect. Examples are measurements of kinetic isotope effects, for which the interpretation is not always straightforward.

In addition to their impressive efficacy, another important aspect of enzymes is related to their functional specificities (i.e., most enzymes act only on certain substrates to optimize the formation of certain products). An understanding of enzyme specificity is of interest also for the rational design of enzymes with desired catalytic properties. In this regard, TIM stands out as a prototypical system, particularly when compared with a related enzyme, methyl glyoxal synthase (MGS). MGS is an enzyme that catalyzes the first reaction in the methylglyoxal bypass of the Embden-Myerhof glycolytic pathway (Copper, 1984); the methylglyoxal metabolism is of biomedical interest because methylglyoxal is implicated in diabetes (Brownlee and Cerami, 1981). Although MGS shares neither sequence nor overall structural similarity (Davenport *et al.*, 1991; Saadat and Harrison, 2000; Marks *et al.*, 2001) with TIM, both enzymes bind DHAP as the natural substrate. However, the two enzymes catalyze different reactions (see Schemes I and II). While TIM catalyzes the isomerization of DHAP and GAP *bidirectionally*, MGS catalyzes the elimination reaction of DHAP, which leads to inorganic phosphate and the enol of methylglyoxal (which subsequently tautomerizes to methylglyoxal in solution), and it does *not* catalyze the elimination reaction with GAP as the substrate (Marks *et al.*, 2001; Tsai and Gracy, 1976). The elimination reaction of enediolate phosphate in solution is faster than isomerization by at least a factor of 100 (Richard, 1984), and it is suppressed by a factor of 10^5 to 10^8 in TIM (Richard, 1991). The specificities of the two enzymes are all the more impressive in view of the striking similarity of their active sites (see below); both have a base (Glu 165 in TIM and Asp 71 in MGS [Saadat and Harrison, 1998]), a histidine (His 95 in TIM and His 98 in MGS) and a lysine (Lys 12 in TIM and Lys 23 in MGS) close to the substrate. Both MGS and TIM catalyze their reactions near the diffusion limit. Therefore, MGS and TIM form an ideal pair of examples for studying both catalytic *efficacy* and *specificity*.

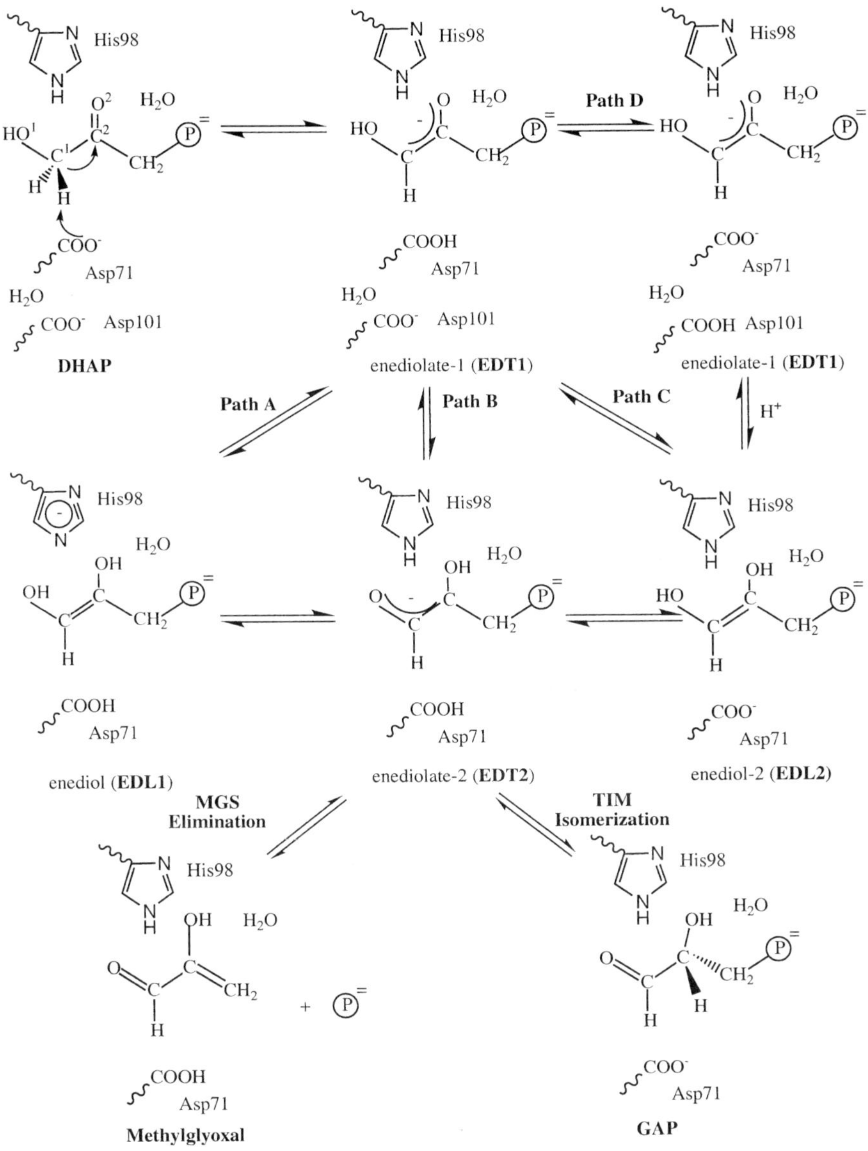
His98
DHAP
Asp71
Asp101
Path D
enediolate-1 (EDT1)
Path A
Path B
Path C
enediol (EDL1)
enediolate-2 (EDT2)
enediol-2 (EDL2)
MGS
Elimination
TIM
Isomerization
Methylglyoxal
GAP

SCHEME II.

In this work, we briefly review the computational method used in our studies and then present the results of a theoretical analysis of catalysis by TIM and a comparison with MGS; these include a comparison of possible catalytic pathways (Cui and Karplus, 2001), the origins of the catalytic power (Cui and Karplus, 2002a,b) and its specificity (Zhang *et al.*, 2002). The results are related to experiments where possible. We emphasize a reinterpretation of mutagenesis studies and a comparison with data (Rozovsky and McDermott, 2001; Kursula *et al.*, 2001; Amyes *et al.*, 2001; Jogl *et al.*, 2003) that have appeared since the publication of our theoretical work.

It is important to remember than an enzyme is a ''reactor,'' which generally operates in a very complex solvent environment (e.g., in the cell) and often has to isolate the active site from the solvent during the reaction. Since it is necessary also for the substrate to enter and the product to leave, the mechanism by which this is accomplished is of considerable interest. A variety of mechanisms, most of which involve a conformational change, exist (Gerstein *et al.*, 1994). TIM is one of the few systems where the nature and dynamics of the conformational change (a ''lid'' closing and opening) have been studied both experimentally and theoretically, as we describe later.

II. Computational Methods

We outline here the methodology used for calculating the contribution to the lowering of $\Delta G^{\dagger}$ and for the effect of tunneling on Eq. (1); other methodological aspects of the work described in this review are mentioned as they are required. The enzyme system was set up using the stochastic boundary condition following the standard procedures (Brooks and Karplus, 1989), the starting coordinates are from the 1.8 Å resolution x-ray structure of the yeast TIM-PGH complex (Davenport *et al.*, 1991) and a sphere of 16 Å radius centered on C2 in the substrate DHAP was used. The potential functions employed in the theoretical analysis are based on the combined quantum mechanical and molecular mechanical (QM/MM) approach (Warshel and Karplus, 1972; Field *et al.*, 1990; Gao, 1996), which allows bond breaking and bond formation in the substrate and the catalytic amino acids. The QM region in our analysis includes the substrate, Glu 165 and His 95. Link atoms were introduced between the Cα and Cβ atoms of the QM sidechain to saturate the valence of the Cβ boundary QM atom. The link atoms interact with the MM atoms, except the ''link host'' MM atom (the Cα atom in this case), through electrostatic terms; no van der Waals interactions are included. This scheme has been shown to be satisfactory for treating the QM/MM interface, particularly when the charges of the atom in the neighborhood of the link atom are small (Reuter *et al.*, 2000); this is true in the present case.

Although fast semi-empirical QM methods (e.g., AM1) can be used for qualitative investigations of enzyme catalysis, a sufficiently high level of QM method is required to distinguish different proposed pathways. In our studies, the B3LYP/6-31+G(d,p) method was chosen as the QM method; it was shown to give satisfactory results for the TIM reactions by examining model system calculations at higher levels, such as MP2 and CCSD with the 6-311+G(d,p) basis set (Cui and Karplus, 2002b). Although B3LYP/MM can be employed to compare different mechanisms with reaction path calculations, the high computational cost does not allow its use for free energy or rate constant calculations including quantum effects on the atomic motions. For these studies, we developed a reaction-specific form of AM1 (AM1-SRP) by adjusting certain parameters (C, N, O, H) to fit B3LYP results (energetics, structure, and dipole moments) for relevant model systems (Cui and Karplus, 2002b). The molecular mechanics force field used throughout our studies is that of CHARMM 22 (Mackerell *et al.*, 1998).

A major difference in the present studies, as compared with earlier QM/MM calculations, is the appropriate (though approximate) treatment of solvation effects; the approach is based on a method developed for free energy calculations of ligand binding (Simonson *et al.*, 1997; Dinner *et al.*, 2003). The partial charges of solvent exposed charged residues were scaled to account for bulk solvent screening effects, which avoids overpolarization of the QM region and structural distortions of the enzyme system. The scaling factors were obtained for each charged residue by use of the Poisson equation (i.e., the Poisson-Boltzmann equation with zero ionic strength). The effects of charge scaling and bulk solvation were accounted for by another set of Poisson calculations (Simonson *et al.*, 1997).

To determine the structures and energetics of critical species during the reaction (e.g., intermediates and transition states), both reaction path and potential of mean force (PMF) calculations have been performed. In both cases, the anti-symmetric stretch involving the proton donor, the transferring proton and the proton acceptor atoms was used as the reaction coordinate. More rigorous saddle point and reaction path calculations using the conjugate peak refinement (CPR) (Fischer and Karplus, 1992) approach confirmed that the antisymmetric stretch is an appropriate choice.

To better understand the origin of the catalysis by TIM, we constructed several model systems including only the active site residues (e.g., Glu 165,

His 95 and Lys 12). The "basic model" includes the substrate, intermediate, etc. and the catalytic residue (Glu 165 or His 95) directly involved (see Fig. 6). The more complete "active site model" includes the substrate and Glu 165 plus His 95. In these models, the phosphate group in the substrate was replaced by $-OCH_3$ to avoid artifacts induced by the large charge of the phosphate, which is stabilized in the enzyme by polar and charge interactions. The catalytic residues were replaced by small organic molecules; His 95 and Glu 165 was modeled by 4-methyl-imidazole and $CH_3CH_2CO_2^-$, respectively. During the geometry optimization, the terminal CH_3 group of the "catalytic residue" was fixed relative to C^1, C^2, and O^2 (see Scheme I) of the substrate in accord with the PGH TIM X-ray structure. All calculations were carried out at the B3LYP/6-31+G(d,p) level in the gas phase, and the PCM model (Miertus *et al.*, 1981; Cossi *et al.*, 1996) with the Pauling set of atomic radii was used to account for solvation effects.

To estimate the contribution of tunneling to the proton transfers in TIM, rate constant calculations were carried out for both the first intermolecular proton transfer step (DHAP→EDT1) and the intramolecular proton transfer in an enediolate involved in path B (Scheme I) (Cui and Karplus, 2002a). Although the latter was shown not to be the dominant pathway in wild type TIM by the present computational analysis (see later), the relatively simple nature of the reaction made possible detailed comparison of tunneling in different environments (i.e., the intramolecular proton transfer was studied in the gas phase, in solution, and in the enzyme [Cui and Karplus, 2002a]). In the current review, we limit ourselves to the results on the first intermolecular proton transfer, because it is rate limiting for k_{cat} in the forward direction and is the step for which kinetic isotope effects have been measured (Alston *et al.*, 1996). The rate constants were calculated using the semi-classical variational transition state theory (VTST) (Truhlar *et al.*, 1985; Truhlar *et al.*, 1996), with the hybrid AM1-SRP/CHARMM potential function mentioned above. In VTST, the effect of barrier recrossing is reduced by letting the transition state deviate from the saddle point *along* the reaction coordinate, and optimizing its position by variationally minimizing the reaction flux. This corresponds to searching for the free energy maximum along the reaction coordinate in the canonical ensemble (CVT). In a complementary approach, we also examine the effect of barrier recrossing by classical activated dynamics (Neria and Karplus, 1997). The effect of tunneling is introduced by computing the one-dimensional WKB integral along a given path. In the small-curvature tunneling (SCT) limit, the tunneling path is taken to be the minimum energy path (MEP), and multi-dimensional corner-cutting (Marcus & Coltrin, 1977) is taken into account by modifying the mass of the tunneling particle based on the reaction path curvature (Miller *et al.*, 1980). For the SCT calculations, normal modes along the MEP are required. Since saving the eigenvectors at many points is very memory intensive, we have implemented an algorithm to evaluate the reaction path curvature components in the TRAVEL module of CHARMM (Brooks *et al.*, 1983). This allows us to use the conjugate peak refinement (CPR) algorithm (Fischer and Karplus, 1992) to optimize the MEP for large systems and reduces the memory requirement of the computation dramatically. Only the frequencies and the curvature components along the MEP have to be saved in disk or memory; they form two arrays that are linear in the size of the system. The required information (geometries, energies, frequencies, and curvature components along the MEP) is then used as input for a modified version (Cui and Karplus, 2002a) of POLYRATE 8.0 (Chuang *et al.*, 1998), which calculates the effective reduced mass along the MEP, the tunneling integrals, and the rate constants. Between 70 and 110 points along the reaction coordinate were used in the VTST-SCT calculations. Of these, at least 50 points were in the critical region with the mass-weighted reaction coordinate in the range of (−3.0,+3.0) $amu^{\frac{1}{2}}$·bohr. Test calculations with 320 points gave transmission coefficients within a factor of 1.2 at 300K, indicating that the results with 50 points are sufficiently accurate for the present analysis.

To analyze the origin of catalytic specificity, combined QM/MM calculations were carried out for the proton transfer and phosphate elimination reactions in methyl glyoxal synthase (MGS), and the results were compared to those for TIM. The QM level employed was the semi-empirical density functional theory (SCC-DFTB) (Elstner *et al.*, 1998), which has shown to be more accurate than the commonly used AM1 and PM3 methods, and about one thousand time faster than DFT methods, such as B3LYP. Test calculations on selected configurations involved in the TIM system indicate that the SCC-DFTB/CHARMM approach gives results similar to those obtained with B3LYP/CHARMM (Cui *et al.*, 2001). MGS was treated using the stochastic boundary model, described above for TIM, and the x ray structure for MGS·PGH at 1.8 Å was used as the starting point (Marks *et al.*, 2001). The SCC-DFTB region included the substrate, Asp 71, His 98, and an active site water molecule (Wat 30) that was believed to be important.

All QM/MM calculations were carried out using an appropriately modified version of the CHARMM program (Brooks *et al.*, 1983). The model systems in the gas phase and solution were studied with Gaussian94 (Frisch *et al.*, 1995).

III. Preparation of the Active Site: The TIM "Lid" Transition

As mentioned in the introduction, TIM has to provide access to its active site to permit entrance of the substrate and exit of the product, while protecting the active site from solvent when the chemical reaction takes place. In TIM, an 11-residue loop region (residues 166–176) moves more than 7 Å and closes over the active site when the substrate binds (Fig. 1) (Lolis *et al.*, 1990; Joseph *et al.*, 1990). Mutagenesis experiments have shown that the loop is essential for catalysis and for preventing the hydrolytic phosphate elimination reaction (Pompliano *et al.*, 1990). When the open and closed structures of TIM are superimposed by least-squares optimization of all α-carbons, the root-mean-square (rms) deviation of the α-carbons is only 0.42 Å, whereas that for the loop residues is 4.8 Å. The α-carbons of Thr-172, which is at the center of the loop, is shifted by 7.1 Å, while the ends of the loop change little; for example, the distance from α-carbon 166 to α-carbon 176 is 7.3 Å in the open structure and 7.5 Å in the closed structure. Most important is the striking similarity between the internal structure of the loop in the open and closed

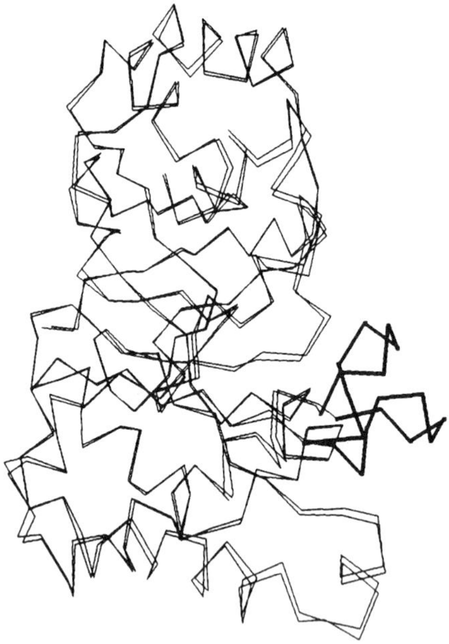

FIG. 1. Comparison of the closed and open structure of subunit 1; only α-carbons are shown. The two structures were superposed by least-squares optimization excluding the loop residues that are shown in thick black lines.

forms; that is, when residues 166–176 in the initial and final states are superposed, the rms deviation for all loop atoms is only 0.73 Å.

These structural results suggest that a rigid-body, hinge-type motion takes when the loop closes over the substrate in the active site; i.e., the so-called 'loop' moves more like a "lid." This suggests that similar behavior will be found in other enzymes that have 'loop' regions involved in substrate binding and release. In fact, 'lid' behavior has now been identified in lactate dehydrogenase (Gerstein and Chothia, 1991), as well as in other enzymes (Gerstein *et al.*, 1994).

To identify the hinge regions for the lid opening and closing, α-carbon pseudo-dihedral angle and mainchain dihedral angle differences between the open and closed loop were plotted against residue number. From the α-carbon plots in Fig. 2a, it is evident that there are two hinges. One of these involves the angles 166–167 and 167–168, and the other the angles 174–175, 175–176, and 176–177. Changing only these angles from their values in the closed structure to the values in the open structure yields an 'open' loop with an α-carbon rms difference of 1.9 Å from the open structure when non-loop α-carbons are superimposed; if the dihedral 164–165 is also rotated, the rms difference is reduced to 1.1 Å. This rather simple picture contrasts with that obtained from the dihedral angles ϕ and ψ

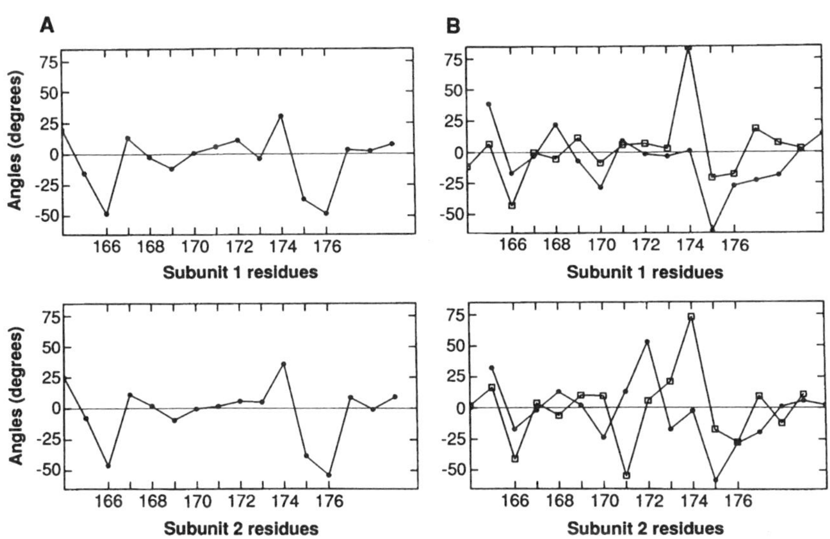

FIG. 2. Plots of closed minus open loop structure dihedral angle differences for subunit 1 and for subunit 2. The α-carbon pseudo-dihedral angle differences and B mainchain dihedral angle differences vs. residue number. In A the letter i corresponds to the pseudo-dihedral angle *i*, *i*+1; in B circles represent relative ϕ and squares relative ψ.

(Fig. 2b). It is evident from the dihedral angle difference plots that motion occurs in the region of residues 174 and 175, but it is less evident that there is also a significant change in the region of residues 166 and 167. Analysis of the van der Waals packing and hydrogen-bonding interactions within the lid provides an explanation for its rigid-body motion and the presence of two hinges (Joseph *et al.*, 1990).

To obtain information about the lid motion, several molecular dynamics simulations were performed. High temperatures were used to decrease the time required for a conformational change. All simulations began with the closed form since an open structure with increased entropy was expected to be favored at high temperatures, even in vacuum. At 298 K, the loop oscillated about the closed position, whereas during the 1000 K molecular dynamics simulation the loop opened and closed repeatedly. It was found that the loop opens and closes in a "jump-like fashion" within approximately 20 ps at this temperature. The 500 K molecular dynamics simulation was also interesting; the loop oscillated about the closed position until 29 ps when it flipped to a more open position; the Thr-172 α-carbon difference from the closed form was *ca.* 5.5 Å, somewhat less than the value of 7.1 Å found experimentally. Plots of the dihedral angle differences between the molecular dynamics structure at 29 ps and the closed and open structures show the largest changes between residues 174 and 175. This suggests that the loop starts to open in the region of 174 and 175.

For most systems, where the occurrence of a conformational change has been demonstrated by the observation of two different crystal structures, there are no measurements of the dynamics. However, for the TIM lid, experimental data on the dynamics exist. In particular, several studies have been made by McDermott and coworkers (Williams and McDermott, 1995; Rozovsky and McDermott, 2001; Rozovsky *et al.*, 2001). Also, in addition to our simulations described above, Wade *et al.* [1993] have used Brownian dynamics to study the lid motion in the presence of the electrostatic field due to the rest of the (rigid) protein. They used a simplified representation of the loop residues, analogous to that employed by McCammon *et al.* (1980) in Brownian dynamics simulations of α-helix unwinding. Such a model permitted simulations as long as 100 ns, which would have been impossible with a more detailed description. The simulations showed that the opening and closing motion occurred on a 1 ns time scale. Moreover, many dihedral transitions within the loop lid were observed, in contrast to the analysis of Joseph *et al.* (1990); these are likely to be a consequence of the simplified model used by McCammon *et al.* (1980). Another simulation of the lid motion (Derreumaux and Schlick, 1998) with an all-atom model using stochastic dynamics supports the rigid-lid description and suggests a sizable activation barrier for the transition.

The dynamics of the loop/lid motion have been studied by several NMR techniques. One of these uses line shape analysis of the quadrupole powder pattern determined by solid state deuterium NMR (Williams and McDermott, 1995; Rozovsky and McDermott, 2001) to probe the motion of the perdeuterated indole sidechain of Trp 168. This residue is near the end of the loop (166–176); in fact, one of the hinges involves motion about the pseudodihedral angles 166–167 and 167–168, though the displacement of the former is larger than the latter. The experimental data were interpreted by assuming displacements of the Trp 168 sidechain that correspond to the magnitude observed in the different crystal structures; with this assumption, a transition rate of about $10^4\ s^{-1}$ was obtained. This is much slower than that observed in the Brownian dynamics simulation. A range of temperatures were studied and an activation barrier between 12 and 14 kcal/mol was estimated. Similar results were obtained with ^{19}F solution NMR spectra when a 5′-fluorotryptophan replaced Trp168 (Rozovsky *et al.*, 2001). An obvious complication in the interpretation is the (reasonable) assumption that the conformational change used to fit the data actually involves the open and closed form of the loop, rather than two unrelated slowly interconverting conformations of Trp 168, such as have been found to exist in studies of tryptophan fluorescence in other proteins. With a barrier of 12 kcal/mol, the 20 ps time for the transition found in the molecular dynamics simulation at 1000 K yields a rate of about $10^4\ s^{-1}$ at 300 K, close to that estimated from the NMR experiment. Another study examined the temperature dependence of a TIM transition by using ^{31}P NMR of Glu 165 modified covalently with the substrate analogue 3-chloroacetol phosphate (Yüksel, 1994). Signals corresponding to two conformations were observed (again there is no direct evidence that they represent the open and closed lid) and the temperature dependence of the transition rate between them was fitted to an Arrhenius plot; it yielded a barrier of 34 kcal/mole. The actual data indicate significant curvature, though it is difficult to determine whether this is meaningful since only three temperature points were measured.

IV. Analysis of Mechanisms of the TIM Catalyzed Reactions: The Effect on the Activation Barrier

In this section, we first discuss the energetics associated with different proposed mechanisms (see Scheme I), and then analyze the various contributions to the catalytic power of TIM. We emphasize the relation between the results of the calculations and experiment, particularly those published after the theoretical studies [Kursula *et al.*, 2001; Amyes *et al.*, 2001; Jogl *et al.*, 2003].

A. *Alternative Catalytic Pathways*

As shown in Scheme I, there is agreement for the first and last steps of the TIM catalyzed reaction, which involve proton exchange between the substrate carbon atoms with Glu 165. For the conversion between the enediolate intermediates (EDT1 and ETD2), however, there have been several proposals. Path A was initially proposed by Knowles and coworkers (Albery *et al.*, 1976a,b), and has been widely accepted, although there is no *direct* evidence for intermediate formation of an *enediol* during the reaction. The energetics related to path A have been analyzed theoretically by Bash *et al.* (1991) with QM/MM calculations at the AM1/CHARMM level, and by Åqvist and Fothergill (1996) with an EVB model using a free energy perturbation technique. Path B, which involves an intramolecular proton transfer leads from EDT1 to EDT2 without proton transfer from His 95, was proposed by Aalgona *et al.* (1995), based on a simplified model of the enzyme active site and a QM/MM type approach at the MP2/3-21+ G level. They obtained a barrier of 12 kcal/mol for the intramolecular transfer and found that path A involved an endothermic process requiring about 20 kcal/mol. The latter result differs drastically from the earlier work of Bash *et al.* [1991] and Åqvist and Fothergill (1996), as well as from the present calculations (see below). Although the origin of this difference is not clear, Aalgona *et al.* (1995) did emphasize "that our [their] model for representing the environment is not quantitatively reliable." The third mechanism, path C in Scheme 1, involves Glu 165 as the only catalytic residue that transfers protons between different oxygen atoms in the substrate. Although this path was originally proposed by Nickbarg *et al.* [1988] only for the His95Gln mutant due to the large displacements of Glu 165 required, several authors have suggested that path C can also contribute in the wild type TIM (Harris *et al.*, 1997; Harris *et al.*, 1998; Peräkylä, 1996, 1997). For example, analysis of the extent of tritium transfer from the *pro*-R position in DHAP to GAP as a function of the substrate concentration11(c) suggests that the "classical" mechanism (path A) contributes at least 3.9% in yeast TIM, and the rest may come from either path A or path C.

Figure 3 shows the QM/MM results for the stable species and transition state barriers along the three different paths; EDT1, which is common to all paths (see Scheme I), is used as the reference of energy. The most significant result is that the intramolecular proton transfer (path B) has a substantially higher barrier between EDT1 and EDT2 (14.0 kcal/mol) than either intermolecular path (path A and C). Analysis of the contribution to the energies by a perturbation method (Bash *et al.*, 1991) (see later) indicates that His 95 destabilizes the intramolecular transition state

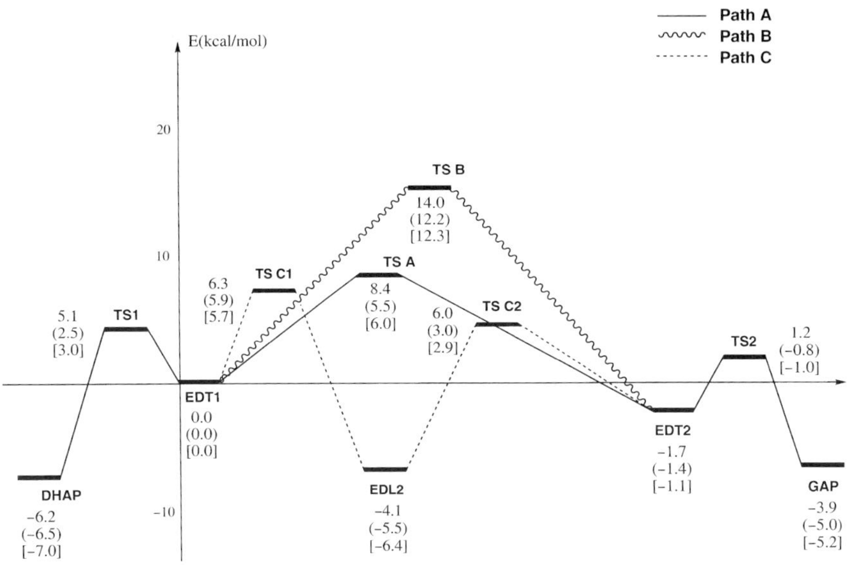

FIG. 3. Schematic potential energy profiles for different paths in the TIM reactions from QM/MM calculations. Numbers are obtained from single point B3LYP/6-31+G(d,p)/CHARMM calculations with a QM partition that includes the substrate, Glu 165 *and* His 95, at the structures optimized with a smaller QM region that includes only the substrate and the catalytic residue (Glu 165 *or* His 95) explicitly involved in a given proton transfer step. The values in the parentheses include zero-point corrections, and the values in the brackets are effective free energies obtained by adding vibrational free energy contributions (see Methods). The absolute energy for the reference structure EDT1 is −1445.91678 hartree, the zero point correction is 3.64467 hartree, and the vibrational free energy is 3.44308 hartree. The dominant contribution that lowers the transition state free energy, relative to the stable states, comes from the zero-point correction. The correction is larger for TSC2, the transition state between EDL2 and EDT2 along path C, because EDL2 has somewhat smaller vibrational contributions compared to EDT1; most of the difference originates from modes with frequencies below $1000 cm^{-1}$.

by about 5 kcal/mol. The two intermolecular proton transfer paths have similar calculated activated energies, in spite of the fact that path C is more complex structurally. In the latter, the sidechain of Glu 165 is required to displace substantially (~0.90 Å for all heavy elements) to permit it to transfer a proton to O^2 of EDT1 (to form the enediol EDL2) and to pick up a proton from O^1 (to form EDT2). The stability of EDL2 along path C and the lack of stability of EDL1 along path A is due to the difference in the deprotonation energy of GluH 165 and His 95.6(b).

To better understand the proton transfer mechanism along path A, a two-dimensional adiabatic map was calculated at the AM1-SRP/CHARMM level (Cui and Karplus, 2002b). The two coordinates used to describe the reaction are the antisymmetric stretch involving the transferring protons (H1 in the substrate and Hϵ2 in His 95) and associated donor and acceptor atoms (the two substrate oxygen atoms involved and the Nϵ in His 95); all degrees of freedoms other than the two mapping coordinates were optimized in the two-dimensional grid scan used to generate the adiabatic map. As shown in Fig. 4, there is only one saddle point in the conversion between EDT1 and EDT2 along path A. The character of this saddle point involves mainly the motion of H^1; i.e. the first proton to be transferred in path A. The motion of the second proton, Hϵ2 in His 95, is significantly involved immediately after the saddle point (i.e., $\delta 1 > 0.2$Å). Since the change from the neutral His 95 to a negative His 95- is not energetically favorable, the second proton transfer is nearly concerted with the first. We note also that the fact that a neutral His 95, rather than the

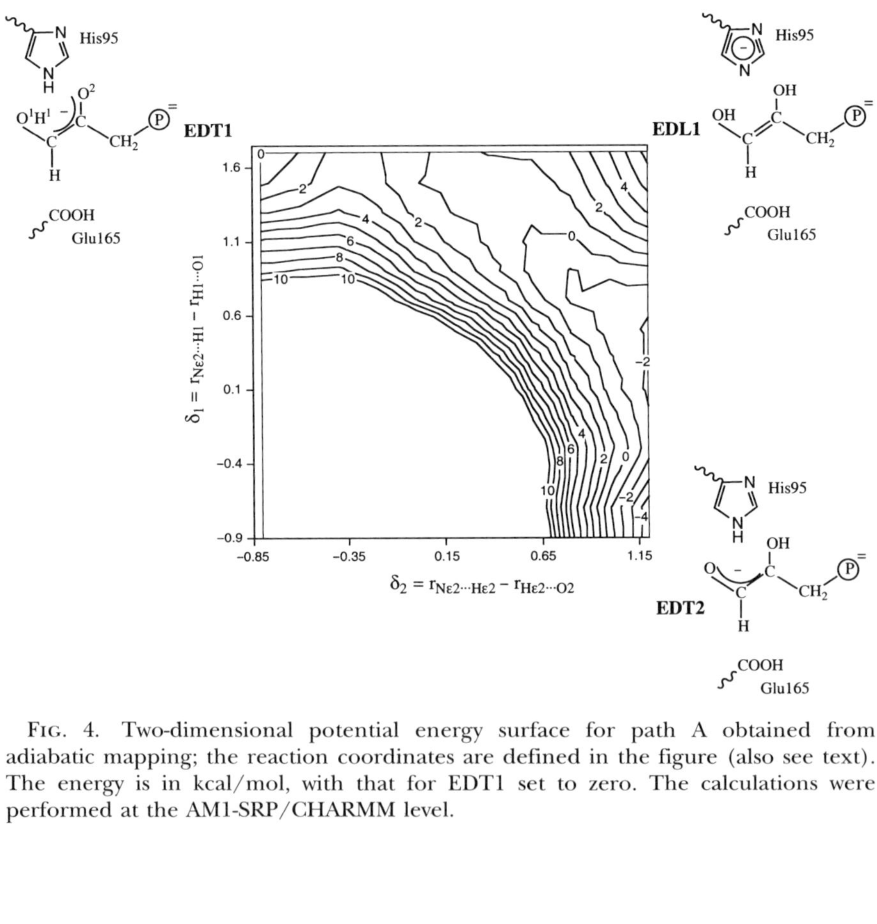

FIG. 4. Two-dimensional potential energy surface for path A obtained from adiabatic mapping; the reaction coordinates are defined in the figure (also see text). The energy is in kcal/mol, with that for EDT1 set to zero. The calculations were performed at the AM1-SRP/CHARMM level.

expected doubly protonated His 95^+, is involved in the reaction, was predicted by the calculation of Bash *et al.* (1991) and confirmed by the experiments of Lodi and Knowles (1991). The latter paper contains an illuminating commentary, which we quote here:

> Yet the reasons why the enzyme has evolved to use a neutral histidine as a general acid are not clear. Perhaps if histidine-95 had carried a positively charged imidazolium side chain, the basicity of glutamate-165 would have been lowered and the catalytic ability of the enzyme would have been impaired. Possibly the enzyme evolved to "match" the pK_a values of the catalytic imidazole nitrogen and the substrate's enediol oxygen in order to allow a rapid proton transfer between these sites. Maybe the use of imidazolium as a general acid would have caused the intermediate to be *too* stable, thereby slowing the reaction catalyzed by the enzyme. Indeed, Bash *et al.* (1991) have performed ab initio and molecular dynamics calculations that suggest that if the sidechain of histidine-95 were positively charged, the imidazole-enediol pair formed during the catalyzed reaction would have a very low free energy. Such a deep thermodynamic well in the free energy profile would obviously be inconsistent with the efficient catalysis that the isomerase mediates. This study further indicated that the imidazol*ate*-enediol pair (resulting from the use of a neutral imidazole sidechain of histidine-95) would have a free energy that is consistent with the known kinetic parameters for triosephosphate isomerase. Whether or not the details of this analysis will turn out to be correct, it is interesting that theory and experiment have here agreed upon a result that runs counter to the initial prejudices of mechanistic chemistry.

B. *Comparison with Model Reactions*

As shown in Fig. 5, the overall energetics in solution are significantly different from those in the gas phase and in the enzyme. Relative to the gas phase, the major solvation effect is to decrease the strength of the hydrogen bonding interaction involving ionic species due to solvent screening, and to stabilize the uncoordinated ionic species and leads to the stabilization of structures such as EDT2. Nevertheless, in the non-enzyme environment (gas phase and solution), the steps with the highest barriers are those associated with proton transfer from and to the carbon atoms of the substrate, i.e., DHAP $\rightarrow$ EDT1 and EDT2 $\rightarrow$ GAP. The calculations indicated that the uncatalyzed proton transfer in solution has a barrier of *at least* 24 kcal/mol (because the donor and acceptor groups were arranged in the favorable orientation found in the enzyme), which is close to the experimental estimate of 26 kcal/mol for the uncatalyzed reaction in solution ($k = 6\times10^{-6}\ s^{-1}$) (Hall and Knowles, 1975; Richard, 1984). The proton transfer processes that do not involve carbon atoms have much lower barriers. Therefore, the primary issue for

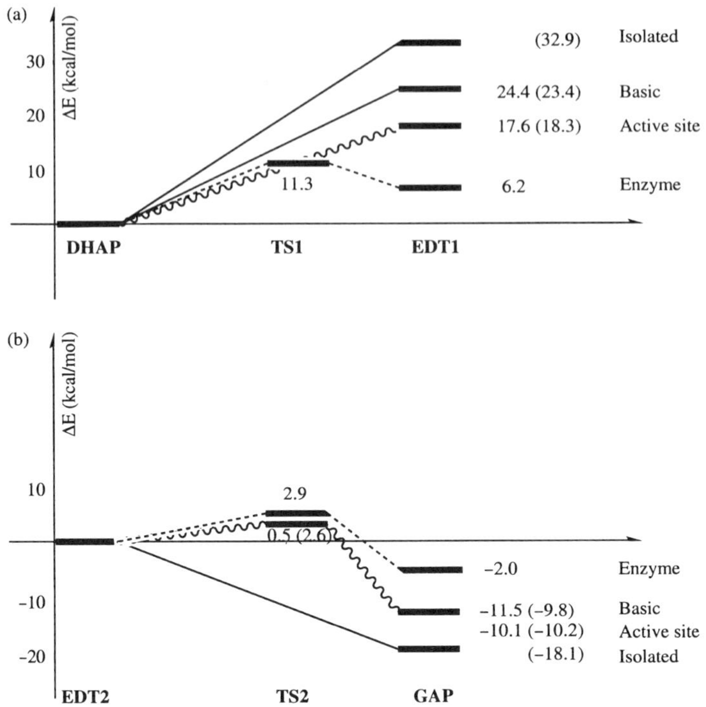

FIG. 5. Comparison of energetics for the reactions (see Scheme I) in the gas phase, in solution (in parentheses) and in TIM. See text for the definition of various models.

the enzyme is to stabilize the enediolate species, EDT1 and EDT2, which is exactly what TIM has achieved. The reaction of DHAP to give EDT1 becomes more facile in the enzyme due largely to the strong stabilization of EDT1 by Lys 12 (but see later). Moreover, along path A, EDT2 becomes very close in energy (−2 kcal/mol) to EDT1 in the enzyme, while the former is less stable by 14.6 and 4.4 kcal/mol in the basic model in the gas phase and solution, respectively.

Since the barrier for proton transfer reactions depends sensitively on the distance between the donor and acceptor atoms, the first task of the enzyme is to provide an active site geometry with the key atoms of the catalytic residues in good positions. The arrangements of the sidechains of Glu 165 and His 95 in TIM clearly illustrate this point. In the B3LYP/CHARMM minimized structure for the substrate(DHAP)-enzyme complex, for example (Fig. 6), the sidechain of Glu 165 is rather well aligned with the substrate; the O_ϵ^1–O^1 distance is about 2.7 Å, and

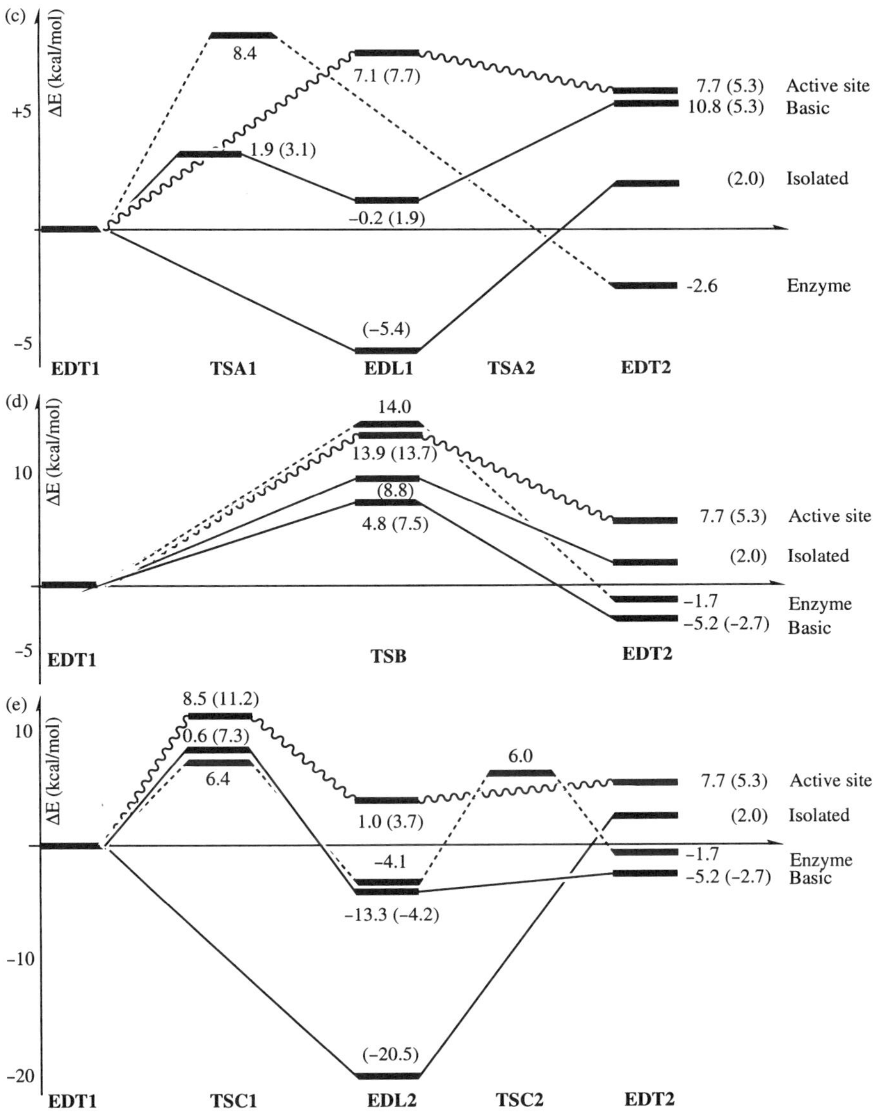

FIG. 5. *Continued*

the $O\epsilon^2$-C^1 distance is about 2.8 Å. The His 95 residue is also well positioned to polarize the carbonyl group in the substrate and to donate a proton in path A; the $N\epsilon$-O^2 and $N\epsilon$-O^1 distances are about 2.8 and 3.0 Å, respectively.

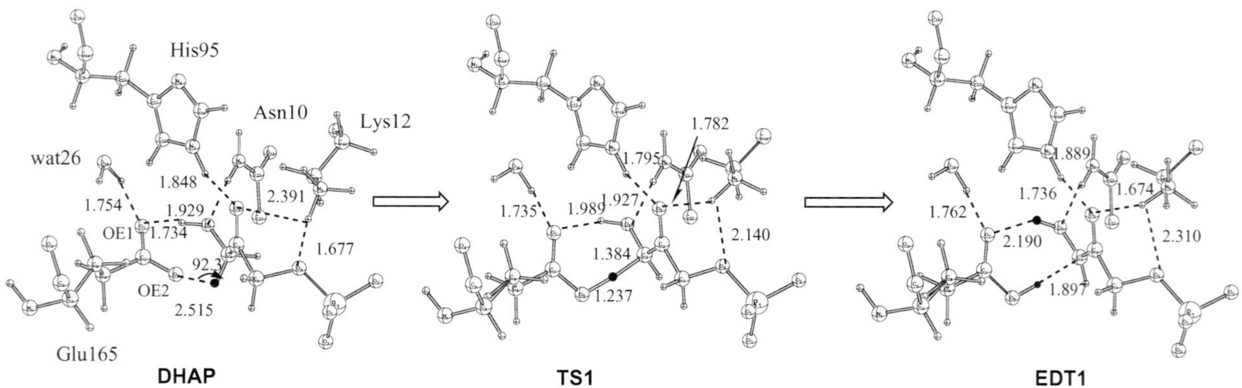

FIG. 6. Active site geometries at the B3LYP/6-31+G(d,p)/CHARMM level for important configurations in the first proton transfer step in the TIM reactions (DHAP→EDT1). The substrate and the side chain of Glu 165 are treated with QM in these calculations (see text). Distances are given in Å. The solid dots represent the photon that is transferred.

As shown in Fig. 5, placing the reacting groups (i.e., CH_3COO- and a model DHAP) in the x-ray positions in a solution environment makes the proton transfer between DHAP and Glu 165 less endothermic. However, this is not sufficient; the proton transfer from DHAP to EDT1 is still very endothermic (~24 kcal/mol!) in the basic model (with only the substrate and model Glu 165). The protein provides electrostatic interactions and polarization by specific residues to stabilize EDT1, which is analyzed best by the perturbation calculations that we turn to in the next section.

Recently, a high-resolution x-ray structure (1.2 Å) of the DHAP-TIM complex has became available (Jogl *et al.*, 2003). It is of interest to compare the results of that structure with the theoretical prediction for the reactant structure (i.e., the Michaelis complex with DHAP in the active site), based on optimization of the crystal structure with the PGH inhibitor (Cui and Karplus, 2001, 2002a). We discuss the structural parameters that were emphasized by Jogl *et al.* (2003) and their interpretation of their relevance to catalysis. In the x-ray structure the $O\epsilon^2$-C^1 and $O\epsilon^2$-C^2 distances are both close to 3.0 Å (calculated values are 2.8 and 2.3 Å, respectively), and the $N\epsilon$-O^2 and $N\epsilon$-O^1 distances are in the range of 2.6-2.7 Å and 3.0 Å, respectively. (Calculated values are 2.8 and 3.0 Å, respectively.) It was then proposed by Jogl *et al.* (2003) that these short, bifurcated hydrogen bonds are essential for the catalysis by TIM. Our calculations support the idea that the hydrogen bond between His 95 and O^2 introduces a favorable polarizing effect in the conversion from DHAP to EDT1 (Cui and Karplus, 2002a). However, it is important to emphasize that this polarization effect contributes to catalysis only if the hydrogen bond becomes stronger (and shorter) as the reaction proceeds, which of course is not demonstrated by the reactant structure of Jogl *et al.* (2003); otherwise, DHAP and EDT1 gets stabilized equally, and the reaction is hardly affected. Their suggestion that the proximity of Glu 165 to both C^1 and C^2 aids in the formation of both EDT1 and EDT2 is in accord with the calculations. However, we did not find any evidence that indicates the Glu 165–C^2 interaction is involved in the formation of EDT1; in fact, the protonated carboxylate group of Glu 165 aligns directly with C^1 in EDT1 (Fig. 6). Several torsional angles associated with the substrate have similar values in the computated and x-ray structure of the Michaelis complex. The $\chi 1$ and $\chi 3$ angles are (see Scheme I) 25° and 64° from B3LYP/CHARMM minimizations, while the x-ray study gave 19–45° and 73°, respectively. The fact that the phosphoryl torsional angle ($\chi 3$) is significantly out of plane, unlike previous observations with substrate analogues (e.g., PGA and PGH), argues that the phosphoryl torsional angle cannot be the major factor that suppresses the phosphate elimination reaction in TIM, in agreement with our recent theoretical analysis (Zhang *et al.*, 2002; see Section VI).

In short, the new data of Jogle *et al.* (2003) confirm the calculated structure of the Michaelis complex. This provides additional support for our analysis of the TIM reaction and the origin of its catalytic power.

C. *Perturbation Analysis of Important Residues*

The perturbation analysis at the AM1-SRP/CHARMM level (Fig. 7) gives more detailed information on the energetic contribution from individual residues in variety steps of the TIM reaction. It shows that Lys 12 and certain polar residues, such as His 95, Asn 10, and Ser 211, plus a number of water molecules in the active site, make the most significant contributions to the energetics of the TIM reactions. However, a few residues that are relatively far from the active site, such as Gly 210 and Ile 127 (Fig. 8), also make significant contributions for certain steps (i.e., the main chain atoms on these non-polar residues have a substantial polarizing effect on the reacting groups in proton transfers along path C [see Fig. 7]). Not all the proton transfer barriers have to be lower in the enzyme than in solution to achieve the overall rate acceleration. For example, the barrier along path A for the transfer from EDT1 to His 95, which is not the rate-determining chemical step, is raised by residues such as Lys 12 and Glu 165 (see Section IV.B.), consistent with the fact that EDL1 is not stabilized in the enzyme.

The major evolutionary pressure on TIM was to lower the barrier between DHAP and EDT1, in part by stabilizing the latter relative to DHAP (or GAP), since this corresponds to the slow step in the uncatalyzed reaction in solution. Among all the residues, Lys 12 makes the most important favorable contribution to the first proton abstraction from DHAP; as a matter of fact, the sidechain of Lys 12 was found to move closer (~0.7 Å) to O^1 in the substrate during the first proton transfer (see Fig. 6). It was known that Lys 12 is crucial for the appropriate binding of the substrate by stabilizing the phosphate group; the mutant Lys 12 Met results in enzymes that do not bind substrate (Lodi *et al.*, 1994), so this mutant gives no information as to its role in catalysis, per se (see later).

In a recent study Amyes *et al.* (2001) used a substrate analogue without the phosphate group ([R]-glyceraldehyde, GA). They found that k_{cat} for the GA isomerization by TIM is drastically lower than that for GAP. In fact, Amyes *et al.* observed that the rate is comparable to that of the nonspecific protein catalyzed degradation of GA, which occurs in the presence of TIM saturated by the potent competitive inhibitor 2-phosphoglycolate and for

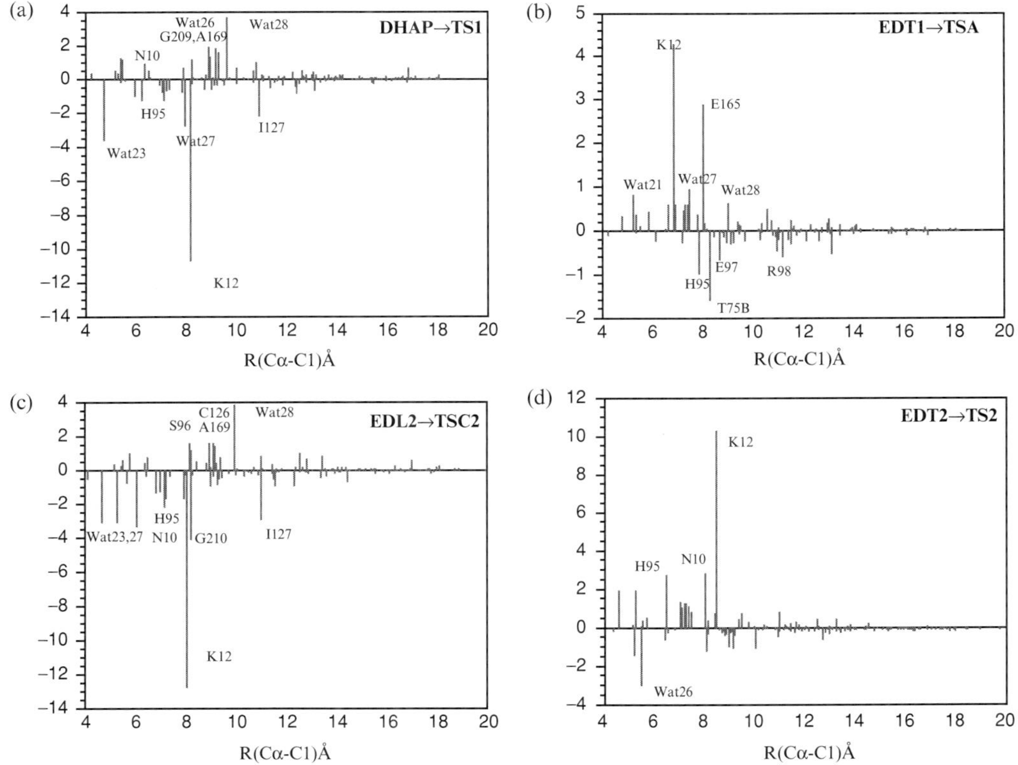

FIG. 7. Perturbational analysis of electrostatic contributions at various stationary points (see Scheme) from the most important residues and water molecules (see text). The contribution to a particular reaction process for each residue is defined as the energy difference caused by zeroing out the partial charges of this residue. A *negative* value indicates that the residue contributes favorably to a particular step of reaction. The difference in energy scale of the figures should be noted.

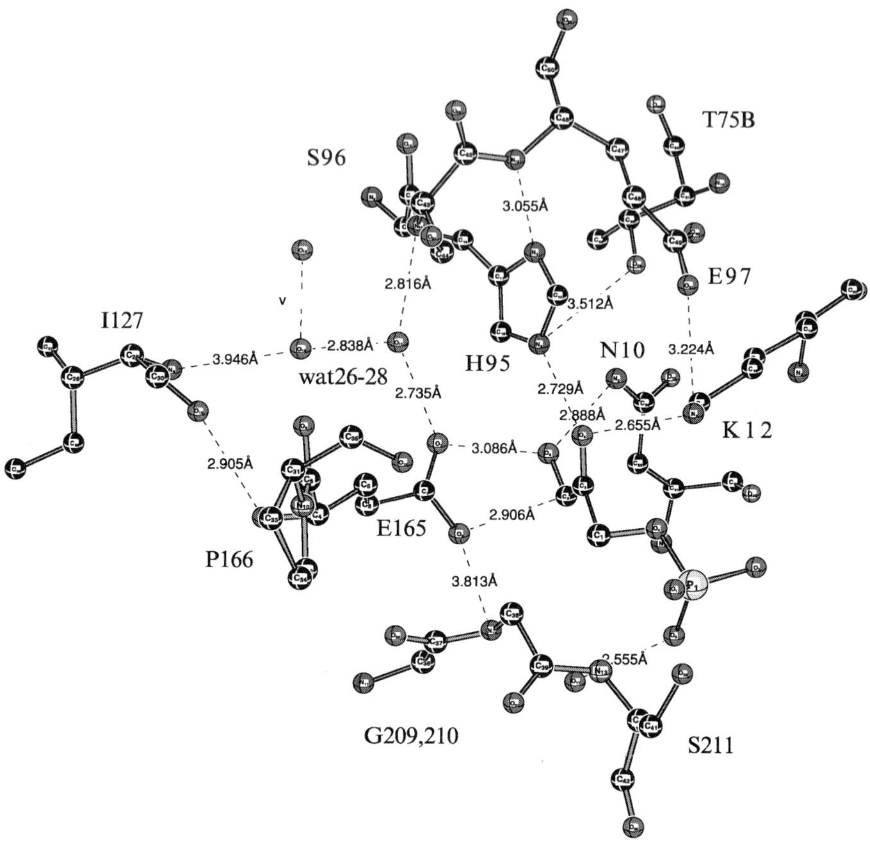

FIG. 8. Position of residues and active site water molecules that were found to be important in the perturbation analysis (see Fig. 7). The structure for EDT1 optimized at the B3LYP/6-31+ G(d,p)/CHARMM level is shown.

which the major product is the side reaction to methyl glyoxal. Since the binding energy of GAP and GA differ by about 4.2 kcal/mol, it was argued by the authors that the dominant contribution to k_{cat} for GAP is due to the preferential binding of the phosphate group to the transition state over the reactant in the enzyme. However, an alternative explanation seems more reasonable to us. That the difference in binding free energy between GA and GAP (4.2 kcal/mol) is similar to the value obtained for inorganic phosphate is what one would expect. Given these data, it is likely that GA binds in a non-specific manner so that catalysis is negligible and/or that the phosphate group is important for positioning certain catalytic groups required for catalysis. The care required in the interpretation of this type of experiment is illustrated by an example (*reducto ad absurdum*) in which a mutation far from the active site leads to a denatured and, therefore, inactive enzyme. This does not permit one to conclude that the mutated residue had an essential role in catalysis!

A residue that was found to play an important role in the perturbation analysis is Asn 10, which contributes favorably to both the first proton abstraction (~1.3 kcal/mol) and the last proton transfer step converting EDT2 to GAP (~2.8 kcal/mol); see also, Alagona *et al.* (1995) and Peräkylä and Pakkanen (1996). A recent study by Kursula *et al.* (2001) reported the synthesis of a new substrate inhibitor (2-(*N*-formyl-*N*-hydroxy)-aminoethyl phosphonate, IPP) that mimics GAP and solved the x-ray structure of IPP complexed to leishmania TIM. It was found that the distance between O^1 in IPP (equivalent to O^2 in GAP, see Scheme I) and Asn 10 is substantially shorter than the corresponding distance in the PGH-TIM complex. Since PGH mimics the substrate DHAP or the enediolate intermediate (EDT1), this result suggests that Asn 10 plays a stabilizing role in the later stage of the catalytic cycle, which is consistent with the computational results that Asn 10 contribute favorably to the conversion from EDT2 to GAP. Kursula *et al.* (2001) also suggested from comparison of different TIM structures, that the active site is rather flexible. The mobility apparent for Glu 165 is in accord with our calculations (see earlier).

Finally, the results from the current calculations serve to suggest why the His 95 is conserved in all known TIM sequences. It plays an essential role along path A (as proton donor and acceptor), and also makes favorable contributions through electrostatic interaction to the first proton transfer and in path C (see Fig. 8). Although the electrostatic effect is relatively small (on the order of 2–3 kcal/mol), the fact that the stabilization is essential for the first proton transfer step (which is the most difficult elementary step during the reaction cycle) makes it favorable to conserve His 95. A simple active-site model calculation (see Section IV.B) indicated that if the His 95 is mutated to a Gln, the enediolate species will be destabilized by about 3 kcal/mol. This is in accord with mutagenesis experiments which found that the H95Q mutant has a lower (~200 fold) catalytic efficiency as compared with the wild type. (Belasco and Knowles, 1980; Komives *et al.*, 1991; Nickbarg *et al.*, 1988).

D. Substrate Polarization Analysis by Mutagenesis

The unusual behavior of His-95 (i.e., that a neutral histidine acts as a generalized acid, as described above) suggested that a further investigation concerning its proposed role in the reaction would be fruitful (Karplus *et al.*, 1992). Fourier-transform infrared (FTIR) spectroscopy was used by Knowles *et al.* (Belasco *et al.*, 1980; Komives *et al.*, 1991) to show that the carbonyl stretching frequency of DHAP bound to TIM is red-shifted by *circa* 19 cm^{-1} relative to solution; the observed spectrum is rather broad so that the quantitative shift is difficult to measure. This was interpreted as an electronic distortion of the carbonyl moiety by which the enzyme-substrate complex destabilizes the ground state of DHAP relative to the transition-state structure for deprotonation of C^1 (see Scheme I). Based on the x-ray structure of the TIM-PGH complex (Davenport *et al.*, 1991), His-95 was thought to have the predominant role in the polarization of DHAP. This conclusion appeared to be confirmed by the demonstration of a reduced red shift in the mutants H95Q and H95N (Komives *et al.*, 1991). We describe here the results of a theoretical analysis of the origin of the observed shifts in the carbonyl frequency of DHAP bound to wild-type TIM and to the mutants H95Q and E165D. It provides an object lesson concerning conclusions based on mutagenesis experiments, particularly those that confirm preconceived notions of what should happen. Such results should be viewed with caution as long as structures for the mutants are not available. Another (simpler) example that occurred with TIM concerns the Lys12/Met mutant (Lodi *et al.*, 1994), which showed that the enzyme was inactive, as expected from the mechanistic analysis. Unfortunately, a subsequent x-ray study found that there was no ligand in the active site, demonstrating that the experimental result gave information about the role of Lys 12 in binding but not in the reaction. Other mutants (e.g., K12R) do show some activity, supporting the importance of Lys12.

To calculate the vibrational frequency of the carbonyl group of DHAP, quantum mechanical studies with *ab initio* (Frisch *et al.*, 1990) and semi-empirical methods (Dewar *et al.*, 1985) were performed. Initial atomic coordinates were obtained from x-ray crystal structures of yeast wild-type TIM-PGH (Davenport *et al.*, 1991), yeast H95Q TIM-PGH (Nickbarg *et al.*, 1988), and chicken E165D TIM-PGH (Joseph-McCarthy *et al.*, 1994). Active-site residues that are implicated in catalyzing the abstraction of a proton from DHAP (see earlier) were examined. The residues are Asn-10, Lys-12, His/Gln-95, Ser-96, Glu-97, and Glu/Asp-165, all of which are conserved in thirteen TIM sequences (Lolis *et al.*, 1990; Joseph *et al.*, 1990). To simplify the calculations, the sidechains of Asn or Gln, Lys, His, Ser and Glu, or Asp were modeled by acetamide, ammonium or methylammonium, imidazole, ethanol and formate, or acetate, respectively; the protein backbone was modeled by acetamide and the essential portion of DHAP was modeled by acetone; neglect of the phosphate group of DHAP has been shown to have a small effect on the carbonyl stretching frequency.

Since the full active site was not included in the calculation and the main purpose of the study was to determine the contribution of individual amino acid residues to substrate polarization, it was necessary to introduce constraints to preserve the geometry corresponding to the x-ray structures. The constraints were applied to appropriately chosen intermolecular distances, angles, and dihedral angles. The internal degrees of freedom of the substrate and amino acid model were minimized in the presence of the geometric constraints. Because of the isolated position of the carbonyl stretching frequency in the vibrational spectrum, the error due to the presence of the constraints is expected to be small; test studies indicate the effect is $< 5\ cm^{-1}$.

The semiempirical AM1 method (Dewar *et al.*, 1985) was used for most of the calculations because it is much less computer intensive than *ab initio* quantum mechanical procedures and it generally yields satisfactory values for ground-state properties. In particular, AM1 vibrational frequencies compare well with *ab initio* theory with medium-sized basis sets (e.g., 3–21G) (Seeger *et al.*, 1991). However, both AM1 and *ab initio* Hartree-Fock results deviate significantly from experiment (Seeger *et al.*, 1991). Since these deviations tend to be systematic, it is possible to scale the computed frequencies by a constant factor that can be transferred among similar molecular systems (Pulay *et al.*, 1983). For isolated acetone the carbonyl stretching frequencies calculated by AM1 ($2063\ cm^{-1}$) and *ab initio* 6–31G* ($2022\ cm^{-1}$) methods can be compared with experiment in the gas phase ($1735, 1735\ cm^{-1}$) (Rogers *et al.*, 1981; Nyquist *et al.*, 1989). Scale factors of 0.84 for AM1 and 0.86 to 6–31G* are required to obtain agreement with experiment. These values are similar to those for other systems at a corresponding level of theory (Seeger *et al.*, 1991). An estimate of the frequency shift expected from bulk solvation by water was based on a molecular dynamics simulation with the CHARMM program (Brooks *et al.*, 1983). The simulation was made for acetone in a box of 211 waters; 50 ps of equilibration and 100 ps of production were performed. For a series of resulting configurations, the structure of acetone was minimized in the presence of the first-shell waters. AM1 calculations for these structures yielded a frequency shift equal to $-73 \pm 16.5\ cm^{-1}$ (or $-61 \pm 14\ cm^{-1}$ with scaling); the range corresponds to inhomogeneous line broadening. The shift is to be compared with the experimental estimate of $-39\ cm^{-1}$ (Rogers *et al.*, 1981; Nyquist *et al.*, 1989).

With acetone placed in the position of the $CH_2OHC{=}OCH_3$ moiety of DHAP in the active site, calculations were performed on the constrained minimum-energy structures for wild-type TIM and the two mutants H95Q and E165D. Individual amino acids and various combinations of them were considered. All the model systems were studied using the AM1 method and a certain subset was checked by *ab initio* calculations. The AM1 results for the wild-type and mutant proteins are given in Table I. It is important to note that for certain cases, the carbonyl stretching frequency of acetone is significantly mixed with one of the Glu-165 carboxyl moiety stretching frequencies. When this occurs, both frequency shifts are listed in the table; if only one frequency shift is listed, it corresponds to the dominant acetone carbonyl stretch. The full four residue calculations (Lys-12, His/Glu-95, Glu-97, Glu/Asp-165) yield unscaled shifts of −53/−60, −16, and −44 cm^{-1} for the wild type, the H95Q mutant and E165D mutant, respectively. This is in accord with the measured values for the wild type (−19 cm^{-1}) and the H95Q mutant (0 cm^{-1}) (Komives *et al.*, 1991); the value for the E165D mutant is reduced from the wild type, but no quantitative value is available. The calculated shifts are with respect to isolated acetone, while the experimental shifts use DHAP in solution as a

TABLE I
Carbonyl Frequency Shift (cm^{-1}) for Acetone Calculated with AM1 for the X-Ray Structure[a,b]

Residue	Wild type	H95Q[c]	E165D
N10	−2	+1	−1
K12[d]	−42	−41	−54
K12[e]	−44	−40	−55
H95	−13	−6	−11
H95[+f]	−33	—	−29
S96	0	0	0
E97	+10	+10	+11
E165	−34/−18	−2	−7
K12 + E97	−17	−12	−26
K12 + E165	−85	−49	−67
S96 + E165	−32	−1	−7
K12 + E97 + E165	−45/−38	−17	−39
K12 + H95 + E165	−90	−41	−73
K12 + H95 + E97 + E165	−60/−53	−16	−44

[a]The shifts are unscaled values relative to the gas phase acetone frequency of 2063 cm^{-1}.

[b]Frequency shifts for the vibration for which the major component is the carbonyl stretch are listed. If there is a second vibration that contributes more than 15%, it is also listed.

[c]The His95 model (imidazole) was replaced with Gln95 (acetamide) in this calculation.

[d]Ammonium was used as the model residue.

[e]Methylammonium was used as the model residue.

[f]A protonated imidazole was used as the model residue.

reference. For a more quantitative comparison the background polarization due to the full active site environment treated by the QM/MM method (Field *et al.*, 1990) would have to be included. Thus, only the direction and qualitative magnitudes of the calculated and experimental shifts can be compared in a meaningful way. The individual amino acid contributions obtained by separate calculations (see Table I) show that for the wild type, Lys-12 (-42 to -44 cm^{-1}) and Glu-165 ($-18/-34$ cm^{-1}) make the dominant contributions; His-95 (-13 cm^{-1}) and Glu-97 (10 cm^{-1}) have smaller effects; and Asn-10 and Ser-96 produce negligible shifts. If His-95 were protonated, it would have a larger effect (see Table I). These results are in accord with the analysis (Bash *et al.*, 1991) of the contributions of the various residues to the catalysis of the proton abstraction from DHAP. For the H95Q mutant, Lys-12 is still very important (-40 cm^{-1}), but the contribution of Glu-165 is very small (-2 cm^{-1}); the effects of the other amino acids are only slightly altered. In particular, the direct effect of Gly-95 vs. His-95 is calculated to be -6 cm^{-1} vs. -13 cm^{-1}. For the E165D mutant, Lys-12 still makes the largest contribution (-54 cm^{-1}), but Asp-165 gives a smaller shift (-7 cm^{-1}) than does Glu-165 ($-18/-34$ cm^{-1}) in the wild type; the effects of the other residues are essentially unperturbed. In addition to the effects of the individual residues, Table I shows that there is significant cooperativity (particularly for Lys-12 and Glu-165) and some anticooperativity (e.g., Lys-12 and Glu-97). The *ab initio* test calculations verify the nature of the frequency shifts obtained with AM1, though the values tend to be somewhat larger.

From the calculated results and the crystal structures, the interpretation of what is happening is clear (Fig. 9). In the wild type, Lys-12 and Glu-165 make the dominant contribution and His-95 plays a smaller role. In the H95Q mutant, it is not the mutation *per se* that is responsible for the observed reduction in the frequency shift, but rather the structural change *induced* by the mutation. In the wild type Fig. 9a, Glu-165 is positioned so that the carboxyl group dipole is oriented so as to polarize the carbonyl group prior to the reaction, as well as to abstract the proton during the reaction (Davenport *et al.*, 1991). It is apparently oriented in this way by direct interaction with the substrate because there is a significant displacement between the unligated and ligated structure (Davenport *et al.*, 1991; Lolis *et al.*, 1990). The positive charge of Lys-12 is also ideally positioned for polarization. The structure (Nickbarg *et al.*, 1988) of the H95Q mutant shows that the Glu-165 has moved significantly so that it is no longer in position to polarize the carbonyl group. Instead, Glu-165 now makes a hydrogen bond to the sidechain NH of Gln-95 and the backbone amide hydrogen of Ser-96. The positions of other residues such as Lys-12 change very little in the H95Q mutant relative to the wild type. In E165D,

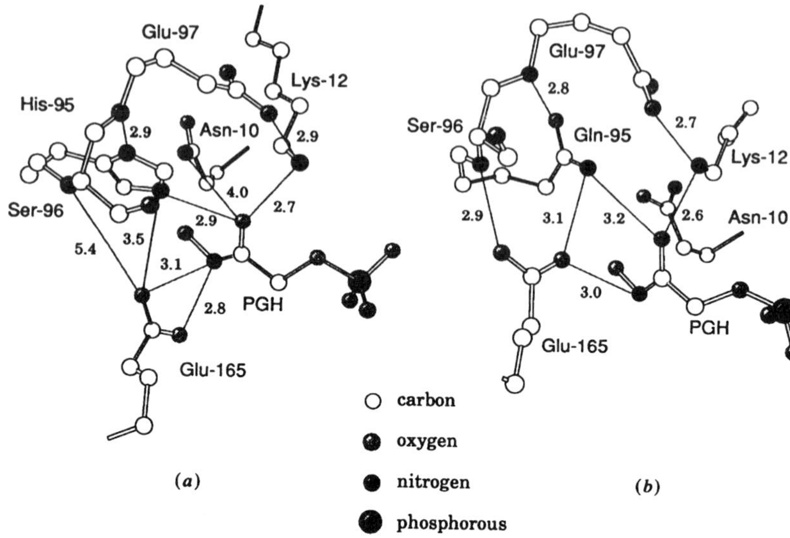

FIG. 9. Simplified view of TIM-PGH active site based on the X-ray structures. (a) Wild type; (b) H95Q mutant. Distances of interest are given in Å.

the mutation itself causes the frequency shift, but it is again the change in the position of the carboxy group relative to the polarizable carbonyl that is responsible. The Asp-165 carboxy group is further from the DHAP carbonyl because of its shorter sidechain, which also has rotated away from the substrate.

V. Dynamic Effects in the TIM Reaction

Although dynamical effects could contribute to all of the steps in the TIM catalyzed chemical reaction, we present here the results for the transfer of a proton from DHAP to the residue Glu 165 to form an enediolate intermediate (EDT1) (see Scheme 1) because this is the rate-determining step in the overall chemical reaction. As described in Section IV, the reduction of the activation barrier is due to interactions of the reactants and the transition state with specific residues of the enzyme. Displacements of atoms in the enzyme by 1 Å or so can have a large effect on the activation barrier (see previous section; Knowles, 1991; Joseph-McCarthy *et al.*, 1995). Such a tight coupling of structure and reactivity raises the question of its possible effect on the dynamics of the reaction; i.e., whether the structural requirements for transition state stabilization result in a significant reduction in the effective rate of crossing the barrier.

Also, there is the possibility that tunneling increases the preexponential factor, A(T) in Eq. (1). We discuss both of these points in the present section.

A. Classical Transmission Coefficient

To determine the transmission coefficient for the DHAP to EDL1 reaction, a reaction coordinate was selected, the transition state was determined by calculating the potential of mean force along the reaction coordinate, and the transmission coefficient was evaluated by classical activated dynamics using the reaction flux method (Chandler, 1978; Karplus, 2000).

In this approach, a set of equilibrated configurations of the system in the transition state region is determined by restraining the reaction coordinate. Each of them is then propagated by molecular dynamics starting with the velocity along the reaction coordinate directed toward product. The trajectories are integrated until they are trapped by loss of the activation energy in either the product or the reactant well. This may involve multiple crossings of the transition state. The transmission coefficient is equal to the ratio of the number of trajectories that end up in the product well to the total number of trajectories.[*]

Although the reaction had been studied by semiempirical and *ab initio* molecular orbital QM/MM methods (see Section III), a simpler representation of the potential energy surface was needed to calculate the large number of trajectories that are required for obtained and converged results for the transmission coefficient. The representation used describes the surface for the reaction in terms of the coupling of two states, one corresponding to the reactants and the other to the products. In each state

In a recent paper, Villa and Warshel [2001] commented on our calculations by saying (see their reference 15): "The fact that the transmission factor depends on the *number of times the trajectory moves back and forth on the transition state* [italics added] was questioned in ref. 157 [Karplus, 2000]. However, this recrossing effect is an integral part of current reaction rate formulations." This comment suggests that Villa and Warshel misunderstand what we did. As stated above, it is not the number of recrossings that is involved in the transmission coefficient, but only the fraction of trajectories that end up as product. Reference 9 [Grimmelmann, E. K., Tully, J. C., and Helfond, E. (1981). *J. Chem. Phys.* **74,** 5300-5310] of Villa and Warshel has a very clear discussion of how the transmission coefficient (p. 5302) is obtained, and why a factor of the type used by Villa and Warshel can be used even if the number of recrossings is not relevant. "The factor is introduced in order that all trajectories which originate in A and end in B be counted only once, no matter how many times they cross S [the dividing surface] and that the trajectories which go from B to A, A to A and B to B are not be counted at all."

the potential surface is represented by a standard molecular mechanics (MM) force field. The possibility of reaction is introduced by coupling the two electronic states (Brumer and Karplus, 1973; Åqvist *et al.*, 1993). The potential function V(**R**) is chosen to have the EVB form (Åqvist *et al.*, 1993; Åqvist *et al.*, 1996):

$$V(\mathbf{R}) = \frac{1}{2}[V_R(\mathbf{R}) + V_P(\mathbf{R})] - \frac{1}{2}\left[(V_R(\mathbf{R}) - V_P(\mathbf{R}))^2 + 4V_{RP}^2\right]^{\frac{1}{2}} \tag{2}$$

where **R** is the set of all coordinates, $V_R(\mathbf{R})$ and $V_P(\mathbf{R})$ are the MM potential energies in the reactants and products states, respectively, and V_{RP} is approximated by a constant coupling, assumed to be independent of **R** for simplicity.

An appropriate reaction coordinate for the proton transfer from carbon C_1 of DHAP to the O of Glu 165 is the asymmetric stretch defined by

$$q_{as} = \frac{1}{m_C + m_0}(m_C r_{CH} - m_O r_{OH}) \tag{3}$$

where r_{CH} and r_{OH} are the distances of the proton from the donor carbon and the acceptor oxygen, respectively, and m_C and m_O are their masses; for a collinear arrangement q_{as} has a reduced mass of 0.97amu. In the reaction, the proton moves from the vicinity of the donor carbon atom ($q_{as} = -1.1$ Å) to the vicinity of the accepting oxygen ($q_{as} = 0.3$ Å). To determine the location of the transition state, the potential-of-mean force (PMF) (McQuarrie, 1976) as a function of q_{as} was determined by use of umbrella sampling (Valleau and Torrie, 1977). As already described, the PMF is an effective potential for the reaction coordinate that is based on an equilibrium average over all other degrees of freedom. The resulting free energy profile for the reaction is shown in Fig. 10.

As a starting point for the transition state trajectory calculations, equilibrium configurations in the transition state were sampled by running a 400ps trajectory; $q_{as}(t)$ was constrained to $q_{as}^{\dagger}$ with a SHAKE-like algorithm (Ryckaert *et al.*, 1977), and 40 configurations at 10ps intervals were saved. For each of these configurations, 100 trajectories were initiated with different values of $\dot{q}_{as}(0)$ sampled from a Maxwell distribution. The function $\kappa(t)$ calculated from these trajectories is shown in Fig. 11. It decays rapidly and has almost reached the plateau value by 10fs. The plateau value is $\kappa = 0.43 \pm 0.08$, a significant reduction from the TST value of unity. However, the reduction by a factor of two is a small effect relative to the specific interactions in the protein that lower the potential of mean force barrier for the reaction by five orders of magnitude. It would be interesting to compare this result for the enzyme catalyzed reaction with that for the reaction in solution.

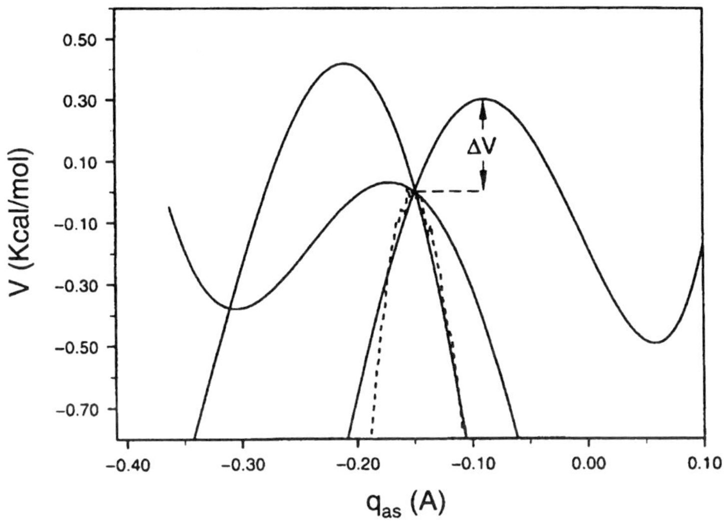

FIG. 10. The potential of mean force near the transition state (---) and the potential energy for three initial configurations used in the calculation of the transmission coefficient. The potential energy was calculated with all degrees of freedom, other than the reaction coordinate, fixed. The zero of energy is at $q_{as}^{\dagger}$and ΔV is the height of the instantaneous barrier that trajectories need to surmount.

The rapid decay of $\kappa(t)$ to the plateau value makes it likely that most degrees of freedom coupled to the reaction coordinate do not change significantly during the time required to trap the trajectory in the reactant or product wells. Defining the substrate and the GLU165 atoms as the "intramolecular" part and the rest of the system as the "environment," the calculation of the transmission coefficient was repeated from the same initial configurations, but with only the intramolecular atoms allowed to move and the environment atoms fixed in their initial (t = 0) positions (see Fig. 11). The behavior of $\kappa(t)$ is very similar to that obtained when all atoms are allowed to move. This demonstrates that the dynamics of the environment atoms is unimportant for the barrier crossing per se and suggests that the structure at t = 0 determines whether or not a reaction takes place.

If all degrees of freedom other then the reaction coordinate are frozen ("frozen bath assumption") and there is a barrier ΔV, in addition to the potential-of-mean force barrier, for a particular set of bath coordinates (a configuration chosen from the 400 ps trajectory), the transition state trajectory has to surmount that barrier to complete the reaction; the range ΔV observed in the calculations is about 0.5 kcal/mol. This reduces the rate from the TST value, which is based on the potential-of-mean force

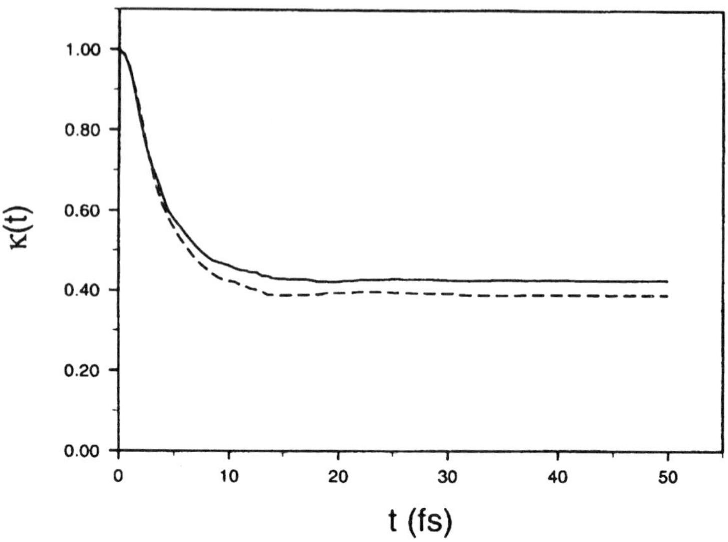

FIG. 11. The time-dependent transmission coefficient, $\kappa(t)$, when all atoms are allowed to move (———), and in a rigid environment (---).

barrier. With such a "frozen bath" assumption, the transmission coefficient is given by (Bergsma *et al.*, 1987)

$$\kappa^{\text{froz}} = < e^{-\Delta V/k_B T} > \tag{4}$$

where k_B is Boltzmann constant, T the is temperature, and the average is over all initial transition state configurations (see Fig. 10). The average over the 40 transition-state configurations gives κ^{froz} 0.40, in good agreement with the simulations. This confirms that the frozen bath approximation is valid (i.e., the surrounding system is fixed within the femtosecond time scale of the barrier crossing).

Two mechanisms were described for the influence of the environment on the instantaneous barrier. The first, found in a study of a symmetric S_N2 reaction in water [Bergsma *et al.*, 1987], corresponds to a nonequilibrium solvation mechanism. In that reaction, the instantaneous barrier arises from different solvation states of the intramolecular subsystem (e.g., a fluctuation in the environment that results in better solvation of the reactant relative to the products introduces an instantaneous barrier located between the transition state and the products state and vice versa). In an alternative mechanism for the environmental influence, the instantaneous barrier is determined by the configuration of the intramolecular subsystem, but the intramolecular dynamics is coupled to low frequency fluctuations of the environment (e.g., intramolecular motions that are required for the reaction are hindered by the environment, which appears rigid on the time scale of the barrier crossing (~10 fs) but fluctuates on a picosecond time scale). The two mechanisms were tested for the proton transfer in the TIM reaction and it was demonstrated that the second mechanism is dominant in introducing the effect of the environmental fluctuations on the barrier modulation and in the reduction of the rate from the transition state limit.

From this analysis, there is a well-defined rate constant for the reaction and there are three time scales that are involved. There is the very fast time scale of relaxation of the high energy transition state trajectories to a thermal distribution (~10 fs); there is the time scale of the motions associated with the enzyme environment that determines the height of the transition state barrier (~1 ps). The dominant modulations of the barrier are of a low frequency character ($\omega < 300\ \text{cm}^{-1}$) and originate from the interaction between the intramolecular subsystem directly involved in the reaction and the rest of the enzyme. Finally there is the time scale associated with the overall reaction rate (~1 ms). Since the third is much longer than the other two, simple behavior, as exemplified by the well-defined plateau value for the transmission coefficient, it expected and was obtained in the calculations. However, it should be noted that the trajectories covered a time scale (400 ps) many orders of magnitude shorter than that of the overall reaction rate (ms). Consequently, it is not excluded that there exist very slow structural fluctuations (on the same time scale or longer than the reaction) that alter the barrier. This would lead to nonexponential behavior, but it appears unlikely at room temperature for such a ''slow'' reaction; at low temperatures, nonexponential kinetics might occur. Single molecule techniques can provide additional information in this area (Xie, 2002).

B. *Contribution of Tunneling in TIM Catalysis*

The possibility that tunneling makes a significant contribution to the rate, even at room temperature, has to be considered for proton- and hydride-transfer reactions (Bell, 1973). A number of experimental studies point to the importance of such tunneling in enzymes (Kohen and Klinman, 1998). In this section, we describe the results obtained for the DHAP to EDT1 reaction (Scheme 1) in TIM (Cui and Karplus, 2002a).

Because of the sensitivity of the tunneling contribution to the shape of the barrier for the proton transfer, meaningful calculations are limited by the accuracy of the potential energy surface. The detailed and relatively high-level QM/MM potential energy surface obtained for the TIM-catalyzed reactions sets the stage for an analysis of the effect of tunneling on the various proton-transfer steps involved and of its relative contribution to the different proposed mechanisms. In both paths A and C (see Scheme 1), the elementary step that has the largest barrier is the first proton transfer from CR of the substrate DHAP to Glu 165, a step common to all three paths. The contribution of tunneling to this step are reported here. For path B, the intramolecular proton transfer in the enediolate species has the highest barrier. If tunneling were significant, it might increase the rate sufficiently so that it could contribute to the overall reaction. We therefore also determined the role of tunneling for this step. In addition, we made a comparison calculations for the intramolecular proton process of a model enediolate in the gas phase and in aqueous solution. The intramolecular proton transfer results are reported in Cui and Karplus (2002a). They show that the contribution of tunneling does not lower the effective barrier for this step sufficiently to make path B contribute significantly to the reaction in the wild type protein.

The methodology used for the tunneling calculation is outlined in Section II. The CVT rate constants differ notably from the TST values, as shown in Table I. The reduction of the CVT rate constant relative to the TST value corresponds to a transmission coefficient of 0.69 at 300K; this is qualitatively similar to the result ($\sim$0.43) from activated dynamics calculations based on an EVB potential (see earlier). The value from the current calculation is larger partly because barrier recrossing induced by the environment (atoms other than the substrate and Glu 165) is already partially accounted for in the TST result by including the environment in the normal mode analysis (i.e., the definition of the dividing surface). The tunneling factor is 1.22 and 1.45 at the CVT-ZCT and CVT-SCT level, respectively, at 300K. The reliability of the estimate is supported by the fact that computed secondary H/T kinetic and equilibrium effects are fairly close to the experimentally measured values; the computed CVT-SCT values are 1.14 and 1.31 for the kinetic and equilibrium isotope effects, respectively, and the corresponding experimental values are 1.12 and 1.27, respectively (Alston *et al.*, 1996). Alston *et al.* were able to make the proton-transfer steps rate limiting by using [1(*R*)-2H]-labeled substrate in D_2O to slow the chemical steps relative to the product release, which is normally rate limiting, as mentioned above (Knowles and Albery, 1977).

It is clear from the calculations that the contribution of tunneling is fairly small for the intermolecular proton transfer in TIM, at least at room temperature. Corresponding behavior has been found in several studies (Alhambra *et al.*, 1999; Billeter *et al.*, 2001), including ours (Cui *et al.*,

2001), of horse liver alcohol dehydrogenase (Bahnson *et al.*, 1997), where the tunneling particle is a hydride rather than a proton. For this reaction the tunneling factors from several studies, using different potential functions and treatments for the quantal nature of the hydride, are all close to be a factor of 2 at 300K. In cases where the tunneling factor is larger (Hwang *et al.*, 1996; Truhlar and Gao, 2002), a similar value is likely to be found for the corresponding solution reaction. It is unlikely, therefore, that tunneling makes a major contribution to the catalytic power of enzymes.

C. *The Effects of Protein Motion*

It is well known from model studies that the energetics and tunneling contribution of proton transfer are sensitive to the distance between the donor and acceptor atoms. It is necessary to examine, therefore, whether the motions of the enzyme make a significant contribution to the proton transfer rate constant, including of tunneling. As shown in Table II, the calculated potential barrier for the intermolecular proton transfer in TIM from DHAP to Glu 165 has a very pronounced dependence on the donor-acceptor distance. The OE2-Cα distances at the beginning of the reaction and in the transition state are 3.0 Å and 2.65 Å, respectively. To determine the effect of this distance in the transition state, it was fixed at 3.0, 2.7, and 2.6 Å, and the resulting barriers were calculated; the values are 24.5, 9.2, and 3.9 kcal/mol, respectively. Furthermore, the barrier frequencies also were very different; the values are 2073*i*, 1778*i*, and 1361*i* cm-1, respectively. Correspondingly, the tunneling probabilities differ dramatically. They are 3×10^4, 36.5, and 2.1, respectively. Thus, counterintuitively, tunneling is much *larger* at the *longer* distance because the barrier is so much higher and it is significantly narrower at the top. Nevertheless, the rate constant for the proton transfer with the shortest donor-acceptor is much larger than that for the longest distance, by 10 orders of magnitude at 300K (see Table II), due to the huge difference in the barrier heights. It should be noted that it takes only 2.9 kcal/mol to bring the donor-acceptor distance from 3.0 to 2.6 Å in the EDT1 species. Therefore, in the actual system the proton transfer is expected to be dominated by over-the-barrier transfer following the donor-acceptor vibration, in correspondence with the small tunneling contribution given above.

Next, we examine modes that have strong coupling to the proton transfer coordinate based on the reaction path Hamiltonian method (Miller *et al.*, 1980). As shown in Figure 12, the reaction path properties for the intermolecular proton transfer are typical for a proton transfer

TABLE II
Rate Constants and Tunneling Factors for the Intermolecular Proton Transfer Reactions in TIM at 300K

	TST	CVT	CVT-ZCT	CVT-SCT
E_a/ω^a	6.1/924.7 (5.8/979.0)			
s^{CVT}	-----	−0.31(−0.30)		
$\mathrm{Log}k_H(s^{-1})$	9.7 (10.4)	9.6 (10.3)	9.7 (10.4)	9.7 (10.5)
$\kappa_H^{CVT\text{-}SCT}$	-----	-----	1.2 (1.2)	1.5 (1.5)

[a]Values without parentheses were obtained with the large model including 445 movable atoms; those with parentheses or brackets were obtained with a smaller model including 106 movable atoms. E_a is the barrier height, in kcal/mol, without zero-point. energy (ZPE) correction; ω is the imaginary frequency (in cm^{-1}) at the saddle point. s^{CVT} is the optimized position of the transition state from the canonical variational transition state theory calculations. The SCT correction factor is defined as $k^{CVT\text{-}SCT}/k^{CVT}$.

reaction (i.e., the three distances that are involved directly [D-H, A-H, and D$\cdots$A] only start to change rapidly close to the saddle point, in the range-1.0 to 2.0 amu$^{\frac{1}{2}}$·bohr^{-1}). Both His 95 and Lys 12 are perturbed substantially by the reaction, as indicated in Fig. 12b; that is, the O1-HNζ of Lys 12 and O2-Hϵ of His 95 increase by nearly 0.2 Å during the critical part of the reaction. The reaction path curvature shows two sets of peak positions, one set with large curvature (>2.0 amu$^{\frac{1}{2}}$·bohr^{-1}) at $s\sim\pm0.5$ and another set with smaller peak values ($\sim$0.7 amu$^{\frac{1}{2}}$·bohr^{-1}) at larger s values, −3.0 and +2.0 amu$^{\frac{1}{2}}$·bohr. As discussed in our publications (Cui and Karplus, 2002a,c), the inner set of peaks is due to modes that are coupled directly to the proton transfer coordinate. The outer set of peaks is due to modes that represent the character of the reaction path in the beginning/final stage of the reaction. They correspond to motions that are relevant to the reorganization of the environment necessary for the reaction; i.e.; the donor-acceptor stretch and transfer of the proton. Only the inner set of peaks is expected to be important for the tunneling contribution. This was confirmed in the TIM model, (i.e., zeroing out the outer-peaks in the reaction path curvature gave virtually identical SCT rate constants). We focus, therefore, on the inner set of peaks in the following analysis of important motions in TIM.

The reaction path curvature at the inner peak positions (s=−0.31 amu$^{\frac{1}{2}}$·bohr) was decomposed into contribution from the generalized normal modes orthogonal to the reaction path (Fig. 12d); the

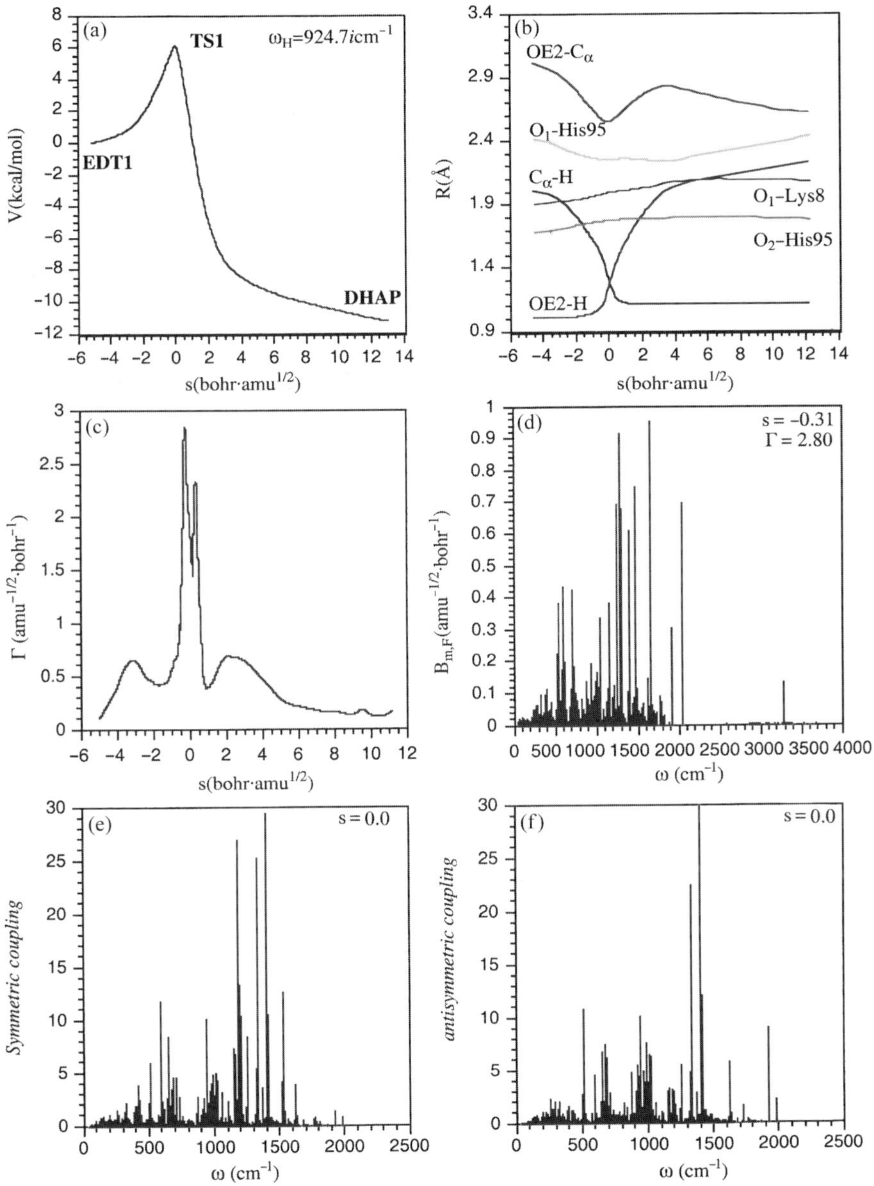

FIG. 12. Energetics (a), important geometrical parameters (b), and reaction path curvature (c) along the reaction coordinate for the first proton transfer step in triosephosphate isomerase, Normal mode decomposition (d) of the reaction path curvature at the peak curvature position (s=−0.31 amu$^{\frac{1}{2}}$·bohr). Approximate symmetric and asymmetric coupling coefficients (see Eq. 14 in Cui and Karplus, 2002c) for modes at the saddle point (e and f).

overlaps between these "path modes" and the eigenvectors at the saddle point were calculated (Table III). The effective coupling coefficients, defined in Cui and Karplus [2002c], were also computed for the eigenvectors ("essential modes") with large overlaps (Fig. 12e and f and Table IV). The normal modes at the saddle points having the largest coupling coefficients are plotted in Fig. 13. As discussed in Cui and Karplus [2002c] and also in related literature (Benderskii *et al.*, 1993; Benderskii *et al.*, 1994; Caratzoulas and Schwartz, 2001; Antoniou and Schwartz, 1997; Antoniou and Schwartz, 2001), modes with large symmetric coupling coefficients have a "promoting effect" in the sense that their excitation lowers the effective barrier along the proton transfer coordinate, relative to the uncoupled system and the proton transfer with this mode fixed at its equilibrium value. By contrast, modes with large antisymmetric coupling coefficients have a "demoting effect" relative to uncoupled proton transfer, in that the coupling lowers the effective barrier frequency, which leads to the decrease of tunneling. More importantly, the antisymmetric coupling gives rise to asymmetry in the effective potential along the proton transfer coordinate, and therefore also reduces tunneling.

As shown in Table III, the "path modes" with large curvature components typically have substantial overlap with many modes at the saddle point ("essential modes"). The barrier mode ($936i$) has large overlaps with "path modes" of relatively high frequency, corresponding to the fact that the character of the reaction coordinate is changing rapidly at the peak curvature position ($s=-0.3$). As shown in Table IV and Fig. 13, most of the essential modes are rather localized and involve mainly the substrate and Glu165 (which contains the reacting group), as reflected by the participation ratios; the barrier mode has a similar character. The donor and acceptor atoms are heavily involved in many modes, especially those at low frequencies (e.g., 514 and 598 cm^{-1}). Other atoms in the substrate/Glu165 can also be involved; e.g., C^1-O^1 and C^3-OP stretches in the substrate (see Scheme I for labels) and $O\epsilon^1$-$C\epsilon$-$O\epsilon^2$ in Glu 165 bending are heavily involved in the modes at 1192 and 598 cm^{-1}, respectively. In some modes, other nearby residues also make notable contributions; Ala 212 and Ile 170 have sizable components in the modes at 1410 and 1417 cm^{-1}, respectively. A residue from the other subunit on the dimer interface, Thr 75B, also has a small component in the mode at 1333 cm^{-1}. This residue has been shown to contribute significantly to the reaction [Borchert *et al.*, 1994; Åqvist and Fothergill, 1996].

The calculations have clearly demonstrated that the DHAP to EDT1 proton transfer is strongly coupled to a large number of vibrations, which in general are localized on atoms in the active site. One of those modes is the donor-acceptor stretch, which modulates the effective barrier for the

TABLE III

Rate Constants and Tunneling Factors for the Intermolecular Proton Transfer Reactions in TIM at 300K with *the Proton Donor and Acceptor Atoms Fixed* at Difference Distances[a]

	TST	CVT	CVT-ZCT	CVT-SCT
E_a/ω	24.5/2073 (9.2/1778) [3.9/1361]			
s^{CVT}	—	−0.01 (−0.03) [+0.00]		
$\mathrm{Log}(k_{\mathrm{H}})$	−2.5 (8.2) [11.9]	−2.5 (8.2) [11.9]	1.3 (9.4) [12.1]	2.0 (9.7) [12.2]
$\kappa_H^{CVT\text{-}SCT}$	—	—	4.9103 (18.5) [1.9]	3.0 104 (36.5) [2.1]

[a]The values were obtained with the small model including 106 movable atoms; the positions of the proton donor (OE2 in Glu 165) and acceptor (C3 in DHAP) were fixed at 3.0, 2.7, and 2.6 for the numbers without parentheses, with parentheses, and with brackets, respectively. The definitions for the quantities are the same as in Table II.

TABLE IV
Overlap among Frequencies with Large Curvature Components and Modes at the Saddle Point in the First Proton Transfer in TIM[a]

Path Modes ω (cm^{-1})	$B_{k,F}$	Overlap	Saddle Point Modes ω (cm^{-1})[c]	Coupling strength[d]
0[b]	—	0.75	936i	---
		0.30	514	5.9 (10.7)
		0.17	651	8.4 (6.8)
		0.11	918	2.6 (5.5)
		0.14	987	4.7 (7.5)
		0.13	1027	4.2 (6.4)
		0.21	1187	26.9 (3.1)
1328	0.75	0.99	1328	0.7 (0.6)
1715	0.70	0.26	935i	---
		0.13	1187	26.9 (3.1)
		0.13	1333	25.1 (22.4)
		0.22	1618	3.1 (5.0)
		0.45	1623	3.8 (5.7)
		0.16	1628	0.8 (1.6)
		0.21	1712	0.1 (0.0)
		0.57	1717	0.1 (0.0)
		0.18	1726	0.0 (1.7)
		0.11	1766	0.6 (0.3)
		0.23	1921	1.3 (8.9)
		0.19	1978	0.8 (2.1)
2093	0.69	0.30	935i	—
		0.12	1187	26.9 (3.1)
		0.18	1920	1.3 (8.9)
		0.88	1978	0.8 (2.1)
1719	0.58	0.18	935i	—
		0.15	1618	3.1 (5.0)
		0.31	1623	3.8 (5.7)
		0.11	1628	0.8 (1.6)
		0.81	1717	0.1 (0.0)
		0.22	1726	0.0 (1.7)
		0.16	1920	1.3 (8.9)
		0.13	1978	0.8 (2.1)
1520	0.55	0.18	935i	—
		0.13	1187	26.9 (3.1)
		0.14	1333	25.1 (22.4)
		0.18	1410	21.0 (22.4)
		0.25	1411	29.3 (31.3)
		0.13	1417	10.5 (12.0)
		0.15	1417	9.9 (11.3)
		0.11	1418	8.0 (9.4)
		0.41	1522	1.0 (0.4)
		0.35	1525	4.1 (0.3)
		0.39	1528	12.6 (0.3)

(*continued*)

TABLE IV (*continued*)

Path Modes ω (cm^{-1})	$B_{k,F}$	Overlap	Saddle Point Modes ω (cm^{-1})[c]	Coupling strength[d]
		0.20	1619	3.1 (5.0)
		0.34	1623	3.8 (5.7)
		0.11	1627	0.8 (1.6)
		0.15	1978	0.8 (2.1)
1328	0.46	0.23	935*i*	---
		0.12	987	4.7 (7.5)
		0.10	1027	4.2 (6.4)
		0.25	1187	26.9 (3.1)
		0.31	1330	4.3 (4.0)
		0.61	1333	25.1 (22.4)
		0.20	1335	4.9 (4.2)
		0.20	1410	21.0 (22.4)
		0.30	1411	29.3 (31.3)
		0.11	1417	10.5 (12.0)
		0.13	1417	9.9 (11.3)
		0.12	1621	3.8 (5.7)
552	0.46	0.12	511	1.1 (2.7)
		0.59	514	5.9 (10.7)
		0.19	547	0.4 (0.1)
		0.12	548	0.4 (0.0)
		0.22	550	0.2 (0.2)
		0.35	553	0.1 (0.1)
		0.22	555	0.2 (0.1)
		0.28	599	11.8 (4.5)
		0.12	644	2.3 (1.8)
		0.17	651	8.4 (6.8)
		0.14	677	3.4 (7.4)
		0.11	691	5.9 (10.7)
				4.5 (5.8)

[a]The frequencies orthogonal to the reaction path were computed as s = 0.32amu$^{\frac{1}{2}}$·bohr, where the reaction path curvature has the peak value (see Fig. 6). The reaction path curvature components, $B_{k,F}$, are given in amu$^{-\frac{1}{2}}$·bohr^{-1}.

[b]The eigenvector associated with the reaction path motion at this configuration.

[c]The letter in parentheses indicate the approximate character (symmetrically or asymmetrically coupled to the proton transfer coordinate).

[d]The coupling strengths are defined in Eq. (14) in Cui and Karplus (2000c) and are given in the unit of kcal/mol·Å. The values without parentheses are the approximate symmetric components, and the values with parentheses are the approximate asymmetric components.

proton transfer and also effects the tunneling. There are other modes that are symmetrically or antisymmetrically coupled to the proton transfer coordinate, and they may involve nearby residues, such as Ala 212 and Ile 170. Their effect is to prepare the enzyme environment for the reaction by lowering the effective proton transfer barrier.

VI. Catalytic Specificity: Comparison of TIM and MGS

TIM and MGS have highly different functional specificities, although their active sites are similar (see earlier section). Two features have been suggested as making significant contributions to this difference (Saadat *et al.*, 2000; Marks *et al.*, 2001). First, the general acid for GAP formation in TIM is a glutamic acid (Glu 165), which is more flexible than the aspartic acid (Asp 71) in MGS. This has been invoked to explain that GAP formation is more strongly favored in TIM than in MGS (i.e., the sidechain of Asp 71 does not have the flexibility to reach C^2 in the substrate to form GAP). To explain the different propensities towards phosphate elimination (i.e., that MGS cleaves the C^3-OP bond in the enediolate intermediate, while TIM does not) the traditional hypothesis is based on a stereoelectronic argument. It was found that the O^2-C^2-C^3-O^3 torsional angle (Fig. 14) of competitive inhibitors are different in x-ray structures of the two enzymes; it is $\sim 15°$ in TIM and $\sim 45°$ in MGS for PGH, and $\sim 5°$ in TIM and $\sim 55°$ in MGS for PGA. Therefore, the orbital overlaps between the breaking C^3-O^3 bond and the enediol π system could be very different, making the elimination reaction less favorable in TIM than in MGS (Lolis & Petsko; 1990; Pompliano *et al.*, 1990). Although those suggestions are physically reasonable, no quantitative estimate of the magnitudes of their effects is available. It seemed appropriate, therefore, to investigate MGS by theoretical methods and to examine what structural and electrostatic effects make contributions to the specificities. SCC-DFTB/CHARMM calculations have been performed on the reaction energetics for different pathways (Scheme II) in the two enzymes (Cui *et al.*, 2002); the main results are summarized in Figs. 15 to 19. There have been several proposals for the proton transfer pathways in MGS (Marks *et al.*, 2001), some of which were inspired by TIM (path B and C in Scheme II). Preliminary reaction path and potential of mean force calculations indicate that the "TIM-like" mechanism (path A) is reasonable in terms of energetics, although the recently introduced path D (Marks *et al.*, 2001), which involves Asp 101 explicitly in the deprotonation of Asp 71 in EDT1 and an active site water molecule to form EDT2 (Scheme II), is also likely to contribute. Path B and C are not expected to be competitive due to high calculated barriers; (i.e., Path B is energetically unfavorable due to the interaction with His 98, similar to the situation in TIM [Cui and Karplus, 2001]), and Path C has a high barrier because Asp 71 does not have the flexibility required to shuttle the proton to O^2. Thus, for the analysis of functional specificities, results for path A were used for both MGS and TIM.

As shown in Fig. 15, the energetics associated with the critical structures (intermediates and transition states) are rather similar in MGS and TIM

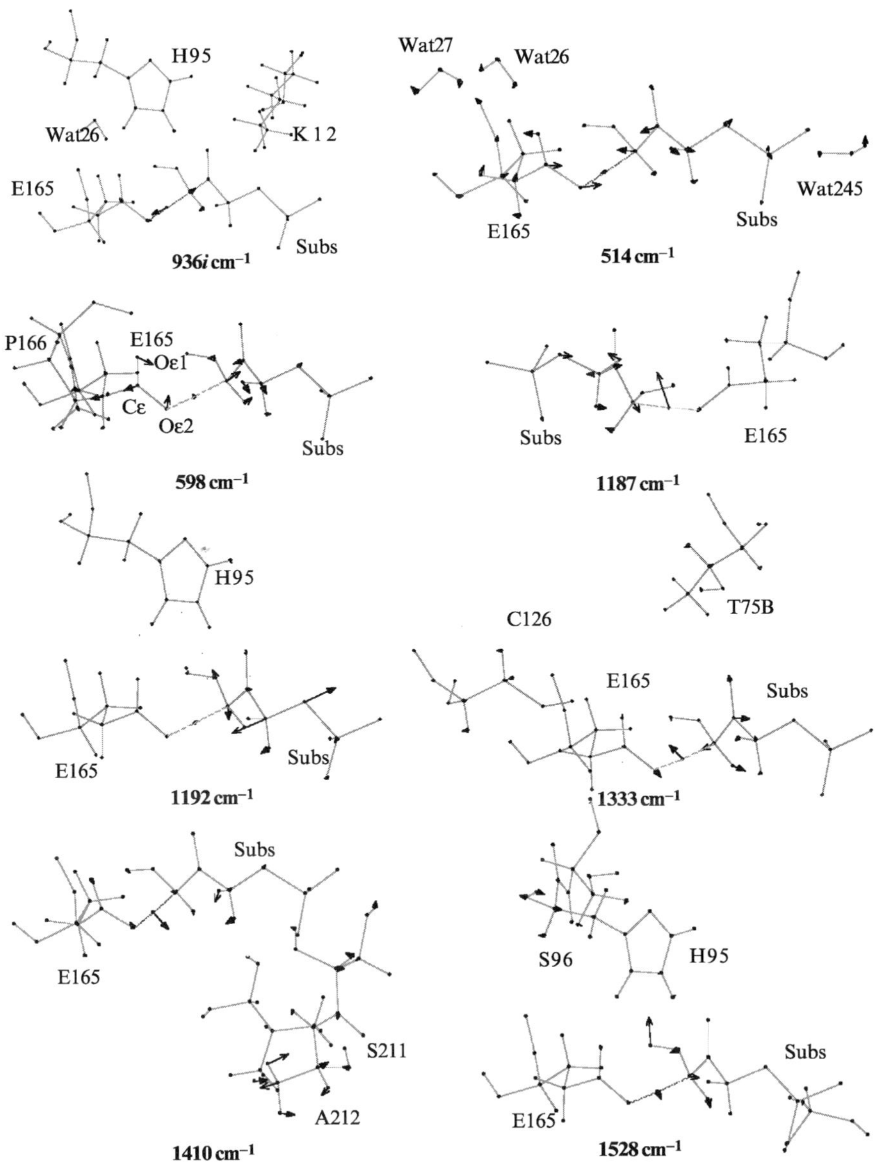

FIG. 13. Normal modes at the saddle point for the first proton transfer step in triosephosphate isomerase; these modes have large overlaps with the modes that have large curvature components. For the overlaps between the path modes and essential modes, the approximate coupling coefficients and their characters, see Table III and IV.

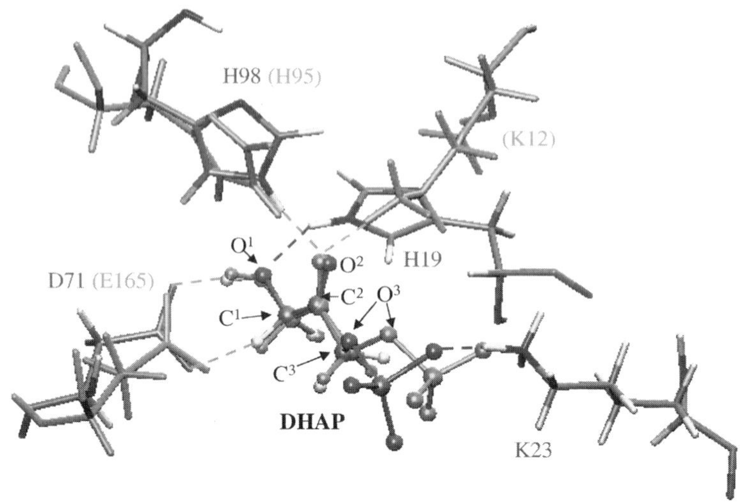

FIG. 14. A comparison of active site arrangements in methylglyoxal synthase (MGS) and triosephosphate isomerase (TIM) based on the x-ray structures with PGH as the substrate analogue (replaced by DHAP in the figure). The substrate is shown in ball-and-stick, and protein residues are shown in line-forms; the MGS residues are colored according to atom types, and those in TIM are shown in orange. The hydrogen bonding interactions found in both enzymes are shown in green, these found specifically in MGS and TIM are shown in blue and orange, respectively. Note the similarity in the orientations of the catalytic base (Asp 71 in MGS and Glu 165 in TIM) and the distal histidine (H98 in MGS and H95 in TIM) relative to the substrate. The phosphate torsional angle (O^2-C^2-C^3-O^3) is rather different in the two enzymes. (See Color Insert.)

for all the proton transfer steps up to the second enediolate species (EDT2). The last isomerization step, which is a proton transfer from the protonated catalytic base (Glu 165 in TIM, Asp 71 in MGS) to the substrate and converts EDT2 to GAP, has dramatically different energetics in the two enzymes. While the barrier is vanishingly small in TIM (i.e., ~0.1 kcal/mol), it is close to 16 kcal/mol in MGS. As shown in Fig. 16a, a perturbation analysis for the barrier to go from EDT2 to GAP did not reveal any residues that make large destabilizing contributions in MGS. Instead, the calculations indicate that the high barrier for GAP formation in MGS is due primarily to the longer separation of the proton donor (carboxylate oxygen) and acceptor (C^2 in EDT2), which is a result of the reduced flexibility of D71 in MGS relative to E165 in TIM. As shown in Fig. 16b, the proton-acceptor distance in EDT2 is less than 1.9 Å in TIM, while it is about 3.4 Å in MGS. The differences in the barriers for the proton transfer

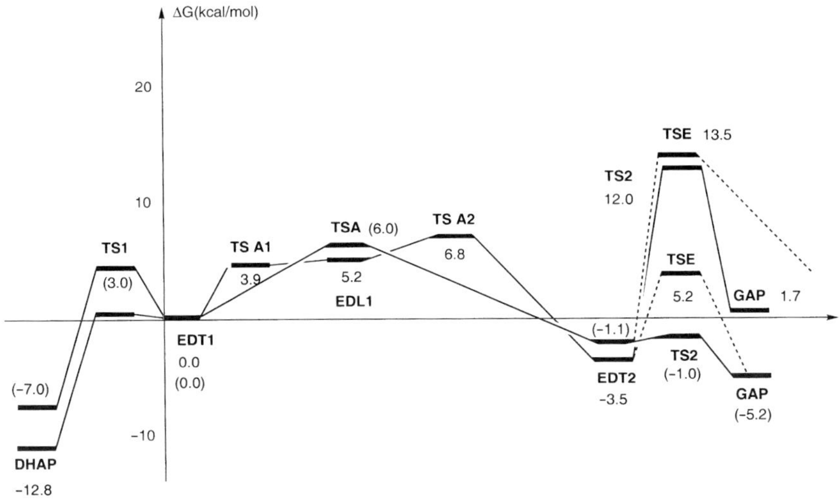

FIG. 15. The schematic effective free energy surfaces (in kcal/mol) for the isomerization (solid line) and phosphate elimination reactions (dotted line) in MGS (values without parentheses) and TIM (values with parentheses). The values without parentheses are from SCC-DFTB/CHARMM calculations, in which the substrate, His 19, His 98, Asp 71, and an active site water (Wat 30) were treated with SCC-DFTB, and the rest protein plus solvent molecules described with CHARMM 22 force field. The numbers with parentheses are from B3LYP/6-31+G(d,p)/CHARMM calculations (see legend Fig. 3). The zero-point energies and vibrational free energy contributions were included with harmonic approximations based on SCC-DFTB/CHARMM normal mode analyses for the critical structures. For the conversion between EDT1 and EDT2, which has several alternatives, the "classical" TIM-like mechanism (Scheme II) was used here because it was found to be energetically reasonable in both enzymes; however, calculations found that other pathways are also likely to contribute in TIM and MGS with similar energetics.

correspond to expectations from the results described above for DHAP to EDT1 in TIM (Section V.C) at fixed donor-acceptor distances.

For the phosphate elimination reaction, the QM/MM calculations also gave results that are consistent with the specificities of the two enzymes. The barrier in TIM is very large as measured from EDT2, on the order of 14 kcal/mol, which is in accord with experimental estimates for the fraction of elimination in TIM (Richard, 1991). For MGS, the corresponding barrier is about 8.7 kcal/mol. Although the traditional explanation by a stereoelectronic effect emphasized the importance of the phosphate torsional angle of the substrate, our computational analysis indicates that there is a different explanation. This is based primarily on two results. First, the gas-phase model calculation for EDT2 described above indicates that the elimination energetics depend only weakly on the phosphate

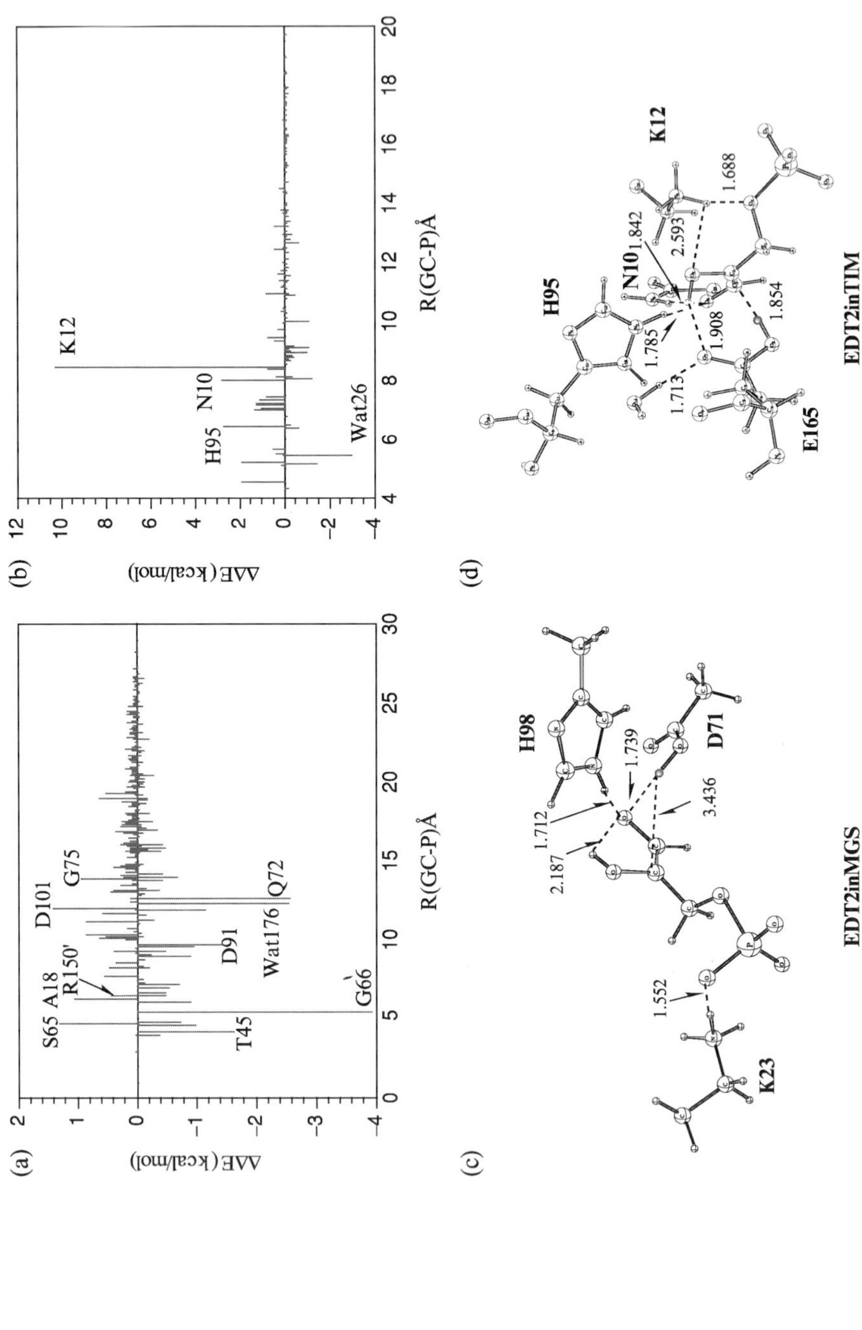
(a)
ΔΔE (kcal/mol)
R(GC-P)Å
S65
A18
R150'
D101
G75
T45
D91
Wat176
Q72
G66
(b)
ΔΔE (kcal/mol)
R(GC-P)Å
H95
N10
K12
Wat26
(c)
H98
D71
K23
1.552
2.187
1.712
1.739
3.436
EDT2inMGS
(d)
H95
N10
K12
E165
1.842
2.593
1.688
1.785
1.713
1.908
1.854
EDT2inTIM

torsional angle; with the torsional angle fixed at 60° (MGS-like) and 0° (TIM-like), the endothermicity of the elimination reaction differ by only about 4 kcal/mol at a C to O distance of 2 Å (Fig. 17a). Second, although the phosphoryl torsional angle is somewhat different in the x-ray structures of TIM and MGS with different *inhibitors* (see above), molecular dynamics simulations with the SCC-DFTB/MM potential functions found very similar average torsional angles for the intermediate (EDT2) in the two enzymes; the average absolute values O^2-C^2-C^3-O^3 torsion (Fig. 17b) of TIM and MGS is 21° and 18°, respectively, with similar magnitudes for the fluctuations (on the order of 12°). Combining these two observations, we conclude that even if the torsional dependence of phosphate elimination were larger than that found for EDT2 in the gas phase (Fig. 17a), it is unlikely that the effect is the primary source of the catalytic specificities of TIM and MGS. Moreover, in the recent x-rays structure for TIM·DHAP of

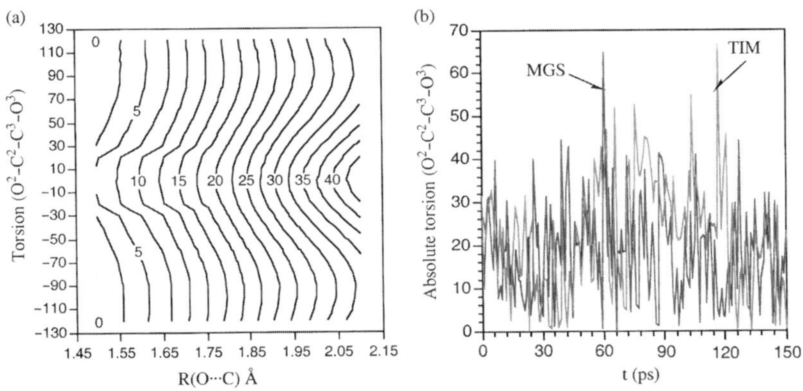

FIG. 17. (a) The two-dimensional potential energy surface associated with the phosphate elimination from EDT2 in the gas phase; the two coordinates are the breaking C-O distance, and the O^2-C^2-C^3-O^3 torsional angle. (b) The substrate O^2-C^2-C^3-O^3 torsional angles in MD simulations of MGS·EDT2 and TIM·EDT2.

FIG. 16. Results from perturbation analysis for the contributions from the protein residues and water molecules to the activation barrier of the EDT2→GAP step in (a) MGS and (b) TIM. Negative values indicate favorable contributions that lower the activation barrier, while positive values indicate unfavorable contributions that increase the barrier. The X-axis indicates the distance between the residues and the substrate atoms. The active site structures with the substrate in the EDT2 state in (c) MGS and (d) TIM from QM/MM minimizations. Note that the distance (in Å) between the catalytic residue (Asp 71 in MGS and Glu 165 in TIM) and the proton acceptor (C^2 in the substrate) is very different in the two enzymes.

Jogl *et al.* (2003; see earlier), the phosphate torsional angle was found to be substantially non-planar (i.e., similar to the MGS·PGH structure), which also argues against the stereoelectronic proposal.

Instead, the current work supports the idea that the electrostatic environment of the leaving phosphate group plays an essential role in MGS. As noticed previously (Saadt *et al.*, 2000; Marks *et al.*, 2001), there are far fewer hydrogen-bonding interactions involving the substrate phosphate group in TIM (~2) than in MGS (~11) (Fig. 19). As shown in Fig. 18, a perturbation analysis clearly illustrates the fact that several polar and charged residues in MGS make substantial contributions to stabilizing the elimination barrier; they include Lys 23 (~2.8 kcal/mol), Arg 150′ (~4.0 kcal/mol), Ser 65 (~2.1 kcal/mol), Thr 45 (~1.5 kcal/mol, and Thr 47 (~1.5 kcal/mol). In TIM, only Lys 12 and 237 make sizable stabilizing contributions, each on the order of 2 kcal/mol. Active site water molecules were found to make substantial contributions in TIM, although their effects tend to cancel out, so that the net contribution is small.

We note that phosphate elimination was found to be the dominant pathway for a mutant of TIM in which the active site lid (see Section III) was deleted (Pompliano *et al.*, 1990). This result is thought to be due to the fact the phosphate group is no longer protected from the bulk solvent. Phosphate elimination is a rapid process for triosephosphates in solution (Richards, 1991), and an important function of the lid is to prohibit the "immature" release of the intermediate(s) before the formation of the final product, GAP. This result *per se* does not directly support either the stereoelectronic model or the current interpretation based on electrostatic interactions, although it underscores the importance of active site structural integrity for maintaining catalytic specificity.

Summarizing, the QM/MM calculations correctly reproduce the functional specificities of TIM and MGS and explain their origin. In agreement with a previous proposal (Marks *et al.*, 2001) GAP formation is more favored in TIM compared to MGS because the catalytic residue, Glu 165, is more flexible than Asp 71 in MGS, so that the proton donor (Glu 165 Oϵ2) and acceptor (C^2 in EDT2) are sufficiently close for a facile proton transfer. Furthermore, the phosphate-binding site of TIM is such that only a few residues are involved in hydrogen bonding interactions with the phosphate, which makes it energetically less favorable to undergo the elimination reaction. The stereoelectronic effect associated with the phosphate torsional angle was found to play a minor role as compared with the electrostatic effects.

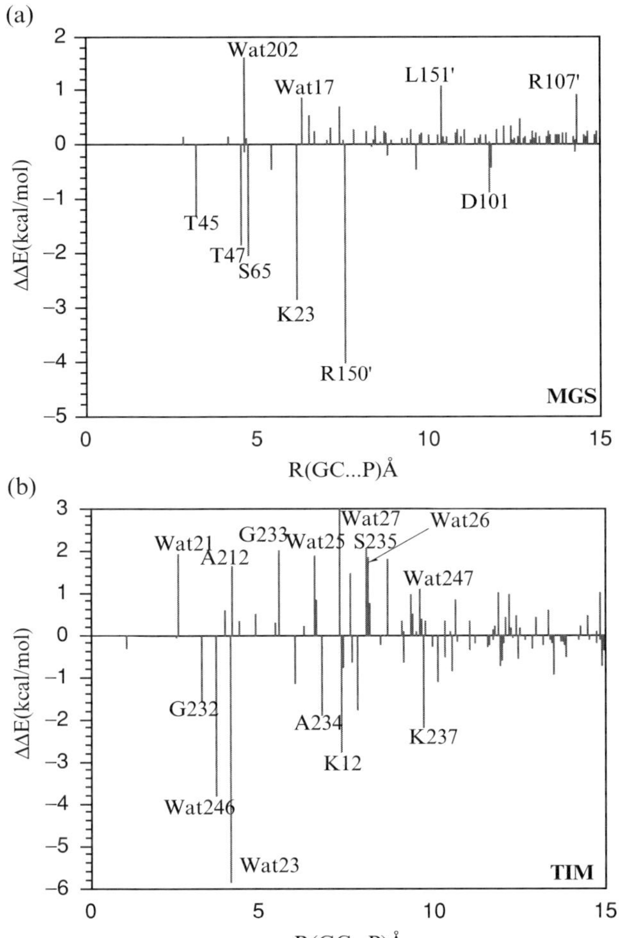

FIG. 18. Results from perturbation analyses for the contributions (in kcal/mol) of protein residues and active-site water molecules to the phosphate elimination barrier in (a) MGS and (b) TIM. Although potential of mean force calculations were made, the perturbation analyses were more straightforwardly done with reaction paths for the elimination reaction. Negative (positive) values indicate stabilizing (destabilizing) contributions. The x-axis is the distance between the center-of-geometry of a particular group to the P atom in the substrate (in Å).

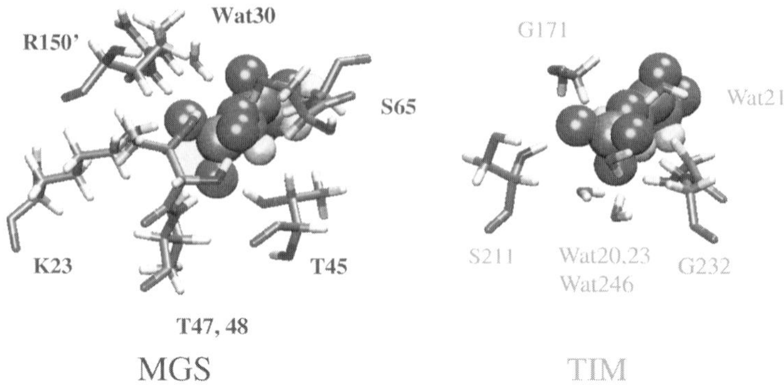

FIG. 19. The environment of the substrate phosphate group in (a) MGS and (b) TIM, based on the relevant x-ray structures with PGH as the substrate inhibitor. Residues and water molecules within 3.5 Å of the substrate P atom are shown. Note that there are many more hydrogen bonding interactions between protein residues and the phosphate group in MGS than in TIM. (See Color Insert.)

VII. Conclusions

The purpose of this review is to use a specific example for a detailed study of how *an* enzyme "works." This is not to say that all enzymes "work" in the same way but rather to illustrate how present-day computational methodologies can be applied, successfully in this case, to provide an understanding of the catalytic function of an enzyme. Triosephosphate isomerase (TIM), the so-called "perfect enzyme," was selected. It accelerates the rate of the DHAP to GAP conversion by a factor of 10^9 relative to the solution reaction. In our study of TIM, we have examined the essential aspects of the function of this enzyme, starting with the preparation of the active site (the TIM "lid"), an important step in many enzymes. We then described the theoretical results for the energetics along different possible pathways of the TIM catalyzed reactions and showed that only certain of these are likely to make a significant contribution to the overall reaction. Most importantly, we have been able to determine the primary contributions to the rate increase in the enzyme by comparing the results with models for the same sets of reactions in the gas phase and solution. By use of a perturbation analysis the (electrostatic) contributions of specific amino acids to the rate enhancement have been identified. Since "predictions" in theoretical chemistry are often made "retrospectively" in the jargon of the field, we have emphasized results that reinterpreted experiments or are related to experiments that were published subsequent to the theoretical work. In particular, we have analyzed certain mutation studies and shown that some of the observed effects have an origin different from

TABLE V
Characters of Normal Modes at the Saddle Point that have Large Overlaps to the Modes with Large Reaction Path Curvature Components for the First Proton Transfer in TIM[a]

ω, R_i^I, R_i^{II}, C	Resi.	Dist.	Contri.	ω, R_i^I, R_i^{II}, C	Resi.	Dist.	Contri.
936*i*	Subs	—	0.89	**1333**	Subs	—	0.87
5.5	E165	5.8	0.09	13.6	C126	9.5	0.03
1.2	K12	5.9	0.01	1.3	T75B	8.7	0.03
—	H95	7.3	0.01	25.1 (22.4)	L230	5.6	0.03
	W26	6.3	0.01		E165	5.8	0.03
514	E165	5.8	0.45	**1410**	A212	7.0	0.56
30.2	Subs	—	0.41	15.4	Subs	—	0.31
2.7	W27	8.2	0.03	2.4	S211	6.8	0.09
5.9 (10.7)	W245	6.0	0.03	21.0 (22.4)	E165	5.8	0.02
	W26	6.3	0.02		V231	5.9	0.01
599	Subs	—	0.52	**1411**	Subs	—	0.60
28.9	E165	5.8	0.35	11.4	A212	7.0	0.29
2.6	P166	9.1	0.02	2.2	S211	6.8	0.05
11.8 (4.5)	W	8.6	0.02	29.3 (31.3)	E165	5.8	0.04
	S235	8.4	0.01		N8	5.5	0.01
1187	Subs	—	0.94	**1528**	Subs	—	0.84
5.4	E165	5.8	0.02	2.6	H95	7.3	0.10
1.1	L230	5.6	0.01	1.4	S96	8.4	0.05
26.9 (3.1)	F229	10.0	0.01	12.6 (0.3)	E165	5.8	0.01
	V231	5.9	0.01		G232	3.9	0.01
1192	Subs	—	0.97	**1417**	I170	4.8	0.34
3.5	H95	7.3	0.01	21.7	Subs	—	0.33
1.1	K12	5.9	0.01	3.6	E165	5.8	0.24
13.2 (3.0)	E165	5.8	0.01	10.5 (12.0)	P166	9.1	0.06
	I170	4.8	0.01		H95	7.3	0.02

[a] Distances are defined as the center of geometry distances between the residues (*including* the main chain atoms) and the substrate. *Ri* is the participation ratios for the *i*th mode, and C is the coupling coefficients.

that assumed in the published interpretation. We have also investigated the dynamics of the reaction, including the transmission coefficient and the role of tunneling. It was found that the contribution of the latter to the first proton transfer (DHAP to EDT1) at room temperature is on the order of a factor two. Therefore, although quantum effects on the nuclear motions are important for kinetic isotope effects, the net contribution to the catalytic power of the enzyme is small for TIM, especially if the comparison is made with the same reaction in solution. Finally, we compared TIM with an enzyme MGS that has a very similar active site, but catalyses the conversion of DHAP to methylglyoxal, a side reaction that is avoided in TIM. The origin of the difference was described.

The series of computational studies described in this review exemplifies the power of the theoretical analyses, particularly when combined with experimental techniques such as x-ray crystallography and kinetic measurements, for an understanding of the detailed mechanism of catalysis by an enzyme. The calculations are capable of providing a semi-quantitative measure of the relative importance of factors that all seem physically reasonable. Although experiments are clearly the essential underpinning of the theoretical approaches, the latter are required for an understanding of the mechanism at an atomic level of detail. Since the importance of theory was demonstrated many years ago for one of the simplest chemical reactions (the $H+H_2$ exchange reaction in the gas phase), it is not surprising that it is all the more true that theoretical analyses are essential for understanding processes as complex as enzyme catalyzed reactions in solution.

Acknowledgments

Much of this review is based on published references, as cited in the text. Both Q.C. and M.K. acknowledge the many coworkers who have contributed to the work presented here. M.K. would like also to acknowledge many discussions with J. Knowles and G. A. Petsko that stimulated his interest in triosephosphate isomerase. We thank J. Knowles for many useful comments on the manuscript. The work at Harvard was supported in part by a grant from the National Institutes of Health.

References

Alagona, G., Desmeules, P., Ghio, C., and Kollman, P. A. (1984). *J. Am. Chem. Soc.* **106,** 3623–3632.

Alagona, G., Ghio, C., and Kollman, P. A. (1986). *J. Mol. Biol.* **191,** 23–27.

Alagona, G., Ghio, C., and Kollman, P. A. (1995). *J. Am. Chem. Soc.* **117,** 9855–9865.

Albery, W. J., and Knowles, J. R. (1976a). *Biochem.* **15,** 5627–5631.

Albery, W. J., and Knowles, J. R. (1976b). *Biochem.* **15,** 5588–5600.

Alhambra, C., Gao, J., Corchado, J. C., Villa, J., and Truhlar, D. G. (1999). *J. Am. Chem. Soc.* **121,** 2253–2258.

Amyes, T. L., O'Donoghue, A. C., and Richard, J. P. (2001). *J. Am. Chem. Soc.* **123,** 11325–11326.

Alston, W. C., Kanska, M., and Murray, C. J. (1996). *Biochem.* **35,** 12873–12881.

Antoniou, D., and Schwartz, S. D. (1997). *Proc. Natl. Acad. Sci. USA* **94,** 12360–12365.

Antoniou, D., and Schwartz, S. D. (2001). *J. Phys. Chem. B* **105,** 5553–5558.

Åqvist, J., and Warshel, A. (1993). *Chem. Rev.* **93,** 2523–2544.

Åqvist, J., and Fothergill, M. (1996). *J. Bol. Chem.* **271,** 10010–10016.

Bahnson, B. J., Colby, T. D., Chin, K. J., Goldstein, B. M., and Klinman, J. P. (1997). *Proc. Natl. Acad. Sci. USA* **94,** 12797–12802.

Bash, P. A., Field, M. J., Davenport, R. C., Petsko, G. A., Ringe, D., and Karplus, M. (1991). *Biochem.* **30,** 5826–5832.
Belasco, J. G., and Knowles, J. R. (1980). *Biochem.* **19,** 472–477.
Bell, R. P. (1973) *The Proton in Chemistry*, Chapman and Hall, London.
Benderskii, V. A., Markarov, D. E., and Grinevich, P. (1993). *Chem. Phys.* **170,** 275–293.
Benderskii, V. A., Markarov, D. E., and Wight, C. A. (1994). *Adv. Chem. Phys.* **LXXXVIII,** 275–293.
Bergsma, J. P., Gertner, B. J., Wilson, K. R., and Hynes, J. T. (1987). *J. Chem. Phys.* **86,** 1356–1376.
Billeter, S. R., Webb, S. P., Agarwal, P. K., Iordanov, T., and Hammes-Schiffer, S. (2001). *J. Am. Chem. Soc.* **123,** 11262–11272.
Borchert, T. V., Abagyan, R., Jaenicke, R., and Wierenga, R. K. (1994). *Proc. Natl. Acad. Sci. USA* **91,** 1515–1518.
Brooks, B. R., Burccoleri, R. E., Olafson, B. D., States, D. J., Swaminathan, S., and Karplus, M. (1983). *J. Comp. Chem.* **4,** 187–217.
Brooks III, C. L., and Karplus, M. (1989). *J. Mol. Biol.* **208,** 159–181.
Brooks, C. L., Karplus, M., and Pettitt, B. M. (1988). *Adv. in Chem. Phys.* **LXXI,** 1–259.
Brownlee, M., and Cerami, A. (1981). *Annu. Rev. Biochem.* **50,** 385–432.
Brumer, P., and Karplus, M. (1973). *Faraday Soc. Discussions* **55,** 80–91.
Caratzoulas, S., and Schwartz, S. D. (2001). *J. Chem. Phys.* **114,** 2910–2918.
Chandler, D. (1978). *J. Chem. Phys.* **68,** 2959.
Chuang, Y., Corchado, J. C., Fast, P. L., Villa, J., Coitino, E. L., Hu, W., Liu, Y., Lynch, G. C., Nguyen, K., Jackels, C. F., Gu, M. Z., Rossi, I., Clayton, S., Melissas, V., Steckler, R., Garrett, B. C., Isaacson, A. D. and Truhlar, D. G. *POLYRATE version 8.0* (1998). Department of Chemistry and Supercomputer Institute, University of Minnesota, Minneapolis, Minnesota.
Cleland, W. W., Frey, P. A., and Gerlt, J. A. (1998). *J. Biol. Chem.* **273,** 25529–25532.
Copper, R. A. (1984). *Annu. Rev. MicroBiol.* **44,** 812.
Cossi, M., Barone, V., Cammi, R., and Tomasi, J. (1996). *Chem. Phys. Lett* **255,** 327–335.
Cui, Q., Elstner, M., and Karplus, M. (2002). *J. Phys. Chem. B* **106,** 2721–2740.
Cui, Q., Elstner, M., Kaxiras, E., Frauenheim, T, and Karplus, M. (2001). *J. Phys. Chem. B* **105,** 569–585.
Cui, Q., and Karplus, M. (2001). *J. Am. Chem. Soc.* **122,** 2284–2290.
Cui, Q., and Karplus, M. (2002a). *J. Am. Chem. Soc.* **124,** 3093–3124.
Cui, Q., and Karplus, M. (2002b). *J. Phys. Chem. B* **106,** 1768–1798.
Cui, Q., and Karplus, M. (2002c). *J. Phys. Chem. B* **106,** 7927–7947.
Davenport, R. C., Bash, P. A., Seaton, B. A., Karplus, M., Petsko, G. A., and Ringe, D. (1991). *Biochem.* **30,** 5821–5826.
Derreumaux, P., and Schlick, T. (1998). *Biophys. J.* **74,** 72–81.
Dewar, M. J. S., Zoebisch, E. G., Healy, E. A., and Stewart, J. J. P. (1985). *J. Am. Chem. Soc.* **107,** 3902–3909.
Dinner, A. R., Lopez, X., and Karplus, M. (2003). *Theor. Chem. Acc.* **109,** 118–124.
Elstner, M., Porezag, D., Jungnickel, G., Elsner, J., Haugk, M., Frauenheim, T., Suhai, S., and Seifert, G. (1998). *Phys. Rev. B* **58,** 7260–7268.
Fersht, A. (1999). *Structure and Mechanism in Protein Science*, W. H. Freeman and Company, New York.
Field, M. J., Bash, P. A., and Karplus, M. (1990). *J. Comput. Chem.* **11,** 700–733.
Fischer, S., and Karplus, M. (1992). *Chem. Phys. Lett.* **194,** 252–261.
Frisch, M. J., Trucks, G. W., Schlegel, H. B., Gill, P. M. W., Johnson, B. G., Robb, M. A., Cheeseman, J. R., Keith, T., Petersson, G. A., Montgomery, J. A., Raghavachari, K.,

Al-Laham, M. A., Zakrzewski, V. G., Ortiz, J. V., Foresman, J. B., Peng, C. Y., Ayala, P. Y., Chen, W., Wong, M. W., Andres, J. L., Replogle, E. S., Gomperts, R., Martin, R. L., Fox, D. J., Binkley, J. S., Defrees, D. J., Baker, J., Stewart, J. P., Head-Gordon, M., Gonzalez, C., and Pople, J. A. (1995). *Gaussian 94*, Revision B.3., Gaussian, Inc., Pittsburgh Pennsylvania.

Frisch, M. J., Head-Gordon, M., Trucks, G. W., Foresman, J. B., Schlegel, H. B., Raghavachari, K., Robb, M. A., Binkley, J. S., Gonzalez, C., Defrees, D. J., Fox, D. J., Whitesides, R. A., Seeger, R., Melius, C. F., Baker, J., Martin, R. L., Kahn, L. R., Stewart, J. J. P., Topiol, S., and Pople, J. A. (1990). *Gaussian 90*. Gaussian Inc., Pittsburgh Pennsylvania.

Gao, J. (1996). In: *Reviews in Computational Chemistry*, (K. B. Lipkowitza and D. B. Boyd, Eds.), VCH Publishers, New York.

Gerstein, M., and Chothia, C. (1991). *J. Mol. Biol.* **220,** 133149.

Gerstein, M., Lesk, A. M., and Chothia, C. (1994). *Biochemistry* **33,** 6739–6749.

Hall, A., and Knowles, J. R. (1975). *Biochem.* **14,** 4348–4352.

Harris, T. K., Abeygunawardana, C., and Mildvan, A. S. (1997). *Biochem.* **36,** 14661–14675.

Harris, T. K., Cole, R. N., Comer, F. I., and Mildvan, A. S. (1998). *Biochem.* **37,** 16828–16838.

Herlihy, J. M., Maister, S. G., Albery, W. J., and Knowles, J. R. (1976). *Biochem.* **15,** 5601–5607.

Hwang, J. K., and Warshel, A. (1996). *J. Am. Chem. Soc.* **118,** 11745–11751.

Jencks, W. P. (1987) *Catalysis in Chemistry and Biology*, Dover, New York.

Joseph, D., Petsko, G. A., and Karplus, M. (1990). *Science* **249,** 1425–1428.

Joseph-McCarthy, D., Rost, L. E., Komives, E. A., and Petsko, G. A. (1994). *Biochem.* **33,** 2824–2829.

Joseph-McCarthy, D., Petsko, G. A., and Karplus, M. (1995). *Protein Engineering* **8,** 1103–1115.

Jogl, G., Rozovsky, S., McDermott, A. E., and Tong (2003). *L. Proc. Natl. Acad. Sci. USA* **100,** 50–55.

Karplus, M., Porter, R. N., and Sharma, R. D. (1965). *J. Chem. Phys.* **43,** 3259–3287.

Karplus, M., Evanseck, J. D., Joseph, D., Bash, P. A., and Field, M. J. (1992). *Faraday Discussion* **93,** 239–248.

Karplus, M. (2000). *J. Phys. Chem. B* **104,** 11–27.

Knowles, J. R. (1991). *Nature* **350,** 121–124.

Knowles, J. R., and Albery, W. J. (1977). *Acc. Chem. Res.* **10,** 105–111.

Kohen, A., and Klinman, J. P. (1998). *Acc. Chem. Res.* **31,** 397–404.

Komives, E. A., Chang, L. C., Lolis, E., Tilton, R. F., Petsko, G. A., and Knowles, J. R. (1991). *Biochem.* **30,** 3011.

Kursula, I., Partanen, S., Lambeir, A., Antonov, D. M., Augustyns, K., and Wierenga, R. K. (2001). *Eur. J. Biochem.* **268,** 5189–5196.

Lodi, P. J., and Knowles, J. R. (1991). *Biochem.* **30,** 6948–6956.

Lodi, P. J., Chang, L. C., Knowles, J. R., and Komives, E. A. (1994). *Biochem.* **33,** 2809–2814.

Lolis, E., Alber, T., Davenport, R. C., Rose, D., Hartman, F. C., and Petsko, G. A. (1990). *Biochem.* **29,** 6609–6618.

MacKerell, Jr., A. D., Bashford, D., Bellott, M., Dunbrack, R. L.Jr., Evanseck, J. D., Field, M. J., Fischer, S., Gao, J., Guo, H., Ha, S., Joseph-McCarthy, D., Kuchnir, L., Kuczera, K., Lau, F. T. K., Mattos, C., Michnick, S., Ngo, T., Nguyen, D. T., Prodhom, B., Reiher, W. E.III, Roux, B., Schlenkrich, M., Smith, J. C., Stote, R.,

Straub, J., Watanabe, M., Wiórkiewicz-Kuczera, J., Yin, D., and Karplus, M. (1998). *J. Phys. Chem. B* **102,** 3586–3616.
Marcus, R. A., and Coltrin, M. E. (1977). *J. Chem. Phys.* **67,** 2609.
Marks, G. T., Harris, T. K., Massiah, M. A., Mildvan, A. S., and Harrison, D. H. T. (2001). *Biochem.* **40,** 6805–6818.
McCammon, J. A., Northrup, S. H., Karplus, M., and Levy, R. M. (1980). *Biopolymers* **19,** 2033–2045.
McQuarrie, D. A. (1976) *Statistical Mechanics*, Harper & Row, New York.
Miertus, S., Scrocco, E., and Tomasi (1981). *J. Chem. Phys.* **55,** 117–129.
Miller, W. H., Handy, N. C., and Adams, J. E. (1980). *J. Chem. Phys.* **72,** 99–112.
Neria, E., and Karplus, M. (1997). *Chem. Phys. Lett* **267,** 23–30.
Nickbarg, E. B., Davenport, R. C., Petsko, G. A., and Knowles, J. R. (1988). *Biochem.* **27,** 5948–5960.
Nyquist, R. A., Chrzan, V., and Houck, J. (1989). *Appl. Spectroscopy* **43,** 981–983.
Peräkylä, M. (1997). *J. Chem. Soc. Perkin Trans.* 2185–2189.
Peräkylä, M., and Pakkanen, T. A. (1996). *Proteins* **25,** 225–236.
Pompliano, D. L., Peyman, A., and Knowles, J. R. (1990). *Biochem.* **29,** 3186–3194.
Pulay, P., Fogarasi, G., Pongor, G., Boggs, J. E., and Vargha, A. (1983). *J. Am. Chem. Soc.* **105,** 7037–7047.
Reuter, N., Dejaegere, A., Maigret, B., and Karplus, M. (2000). *J. Phys. Chem. B* **104,** 1720–1735.
Richard, J. P. (1984). *J. Am. Chem. Soc.* **106,** 4926.
Richard, J. P. (1991). *Biochem.* **30,** 4581.
Rogers, J. D., Rub, B., Goldman, S., and Person, W. B. (1981). *J. Phys. Chem.* **85,** 3727–3729.
Rozovsky, S., and McDermott, A. E. (2001). *J. Mol. Biol.* **310,** 259–270.
Rozovsky, S., Jogl, G., Tong, L., and McDermott, A. E. (2001). *J. Mol. Biol.* **310,** 271.
Ryckaert, J.-P., Ciccotti, G., and Berendsen, H. J.-C. (1977). *J. Comp. Phys.* **23,** 327–341.
Saadat, D., and Harrison, D. H. T. (1998). *Biochem.* **37,** 10074–10086.
Saadat, D., and Harrison, D. H. T. (2000). *Biochem.* **39,** 2950–2960.
Seeger, D. M., Korzeniewski, C., and Kowalchyk, W. (1991). *J. Phys. Chem.* **95,** 6871–6879.
Simonson, T., Archontis, G., and Karplus, M. (1997). *J. Phys. Chem. B* **101,** 8347–8360.
Truhlar, D. G., Isaacson, A. D., and Garrett, B. C. (1985). In: *Theory of Chemical Reaction Dynamics* (M. Bear Ed.), CRC Press, Boca Raton, Forida.
Truhlar, D. G., Garrett, B. C., and Klippenstein, S. J. (1996). *J. Phys. Chem.* **100,** 12771–12800.
Truhlar, D. G., and Gao, J. (2002). *Ann. Rev. Phys. Chem.* **53,** 467–505.
Truhlar, D. G., and Wyatt, R. E. (1976). *Ann. Rev. Phys. Chem.* **27,** S1–S43.
Tsai, P., and Gracy, R. W. (1976). *J. Biol. Chem.* **251,** 364–367.
Valleau, J. P., and Torrie, G. M. (1977) In: *Statistical Mechanics*, (B. J. Berne, Ed.), Plenum Press, New York.
Wade, R. C., Davis, M. E., Luty, B. A., Madura, J. D., and McCammon, J. A. (1993). *Biophysical J* **64,** 9–15.
Villa, J., and Warshel, A. (2001). *J. Phys. Chem.* **105,** 7887–7907.
Warshel, A., and Karplus, M. (1972). *J. Am. Chem. Soc.* **94,** 5612–5625.
Williams, J. C., and McDermott, A. E. (1995). *Biochemistry* **34,** 8309–8319.
Wolfenden, R., and Snider, M. J. (2001). *Acc. Chem. Res* **34,** 938–945.
Xie, X. S. (2002). *J. Chem. Phys.* **117,** 11024–11032.
Xie, X. S., and Lu, H. P. (1999). *J. Biol. Chem.* **274,** 15967–15970.

Yüksel, K. Ü., Sun, A.-Q., Gracy, R. W., and Schnackerz, K. D. (1994). *J. Biol. Chem.* **269,** 5005–5008.
Zhang, X., Harrison, D. H. T., and Cui, Q. (2002). *J. Am. Chem. Soc.* **124,** 14871.
Zhang, Z., Komives, E. A., Sugio, S., Blacklow, S. C., Narayana, N., Xuong, N. H., Stock, A. M., Petsko, G. A., and Ringe, D. (1999). *Biochemistry* **38,** 4389–4397.

ALL-ATOM SIMULATIONS OF PROTEIN FOLDING AND UNFOLDING

By RYAN DAY AND VALERIE DAGGETT

Department of Medicinal Chemistry, University of Washington, Seattle, WA

I. Background ... 373
A. The Folding Problem and Minimal Models ... 373
B. Minimal Models ... 375
C. All-Atom Approaches ... 375
II. Overview of Results ... 377
A. The First Systems ... 377
B. Modern Simulations ... 381
C. Other Fast Folding Proteins ... 393
III. Conclusions/Summary ... 395
A. Proteins Unfold by Pathways ... 395
B. Direct Simulation of Protein Folding Is in Its Infancy ... 397
C. The Future of Unfolding Simulations ... 398
References ... 399

I. BACKGROUND

A. *The Folding Problem and Minimal Models*

The question of how a given sequence of amino acids assumes a unique three-dimensional structure has sparked considerable debate since its framing. In the early 1960s, Anfinsen and co-workers firmly established the reversibility of protein unfolding and the idea that a protein's structure was encoded solely in its amino acid sequence (Anfinsen, 1973). This reversibility was formally recognized as a problem by a calculation that would later come to be known as Levinthal's paradox (Levinthal, 1968). The calculation simply points out that a 100 amino acid protein has about 10^{49} conformations available to it, and that finding the single native conformation would take on the order of 10^{29} years, assuming that only 10^{-11} s are required to change from one conformation to another (Dobson *et al.*, 1998). The paradox is that proteins fold on the order of minutes or less, not on the order of lifetimes of universes.

The trivial solution to this paradox is that a given peptide chain only searches a small region of conformational space before arriving at the native state. The problem then is which region of conformational space does the peptide chain explore, and can we predict that region from the

ADVANCES IN PROTEIN CHEMISTRY, Vol. 66

0065-3233/03 $35.00

amino acid sequence? The first solution to this problem was proposed, unsurprisingly, by Levinthal (1968). He proposed that protein folding occurs by the sequential formation and condensation of smaller three-dimensional structures. In what is likely the first example of a computational study of protein folding, he used computer-aided model building to generate a "plausible structure" of cytochrome c. Levinthal (1968) also pointed out that "the uniqueness of the proposed folding process has not been determined." This caveat was largely ignored in later formulations of the pathway hypothesis.

In recent years, a more physically reasonable description of conformation space has given rise to a "new view" of the solution to the Levinthal paradox (Baldwin, 1995; Dill and Chan, 1997; Dobson *et al.*, 1998). Where the formulation of the Levinthal paradox puts all 10^{49} possible conformations of a 100 residue peptide on equal footing, the funnel view recognizes that these conformations will have different energies and different probabilities of being accessed. The funnel is best envisioned by imagining the high dimensional conformational space available to a peptide chain projected into two dimensions. Each conformation of the chain can then be imagined as a point in the two dimensional space, with the distance between any two points approximating the similarity between the two conformations. Each conformation has what Dill and Chan (1997) term a "conformational free energy," consisting of its internal energy plus the solvation free energy. If this conformational free energy is plotted on a third axis, the resulting surface is imagined to be broadly funnel shaped, with the native ensemble at the lowest point of the funnel. The conformational space that is searched in protein folding is limited by the slope of the funnel as a given conformation and is more likely to change into a conformation having a lower free energy than one with higher free energy. As a complete enumeration of conformational space is prohibitively time consuming, the free energy surfaces are typically envisioned in terms of simple progress variables. The funnels arose from minimal models of protein folding in which a conformation's internal energy is based solely on native contacts. As such, another common visualization of folding funnels groups conformations according to the fraction of native contacts formed, which is Q. The funnel arises when the internal energy of each value of Q is plotted on one axis and the volume of conformational space available at each value is plotted on another (see Dobson *et al.*, 1998).

As we shall see in this review, the pathway and funnel ideas are not mutually exclusive. A pathway view is essentially a subset of the funnel view. It imagines that there are common features of the conformations that a peptide chain assumes as it travels different routes down the funnel.

These common features can then be used to describe a pathway by which an unfolded structure finds its way to the native structure.

B. *Minimal Models*

Early simulations of protein folding necessarily relied on minimal models of the protein chain. Simplified models still find extensive use today, as they have far fewer degrees of freedom than all-atom representations and their conformation space can be more fully explored. Early models were developed by Levitt and Warshel, Kuntz, Crippen, and Kollman, Scheraga and co-workers, and Gō and co-workers (see Gō, 1983, and references therein). All of these models represented amino acids by one or two "beads." The models of Gō and co-workers took the additional step of restricting these beads to points on a two-dimensional lattice and used only two types of beads. The bead types were hydrophobic, which were attracted to each other, and hydrophilic, which were not. Sequences of such beads could be designed that had a unique low energy structure and underwent a two-state transition much like that in real proteins. These lattice models had the additional advantage that their conformational space could be completely enumerated in a relatively short amount of time. As computer speeds have increased, minimal models have become less minimal. They have evolved onto three-dimensional lattices, then off lattices completely, and are now allowed to have more than two flavors of bead, although the formation of three bead types seems to be a common maximum.

Minimal models have greatly enhanced our understanding of protein folding. The funnel view and energy landscapes arose directly from the exhaustive sampling allowed by minimal models (Onuchic *et al.*, 1997). Nevertheless, there is much that is left out or approximated by even the most complex minimal models. A funnel-shaped landscape is a general property of proteins that fold. In order to determine how the details of that landscape are encoded by specific protein sequences, we need to add more detail to our models. We need to capture the unique hydrogen-bonding requirements of the backbone and the multitude of ways that twenty different side chains can interact with one another, with the protein backbone, and with the solvent. We need atoms.

C. *All-Atom Approaches*

The force fields and dynamics protocols used in all-atom simulations of protein folding and unfolding are essentially the same as those used in other simulation studies of proteins and are detailed elsewhere in this volume. The problem encountered in simulating protein folding and unfolding is similar to that of the Levinthal paradox. There are many possible conformations of the protein chain and a limited amount of time in which to sample them. Protein folding takes place in timescales ranging from microseconds to minutes, while simulations were, in their early days, limited to hundreds of picoseconds (ps) and are only recently beginning to regularly extend beyond a few nanoseconds (ns).

The simplest method of increasing sampling is by increasing the temperature of the simulation. Simulations of protein unfolding are regularly run at 498 K and above. At these temperatures, the native structure of the protein is typically lost in a few nanoseconds, with most unfolding events taking place inside of the first nanosecond. In early simulations of protein unfolding, these temperatures were necessary to see even early unfolding events. Although lower temperature simulation is now possible, high temperature simulations are still used extensively. Many high temperature simulations can be run in the same amount of time as one lower temperature simulation, leading to more extensive sampling of the energy landscape. Although they sample only a limited portion of the total conformational space, features of the unfolding process such as the transition state ensemble of structures and the unfolded ensemble have shown remarkable agreement with experimentation. This will be detailed on a case-by-case basis below.

Brooks and co-workers developed another method to increase the conformational sampling allowed by all-atom simulation (Shea and Brooks, 2001). The essence of the method is to generate a large number of conformations from a variety of points in the folding/unfolding process and to use them to create a free energy surface. Initial conformations are generated by the standard thermal unfolding simulations previously described. These conformations are grouped according to the fraction of the native contacts remaining in each. Representative structures from each group are then used as starting points to generate more conformations with a harmonic biasing potential based on the fraction of native contacts. These structures are combined using a weighted histogram and then used to construct free energy surfaces for a variety of simple reaction coordinates. While this method is quite effective at generating a large number of conformations, the folding behavior predicted by these conformations does not appear to agree well with available experimental data.

Parallel implementations of molecular dynamics code hold the potential to greatly increase the timescales that are accessible by all-atom simulation. In such implementations, the computational load of determining the positions and interactions between the atoms in a system is spread out over many processors. Direct speed increases from these types of implementations require that the processors being used can exchange information very rapidly. Processors with slower connections can still be used in a couple of ways. If very large systems are divided among a smaller number of processors, the data exchange requirements will not be as high, and the processing time will not be severely affected by the slow connection. Another way to take advantage of multiple processors is to run multiple trajectories (replicas) with slightly different starting conditions and monitor them for significant barrier crossing or folding events. When such an event occurs in one trajectory, one should replicate that trajectory to the other processors and restart them all, again using slightly different starting conditions in each. The most difficult part of these types of methods is identifying events. If the structure of the native state is known, it can be used for comparison. This is, however, somewhat unsatisfying and may lead to bias in the conformations explored. Another method, developed by Pande and co-workers (Zagrovic *et al.*, 2001), uses the sharp spike in heat capacity that accompanies folding in identifying folding events. This method has the advantage of not requiring any knowledge of the native state. It is still somewhat dissatisfying in that the primary contributor to the heat capacity change of folding is the burial of hydrophobic groups, so the folding may be biased towards hydrophobic collapse mechanisms.

The first all-atom simulations of the protein folding/unfolding reaction appeared in the literature a decade ago. Since then, the methods described have been applied to many more proteins, and experimental and simulation studies are being performed in concert, yielding a much richer view of the protein folding/unfolding process at atomic resolution (Fersht and Daggett, 2002). In this review, we will discuss the results of many of these simulations and, briefly, how they compare with experimental results. In the end, we will see how these simulations have created a view of the conformational landscape for real protein sequences and that it is not all that different from what Levinthal predicted 35 years ago.

II. Overview of Results

A. *The First Systems*

1. *Bovine Pancreatic Trypsin Inhibitor*

The first all-atom simulations of protein unfolding in explicit solvent began to appear in 1992. Bovine pancreatic trypsin inhibitor (BPTI) (Fig. 1) was simulated at 423 K for 550 ps in both its oxidized and reduced forms and at 498 K for 284 ps in the reduced form (Daggett and Levitt,

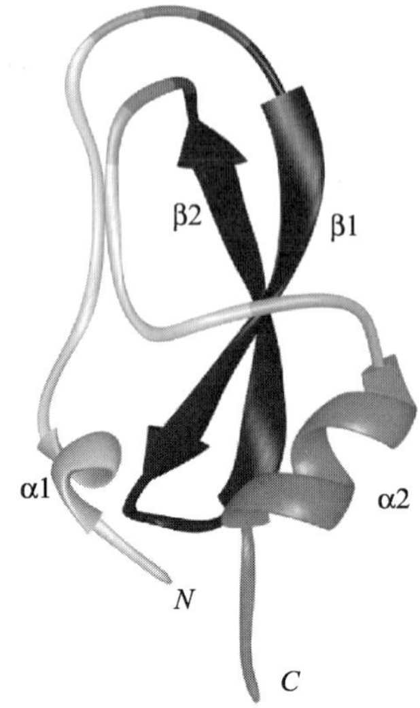

FIG. 1. The structure of bovine pancreatic trypsin inhibitor (Deisenhofer and Steigemann, 1975) colored by the order of structure loss seen in the dominant unfolding pathway at neutral pH (Kazmirski and Daggett, 1998a). Here and in Figs. 2 to 13, the shading ranges from white (early loss) to black (late loss or no loss) and is necessarily quite approximate.

1992, 1993). The oxidized form was constrained to a smaller volume than the reduced form at higher temperatures and most of the discussion of unfolding events followed this latter form. Disruption is seen primarily in turns and loops, as well as in the specific packing of the protein core. Additionally, the protein's α-helix appeared less stable than the β-sheet at high temperature; the sheet gained some secondary structure while the helix lost some structure.

BPTI was revisited in 1998 in a series of 2.5 ns simulations at 498 K (Kazmirski and Daggett, 1998a). Four simulations were performed of the native structure at neutral pH and three at low pH. Four additional simulations were performed at neutral pH, each with the charge on one lysine reduced to approximate chemically modified lysines used as probes in fluorescence energy transfer experiments. Two general pathways to the denatured state were observed. First, the protein expanded to the transition state, losing specific core packing, after which a secondary structure was lost. In one pathway that was observed in three of the four simulations at neutral pH, unfolding of the α-helix preceded the unfolding of the β-sheet. This order of loss was reversed in the other pathway, which was observed in all three low pH simulations and one of the neutral pH simulations. The denatured state is essentially a random or "molten" coil, with no real residual secondary structure and fluctuating hydrophobic contacts. Three of the four simulations using modified lysines also unfolded by this latter pathway.

2. *Hen Egg White Lysozyme*

Appearing later in 1992 was a study of the unfolding of hen egg white lysozyme (HEWL) (Fig. 2) (Mark and van Gunsteren, 1992). The study simulated HEWL at 500 K for about 200 ps in order to characterize its unfolding. An initial expansion of the protein with an associated increase in solvent exposed surface area was observed. A sharp decrease in helical hydrogen bonding was observed to occur in this initial solvent expansion. The only portions of the protein that maintained significant secondary structure through this initial expansion were helices A and C.

van Gunsteren and co-workers revisited HEWL in 1995 to compare different methods of unfolding proteins (Hunenberger *et al.*, 1995). They describe the thermal unfolding of HEWL in more detail. Helix D and the C-terminal 3_{10} helix are rapidly distorted. Helix C shrinks slowly and steadily from its ends. Helices A and B fluctuate in and out of existence over the course of the simulation. The β-domain is stable until about 120 ps, at which point the sheet begins to fray and melt, beginning with loss of strand 3. The β-domain appears much more stable than the α-domain in this simulation, in apparent disagreement with experimental amide proton protection (Dobson *et al.*, 1994). Other biasing forces were also explored in this work, including pressure, a radially directed artificial force, and a velocity-rescaling scheme in which inwardly moving protein atoms are slowed and outwardly moving protein atoms are accelerated.

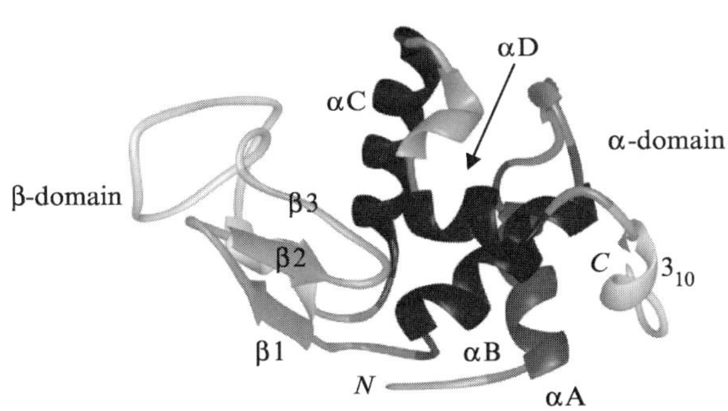

FIG. 2. The structure of hen egg white lysozyme (Wilson *et al.*, 1992). Coloring ranges are as in Fig. 1 and are based on the simulations of Kazmirski and Daggett (Kazmirski and Daggett, 1998b).

The pressure denaturation yielded little change from the native state in 210 picoseconds of simulation. The other two forces led to rapid, fairly uniform radial expansion from which little structural information about the unfolding process was obtained.

Hen lysozyme also served as the model for a denaturation technique that involved simulation at 353 K and accelerated solvation of the protein core (Williams *et al.*, 1997). The method involved simulating for very short (1 ps) times and then filling any core cavities that had opened with water molecules. Denaturation occurred early and catastrophically when the number of core waters placed went from ten to eighteen in a single step, leading to a rapid disruption of the protein. The helices denature first, while the sheet remains essentially fully formed throughout the simulation.

More recently, HEWL was the subject of two additional high temperature denaturation simulations, each of which lasted for 9 ns (Kazmirski and Daggett, 1998b). These simulations saw an initial loss of tight packing, followed by loss of secondary and tertiary structure in the β-domain. The native packing of the helices in the α-domain is lost, but the helical secondary structure is maintained by non-native contacts. While different from the earlier simulations, this order of events is similar to that seen experimentally in the major folding pathway (Dobson *et al.*, 1994). Lysozyme does, however, have multiple folding pathways, and it is possible that the different simulations are simply exploring different pathways.

3. *Myoglobin*

While more recent work tends to focus on the solvation and functional aspects of the native state of myoglobin (see Makarov *et al.*, 2000, Kiefl *et al.*, 2002), it was also among the first proteins to have its denaturation simulated (Tirado-Rives and Jorgensen, 1993). The simulations, run at 358 K, were quite short by today's standards, and the 298 K control deviated significantly from the crystal structure. An earlier simulation of apomyoglobin at 298 K, using a different parameter set, maintained the crystal structure to a larger degree (Brooks, 1992). The deviations in the control were ascribed to the removal of the heme group, which drops the amount of helical structure measured by far-UV circular dichroism from 85% to 55% (Griko *et al.*, 1988), and the authors made observations of the early events in myoglobin unfolding (Fig. 3). All the helices frayed somewhat over the course of the simulations, with helices A to E unfolding the most. Helix F remains relatively well formed at neutral pH but unfolds completely at lower pH. Helix H frays from the C-terminus, leading to a corresponding loss of structure in the N-terminal end of helix G. Generally, it

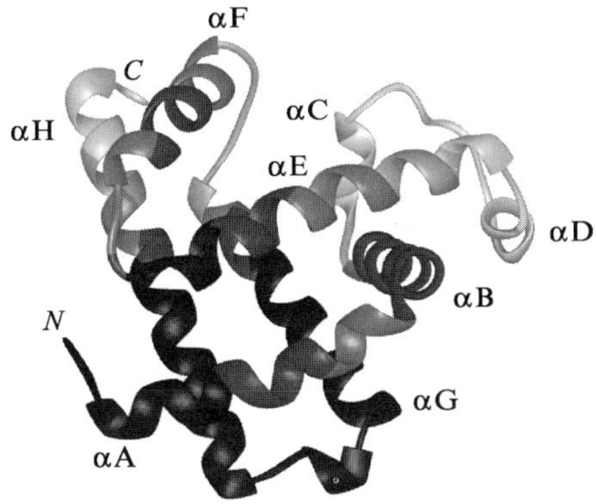

FIG. 3. The structure of apo-myoglobin (Vojtechovsky *et al.*, 1999). The order of structure loss seen here is based on only one unfolding trajectory (Tirado-Rives and Jorgensen, 1993).

seems that helical structure is better maintained in areas with more tertiary contacts, such as where helices A and E cross G and H.

B. Modern Simulations

1. Barnase

Barnase's unfolding (Fig. 4) was first studied by molecular dynamics simulation by Caflisch and Karplus (1994, 1995). The protein was denatured thermally at 600 K in a series of simulations. In one of these simulations, starting with a structure that had been first simulated at 300 K for 100 ps, the mainchain becomes fully solvated and all secondary structure is lost in the time course of the simulation. Despite these changes, the protein is still described as having the same overall fold. The order of events of unfolding is conserved across several simulations using different starting conditions. First to unfold were the N-terminus and loops. This was followed by water penetration and subsequent unfolding in both cores 1 and 2, with core 2 becoming solvated somewhat before core 1. Caflisch and Karplus verified these results by investigating the unfolding of barnase at a much lower temperature, 360 K, and protonating acidic residues to simulate low pH conditions. The chain termini and loops are again seen to lose structure first, followed by helix 1 separating from the β-sheet to expose core 1 to solvent and allow its unfolding. The largest difference in the unfolding between the two conditions is that helices 2 and 3, which make up core 2, remain stable in the lower temperature, low pH simulation but are rapidly disrupted at the higher temperature.

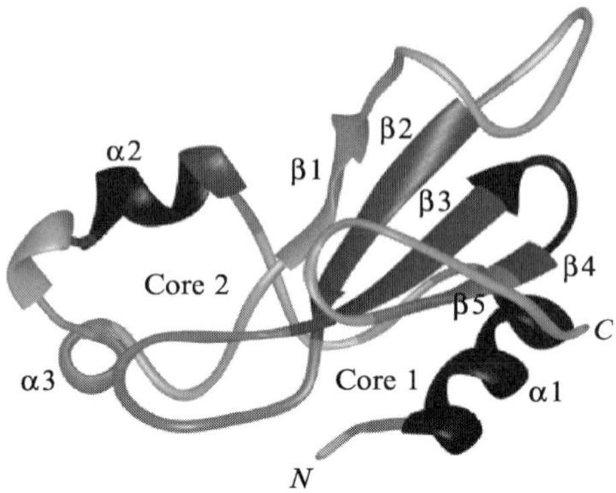

FIG. 4. The structure of barnase (Martin *et al.*, 1999) colored by the order of structure loss.

Daggett and co-workers used molecular dynamics simulation at 498 K to determine atomic information about the unfolding transition, major intermediate, and denatured states of barnase (Li and Daggett, 1998; Daggett *et al.*, 1998; Bond *et al.*, 1997; Wong *et al.*, 2000). The transition and intermediate states were identified by conformational clustering, and the denatured state was defined to span the last 2 ns of the 4 ns trajectories in which the average α-carbon RMSD to native was greater than 10 Å. The structure of the transition state found in the molecular dynamics simulations agrees quite well with the experimentally determined Φ-values for all residues except those found in core 2, which is more structured in the simulation than the Φ-values seem to indicate. An alternate interpretation of the Φ-values was suggested in which core 2 folds semi-autonomously and does not contribute significantly to the rest of the protein's folding free energy until after the transition state. The intermediate was found to have disrupted core packing and with an increase in mobility of the termini and loops. Core 2 is found to be most significantly disrupted, though helix 2 maintains most of its structure. The final denatured state was conformationally diverse, though some residual structure was observed. Specifically, helices 1 and 2 fluctuated in and out of existence and a small hydrophobic core was transiently formed between strands 3 and 4 of the β-sheet. The overall picture of barnase's unfolding that emerges from these studies is quite similar to that provided by Caflisch and Karplus. Namely, the unfolding begins with distortion of the loops and N-terminus and proceeds through an intermediate in which the hydrophobic cores are quite disrupted, though significant secondary structure remains and the general topology is native-like. Finally, the protein moves into a denatured state that is quite conformationally diverse with some transient secondary structure.

Barnase has also been the subject of simulated chemical denaturation. Two such studies have been carried out, both using 8 M urea and elevated temperature to effect the denaturation. In the first such study, carried out by Jorgensen and co-workers (Tirado-Rives *et al.*, 1997) the protein's helical content is largely maintained, while the β-sheet structure frays rapidly with a concomitant loss of core 1. The protein is not, however, fully unfolded by the end of the 2-ns simulation. They conclude that the primary effect of the urea is to stabilize the denatured state by excluding water from the protein surface, hydrogen bonding to the main chain, and scaffolding remaining waters in the protein's solvation shell. Caflisch and Karplus (1999) also simulated barnase in urea, though for shorter time-scales. They do not observe any significant unfolding events at 360 K, 8 M urea, and neutral pH, though they do observe some early events such as increased mobility of the N-terminus and disruption of core 2 when low pH conditions are employed. Caflisch and Karplus note differences between the behavior of their urea model and that of Tirado-Rives *et al.* These include a lower degree of hydrogen bonding to the protein and a lack of change in the solvent-accessible surface area going from water to a water-urea mix. It seems possible that these differences could be due to the greater degree of unfolding observed in the simulations performed by Tirado-Rives *et al.*

2. *Chymotrypsin Inhibitor 2*

Chymotrypsin inhibitor 2 (CI2) was among the first proteins to have its folding/unfolding transition state extensively characterized by the protein engineering method (Jackson *et al.*, 1993; Itzhaki *et al.*, 1995) and its denatured state and folding were subsequently characterized by NMR and hydrogen exchange (Killick *et al.*, 1998; Itzhaki *et al.*, 1997; Kazmirski *et al.*, 2001). This wealth of experimental information has been exploited by Daggett and co-workers in validating extreme temperature simulation as a method of exploring the atomic details of protein unfolding (Li and Daggett, 1994, 1996; Daggett *et al.*, 1996; Kazmirski *et al.*, 2001; De Jong *et al.*, 2002; Day *et al.*, 2002). A recent study of CI2's unfolding at temperatures ranging from 373 K to 498 K (Day *et al.*, 2002) suggests a general description of the unfolding pathway of CI2 that is largely independent of temperature (Fig. 5). At all temperatures, the protein began unfolding by expanding slightly with a corresponding disruption of the core packing. The initial expansion of the protein was followed by either a sliding or breathing motion of the helix and N-terminal strand relative to the sheet and active site loop. After undergoing this motion for some time, the

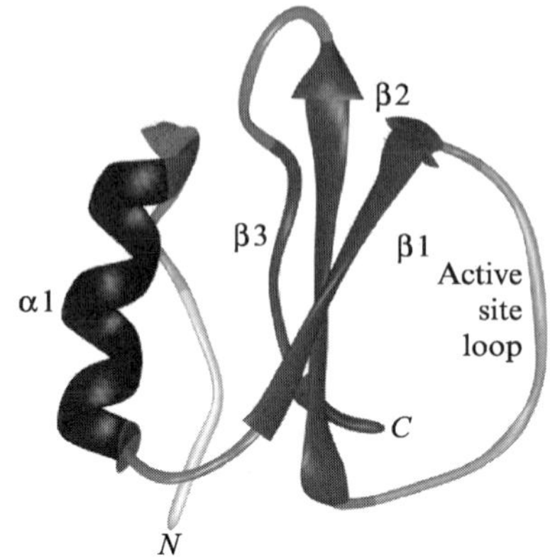

FIG. 5. The structure of chymotrypsin inhibitor 2 (Harpaz *et al.*, 1994) colored by the order of structure loss.

protein reached the unfolding transition state, characterized by a weakened hydrophobic core and some loss of secondary structure. Once it passed through the transition state, the protein core rapidly became fully disrupted and the active site loop was highly distorted. In time, the protein reached a fully denatured state in which there was virtually no native structure, although fluctuating secondary and non-native tertiary structures were populated. At lower temperatures, a sliding motion of the helix relative to the sheet became more prevalent in the pre-transition state ensemble, but the primary effect of lowering the temperature was to increase the length of time that the protein spends in each of these states. Multiple 20-ns simulations of the denaturation of CI2 in 8 M urea are consistent with the thermal denaturation simulations (Bennion and Daggett, 2003).

The earlier work simulating wild type CI2 at extreme temperatures (498 K) showed extensive agreement between the denatured and transition states of CI2 by simulation and experiment. The denatured state was seen to be highly disordered, in agreement with NMR studies, although portions of the native helix were transiently formed (Kazmirski *et al.*, 2001), suggesting that folding does not take place from a completely random coil. The transition state, identified from conformational clustering of the denaturation trajectory, is quite native-like. A simulated Φ-value, Φ_{MD}, can be calculated based on the ratio of the number of contacts made with a given residue in the native and transition states (Li and Daggett, 1994, 1996). This value was calculated for the transition states of four unfolding trajectories and compared to the experimental Φ-values. The correlation coefficients for the individual trajectories ranged from 0.80

to 0.93, with the average values from all four trajectories giving an even better correlation of 0.94. While the time scale of protein folding from a random coil precludes direct simulation, refolding from this transition state takes place quite rapidly (De Jong *et al.*, 2002). Structures taken before the transition state became significantly more native-like within 3 ns when simulated at a temperature below the protein's melting temperature, whereas structures taken from after the transition state became less native under the same conditions. The principle difference between the two types of structures was seen to be the extent and nature of the protein's hydration. While similar numbers of waters were present in the protein core both before and after the transition state, the waters present after the transition state were strongly bonded to the protein backbone and therefore harder to expel than those present before the transition state.

In 1997, Lazaridis and Karplus performed a set of simulations of C12 that serve as a bridge between highly simplified models and all-atom, explicit solvent simulations (Lazaridis and Karplus, 1997). This set of 24 simulations was performed using an implicit solvent representation at 500 K, allowing the protein to denature in less than 200 ps of simulation time. The simulations show that, while the specific contacts lost as the protein progresses from the native to denatured state appear random, there is an order to which regions of the protein become unstructured first. It was observed that the N-terminus of the protein pulls away first, followed by disruption of the core with an associated loss of secondary structure in the helix and sheet. This is essentially consistent with the observations of the unfolding of C12 using explicit water at a range of temperatures.

3. Protein A

Fragment B of protein A (Fig. 6) served as the initial test system for a method of obtaining thermodynamic levels of sampling developed by Brooks and co-workers (Boczko and Brooks, 1995; Guo *et al.*, 1997) and described earlier. Boczko and Brooks describe folding using the radius of gyration as a reaction coordinate. It should be noted that, while this description is in the folding direction, the simulations essentially start from the native state. At high radii of gyration (unfolded), they see helices I and II as partially formed. As the radius of gyration decreases, helices I and II begin to pack together and the extent of secondary structure increases. The native topology and some fluctuating structure in helix III are observed with further decreases of radius of gyration. Finally, in the most compact regime, helix III is seen to be docked to helices and I and II, with helix I assuming its native orientation. A later study of the same

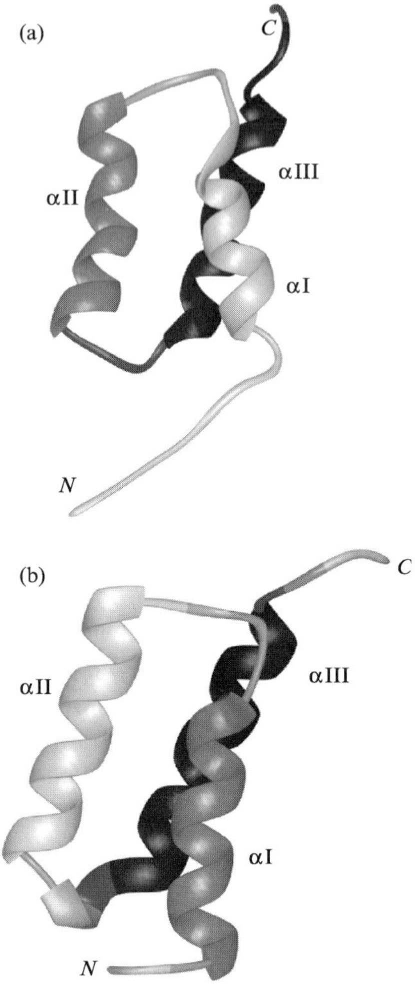

FIG. 6. The structure of staphylococcal protein A colored by the order of loss of structure seen in the simulations of Alonso and Daggett (2000). (A) Fragment B of protein A (Gouda *et al.*, 1992). (B) Fragment E of protein A (Starovasnik *et al.*, 1996).

trajectories (Guo *et al.*, 1997) defines the transition state as occurring after the initial collapse and formation of helices I and II. This transition has about 30% of the native tertiary structure formed and about 50% to 70% of the native secondary structure formed, mostly in helices I and II.

The unfolding of the B and E fragments of protein A was also studied by high temperature molecular dynamics simulation by Alonso and Daggett

(2000). The E fragment has essentially the same structure as the B fragment, and about 67% sequence identity exists between the two. Two simulations of each fragment were carried out at 498 K for 6 ns each. The observed unfolding pathway is quite different from that described by Brooks and co-workers. There is an initial loss of specific tertiary contacts, followed closely by a loss of secondary structure in helix I. Helix II loses its secondary structure at the same time as or slightly later than helix I. Helix III maintains the most structure and some residual structure is observed in the denatured state. The unfolding of the E domain is essentially the same, except that helix II tends to lose structure before helix I. Experimental studies of the helices in isolation and pairs supports this picture of the relative stabilities of the helices (Bai *et al.*, 1997). Hydrogen exchange studies indicate that helix I exchanges more rapidly than II or III under native conditions. This finding is consistent with the simulations of Alonso and Daggett but not those of Brooks and co-workers. The transition state as defined in these simulations is much more structured that that of Brooks and co-workers, with 63% to 76% of the native contacts formed. Experimental determination of the structure of the transition state (Φ-value analysis) and more definitive experiments could be quite useful in resolving and explaining these differences.

4. Ubiquitin

Ubiquitin (Fig. 7) was the subject of the first simulations exploring very early refolding events from a number of denatured and partially denatured structures (Alonso and Daggett, 1995, 1998). Two independent unfolding simulations were run at 498 K. The first unfolding event was a loss of core contacts. This was followed by the dissolution of the β-sheet. The order of loss of specific strands within the β-sheet was different in the two simulations, though early fraying of strands 1 and 2 was observed in both. The α-helix was the last element of structure to be lost. Structures taken from both simulations were simulated at 335 K to investigate refolding. Structures from early in the simulation, after the initial loss of core structure but before significant secondary structure loss, regained core contacts. In addition, the contacts stabilizing the secondary structure became more native-like. The α-carbon RMSD to the native structure decreased from about 5 Å to about 3.5 Å. Structures taken from after the β-sheet had begun to dissolve rapidly collapsed to the same radius of gyration as that of the native state. This collapse brought the N- and C-terminal strands back together and increased the number of contacts between the other strands. The β-sheet is still greatly expanded relative to the native state, however. The final class of starting structures contain a

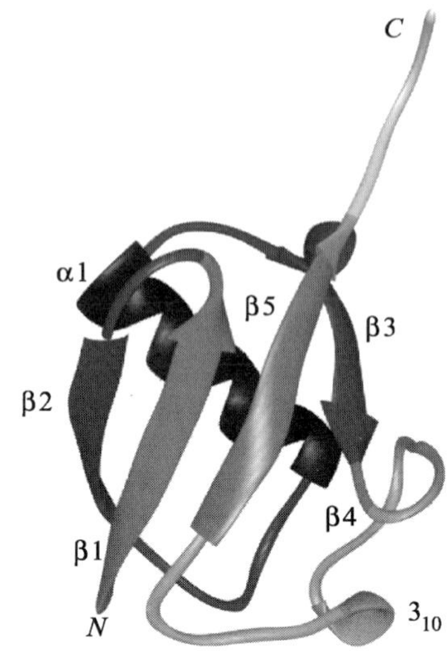

FIG. 7. The structure and order of unfolding of ubiquitin (Alexeev *et al.*, 1994).

highly disordered β-sheet, and in two cases, the helix is highly disrupted as well. These structures all showed rapid collapse as well, but this initial collapse was followed by repeated expansion and recollapse, with little or no native structure and contacts regained.

5. Protein G

The important sampling methods of Brooks and co-workers were also used in exploring the folding/unfolding landscape of segment B1 of streptococcal protein G (GB1) (Fig. 8a) (Sheinerman and Brooks, 1998a,b). Conformations were generated and projected into a two-dimensional reaction coordinate defined by the radius of gyration and fraction of native contacts. This surface suggests that the protein undergoes an early collapse, followed by a conformational search to form correct native contacts. The radius of gyration is constrained to near-native levels in all structures with more than 50% of the native contacts. The order of secondary structure loss is determined by looking in more detail at the fraction of native contacts formed. The specific contacts that tend to make up a given fraction are determined. In structures with a low fraction of native contacts formed, only local contacts are formed. The helix is the only element of secondary structure present. In structures with at least 50% of the native contacts formed, native contacts between the N- and C-terminal β-strands are formed and probably help constrain the protein's radius of gyration. The contacts corresponding to the rest of the β-sheet are formed in the

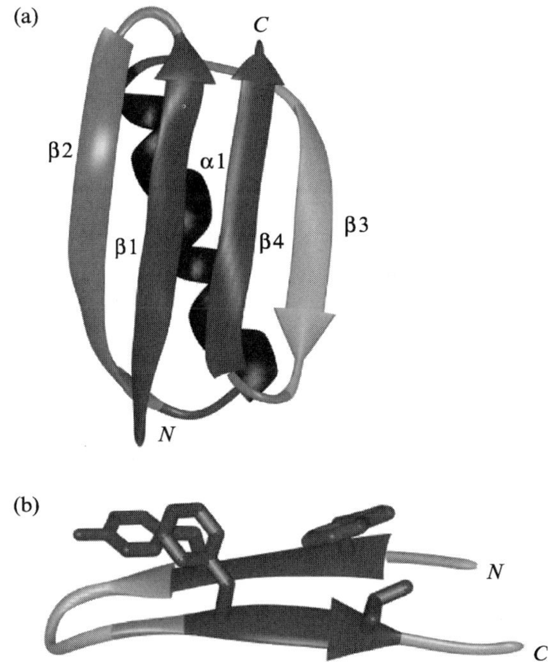

FIG. 8. The structure and order of unfolding of protein G (Gallagher *et al.*, 1994). (A) Order of loss seen in the full protein by simulation (Sheinerman and Brooks, 1998a,b). The order of loss suggested by experiment is quite different (see text and McCallister *et al.*, 2000) (B) Order of loss seen in the C-terminal hairpin. The sidechains making up the hydrophobic core are shown as sticks.

regime having at least 60% of the native contacts formed. Native tertiary contacts between the helix and sheet are formed only in the structures with very high fractions of native contacts. The helix makes more contacts with the N-terminal strands of the β-sheet, and these contacts are formed at a lower fraction of native contacts than the contacts between the helix and C-terminal strands. The formation of these last contacts requires the expulsion of all water molecules from the protein core. Experimental Φ-values paint a somewhat different picture (McCallister *et al.*, 2000). High Φ-values; suggesting native-like structure in the transition state, are observed primarily in the C-terminal hairpin. Next highest are those between the termini, followed by the helix and strand 2. This suggests that the C-terminal hairpin forms first, then the N-terminus and helix dock to it, with strand 2 becoming structured only very late in the folding.

Due to its experimentally observed stability, the second β-hairpin from protein G (Figure 8b) has become a popular subject of all-atom unfolding and folding studies. The fragment's short length makes it quite attractive for simulation because the conformational space that it can sample is much smaller than that of a full-sized protein. The fragment's unfolding was studied in detail by several groups (Roccatano *et al.*, 1999; Pande and Rokhsar, 1999; Lee and Shin, 2001). While the crystallographic coordinates from the full protein are not maintained in any simulations, low temperature controls show that the general hairpin structure is maintained. There is significant movement at the termini of the fragment and in the turn itself, but the small hydrophobic core and inner hydrogen bonds are sufficient to maintain the structure. Pande and Rokhsar characterize the unfolding as passing through four discrete intermediates. First is the folded form, the frayed hairpin described previously. This passes into an "H" state in which the secondary structure is lost, but the core remains intact. In the "S" state, the core becomes partially solvated, the hairpin topology remains, and the "U" state corresponds to a completely unfolded fragment. Multiple transitions are observed between these states, especially between H and S. F has the ability to transition directly to S, but this was an exceedingly rare event. Replica exchange methods are used to determine free energy surfaces for the folding/unfolding of the hairpin (Garcia and Sanbonmatsu, 2001; Zhou *et al.*, 2001). Both of these energy landscapes are quite rugged at low temperatures but smooth as the temperatures increase. Minima are seen corresponding to the F, H, and U state of Pande and Rokhsan (1999). Zhou *et al.* find an additional "P" minimum between F and H, which has fewer hydrogen bonds than F but still maintains more secondary structure than H. Finally, Pande and co-workers refolded the hairpin from a linear structure using a massively distributed approach and implicit solvent (Zagrovic *et al.*, 2001).. Their results again suggest that the hydrophobic core must be formed before a significant secondary structure is formed in the hairpin. It should be noted, however, that of eight putatively refolded structures, five fail to reproduce more than eight of thirteen structurally informative NOEs, with the other three reproducing all but two, three, and four NOEs respectively. Additionally, as discussed earlier, their method of refolding is biased toward a hydrophobic collapse mechanism.

6. *SH3 Domain*

The src SH3 domain (Fig. 9a) was the subject of an extensive molecular dynamics study (Tsai *et al.*, 1999) to complement experimental studies carried out by Baker and co-workers (Grantcharova and Baker, 1997; Grantcharova *et al.*, 2000). In this study, thirty independent simulations of 1 ns each were carried out at 498 K. It was seen that the order of loss and occupancy of individual contacts was quite different among the thirty

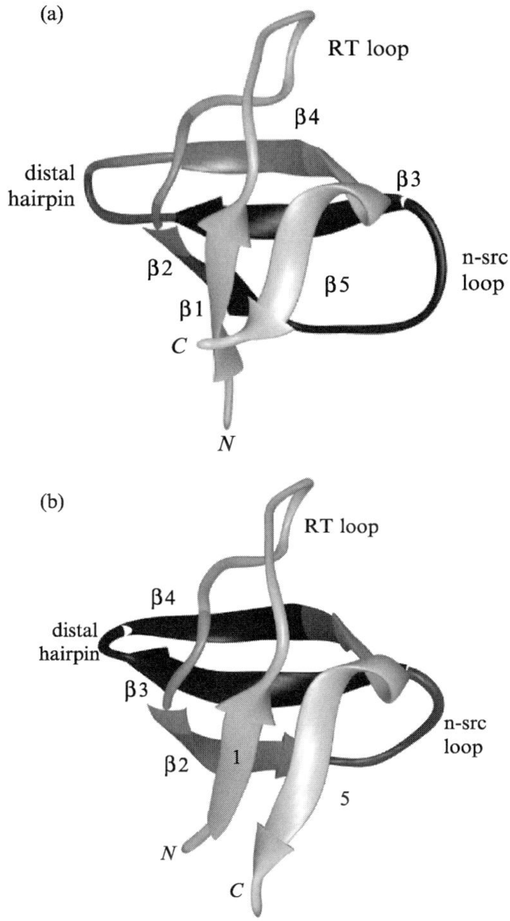

FIG. 9. The structure and order of unfolding of SH3 domains. (A) Src SH3 (Xu *et al.*, 1997). (B) α-spectrin SH3 (Musacchio *et al.*, 1992). The circular permutants discussed in the text are formed by joining the N- and C-termini with a short linker and breaking the chain at either the RT loop (s19) or the distal hairpin (n47).

trajectories. Thus, rather than analyze each of these simulations individually, the conformations generated in them were clustered according to the fraction of native contacts present. When this clustering was performed, a hierarchy of structure loss became apparent. When more than 75% of the native contacts are present, the contact loss is still fairly minimal and all major structures are still apparent. When the fraction of native contacts drops to the range of 50% to 75%, however, most of the contact loss is in the core of the protein and between the N and C termini. Dropping to 25% to 50% of the native contacts brings about a loss of contacts in the RT loop between strands 1 and 2, as well as some weakening of the n-src loop and distal hairpin. Finally, in the range of 0% to 25% of native contacts remaining, only the diverging turn and distal hairpin are significantly populated. These results agree well with the areas of structure seen in the Φ-value analysis of src SH3, though the stabilities of structural elements appear more polarized in the Φ-value analysis than the simulations suggest. It is suggested that this difference could be due to approximations in the potential function, insufficient sampling, or simply that the transition state was not singled out for analysis.

In order to more fully study the relationship between native topology and folding mechanism, Gsponer and Caflisch (2001) simulated the unfolding of both the src SH3 domain and α-spectrin SH3 domain (Fig. 9b), as well as two circular permutants of the α-spectrin SH3 domain. Denaturations were carried out at 375 K for six nanoseconds in implicit solvent. Forty such simulations were performed for each of the two native SH3 domains and twenty for each of the two circular permutants. As in the work of Tsai *et al.*, the number of native contacts was used as a reaction coordinate due to differences in the time scale of unfolding in individual trajectories. The description of the unfolding of src SH3 is also essentially the same as that given by Tsai *et al.* Contacts between the termini and in the core are the first to be lost, followed by the RT loop, with those in the three stranded β-sheet lost last. The unfolding of the α-spectrin SH3 domain follows the same order of events. The principle difference between the two is that in src SH3, contacts between strands 3 and 4 of the β-sheet were lost before those between 2 and 3, whereas in α-spectrin SH3 this order is reversed.

The circular permutants studied here are formed by joining the N- and C-termini with a short linker and breaking the protein in one of the loops. s19 is formed by breaking the protein between residues 19 and 20 in the RT loop. n47 is formed by breaking the protein between residues 47 and 48 in the distal hairpin. In both cases, the effects of these changes are primarily local, affecting only those β-strands that they directly join. Thus, the contacts between strands 1 and 5 are broken much later in both permutants. The cleavage of the RT loop in s19 did not lead to observable changes in the loss of the β-structure that it joins. Cleavage of the distal hairpin in n47 only led to changes in the order of loss of contacts up to two residues away from the site of cleavage. This second permutant also added some apparent stability to the contacts between strands 2 and 3. Thus, the overall picture that emerges from this work is that proteins with similar topologies unfold via similar mechanisms, but that those similarities are not strictly dependent upon the connectivity of the chain.

C. *Other Fast Folding Proteins*

1. *Villin*

Experimentalists' discovery of several small, very fast folding and unfolding protein domains, combined with faster computers and simulation techniques, has led to a significant narrowing of the gap between experimental timescales and those accessible to simulation. The first and best known of these was a full microsecond simulation of the villin headpiece subdomain (Fig. 10) in the folding direction (Duan and Kollman, 1998). The simulation was at least one order of magnitude slower than the 10 to 100 microseconds that the protein was believed to fold in, but it was two orders of magnitude longer than any other simulation at that time. It is still about ten times longer than the second longest single simulation reported. The simulation was started with an expanded conformation that was generated by a 1 ns simulation of the NMR structure at 1000 K. The simulation began with an initial collapse with concomitant formation of helical structures. This collapse took place over about 60 ns of simulation time and led to the formation of about 60% of the native helix and 45% of the native contacts. After this initial collapse, the protein samples a large number of conformations in an attempt to optimize tertiary packing. Most of these clusters are highly interconnected, but some, including a highly

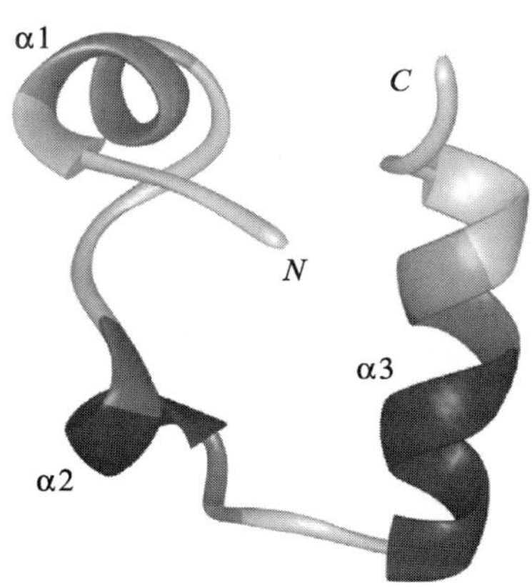

FIG. 10. Structure and order of formation of early structure in the villin headpiece subdomain (McKnight *et al.*, 1997).

populated apparent intermediate, are accessible only by a couple of pathways. As the simulation did not reach the native state, it is not clear whether this intermediate is on or off the folding pathway. Later calculations aimed at determining the volume of the conformational subspace that the protein was exploring suggested that the folding mechanism was a near-random search of this space (Sullivan and Kuntz, 2001).

2. *Engrailed Homeodomain*

The Engrailed Homeodomain protein (Fig. 11) is a 61 residue helical protein that has been observed to unfold with a rate constant of about 205,000 s^{-1} at 336 K. The unfolding half-life at 373 K is extrapolated from this value and others at lower temperature to be ~5 ns, which is well within the reach of molecular dynamics simulation (Mayor *et al.*, 2000). Simulations at 373 K yielded large structural deviations within 10 ns and an unfolding transition state at about 2 ns. In simulations at 373 K and 498 K, the unfolding was seen to begin with Helix III pulling away from the core, followed by a separation of Helices I and II. Much of the native helical secondary structure was maintained at both temperatures. The 498 K simulation is much less compact over the 40 ns of simulation time, and the helices fluctuate in and out of existence much more. Peptides based on these helices were observed to be remarkably stable (Mayor *et al.*, 2000). More recent work confirms the insensitivity of the pathway to temperature (Mayor *et al.*, 2003). Also, the unfolding time for the protein in 348 K simulations is in excellent agreement with the experiment, showing that the timescale separating theory and experiment has finally been bridged.

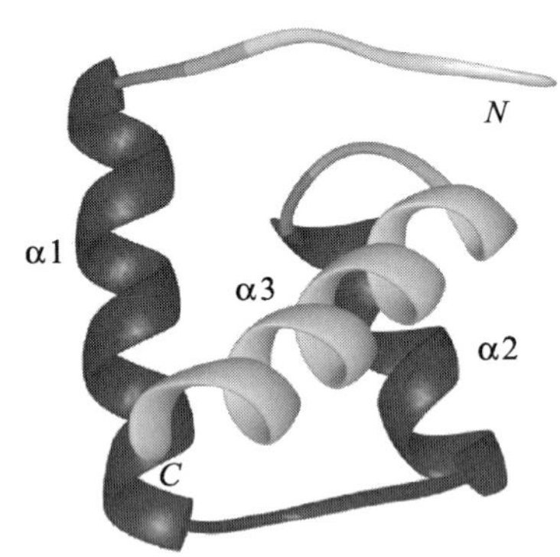

FIG. 11. The native structure and order of structure loss in the Engrailed Homeodomain protein (Clarke *et al.*, 1994). The coloring is based on the order of tertiary contact loss seen at both 373 K and 498 K.

3. WW Domain

The WW domain, a 44 residue, three strand β-sheet protein (Fig. 12), also unfolds rapidly enough to be studied by molecular dynamics at experimentally accessible temperatures. Its unfolding was simulated at 333 K, 348 K, and 373 K (Ferguson *et al.*, 2001). Unfolding events were essentially the same at all temperatures. The C-terminal strand was the first to become unstructured and pull away from the other two strands. Strands 1 and 2 fray and reform at their ends before finally pulling away, leading to what is, essentially, a random coil. This last step in the dissolution of the protein's secondary and tertiary structure is not observed in the 34 ns of simulation time at 333 K.

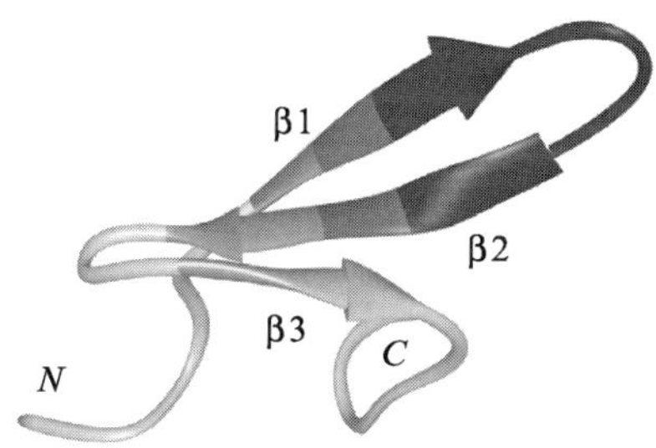

FIG. 12. Structure and order of unfolding of the WW domain (Macias *et al.*, 2000).

III. CONCLUSIONS/SUMMARY

A. Proteins Unfold by Pathways

The results of molecular dynamics simulations of protein unfolding suggest a unification of the funnel and pathway views of protein folding (Lazaridis and Karplus, 1997; Sheinerman and Brooks, 1998b; Tsai *et al.*, 1999; Day *et al.*, 2002). The funnel-shaped free energy surface described by statistical mechanics and simplified protein models is a robust and accurate model for the energy surface of random heteropolymers. What is seen in the molecular dynamics simulations is that there is a progression of structure gain/loss not unlike the pathways originally described by Levinthal (Levinthal, 1968). Some regions of the native protein are very likely to become disordered before others. This observation is entirely consistent with the funnel view. For instance, as chymotrypsin inhibitor 2 is pushed up the free energy funnel towards less native-like conformations, it is very likely that the first conformations that it will sample will be quite native-like, having only lost the interactions between the N- and C-termini. It will then proceed through a series of conformations in which the hydrophobic core has been significantly weakened or lost before losing its secondary structure. While the funnel landscape allows for conformations in which the loss of the β-sheet precedes significant weakening of the core, the topography of the surface is such that these conformations are only rarely, if ever, explored. It is worth noting that no such hierarchy is apparent for the loss of specific contacts (Tsai *et al.*, 1999, Day *et al.*, 2002). Taking again the example of CI2, while the β-sheet will not become significantly disrupted until after the core is lost, a given contact between the first two strands may be lost before a given contact between the β-sheet and α-helix. Thus, the order of loss of structure has less to do with the strength of a specific contact and more to do with the likelihood that all the contacts within a region will be broken. It is as much a function of the number of contacts as the type of contacts.

A key assumption that is typically invoked in describing molecular dynamics simulations of protein unfolding is a ''principle of microscopic reversibility,'' stating that unfolding is essentially the reverse of the folding process. Describing unfolding as taking place in a statistically preferred pathway on an energy funnel allows us to formulate a physical rationale for this principle. The energy surface is a map of conformational space and is the same whether it is being traversed from the folded state to the unfolded state. The statistically favored sequence of unfolding events is a preferred minimum route across this landscape. If the landscape does not change, the route does not change either. In many simulations, native or native-like secondary structures and contacts are present in the structures of the denatured ensemble. This residual structure likely serves as a nucleation point for the early steps in the folding process and limits the conformational space available to the unfolded chain. As Duan and Kollman (1998) illustrated, however, the conformational search from such structure is still a time-consuming process.

The objection has been raised that the extreme temperatures used in many unfolding simulations do in fact distort the landscape. Additionally, simulations using a simplified model showed that high temperature simulations tended to explore conformations with significantly fewer native contacts and a higher radius of gyration than the conformations populated at lower temperatures (Shea and Brooks, 2001). Analysis of the order of events in unfolding, however, suggests that the process itself is essentially conserved across temperature (Mayor *et al.*, 2000, 2003; Ferguson *et al.*, 2001; Day *et al.*, 2002). It comes as no surprise that a greater number of contacts are broken and the protein is more expanded at higher temperatures. The similarities between the preferred pathways of unfolding at different temperatures, however, point to the limitations of globally defined reaction coordinates.

Two caveats should be kept in mind when considering the folding/unfolding pathways described by simulation. The first is that these pathways are inherently statistical in nature. Thus, multiple simulations are necessary in order to fully elucidate the pathway. In the case of BPTI, the simulations revealed two distinct pathways, one of which was preferred at neutral pH and the other at low pH. Both were present under both conditions, however, and multiple simulations were necessary to estimate which was preferred. In other cases, such as those of protein A and hen egg white lysozyme, different pathways were determined by different methods. The second caveat is that a pathway is only as good as the force field and protocols used to generate it. A force field and protocol must be shown to give results that are in agreement with available experimental information before more speculative ventures are undertaken. Detailed information regarding the structure of the transition state of folding and unfolding is now available for several proteins. Additionally, the results of simulations can be compared with information determined in hydrogen exchange and NMR experiments on full proteins, as well as experiments based on protein fragments. A suitably sized ensemble of simulations should reproduce and explain the observations made in these types of experiments.

B. *Direct Simulation of Protein Folding Is in Its Infancy*

Duan and Kollman's ambitious protein refolding simulation is still by far the longest continuous all-atom molecular dynamics simulation published. It is unlikely that it will lose that status in the near future. While processor speeds have increased steadily since 1998 and show signs that they will continue to do so, the simulation was one or two orders of magnitude too short to observe the folding process in its entirety. Additionally, as discussed earlier, truly meaningful data are best obtained from several simulations of the same system. A single folding trajectory that reached the true native ensemble from an extended or expanded conformation would go far towards justifying molecular dynamics force fields and protocols. Nevertheless, multiple simulations would have to be performed, however, before a believable folding pathway could be constructed. Such a feat is not feasible, given current algorithms and available processor speeds.

The first method used to get around this limitation in simulation time while still maintaining all-atom representations of the protein and solvent is to simulate refolding from various locations on the unfolding pathway.

Rapid, non-specific collapse is seen in structures from all points in the unfolding trajectory. Starting structures that are closer to the native state have longer-lived collapsed states (Alonso and Daggett, 1995, 1998). Structures taken from before the unfolding transition refold to the native structure, while those taken from after the unfolding transition do not (De Jong *et al.*, 2002). Structures taken from within a transition state ensemble have about a 50% chance of refolding rapidly (Gsponer and Caflisch, 2002). Expelling water from water-mediated contacts in native-like structures appears to be a significant hindrance to rapid refolding (De Jong *et al.*, 2002).

Another method of discontinuously refolding proteins that is gaining considerable attention is the massively distributed method employed by Pande and co-workers (Shirts and Pande, 2001; Zagrovic *et al.*, 2001). As this method is started from an extended structure, it can potentially be used in structure prediction. It also makes remarkably good use of otherwise underutilized resources. Care must be taken in interpreting the folding pathways from these simulations, however. The heat capacity changes in protein folding, used to detect transitions in the simulations, are primarily dependent on hydrophobic burial. Thus, it is probable that the pathways from these simulations will be biased towards hydrophobic collapse and that the simulations have not been adequately validated by comparison with experimentation. In addition, Fersht (2002) has recently pointed out other fundamental flaws to this approach.

C. *The Future of Unfolding Simulations*

Using modern technology, an unfolding simulation can be carried out on a single processor in a week or less. This provides us with the opportunity to truly begin to explore fold space. Rather than enumerating the unfolding pathways of single proteins, we are now poised to compare the unfolding pathways of proteins of varying degrees of similarity. By comparing the data from multiple simulations of proteins with the same fold but different sequences, we can hope to gain further insight into the sequence-structure relationships that are the essence of the protein folding problem. Having established that there is a progression of structure loss in protein unfolding, we must now attempt to understand and predict that progression. Simple topological constraints such as reduced contact order and number of long- and short-range contacts have provided quantities that correlate fairly well with folding rates and experimentally determined Φ-values. Topological constraints will no doubt also prove useful in predicting the relative stabilities of regions of structure within a native fold. In order to fully predict structural hierarchies, it will be necessary to consider the types and number of contacts, both native and non-native, that stabilize that piece of structure. In order to do this in a statistically relevant fashion, it will be necessary to simulate many examples of the same substructure with different sequences and from proteins with different folds.

Acknowledgments

We are grateful for financial support from the National Institutes of Health (GM 50789 to VD and GM 08268 via an Institutional Molecular Biophysics Training Grant to RD).

References

Alexeev, D., Bury, S. M., Turner, M. A., Ogunjobi, O. M., Muir, T. W., Ramage, R., and Sawyer, L. (1994). Synthetic, structural and biological studies of the ubiquitin system: Chemically synthesized and native ubiquitin fold into identical three-dimensional structures. *Biochem. J.* **299,** 159–163.

Alonso, D. O. V., and Daggett, V. (1995). Molecular dynamics simulations of protein unfolding and limited refolding: Characterization of partially unfolded states of ubiquitin in 60% methanol and in water. *J. Mol. Biol.* **247,** 501–520.

Alonso, D. O. V., and Daggett, V. (1998). Molecular dynamics simulations of hydrophobic collapse of ubiquitin. *Prot. Sci.* **7,** 860–874.

Alonso, D. O. V., and Daggett, V. (2000). Staphylococcal protein A: Unfolding pathways, unfolded states, and differences between the B and E domains. *Proc. Natl. Acad Sci. USA* **97,** 133–138.

Anfinsen, C. B. (1973). Principles that govern the folding of protein chains. *Science* **181,** 223–230.

Bai, Y., Karimi, A., Dyson, J., and Wright, P. E. (1997). Absence of a stable intermediate on the folding pathway of protein *A.*. *Prot. Sci.* **6,** 1449–1457.

Baldwin, R. L. (1995). The nature of protein folding pathways: The classical versus the new view. *J. Biomol. NMR.* **5,** 103–105.

Bennion, B. J., and Daggett, V. (2003). The molecular basis for the chemical denaturation of proteins by urea. *Proc. Natl. Acad. Sci. USA* **100,** 5142–5147.

Boczko, E. M., and Brooks, C. L.III (1995). First-principles calculation of the folding free energy of a three-helix bundle protein. *Science* **269,** 393–396.

Bond, C. J., Wong, K. B., Clarke, J., Fersht, A. R., and Daggett, V. (1997). Characterization of residual structure in the thermally denatured state of barnase by simulation and experiment: description of the folding pathway. *Proc. Natl. Acad. Sci. USA* **94,** 13409–13413.

Brooks, C. L.III (1992). Characterization of ''native'' apomyoglobin by molecular dynamics simulation *J. Mol. Biol.* **227,** 375–380.

Caflisch, A., and Karplus, M. (1994). Molecular dynamics simulation of protein denaturation: Solvation of the hydrophobic cores and secondary structure of barnase. *Proc. Natl. Acad. Sci. USA* **91,** 1746–1750.

Caflisch, A., and Karplus, M. (1995). Acid and thermal denaturation of barnase investigated by molecular dynamics simulations. *J. Mol. Biol.* **252,** 672–708.

Caflisch, A., and Karplus, M. (1999). Structural details of urea binding to barnase: A molecular dynamics analysis. *Structure* **7,** 477–488.

Clarke, N. D., Kissinger, C. R., Desjarlais, J., Gilliland, G. L., and Pabo, C. O. (1994). Structural studies of the engrailed homeodomain. *Protein Sci.* **3,** 1779–1787.

Daggett, V., and Levitt, M. (1992). A model of the molten globule state from molecular dynamics simulations. *Proc. Natl. Acad. Sci. USA* **89,** 5142–5146.

Daggett, V., and Levitt, M. (1993). Protein unfolding pathways explored through molecular dynamics simulations. *J. Mol. Biol.* **232,** 600–619.

Daggett, V., Li, A., Itzhaki, L. S., Otzen, D. E., and Fersht, A. R. (1996). Structure of the transition state for folding of a protein derived from experiment and simulation. *J. Mol. Biol.* **257**(2), 430–440.

Daggett, V., Li, A., and Fersht, A. R. (1998). Combined molecular dynamics and Φ-value analysis of structure-reactivity relationships in the transition state and unfolding pathway of barnase: Structural basis of Hammond and anti-Hammond effects. *J. Amer. Chem. Soc.* **120,** 12740–12754.

Day, R., Bennion, B. J., Ham, S., and Daggett, V. (2002). Increasing temperature accelerates protein unfolding without changing the pathway of unfolding. *J. Mol. Biol.* **322,** 189–203.

De Jong, D., Riley, R., Alonso, D. O. V., and Daggett, V. (2002). Probing the energy landscape of protein folding/unfolding transition states. *J. Mol. Biol.* **319,** 229–242.

Deisenhofer, J., and Steigemann, W. (1975). Crystallographic refinement of the structure of bovine pancreatic trypsin inhibitor at 1.5 Å resolution. *Acta Cryst. B.* **31,** 238–250.

Dill, K. A., and Chan, H. S. (1997). From Levinthal to pathways to funnels. *Nat. Struct. Biol.* **4,** 10–19.

Dobson, C. M., Evans, P. A., and Radford, S. E. (1994). Understanding how proteins fold: The lysozyme story so far. *Trends Biochem. Sci.* **19,** 31–37.

Dobson, C. M., Sali, A., and Karplus, M. (1998). Protein folding: A perspective from theory and experiment. *Angew. Chem. Int. Ed.* **37,** 868–893.

Duan, Y., and Kollman, P. A. (1998). Pathways to a protein folding intermediate observed in a 1-microsecond simulation in aqueous solution. *Science* **282,** 740–744.

Ferguson, N., Pires, J. R., Toepert, F., Johnson, C. M., Pan, Y. P., Volkmer-Engert, R., Schneider-Mergener, J., Daggett, V., Oschkinat, H., and Fersht, A. R. (2001). Using flexible loop mimetics to extend Φ-value analysis to secondary structure interactions. *Proc. Natl. Acad. Sci. USA* **98,** 13008–13013.

Fersht, A. R., and Daggett, V. (2002). Protein folding and unfolding at atomic resolution. *Cell* **108,** 573–582.

Gallagher, P., Alexander, P., Bryan, P., and Gilliland, G. L. (1994). Two crystal structures of the B1 immunoglobulin-binding domain of streptococcal protein G and comparison with NMR. *Biochemistry* **33,** 4721–4729.

Garcia, A. E., and Sanbonmatsu, K. Y. (2001). Exploring the energy landscape of a β-hairpin in explicit solvent. *Proteins* **42,** 345–354.

Gō, N. (1983). Theoretical studies of protein folding. *Annu. Rev. Biophys. Bioeng.* **12,** 183–210.

Gouda, H., Torigoe, H., Saito, A., Sato, M., Arata, Y., and Shimada, I. (1992). Three-dimensional solution structure of the B domain of staphylococcal protein A: comparisons of the solution and crystal structures. *Biochemistry* **31,** 9665–9672.

Grantcharova, V. P., and Baker, D. (1997). Folding dynamics of the src SH3 domain. *Biochemistry* **36,** 15685–15692.

Grantcharova, V. P., Riddle, D. S., and Baker, D. (2000). Long-range order in the src SH3 folding transition state. *Proc. Natl. Acad. Sci. USA* **97,** 7084–7089.

Griko, Y. V., Privalov, P. L., Venyaminov, S. Y., and Kutyshenko, V. P. (1988). Thermodynamic study of the apomyoglobin structure. *J. Mol. Biol.* **202,** 127–138.

Gsponer, J., and Caflisch, A. (2001). Role of native topology investigated by multiple unfolding simulations of four SH3 domains. *J. Mol. Biol.* **309,** 285–298.

Gsponer, J., and Caflisch, A. (2002). Molecular dynamics simulations of protein folding from the transition state. *Proc. Natl. Acad. Sci. USA* **99,** 6719–6724.

Guo, Z., Brooks, C. L.III, and Boczko, E. M. (1997). Exploring the folding free energy surface of a three-helix bundle protein. *Proc. Natl. Acad. Sci. USA* **94,** 10161–10166.

Harpaz, Y., elMasry, N., Fersht, A. R., and Henrick, K. (1994). Direct observation of a better hydration at the N terminus of an α-helix with glycine rather than alanine as the N-cap residue. *Proc. Natl. Acad. Sci. USA* **91,** 3–15.

Hünenberger, P. H., Mark, A. E., and van Gunsteren, W. F. (1995). Computational approaches to study protein unfolding: Hen egg white lysozyme as a case study. *Proteins* **21,** 196–213.

Itzhaki, L. S., Otzen, D. E., and Fersht, A. R. (1995). Detailed structure of the transition state for folding of chymotrypsin inhibitor 2 analyzed by protein engineering. *J. Mol. Biol.* **254,** 260–288.

Itzhaki, L. S., Neira, J. L., and Fersht, A. R. (1997). Hydrogen exchange in chymotrypsin inhibitor 2 probed by denaturants and temperature. *J. Mol. Biol.* **270,** 89–98.

Jackson, S. E., elMasry, N., and Fersht, A. R. (1993). Structure of the hydrophobic core in the transition state for folding of chymotrypsin inhibitor 2: A critical test of the protein engineering method of analysis. *Biochemistry* **32,** 11270–11278.

Kazmirski, S. L., and Daggett, V. (1998). Simulations of the structural and dynamical properties of denatured proteins: The "molten coil" state of bovine pancreatic trypsin inhibitor *J. Mol. Biol.* **277,** 487–506.

Kazmirski, S. L., and Daggett, V. (1998). Non-native interactions in protein folding intermediates: Molecular dynamics simulations of hen lysozyme. *J. Mol. Biol.* **284,** 793–806.

Kazmirski, S. L., Wong, K.-B., Freund, S. M. V., Tan, Y. J., Fersht, A. R., and Daggett, V. (2001). Protein folding from a highly disordered denatured state: The folding pathway of chymotrypsin inhibitor 2 at atomic resolution. *Proc. Natl. Acad. Sci. USA* **98,** 4349–4354.

Kiefl, C., Sreerama, N., Haddad, R., Sun, L., Jentzen, W., Lu, Y., Qiu, Y., Shelnutt, J. A., and Woody, R. W. (2002). Heme distortions in sperm-whale carbonmonoxy myoglobin: Correlations between rotational strengths and heme distortions in MD-generated structures. *J. Am. Chem. Soc.* **124,** 3385–3394.

Killick, T. R., Freund, S. M. V., and Fersht, A. R. (1998). Real-time NMR studies on folding of mutants of barnase and chymotrypsin inhibitor 2. *FEBS Lett.* **423,** 110–112.

Lazaridis, T., and Karplus, M. (1997). "New view" of protein folding reconciled with the old through multiple unfolding simulations. *Science* **278,** 1928–1931.

Lee, J., and Shin, S. (2001). Understanding β-hairpin formation by molecular dynamics simulations of unfolding. *Biophys. J.* **81,** 2507–2516.

Levinthal, C. (1968). Are there pathways for protein folding? *J. Chim. Phys.* **65,** 44–45.

Li, A., and Daggett, V. (1994). Characterization of the transition state of protein unfolding by use of molecular dynamics: Chymotrypsin inhibitor 2. *Proc. Natl. Acad. Sci. USA* **91**(22), 10430–10434.

Li, A., and Daggett, V. (1996). Identification and characterization of the unfolding transition state of chymotrypsin inhibitor 2 by molecular dynamics simulations. *J. Mol. Biol.* **257**(2), 412–429.

Li, A., and Daggett, V. (1998). Molecular dynamics simulation of the unfolding of barnase: Characterization of the major intermediate. *J. Mol. Biol.* **275,** 677–694.

McCallister, E. L., Alm, E., and Baker, D. (2000). Critical role of β-hairpin formation in protein G folding. *Nat. Struct. Biol.* **7,** 669–673.

Macias, M. J., Gervais, V., Civera, C., and Oschkinat, H. (2000). Structural analysis of WW domains and design of a WW prototype. *Nat. Struct. Biol.* **7,** 375–379.

McKnight, C. J., Matsudaira, P. T., and Kim, P. S. (1997). NMR structure of the 35-residue villin headpiece subdomain. *Nat. Struct. Biol.* **4,** 180–184.

Makarov, V. A., Andrews, B. K., Smith, P. E., and Pettitt, B. M. (2000). Residence times of water molecules in the hydration sites of myoglobin. *Biophys. J.* **79,** 2966–2974.

Mark, A. E., and van Gunsteren, W. F. (1992). Simulation of the thermal denaturation of hen egg white lysozyme: Trapping the molten globule state. *Biochemistry* **31,** 7745–7748.

Martin, C., Richard, V., Salem, M., Hartley, R., and Mauguen, Y. (1999). Refinement and structural analysis of barnase at 1.5 Å resolution. *Acta. Cryst. D.* **55,** 386–398.

Mayor, U., Johnson, C. M., Daggett, V., and Fersht, A. R. (2000). Protein folding and unfolding in microseconds to nanoseconds by experiment and simulation. *Proc. Natl. Acad. Sci. USA* **97,** 13518–13522.

Mayor, M., Guydosh, N. R., Johnson, C. M., Grossman, J. G., Sato, S., Jas, G. S., Freund, S. M. V., Alonso, D. O. V., Daggett, V., and Fersht, A. R. (2003). The complete folding pathway of a protein from nanoseconds to microseconds. *Nature* **421,** 863–867.

Musacchio, A., Noble, M., Pauptit, R., Wierenga, R., and Saraste, M. (1992). Crystal structure of a Src-homology 3 (SH3) domain. *Nature* **359,** 851–855.

Onuchic, J. N., Luthey-Schulten, Z., and Wolynes, P. G. (1997). Theory of protein folding: The energy landscape perspective. *Annu. Rev. Phys. Chem.* **48,** 545–600.

Pande, V. S., and Rokhsar, D. S. (1999). Molecular dynamics simulations of unfolding and refolding of a β-hairpin fragment of protein G. *Proc. Natl. Acad. Sci. USA* **96,** 9062–9067.

Roccatano, D., Amadei, A., di Nola, A., and Berendsen, H. J. C. (1999). A molecular dynamics study of the 41–56 β-hairpin from B1 domain of protein G. *Prot. Sci.* **8,** 2130–2143.

Shea, J. E., and Brooks, C. L.III (2001). From folding theories to folding proteins: A review and assessment of simulation studies of protein folding and unfolding. *Annu. Rev. Phys. Chem.* **52,** 499–535.

Sheinerman, F. B., and Brooks, C. L.III (1998). Calculations on folding of segment B1 of streptococcal protein G. *J. Mol. Biol.* **278,** 439–456.

Sheinerman, F. B., and Brooks, C. L.III (1998). Molecular picture of folding of a small α/β protein. *Proc. Natl. Acad. Sci. USA* **95,** 1562–1567.

Shirts, M. R., and Pande, V. S. (2001). Mathematical analysis of coupled parallel simulations. *Phys. Rev. Lett.* **86,** 4983–4987.

Starovasnik, M. A., Skelton, N. J., O'Connell, M. P., Kelley, R. F., Reilly, D., and Fairbrother, W. J. (1996). Solution structure of the E-domain of staphylococcal protein A. *Biochemistry* **35,** 15558–15569.

Sullivan, D. C., and Kuntz, I. D. (2001). Conformation spaces of proteins. *Proteins* **42,** 495–511.

Tirado-Rives, J., and Jorgensen, W. L. (1993). Molecular dynamics simulations of the unfolding of apomyoglobin in water. *Biochemistry* **32,** 3175–3184.

Tirado-Rives, J., Orozco, M., and Jorgensen, W. L. (1997). Molecular dynamics simulation of the unfolding of barnase in water and 8 M aqueous urea. *Biochemistry* **36,** 7313–7329.

Tsai, J., Levitt, M., and Baker, D. (1999). Hierarchy of structure loss in MD simulations of src SH3 domain unfolding. *J. Mol. Biol.* **291,** 215–225.

Vojtechovsky, J., Chu, K., Berendzen, J., Sweet, R. M., and Schlichting, I. (1999). Crystal structures of myoglobin-ligand complexes at near-atomic resolution. *Biophys. J.* **77,** 2153–2174.

Williams, M. A., Thornton, J. M., and Goodfellow, J. M. (1997). Modelling protein unfolding: Hen egg-white lysozyme. *Protein Eng.* **10,** 895–903.

Wilson, K. P., Malcolm, B. A., and Matthews, B. W. (1992). Structural and thermodynamic analysis of compensating mutations within the core of chicken egg white lysozyme. *J. Biol. Chem.* **267,** 10842–10849.

Wong, K., Clarke, J., Bond, C. J., Neira, J. L., Freund, S. M. V., Fersht, A. R., and Daggett, V. (2000). Towards a complete description of the structural and dynamic properties of the denatured state of barnase and the role of residual structure in folding. *J. Mol. Biol.* **296,** 1257–1282.

Xu, W., Harrison, S. C., and Eck, M. J. (1997). Three-dimensional structure of the tyrosine kinase c-Src. *Nature* **385,** 595–602.

Zagrovic, B., Sorin, E. J., and Pande, V. (2001). β-hairpin folding simulations in atomistic detail using an implicit solvent model. *J. Mol. Biol.* **313,** 151–169.

Zhou, R., Berne, B. J., and Germain, R. (2001). The free energy landscape for β-hairpin folding in explicit water. *Proc. Natl. Acad. Sci. USA* **98,** 14931–14936.

AUTHOR INDEX

A

Abagyan, R. A., 77, 90, 352
Abashkin, Y. G., 280
Abele, U., 175, 177
Abeygunawardana, C., 314, 315, 326
Abildgaard, F., 183, 212
Abola, E. E., 11
Abraham, F., 98
Abrahams, J., 213
Abseher, R., 75, 76
Acharya, P., 134
Achouak, W., 175
Achtman, M., 175
Adachi, K., 227
Adams, J. E., 321, 349
Adams, P. D., 3, 4, 6
Agard, D. A., 17, 18
Agarwal, P. K., 348
Aggeler, R., 230
Agrafiotis, D. K., 14
Agre, P., 201, 202, 205, 209
Ahlsen, G., 141
Ahn, K.-H., 298, 299
Akamine, P., 90, 106
Akiyama, S., 236
Aksimentiev, A., 216
Alagona, G., 31, 314, 326, 337
Alber, T., 314, 322, 338, 341, 356
Albertí, S., 177
Albery, W. J., 314, 326, 338, 348
Alden, R. A., 294
Alden, R. G., 257, 260
Alderton, M., 49
Alexander, P., 387
Alexeev, D., 386
Alexov, E., 90, 91
Alhambra, C., 268, 291, 348
Al-Laham, M. A., 321
Allen, M. P., 251, 252, 253
Allen, T. W., 169, 170, 171
Allinger, N. L., 28, 30, 39, 40, 45, 251, 257
Allison, W. S., 213
Alm, E., 387
Almagro, J. C., 75, 107, 108, 114
Almlf, M., 146, 147
Alonso, D. O. V., 3, 8, 381, 383, 384, 385, 392, 394, 396
Alston, W. C., 316, 320, 348
Altendorf, K., 212
Alterman, M., 141
Amadei, A., 388
Ambudkar, S. V., 202
Amicangelo, J. C., 47
Amyes, T. L., 318, 325, 334
Anderson, S., 18
Andersson, H. O., 141
Andres, J. L., 321
Andrews, B. K., 378
Andzelm, J., 67
Anfinsen, C. B., 371
Angevine, C. M., 219
Angyan, J. G., 46
Antoniou, D., 352
Antonov, D. M., 318, 325, 337
Antosiewicz, J., 90, 91, 94, 109
Aota, S., 236
Apostol, I., 18
Appleby, T. C., 291
Applequist, J., 49, 52, 54
Åqvist, J., 14, 110, 111, 128, 129, 133, 134, 135, 136, 137, 138, 139, 140, 141, 142, 144, 145, 146, 147, 154, 169, 254, 255, 256, 257, 262, 269, 281, 283, 291, 298, 299, 315, 326, 344, 352
Archontis, G., 110, 319
Ardelt, W., 18
Arinaminpathy, Y., 172
Armentrout, P. B., 47
Arnaout, M. A., 237

Arora, A., 174, 181, 182, 183
Artacho, E., 185
Ash, E. L., 297
Ashcroft, F. M., 160, 173
Askari, J. A., 233
Astrand, P. O., 55
Augustyns, K., 318, 325, 337
Aukhil, I., 233
Auton, T. R., 124, 142
Ayala, P. Y., 321

B

Baaden, M., 174
Bader, R. F. W., 49
Badilla-Fernandez, C., 232, 234
Bagheri, B., 90, 91, 109
Bahnson, B. J., 349
Bai, Y., 385
Baker, D., 4, 5, 387, 388, 390, 393, 394
Baker, J., 321, 338
Baker, N. A., 90, 91, 114
Bakowies, D., 266, 267, 268
Balabin, I., 216
Baldridge, K., 67
Baldwin, R. L., 372
Ban, N., 198
Banavali, N., 15
Banba, S., 99, 153
Bandyopadhyay, S., 164
Baneyx, G., 232
Banks, J. L., 40, 60
Baptista, A. M., 16
Barclay, M. T., 173
Barford, R. A., 136
Barnes, P., 58
Barone, V., 67, 320
Bartels, C., 64, 75
Bash, P. A., 88, 125, 266, 281, 313, 314, 316, 318, 326, 329, 338, 341
Bashford, D., 4, 9, 10, 14, 34, 36, 65, 68, 69, 70, 74, 75, 89, 94, 199, 260, 262, 319
Basran, J., 289
Bassolino, D., 165
Baudry, J., 197, 199
Baugh, L., 232
Bayas, M. V., 197
Bayly, C. I., 14, 15, 32, 33, 46, 134, 145
Beachy, M. D., 34, 38, 41, 61, 69, 266
Becker, O., 28
Beckstein, O., 172
Beglov, D., 68, 109
Belasco, J. G., 314, 337, 338
Belew, R. K., 102, 113, 114, 139
Belhadj, M., 105, 135
Bell, R. P., 347
Bellott, M., 36, 199, 319
Benderskii, V. A., 352
Benedí, V. J., 177
Benkovic, S. J., 276, 289
Bennett, C. H., 152
Bennion, B. J., 381, 382, 393, 394
Bentzien, J., 268
Berendsen, H. J. C., 39, 40, 58, 62, 63, 98, 124, 127, 134, 142, 145, 164, 165, 176, 179, 289, 344, 388
Berendzen, J., 379
Berens, P. H., 40
Berg, U., 40, 61
Berger, A., 144
Bergsma, J. P., 346
Berkowitz, M. L., 165
Berman, H. M., 170
Bernal, J. D., 62
Bernal-Uruchurtu, M. I., 58
Bernardo, D. N., 58
Berne, B. J., 10, 14, 16, 40, 50, 58, 60, 388
Bernèche, S., 47, 160, 164, 168, 169, 170
Bernstein, F., 170
Beroza, P., 64, 73, 95
Berry, E. A., 197
Bershadsky, A., 232
Beutler, T. C., 131
Beveridge, A. J., 279
Beveridge, D. L., 13, 72, 88, 98, 124, 125, 135
Bhandarkar, M., 199, 233
Bhat, T. N., 141, 143, 170
Biggin, P. C., 160, 163, 167, 168, 169, 170, 171, 172, 174
Bigler, T., 18
Billeter, S. R., 39, 40, 291, 348
Binkley, J. S., 321, 338
Birkenhager, R., 212
Bishop, T. C., 197, 198
Bjelic, S., 141, 154
Bjoersne, M., 141
Bjrkstén, J., 176, 178
Blackburn, G. M., 281
Blacklow, S. C., 314

Blaney, J. M., 12
Bleil, R. E., 103, 104
Bliznyuk, A., 169
Bloomberg, M. R. A., 264
Blouin, C., 20
Blow, D., 283
Board, J., 52
Bocker, J., 164
Bckmann, R. A., 223, 224
Boczko, E. M., 383, 384
Bode, K. A., 52
Boggs, J. E., 339
Bogusz, S., 164, 255
Bhm, H. J., 144, 145, 146, 202
Bolger, M. B., 135
Bond, C. J., 380
Bond, P., 160, 176, 181, 182, 183, 380
Bonham, N. M., 138
Bonhivers, M., 201, 202
Bonhomme, F., 134
Bonneau, R., 4, 5
Bopp, P., 164
Borchert, T. V., 352
Boresch, S., 129, 131
Borgis, D., 271, 290
Borgnia, M. J., 201, 202, 209
Born, M., 69
Bosshard, H. R., 7
Bourne, P. E., 170
Bouzida, D., 43, 152, 153
Bouzide, A., 138
Boxer, S. G., 16
Boyer, P. D., 211, 213
Braisted, A. C., 112
Brandsdal, B. O., 137, 141, 144, 145, 146, 147
Brdarski, S., 40, 55, 61
Breneman, C. M., 46
Breuker, K., 18
Brew, S., 236
Brewer, M. L., 209
Brice, M., 170
Brickmann, J., 164
Briggs, J. M., 90, 91, 109, 114, 115
Bright, J. N., 160, 163, 168, 169
Brobjer, J. T., 48
Brooks, B. R., 35, 37, 164, 199, 215, 255, 316, 321, 339
Brooks, C. L., 3, 47, 94, 99, 100, 109, 151, 153
Brooks, C. L. III, 18, 34, 44, 68, 69, 70, 72, 74, 75, 254, 289, 290, 318, 374, 378, 383, 384, 386, 387, 393, 394
Brown, F. K., 125
Brownlee, M., 316
Bruccoleri, R. E., 35, 164, 199
Bruice, T. C., 276, 285
Brumer, P., 344
Bruneel, J. L., 134
Brünger, A. T., 3, 4, 6, 11, 65, 67, 105, 199
Brunner, R., 199, 233
Brusilow, W. S., 214
Bryan, P., 387
Buchanan, S. K., 175, 181
Buckingham, A. D., 46
Buczek, O., 143
Buning, C., 112
Buono, G. S., 260, 262
Burccoleri, R. E., 316, 321, 339
Burgess, A. W., 30
Burgi, R., 5, 6, 7
Burkert, U., 28, 30, 251, 257
Burnham, C. J., 55, 58
Bursulaya, B. D., 47, 75
Burt, S. K., 280
Bury, S. M., 386
Burykin, A., 258
Bush, B. L., 103, 113
Bushman, F. D., 114, 115
Bushnell, D. A., 198
Bushweller, J. H., 183

C

Cabral, J. M., 133, 167, 168
Cachau, R. E., 141, 143
Cadene, M., 167, 171
Caflisch, A., 7, 379, 381, 390, 396
Cahit, B. T., 167, 168
Calamita, G., 209
Caldwell, J. W., 10, 14, 15, 16, 32, 53, 56, 60, 134, 145, 164
Calimet, N., 72, 75
Cameron, C. E., 289
Cammi, R., 320
Campbell, J. D., 174
Cao, Y. X., 10, 14, 16, 40
Capaldi, R. A., 230
Capel, M. S., 198

Capener, C. E., 160, 163, 168, 169, 172, 173
Car, R., 185, 273
Caratzoulas, S., 352
Careri, G., 289
Carl, J. R., 52
Carloni, P., 169, 170, 185, 279, 280
Carlson, H. A., 110, 111, 114, 115, 140, 141
Carlson, M. J., 40, 50
Carrion-Vazquez, M., 232, 234
Carronde, M. A., 16
Carter, P., 294
Case, D. A., 4, 9, 10, 14, 31, 32, 34, 44, 64, 65, 68, 69, 70, 72, 73, 74, 75, 76, 89, 147, 148, 149, 164, 260
Casewit, C. J., 40, 50
Cassidy, C. S., 297, 298
Cavalli, A., 280
Ceccarelli, M., 44
Cedeno, W., 14
Cerami, A., 316
Chait, B. T., 133, 167, 171
Chalaoux, F. R., 11
Chan, H. S., 3, 372
Chandler, D., 231, 343
Chandrasekhar, J., 62, 63, 264
Chang, G., 173, 174
Chang, L. C., 314, 334, 337, 338, 340
Chang, T.-M., 53
Chapman, D., 34, 38, 41, 61
Charifson, P. S., 8
Chattopadhyaya, J., 134
Cheatham, T. E., 10, 147, 148, 149, 164
Cheatham, T. E. III, 255
Cheeseman, J. R., 321
Chen, J., 167, 171
Chen, J. J., 4, 18
Chen, J. L., 65
Chen, K., 30, 45
Chen, S. H., 164
Chen, W., 321
Chen, X., 18
Cheng, A., 205
Chepelinsky, A. B., 202
Cherepanov, D. A., 213, 214, 219, 224, 225, 231
Chessari, G., 48
Chiabrera, A., 90, 91
Chiang, Y. W., 18
Chimento, D. P., 175
Chin, K. J., 349
Chipot, C., 5, 34, 46
Chiu, S. W., 165
Choe, S., 167, 169, 170
Chong, L., 147, 148, 149
Chothia, C., 89, 318, 323
Chowdhury, S., 165
Christoffersen, R. E., 40
Chrzan, V., 339
Chu, K., 379
Chu, Z. T., 135, 259, 260, 262
Chuang, Y., 321
Chung, S. H., 163, 169, 170, 171
Churg, A. K., 258
Ciccotti, G., 344
Cieplak, P., 5, 10, 14, 15, 16, 32, 33, 34, 46, 53, 56, 60, 134, 145, 147, 148, 149, 150
Cipriano, D. J., 214
Civera, C., 393
Clark, M., 40
Clarke, J., 380
Clarke, N. D., 392
Classon, B., 138, 141
Claussen, H., 112
Clayton, S., 321
Cleland, W. W., 293, 294, 296, 297, 299, 315
Clementi, E., 58
Clemons, W. M., Jr., 198
Clore, G. M., 3, 4, 6
Cohen, B. E., 16
Cohen, S. L., 133, 167, 168
Coitino, E. L., 321
Colby, T. D., 349
Collins, J., 141, 143
Collins, P. J., 141, 143
Colonna, F., 46
Coltrin, M. E., 321
Colwell, K. S., 40, 50
Connolly, M. L., 68
Cook, J., 18
Cooper, T. M., 4
Copper, R. A., 316
Corchado, J. C., 321, 348
Cornell, W., 5, 14, 15, 32, 33, 34, 46, 75, 76, 134, 145
Corongiu, G., 58
Cortes, D. M., 171
Cortis, C. M., 69

Cortona, P., 272
Cossi, M., 67, 320
Coulson, C. A., 294
Cowan, S. W., 177
Craig, D., 197, 232, 233, 234, 236
Cramer, C. J., 28, 64, 67, 68, 72
Cramer, P., 198
Cramer, R. D., 40
Crofts, A. R., 197
Crooks, G. E., 100
Crosby, J., 291
Cuello, L. G., 171
Cui, Q., 224, 266, 314, 318, 319, 320, 321, 328, 333, 334, 347, 348, 349, 350, 351, 352, 356
Cunningham, M. A., 268, 281

D

d. Grotthuss, C. J. T., 202
Dadlez, M., 143
Daggett, V., 3, 5, 40, 63, 375, 376, 377, 378, 380, 381, 382, 383, 384, 385, 392, 393, 394, 396
Dalke, A., 199, 201
Damjanovic, A., 197, 238
Damm, W., 28, 35, 50
Damodaran, K. V., 99
Dang, L. X., 53
Danielson, H., 141
Danielson, U. H., 138, 141
Darden, T., 3, 52, 199
Darden, T. A., 37
Dauber-Osguthorpe, P., 145
Daura, X., 4, 7, 8, 39
Davenport, R. C., 314, 316, 318, 322, 326, 329, 337, 338, 341, 356
David, C. W., 58
David, L., 68, 69, 94, 109
Davis, M. E., 90, 91, 109, 324
Day, R., 381, 393, 394
Debolt, S., 34, 164
Deckers-Hebestreit, G., 212
Deen, P. M. T., 202
Defrees, D. J., 321, 338
de Groot, B. L., 160, 163, 174, 202, 205, 209
Deisenhofer, J., 376
Dejaegere, A., 319
De Jong, D., 381, 383, 396
Dekker, N., 175
DeLano, W. L., 3, 4, 6
Del Buono, G. S., 105, 255
Derreumaux, P., 40, 324
Derrick, J. P., 175
Desjarlais, J., 392
Desmeules, P., 314
deSolms, S. J., 18
DeTar, D. F., 17
Devaney, J., 68
Devidas, S., 202
de Vries, A. H., 55
Dewar, M. J. S., 338, 339
DiCapua, F. M., 88, 125
Diederichs, K., 175
Diefenbach, B., 237
Dill, K. A., 3, 8, 13, 77, 372
Diller, D. J., 8
Diller, T., 90, 106
Ding, Y., 58
Dinner, A. R., 231, 281, 319
Di Nola, A., 111, 388
Dinur, U., 45
Dittrich, M., 238
Dixon, D. A., 59, 60
Dixon, R. W., 34, 48, 145, 146
Dmitriev, O. Y., 214, 219
Dobson, C. M., 371, 372, 377, 378
Dollinger, G., 18
Dolney, D. M., 67
Domene, C., 169, 170, 171, 173
Dominy, B. N., 75, 109
Dong, F., 91
Donini, O., 147, 148, 149
Dorsey, B. D., 18
Dougherty, D. A., 31, 174
Doyle, D. A., 133, 167, 168, 173
Duan, Y., 4, 107, 108, 114, 147, 148, 149, 165, 391, 394
Ducasse, L., 134
Dudek, M. J., 72
Dunbrack, R. L., Jr., 36, 319
Dunietz, B., 264
Dunker, R., 237
Dunn, S. H., 214
Dunning, T. H., 48
Durell, S. R., 173
Dutzler, R., 177, 180
Dykstra, C. E., 47, 49, 51, 52
Dyson, J., 385

E

Eastwood, M. P., 3
Eaton, W. A., 3
Eck, M. J., 389
Egert, E., 49
Egmond, M. R., 175
Eichinger, M., 273
Eisenberg, D., 3
Eisenmesser, E. Z., 289, 290
Elber, R., 98, 99, 151
Eldridge, M. D., 124, 142
Ellman, J. A., 8, 18
elMasry, N., 381, 382
Elmore, D. E., 174
Elsner, J., 319
Elstner, M., 40, 43, 45, 319, 321, 349, 356
Elston, T., 214, 216
Endo, H., 213, 225
Engel, A., 163, 174, 201, 202, 205, 209
Engelbrecht, S., 213, 214, 216, 219, 225, 231
Engelhardt, H., 177
Engelman, D. M., 163, 165
Engels, M., 223
Erickson, H. P., 232, 233
Erickson, J. W., 141, 143, 280
Eriksson, M. A. L., 150, 153
Erlanson, D. A., 112
Ermolaeva, M. D., 271
Ersmark, K., 141, 154
Essex, J. W., 46, 141
Euler, E., 30, 31
Evans, P. A., 377, 378
Evanseck, J. D., 36, 313, 319
Evleth, E., 46
Ewing, T. J. A., 8, 18

F

Faraldo-Gomez, J., 160, 165, 166, 176, 181, 182
Farid, F. F., 98
Fasella, P., 289
Fast, P. L., 321
Fauman, E. B., 10, 11
Feher, G., 95
Feierberg, I., 141, 142, 154, 291, 298, 299
Feig, M., 75
Feller, S. E., 164
Felts, A. K., 73, 75
Feng, Z., 170
Fennen, J., 39, 40
Ferenczy, G. G., 46
Ferguson, D. M., 32, 34, 63, 134, 145, 164
Ferguson, J., 14, 15
Ferguson, N., 393, 394
Fernandez, J., 232, 234
Ferrara, P., 7
Ferretti, J. A., 12, 15
Fersht, A. R., 3, 5, 274, 294, 313, 314, 375, 380, 381, 382, 392, 393, 394, 396
Fesik, S. W., 112, 114
Field, M. J., 36, 266, 268, 313, 314, 318, 319, 326, 329, 341
Figueirido, F., 255, 260, 262
Fillingame, R. H., 212, 214, 216, 219
Fincham, D., 51
Finney, J. L., 58
Fischer, S., 35, 36, 180, 319, 321
Fitzgerald, P. M., 18
Flament, J.-P., 61
Florian, J., 262, 271, 280, 284, 292
Flynn, T. C., 224
Fogarasi, G., 339
Foloppe, N., 15, 37
Ford, L. O. J., 285
Ford, M. G., 141
Foresman, J. B., 321, 338
Forrest, L. R., 164, 173, 179, 209
Forst, D., 175
Fothergill, M., 270, 281, 283, 314, 315, 326, 344, 352
Fotiadis, D., 163
Fowler, P. W., 46
Fowler, R. H., 62
Fox, D. J., 321, 338
Fox, T., 14, 15, 32, 34, 99, 134, 145
Francis, S. E., 141, 143
Franck, P., 103
Frankel, A., 5
Frauenheim, T., 319, 321, 349
Frech, M., 237
Fredkin, D. R., 95
Freire, E., 8, 18
Freire, S., 105
Freisner, R. A., 10, 14, 50, 69, 94
Freund, S. M. V., 380, 381, 382, 392, 394
Frey, P. A., 293, 294, 297, 298, 299, 315
Friedman, R. A., 65

Friesner, R. A., 7, 10, 14, 16, 34, 38, 40, 41, 43, 50, 58, 60, 61, 72, 110, 147, 164, 264, 266
Frisch, M. J., 321, 338
Froemmel, C., 72
Fu, D., 202, 205, 208
Fu, J. H., 198
Fujiyoshi, Y., 174, 202, 205
Fung, K.-K., 52
Futai, M., 210, 212, 213, 214, 224
Futatsugi, N., 279
Fuxreiter, M., 16, 281

G

Gale, J. D., 185
Gallagher, P., 387
Gallant, J., 214
Gallicchio, E., 65, 73, 75, 110
Gao, H., 36
Gao, J., 36, 266, 318, 319, 348, 349
Gao, M., 163, 197, 232, 233, 234, 236
Garcia, A. E., 35, 75, 115, 116, 185, 255, 388
Garcia-Viloca, M., 266, 268, 299
Garg, N., 141
Garrett, B. C., 320, 321
Garza, J., 59, 60
Gaub, H., 232
Gautel, M., 232
Gee, P., 4
Geiger, B., 232
Gelin, B. R., 198, 253
Genest, M., 145
Gennis, R. B., 165
Gerber, P. R., 131, 135, 136
Gerlt, J. A., 293, 294, 296, 299, 315
Germain, R., 388
German, E. K., 290
Gerstein, M., 318, 323
Gertner, B. J., 346
Gervais, V., 393
Ghio, C., 31, 314, 326, 337
Ghosh, A., 10, 14, 72, 94, 147
Ghosh, R., 202
Gibbons, C., 213, 225, 230
Gibson, B. W., 18
Gibson, K. D., 44
Gidofalvi, G., 102
Gill, P. M. W., 321
Gillan, M. J., 291
Gilliland, G. L., 170, 387, 392
Gilson, M. K., 9, 18, 68, 69, 90, 91, 94, 95, 103, 109, 113, 148
Girvin, M. E., 212, 214, 219
Given, J. A., 9, 103, 113
Glattli, A., 4
Glennon, T. M., 279, 280, 281
Gluzman, I. Y., 8, 141, 143
Gnatt, A. L., 198
Go, N., 31, 373
Goddard, W. A., 40, 50
Gogonea, V., 4, 5
Gohlke, H., 124
Goldberg, D. E., 8, 141, 143
Goldman, S., 339
Goldstein, B. M., 349
Gomperts, R., 321
Gonzalez, C., 321, 338
Gonzalez-Lafont, A., 266, 268, 299
Goodfellow, J. M., 378
Goodman, M. F., 262, 280
Goodman, S. L., 237
Goodsell, D. S., 102, 113, 114, 139
Gordon, E. M., 112
Gorfe, A. A., 7
Gorler, A., 17
Gouda, H. I., 9, 10, 384
Gough, C., 34
Gould, C., 106
Gould, I. R., 14, 15, 32, 134, 145
Gracy, R. W., 316, 325
Graffner-Nordberg, M., 111, 139
Grantcharova, V. P., 388
Gratton, E., 289
Grayson, P., 238
Gresh, N., 40, 60, 61, 62
Grigera, J. R., 63
Griko, Y. V., 378
Grimmelman, E. K., 343
Grinevich, P., 352
Gronbeck-Jensen, N., 17
Grondin, J., 134
Gros, P., 3, 4, 6, 175
Grosse-Kunstleve, R. W., 3, 4, 6
Grossfield, A., 165
Grossman, J. G., 392, 394
Groth, G., 214
Grubmüller, H., 160, 163, 174, 202, 205, 209, 223, 224

Gruebele, M., 3, 4, 5, 9, 12
Gsponer, J., 390, 396
Gu, M. Z., 321
Guarnieri, F., 98
Guggino, W. B., 202
Guidoni, L., 169, 170, 185
Guillemette, J. G., 20
Guillot, B., 49
Guissani, Y., 49
Gulbis, J. M., 133, 167, 168
Gullingsrud, J. R., 173, 197, 199
Gulnik, S. V., 141, 143
Gumbiowski, K., 214, 219, 225, 231
Gunner, M. R., 94
Guo, H., 40, 61, 62, 319
Guo, Z., 383, 384
Guo, Z. Y., 99, 100
Gursoy, A., 199, 233
Guthrie, J. P., 294, 296
Guy, H. R., 173
Guydosh, N. R., 392, 394

H

Ha, S., 36, 319
Haak, J. R., 98, 124
Habashita, H., 8
Haddah, R., 378
Hagler, A. T., 7, 17, 30, 31, 39, 40, 45, 103, 145
Hajduk, P. J., 112, 114
Halgren, T. A., 10, 14, 16, 28, 34, 38, 39, 40, 41, 50, 61
Hall, A., 329
Hallberg, A., 111, 138, 139, 141, 154
Halliday, R. S., 102, 113, 114, 139
Halperin, I., 155
Halren, T. A., 18
Ham, S., 381, 393, 394
Hamamoto, T., 213, 225
Hammes-Schiffer, S., 348
Han, H. G., 202, 205
Handy, N. C., 321, 349
Hansch, C., 14
Hansen, J., 198
Hansson, T., 58, 111, 129, 136, 137, 138, 139, 141, 142, 146, 147, 163, 257, 262, 281
Haque, T. S., 8
Hara, K. Y., 213
Hardin, C., 3
Hardmeyer, A., 179
Harpaz, Y., 382
Harris, P., 291
Harris, T. K., 314, 315, 316, 321, 326, 356, 362
Harrison, D. H. T., 316, 318, 321, 334, 356, 362
Harrison, S. C., 389
Hart, W. E., 102, 113, 114, 139
Hartley, R., 380
Hartman, F. C., 314, 322, 338, 341, 356
Harvey, S., 28
Hashitsume, N., 256
Häsler, K., 213
Haudenschild, C., 236
Haugk, M., 319
Havel, T. F., 6
Hawkins, G. D., 67, 72
Hawley, R. C., 16, 69, 70, 72, 92, 147, 152
Hay, B. P., 59, 60
Hayashi, S., 197
Haymet, A. D. J., 3, 13
Hayward, S., 289
Head, M. S., 9
Head-Gordon, M., 321, 338
Healy, E. A., 338, 339
Heginbotham, L., 169
Helfond, E., 343
Heller, H., 197
Heller, K. B., 202
Helms, V., 2, 4, 90
Hempel, J. C., 18
Hendlich, M., 124
Hendrickson, J. B., 251
Hendrickson, T., 16, 69, 70, 72, 92, 147, 152
Hendrickson, W. A., 11
Henrick, K., 382
Herlihy, J. M., 314
Hermans, J., 40, 43, 45, 62, 75, 107, 108, 114, 130, 225, 284
Hermida-Ramón, J. M., 40, 61
Hermolin, J., 212, 214
Hernández-Allés, S., 177
Hernández-Cobos, J., 58
Hess, B., 199
Heulin, T., 175
Heymann, J. B., 174, 201, 202, 205, 209
Heywood, G. C., 279
Higo, J., 4, 111

Hill, H. A. O., 16
Hille, B., 167
Hirai, T., 202, 205
Hirono-Hara, Y., 213
Hirshberg, M., 40, 63
Hoffman, L. R., 8
Hollinger, F. P., 92, 93, 109
Holloway, M. K., 18
Holst, M. J., 90, 91, 114
Holyoake, 171, 173
Hong, G., 268, 272, 273
Honig, B., 4, 7, 65, 68, 89, 90, 91, 94, 148, 258
Hopkins, S. C., 8
Hoppert, M., 212
Horn, D. M., 18
Houck, J., 339
Houk, K. N., 4, 291, 292
Hoyles, M., 170
Hu, H., 40, 43, 45
Hu, W., 321
Huang, N., 48
Huang, S. X., 175
Huang, Z. W., 16
Huber, T., 39, 40
Huey, R., 102, 113, 114, 139
Hughes, C., 175
Hultén, J., 138, 141, 154
Hummer, G., 209, 255
Humphrey, W., 199, 201
Humphries, M. J., 233
Hünenberger, P. H., 28, 30, 39, 40, 90, 106, 382
Hung, H., 68
Hunter, C. A., 48
Huo, S., 5, 10, 147, 148, 149
Huo, S. H.., 148, 149, 150
Hurley, J. H., 281
Husslein, T., 164
Hutchinson, K., 3, 8
Hwang, J.-K., 125, 257, 264, 270, 271, 281, 289, 291, 294, 349
Hwang, M.-J., 39, 40, 45
Hynes, J. T., 290, 346
Hynes, R. O., 232

I

Ilin, A., 90, 91, 109
Im, W., 68, 109, 160, 176, 178, 179, 180
Impey, R. W., 62, 63
Ingham, K., 236
Iordanov, T., 348
Isaacson, A. D., 313, 320, 321, 349
Israelachvili, J., 17
Isralewitz, B., 163, 197, 199, 232, 233, 234, 236
Itoh, H., 213
Itzhaki, L. S., 381
Iwamoto-Kihara, A., 212
Izrailev, S., 197

J

Jackson, J. D., 71
Jackson, S. E., 381
Jacobson, M. P., 7, 43
Jacoby, E., 223
Jaenicke, R., 352
Jain, A. N., 124
Jakobsson, E., 165
Jalkanen, K. J., 39, 40
James, M. N. G., 18
James, T. L., 7, 17
Jan, L. Y., 16, 173
Jan, Y. N., 16, 173
Janssen, P. A., 111
Jap, B. K., 202, 205
Jarzynski, C., 100, 208
Jas, G. S., 392, 394
Jayaram, B., 72, 135
Jelesarov, I., 7
Jencks, W. P., 278, 283, 313
Jensen, J. H., 296, 299
Jensen, M. O., 160, 174, 197, 198, 202, 205, 206, 207, 208, 209
Jentzen, W., 378
Jiang, J.-S., 3, 4, 6
Jiang, W., 212, 214, 219
Jiang, Y., 167, 171
Joachimiak, A., 237
Jogl, G., 318, 324, 325, 333, 334, 362
Johnson, B. G., 321
Johnson, C. M., 392, 393, 394
Johnson, E. T., 260
Jonas, A., 197
Jones, E. Y., 233
Jones, P. C., 212, 214, 219
Jones-Hertzog, D. K., 140, 141

Jonsson, B., 92, 103
Jorgensen, W. L., 3, 4, 10, 14, 31, 38, 39, 43, 45, 62, 63, 88, 110, 111, 114, 115, 135, 140, 141, 147, 164, 264, 378, 379, 381
Joseph, D., 313, 322, 324, 338
Joseph, S., 90, 91, 114
Joseph-McCarthy, D., 36, 319, 338, 342
Joubert, L., 49
Jung, J. S., 202
Junge, W., 213, 214, 216, 219, 224, 225, 231
Jungnickel, G., 319
Junquera, J., 185

K

Kadner, R. J., 175
Kafafi, S. A., 61, 62
Kagawa, Y., 213, 225
Kahn, L. R., 338
Kalé, L., 199, 200, 233
Kaminski, G. A., 10, 14, 16, 38, 39, 40, 43, 45, 60, 164
Kanska, M., 316, 320, 348
Karimi, A., 385
Karlen, A., 138
Karlstrm, G., 40, 55, 61
Karplus, M., 15, 35, 36, 64, 67, 68, 72, 75, 94, 98, 110, 129, 131, 151, 164, 176, 179, 180, 198, 199, 215, 224, 233, 234, 253, 254, 266, 281, 289, 290, 299, 313, 314, 316, 318, 319, 320, 321, 322, 324, 326, 328, 329, 333, 337, 338, 339, 341, 342, 343, 344, 347, 348, 349, 350, 351, 352, 356, 371, 372, 379, 381, 383, 393
Karshikoff, A., 177
Kato, I., 18
Kato-Yamada, Y., 213
Kaufman, A., 163, 168, 169
Kauzmann, W., 3
Kaxiras, E., 321, 349
Kazmirski, S. L., 376, 377, 378, 381, 382
Kebarle, P., 47
Keith, T., 321
Keller, D. A., 46
Keller, T. A., 175
Kelly, C., 18
Kelly, J. A., 17
Kennard, O., 170
Keseru, G. M., 112
Khaliq, S., 75
Khanjin, N. A., 285
Kick, E. K., 8, 18
Kidera, A., 111
Kiefl, C., 378
Kiehart, D. P., 232
Killick, T. R., 381
Kim, D., 280, 299
Kim, H. J., 172
Kim, K., 50, 280, 299
Kim, P. S., 391
Kim, Y., 298, 299
Kimatian, S. J., 164
King, G., 128, 136, 251, 254, 257, 259, 262
Kinosita, K., Jr., 213, 225, 227
Kirkwood, J. G., 73, 97, 124, 258
Kirsch, J. F., 299
Kissinger, C. R., 392
Kitao, O., 50
Kitchen, D. B., 105, 135
Kitson, D. H., 7, 17, 40
Kiyama, R., 18
Klamt, A., 65, 67
Klebe, G., 124
Klein, M. L., 58, 62, 63, 164, 185
Kleinekathoefer, U., 197, 238
Kleinschmidt, J. H., 174
Kleywegt, G. J., 11
Klinman, J. P., 289, 290, 347, 349
Klippenstein, S. J., 320
Klüner, T., 273
Knegtel, R. M., 8, 112
Knowles, J. R., 314, 326, 329, 334, 337, 338, 340, 341, 342, 348
Koch, U., 49
Koebnik, R., 174, 175, 177
Koetzle, T., 170
Kohen, A., 289, 290, 347
Kohr, W., 18
Kollman, P. A., 3, 4, 5, 7, 9, 10, 14, 15, 16, 31, 32, 33, 34, 43, 46, 48, 53, 56, 60, 88, 99, 107, 108, 110, 114, 125, 131, 134, 136, 138, 140, 145, 146, 147, 148, 149, 150, 152, 153, 164, 199, 251, 256, 259, 261, 262, 268, 283, 289, 296, 314, 326, 337, 391, 394
Klmel, C., 67

Kolmodin, K., 139, 142
Kolossvary, I., 112
Komeiji, Y., 176, 179
Komives, E. A., 314, 334, 337, 338, 340
Komiyama, T., 18
Kong, X. J., 151
Kong, Y., 54
Konnert, J. H., 11
Kornberg, R. D., 198
Koronakis, E., 175
Koronakis, V. A. S., 175
Korzeniewski, C., 339
Koshland, D. E., Jr., 300
Kosov, D. S., 49
Kosztin, D., 173, 197, 198, 199
Kosztin, I., 197, 238
Koubi, L., 164
Koumanov, A., 177
Kowalchyk, W., 339
Kozono, D., 209
Kramer, B., 112
Kramer, R. A., 175
Krammer, A., 232, 233, 234, 236
Krauss, M., 263
Krawetz, N., 199, 233
Kreevoy, M. M., 293, 294, 296, 297
Krimm, S., 40, 59, 61
Krishnan, S., 200
Krishtalik, L. I., 283
Kroeger Smith, M. B., 111
Krogh-Jespersen, K., 58
Krokoszynska, I., 143
Kroon, J., 175
Krowarsch, D., 143
Krucinski, J., 202, 205, 208
Krüger, P., 39, 40, 223
Kubo, R., 256
Kuchnir, L., 36, 319
Kuczera, K., 319
Kuharski, R. A., 257
Kuhn, B., 9, 10, 14, 107, 108, 110, 114, 147, 148, 149, 289
Kumar, S., 43, 65, 152, 153
Kuntz, D., 9, 10
Kuntz, I. D., 5, 8, 9, 10, 14, 18, 112, 148, 149, 150, 392
Kuo, A., 167, 168
Kuo, A. L., 133
Kursula, I., 318, 325, 337
Kuszewski, J., 3, 4, 6
Kutyshenko, V. P., 378
Kuwajima, S., 255
Kuyucak, S., 163, 169, 170, 171
Kvarnstrom, I., 138, 141

L

Laason, K., 185
LaBean, T. H., 11
Ladenstein, R., 177
Laidig, K. E., 40, 63
Lamb, M. L., 111, 135, 140, 141
Lambeir, A., 318, 325, 337
Lambert, M. H., 44
Langely, C. H., 39, 40
Langen, R., 280
Langlois, J.-M., 69
Langridge, R., 88, 125
Lappicirella, A., 265
Laskowski, M., 18
Laskowski, M., Jr., 18
Laskowski, R. A., 11
Lasségues, J. C., 134
Lau, F. T. K., 36, 319
Law, R. J., 202
Lazaridis, T., 15, 63, 68, 383, 393
Leach, A. R., 6, 13, 28, 112, 141
Leahy, D. J., 233
Leatherbarrow, R. J., 274, 294
Leavitt, S., 8, 18
Leckband, D., 17, 197
Leclerc, F., 75
Lecomte, C., 18
Lee, A., 167, 171
Lee, A. T., 173, 174
Lee, A. Y., 141, 143
Lee, C. E., 8
Lee, F. S., 128, 135, 255, 256, 259, 261, 262, 268
Lee, J., 388
Lee, J. K., 202, 205, 291, 292
Lee, J. Y., 280, 299
Lee, M. C., 165
Lee, M. D., 202
Lee, M. R., 107, 108, 114, 147, 148, 149
Lee, M. S., 70, 72, 74, 75, 94
Lee, T., 147, 148, 149
Lee, T. S., 271, 293
Lee, Y. S., 281

LeMasurier, M., 169
Lemon, A. P., 112
Lengauer, T., 112
Leplae, R., 5
Lesk, A. M., 318, 323
Leslie, A. G. W., 213, 214, 224, 225, 229, 230
Leslie, M., 55
Leulliot, N., 300
Levin, R. B., 18
Levinthal, C., 371, 372, 393
Levitt, M., 40, 63, 251, 253, 258, 264, 266, 268, 281, 285, 286, 375, 388, 390, 393, 394
Levy, R. M., 10, 14, 58, 65, 73, 75, 105, 110, 135, 147, 255, 260, 262, 324
Li, A., 380, 381, 382
Li, J., 55, 202
Li, Z., 197
Liao, D. I., 281
Libson, A., 202, 205, 208
Lienhard, G. E., 291
Lifson, S., 30, 31, 251, 252
Lightstone, F. C., 291
Lii, J.-H., 30, 39, 40, 45
Lill, H., 216
Lim, C., 68
Lin, E. C., 202
Lin, E. E. C., 202
Lin, J., 297, 298
Lin, J. H., 112, 113, 114, 175
Lin, T.-Y., 18
Lindahl, E., 199
Linge, J. P., 3, 11
Lins, R. D., 114, 115, 184
Linse, P., 52
Linse, S., 92, 103
Liotard, D. A., 67
Lipman, E. A., 3
Litvinovich, S. V., 236
Liu, G., 8, 18
Liu, H. Y., 106, 136, 266, 267, 268, 281
Liu, Y., 72, 171, 321
Liu, Y.-P., 50
Liu, Y. S., 171
Liwo, A., 69
Ljungberg, K. B., 139, 154
Lluch, J. M., 266, 268, 299
Lo, K. K.-W., 16
Lobanov, V. S., 14
Locher, K. P., 174, 175, 177
Lodi, P. J., 314, 329, 334, 338
Lolis, E., 314, 322, 337, 338, 340, 341, 356
Lopez, C. F., 164
Lou, K. L., 179
Louise-May, S., 5
Louro, R. O., 16
Lovas, S., 98
Lovell, S. C., 11
Lovgren, S., 141
Lu, H., 197, 233, 234, 236
Lu, H. P., 347
Lu, S. M., 18
Lu, W., 18
Lu, Y., 378
Luisi, B., 175
Luker, K. E., 141, 143
Luo, H., 9, 103
Luo, R., 68, 69, 94, 109
Luo, Y.-P., 50
Luque, F. J., 89
Luthey-Schulten, Z., 3, 373
Lutter, R., 213
Luty, B. A., 90, 91, 109, 324
Luzatti, V., 10
Luzhkov, V. B., 133, 134, 137, 139, 141, 169, 260, 265, 266, 291
Lybrand, T. P., 60
Lynch, G. C., 321
Lynch, T. W., 197
Lyne, P. D., 266, 281, 299

M

Ma, B., 155
Ma, J., 224
Mace, J. E., 18
Machin, P. A., 285
Macias, M., 393
Mackay, D. H. J., 40
MacKenzie, K. R., 165
MacKerell, A. D., Jr., 15, 28, 30, 35, 36, 37, 48, 199, 319
MacKinnon, R., 133, 163, 165, 167, 168, 169, 170, 171
Madura, J. D., 62, 63, 69, 90, 91, 109, 324
Maeda, M., 210, 214

Maggiora, G. M., 40
Mahoney, M. W., 3, 14, 63
Maigret, B., 46, 319
Mainz, D. T., 60
Maister, S. G., 314
Majer, P., 141, 143
Majeux, 14
Makarov, V. A., 378
Malcolm, B. A., 377
Maloney, P. C., 209
Mangoni, R., 111
Mannfors, B. E., 40, 59, 61
Manning, N. O., 11
Maple, J. R., 39, 40, 45
Marchese, F. T., 13
Marchi, M., 44
Marcus, R. A., 283, 321
Marelius, J., 111, 137, 138, 139, 140, 141, 142, 146, 147, 154
Mark, A. E., 8, 39, 40, 106, 131, 135, 136, 165, 377, 382
Mark, P., 37
Markarov, D. E., 352
Markgren, P. O., 141
Markley, J. L., 212
Marks, G. T., 316, 321, 356, 362
Marrink, S. J., 164, 165
Marszalek, P. E., 232
Martell, P. J., 16
Marti, D. N., 7
Marti, S., 266, 281
Martin, C., 380
Martin, E. J., 12
Martin, R. L., 321, 338
Martin, Y. C., 124
Masaike, T., 213, 225
Masella, M., 61
Masseling, S. J., 173
Massiah, M. A., 316, 321, 356, 362
Massova, I., 5, 107, 108, 110, 114, 147, 148, 149
Masukawa, K. M., 114, 115
Masunov, A., 63, 68
Matsudaira, P. T., 391
Matsuoka, O., 58
Matta, C. F., 49
Matthews, B. W., 377
Mattos, C., 36, 319
Mauguen, Y., 380
Maxwell, D. S., 38, 110, 147
Maxwell, D. W., 38
Maxwell, E. S., 4
May, J. L. C., 198
Mayer, F., 212
Mayor, M., 392, 394
Mayor, U., 392, 394
McAllister, M. A., 295, 299
McAnaney, T. B., 16
McCallister, E. L., 387
McCammon, J. A., 3, 28, 69, 88, 90, 91, 94, 96, 102, 103, 105, 106, 109, 112, 113, 114, 115, 124, 125, 129, 131, 164, 198, 253, 255, 256, 289, 290, 324
McCleskey, E. W., 253
McCutcheon, J. P., 198
McDermott, A. E., 318, 324, 325, 333, 334, 362
McDonald, N. A., 111, 140
McGuire, R. F., 30
McKnight, C. J., 391
McLafferty, F. W, 18
McLean, M. A., 197
McQuarrie, D. A., 344
Meadows, R. P., 112, 114
Medina, C., 14, 110, 135, 136, 137, 138, 140, 145
Mee, R. P., 124, 142
Meerman, J. H. N., 137
Meguro, T., 176, 179
Mehrotra, P. K., 13
Melissas, V., 321
Melius, C. F., 338
Melnyk, P., 205
Meng, E. C., 112
Menger, F. M., 285
Menz, R. I., 213, 229
Merz, K. M., Jr., 4, 5, 8, 14, 15, 32, 47, 134, 145, 266, 268, 271
Meuser, D., 134
Meyer, E., 170
Mezei, M., 13, 88, 97, 98, 124, 128, 129
Michejda, C. J., 111
Michnick, S., 36, 319
Miercke, L. J. W., 160, 174, 197, 198, 202, 205, 208, 209
Miertus, S., 251, 320
Mildvan, A. S., 314, 315, 316, 321, 326, 356, 362
Miller, B., 293
Miller, B. B., 293

Miller, B. G., 291, 292
Miller, C., 169
Miller, K. J., 52
Miller, W. H., 321, 349
Minor, D. L., 173
Mirkin, N. G., 40, 59, 61
Mitchell, P. J., 51
Mitome, N., 213, 225
Mitra, A. K., 202, 205
Mitsuoka, K., 174, 202, 205
Mitsutake, A., 35, 75, 115, 116, 117
Mogilner, A., 216
Mohanty, A. K., 175
Molina, P. A., 296, 299
Molnar, F., 197
Momany, F. A., 30
Monard, G., 266, 268
Montgomery, J. A., 321
Montgomery, M. G., 213, 225, 230
Monticello, R. A., 214
Moore, P. B., 164, 198
Moos, W. A., 12
Morais-Cabral, J. H., 163, 168, 169
Moras, D., 110
Mordasini, T. Z., 106
Morea, V., 5
Morin, P., 148, 149
Morokuma, K., 267
Morris, G. M., 102, 113, 114, 139
Moss, D. S., 11
Mowbray, S. L., 138
Mruzik, M. R., 98
Muegge, I., 124, 260
Muehlman, A., 141
Muir, T. W., 386
Mulholland, A. J., 266, 281, 299
Mulkidjaninan, A. Y., 213, 224
Müller, M., 214, 219, 231
Muller, R. P., 272, 273
Muneyuki, E., 213, 225
Murakami, S., 174
Murata, K., 202, 205
Murphy, R. B., 10, 14, 16, 34, 38, 40, 41, 61
Murphy, R. F., 98
Murray, C. J., 316, 320, 348
Murray, C. W., 124, 142
Murrell, J. N., 48
Musacchio, A., 389
Musil, D., 139, 154

N

Nachbar, R. B., 39
Nachbar, R. B., Jr., 103
Nadanaciva, S., 225
Nakajima, N., 111
Nakamura, H., 111
Nakasako, M., 4
Nakashima, R., 174
Narayana, N., 90, 106, 314
Naray-Szabo, G., 16
Naudin, C., 134
Neira, J. L., 380, 381
Némethy, G., 30, 40
Neria, E., 35, 289, 320
Nettikadan, S., 212
Network, E. U., 11
Newns, D. M., 164
Ngo, T., 36, 319
Nguyen, D. T., 32, 36, 319
Nguyen, H., 3, 4, 5, 9, 12
Nicholas, J. D., 58
Nicholls, A., 4, 89, 90, 91, 109, 148
Nickbarg, E. B., 314, 326, 337, 338, 341
Nielsen, S., 201, 202, 209
Nifosi, R., 5
Niketic, S. R., 30
Nilges, M., 3, 4, 6, 11, 75, 76
Nillroth, U., 138, 141
Nilsson, L., 37, 215
Nilsson, O., 176, 178
Nina, M., 68
Nishiura, M., 213
Nishizawa, K., 47
Nissen, P., 198
Noble, M., 389
Noji, H., 213, 225
Nollert, P., 160, 174, 197, 198, 202, 205, 208, 209
Noodleman, L., 65
Norden, B., 139, 154
Nordlund, P., 281
Northrop, D. B., 279
Northrup, S. H., 324
Noumi, T., 210, 214
Nowak, W., 98, 99
Noworyta, J. P., 209
Nozaki, Y., 89
Nussinov, R., 155

Nymand, T. M., 52
Nyquist, R. A., 339

O

Oberhauser, A., 232, 234
O'Connell, J., 160, 174, 197, 198, 205, 209
O'Donoghue, A. C., 318, 325, 334
O'Donoghue, S. I., 11
Ogawa, M., 18
Ogawa, T., 50
Ogin, E., 18
Ogunjobi, O. M., 386
Oh, K. S., 280, 299
Ohashi, T., 232
Oie, T., 40
Okamoto, Y., 35, 75, 115, 116, 117
Okamura, M. Y., 95
Oksman, A., 141, 143
Olafson, B. D., 35, 164, 199, 316, 321, 339
Olson, A. J., 102, 113, 114, 139
Omote, H., 212, 213, 224
Onuchic, J. N., 373
Onufriev, A., 14, 70, 74, 75
Oostenbrink, B. C., 137
Oostenbrink, C., 163
Oppenheimer, N. J., 7
Ordejon, P., 185
Ornstein, R. L., 46, 110, 141, 145, 147
Orozco, M., 89, 381
Ortega-Blake, I., 58
Ortiz, J. V., 321
Ösapay, K., 34, 68, 69
Oschkinat, H., 393, 394
Osguthorpe, D. J., 145
Oshiro, C. M., 18, 112
Oster, G., 213, 214, 216, 224, 231
Österberg, F., 134
Otlewski, J., 18, 143
Otzen, D. E., 381

P

Pabo, C. O., 392
Paci, E., 233, 234
Packer, M. J., 48
Pages, J. M., 175
Pakkanen, T. A., 314, 326, 337
Palmer, A. G. III, 17
Palmo, K., 40, 59, 61
Pan, Y., 295, 299
Pan, Y. P., 393, 394
Pande, V. S., 3, 4, 5, 9, 12, 75, 375, 388, 396
Pänke, O., 214, 219, 225, 231
Pankov, R., 232
Pannu, N. S., 3, 4, 6
Paolini, G. V., 124, 142
Papazyan, A., 89, 258, 261, 294, 295, 296, 297, 298, 299
Park, E. S., 16
Park, S.-J., 18, 207, 208
Parrinello, M., 185, 273
Parson, W. W., 253, 260, 283, 284, 285, 289
Partanen, S., 318, 325, 337
Pastor, R. W., 164
Pattnaik, P. C., 164
Pauling, L., 276, 278
Paulsen, M. D., 110, 141, 145, 147
Pauptit, R., 389
Pautsch, A., 181
Pearlman, D. A., 37, 131, 164
Pedersen, L., 3, 199
Pedersen, L. G., 37
Pellegrini, M., 17
Pelton, J. G., 299
Peng, C. Y., 321
Peräkylä, M., 268, 314, 326, 337
Perokyla, M., 289
Perozo, E., 171
Perryman, A. L., 112, 113, 114
Persico, M., 64, 68
Person, W. B., 339
Peter, C., 4
Petersson, G. A., 321
Petrache, H. I., 165
Petsko, G. A., 314, 316, 318, 322, 324, 326, 329, 337, 338, 340, 341, 342, 356
Pettitt, B. M., 378
Peyerimhoff, S. D., 2
Pfluegl, G. M. V., 17
Pfuetzner, R. A., 167, 168
Pfuetzner, T. A., 133
Phale, P. S., 175, 177, 178, 179
Phale, V. P., 178
Philippsen, A., 163, 178

Phillips, D. C., 285
Phillips, J., 197, 199, 233
Phillips, J. C., 197
Phillips, R., 238
Piana, S., 279, 280
Pires, J. R., 393, 394
Pisano, M. M., 202
Pitera, J. W., 5, 6, 130, 131, 137, 150, 152, 153
Pohorille, A., 34
Poirier, R. A., 46
Polanyi, M., 276
Pomès, R., 209
Pompliano, D. L., 314, 326, 337
Ponder, J. W., 39, 47, 52, 55, 56, 57, 58, 72
Pongor, G., 339
Popelier, P., 49
Pople, J. A., 263, 321, 338
Popot, J. L., 163
Porezag, D., 319
Porter, R. N., 313
Postma, J. P. M., 40, 62, 98, 124
Pottle, M. S., 30, 40
Pound, G. M., 98
Pratt, L. R., 255
Prentiss, M., 3
Presley, B. K., 11
Preston, G. M., 201, 202
Price, D. J., 44
Price, S. L., 47, 49
Prilipov, A., 179
Prilusky, J., 11
Prince, S. M., 175
Privalov, P. L., 378
Prodhom, B., 36, 319
Profeta, S., Jr., 31
Proks, P., 173
Pulay, P., 339

Q

Qasim, M. A., 18
Qasim, S., 18
Qian, W., 16
Qiu, D., 92, 93, 109
Qiu, Y., 378
Queener, S. F., 139
Quinn, J. E., 58

R

Rabitz, H., 12, 103, 104
Rader, S. D., 17
Radford, S. E., 377, 378
Radkiewicz, J. L., 289, 290
Radmer, R. J., 136, 150
Raghavachari, K., 321, 338
Rahman, A., 62
Ramage, R., 386
Ramakrishnan, V., 198
Ramnarayan, K., 72
Ranatunga, K. M., 160, 169, 173
Randa, H. S., 209
Randal, M., 112
Ranjbar, M., 18
Rao, S. N., 281, 294
Raphael, D. R., 112
Rapp, C. S., 43, 72, 94
Rappé, A. K., 40, 50
Rarey, M., 112
Rasaiah, J. C., 209
Rasmussen, K., 30
Rastogi, V. K., 212, 214, 219
Ravi-Shanker, G., 13
Read, R. J., 3, 4, 6
Reddy, V. S., 205
Rees, D. C., 173, 174
Reggiani, A., 162
Reiher, W. E. III, 35, 36, 319
Rein, R., 46
Remigy, H. W., 163
Remington, S. J., 281
Ren, G., 205
Ren, H., 213
Ren, P., 39, 47, 52, 55, 56, 57, 58
Rendell, A. P., 169
Replogle, E. S., 321
Resat, H., 97, 98
Reuss, L., 202
Reuter, N., 319
Revington, M., 214
Reyes, C., 5, 147, 148, 149
Reynolds, C. A., 46
Rice, L. M., 3, 4, 6
Richard, J. P., 316, 318, 325, 329, 334, 359, 362
Richard, V., 380
Richards, W. G., 46, 281
Richardson, D. C., 11

Richardson, J. S., 11
Rick, S. W., 50
Rickey, D. P., 202
Riddle, D. S., 388
Rief, M., 232
Riley, R., 381, 383, 396
Rinehart, D., 181, 182
Ringe, D., 314, 316, 318, 326, 329, 338, 341
Ripoll, D. R., 69
Rittner, F., 58
Rizzo, R. C., 10, 14, 38, 147
Robb, M. A., 321, 338
Roberts, V. A., 145
Robertson, K. M., 176, 180
Roccatano, D., 111, 388
Rocchia, W., 90, 91
Roche, O., 18
Rodgers, J., 170
Roe, D. C., 8, 18
Rogers, J. D., 339
Roitberg, A. E., 35, 75, 165
Rokhsar, D. S., 75, 375, 388
Roosild, T., 167
Roques, B. P., 40, 61, 62
Rose, D., 314, 322, 338, 341, 356
Rosenberg, J. M., 43, 152, 153
Rosenbusch, J. P., 175, 176, 177, 178, 179
Ross, W. S., 164
Rossi, I., 321
Rossky, P. J., 65
Rost, L. E., 338
Roterman, I. K., 44
Roth, C. B., 174
Rothberg, I., 18
Rothlisberger, U., 185, 273
Rouhi, A. M., 292
Roux, B., 28, 36, 47, 64, 68, 109, 160, 163, 164, 165, 168, 169, 170, 176, 178, 179, 180, 209, 215, 257, 319
Rozovsky, S., 318, 324, 325, 333, 334, 362
Rub, B., 339
Rubins, K., 114, 115
Rumberg, B., 231
Rummel, G., 177
Russell, S. T., 251, 257, 258, 260, 282, 297, 298
Ryan, K., 18
Ryckaert, J.-P., 344
Ryde, U., 300
Ryohei Yasuda, K. K. J., 213

S

Saadat, D., 316, 356, 362
Saam, J., 197
Safront, V. S., 285
Sagnella, D. E., 185
Sagui, C., 52
Saint-Martin, H., 58
Saito, M., 176, 179, 255
Saiz, L., 164
Salahub, D. R., 40, 61, 62
Salem, M., 380
Salemme, F. R., 164
Sali, A., 5, 371, 372
Salomaa, A., 19
Salsbury, F. R., Jr., 70, 72, 74, 75, 94
Salt, D. W., 141
Sambongi, Y., 213, 224
Sampogna, R., 94
Samuelsson, B., 135, 136, 137, 138, 140, 141, 145, 154
Samuelsson, J.-E., 14, 110
Sanbonmatsu, K. Y., 35, 75, 115, 116, 388
Sanchez-Portal, D., 185
Sansom, M. S., 160, 169, 170, 171, 172, 173, 174, 176, 181, 182, 183
Sansom, M. S. P., 160, 163, 164, 165, 166, 168, 169, 170, 171, 172, 176, 179, 202, 209
Saraste, M., 389
Sato, S., 392, 394
Sawyer, L., 386
Schaefer, M., 64, 72, 74, 75, 94, 197
Schäfer, H., 136
Schames, J. R., 112, 113, 114
Schapira, M., 90
Scharf, D., 164
Schechter, I., 144
Schemmel, A., 232
Scheraga, H. A., 30, 40, 44, 46, 69
Schiltz, E., 175, 177
Schiøtt, B., 296
Schirmer, T., 175, 176, 177, 178, 179, 180
Schlegel, H. B., 321, 338
Schlenkrich, M., 36, 319
Schleyer, P. von Ragué, 2
Schlichting, I., 379
Schlick, T., 28, 324
Schlitter, J., 223
Schmitt, U. W., 209, 271

Schnackerz, K. D., 325
Schneider, T. R., 17
Schneider-Mergener, J., 393, 394
Schreiber, D. E., 98
Schrempf, H., 134
Schuler, B., 3
Schulten, K., 160, 163, 173, 174, 197, 198, 199, 201, 202, 205, 206, 207, 208, 209, 211, 216, 232, 233, 234, 236, 238
Schulz, G. E., 175, 177, 181
Schutt, C., 103
Schutt, C. E., 164
Schutz, C. N., 259, 260, 262, 298
Schüürmann, G., 65
Schwartz, S. D., 352
Schwarzbauer, J. E., 232
Schweins, T., 270, 280
Scott, D. L., 237
Scott, H. L., 165
Scott, L. R., 90, 91, 109
Scott, W. R., 39, 40
Scrocco, E., 251, 320
Scrutton, N. S., 289, 290
Seaton, B. A., 314, 316, 318, 338, 341
Sechler, J. L., 232
Seefeld, S., 178
Seeger, D. M., 339
Seeger, R., 338
Seibel, G., 164
Seifert, G., 319
Senior, A. E., 213, 225
Sept, D., 90, 91, 114
Servant, L., 134
Shafer, H., 8
Sham, Y. Y., 255, 257, 259, 260, 261, 262
Sharma, A., 233
Sharma, R. D., 313
Sharon, R., 40, 63
Sharp, K. A., 9, 65, 68, 69, 94, 103, 109, 148, 258
Shea, J. E., 3, 374, 394
Sheinerman, F. B., 386, 387, 393
Sheldrick, G. M., 17
Shelnutt, J. A., 378
Shen, L., 165
Shen, T. Y., 109
Shenkin, P. S., 92, 93, 109
Sherwood, P., 185
Shi, L. B., 202
Shilton, B. H., 214
Shimanouchi, T., 170
Shin, S., 388
Shindyalov, I. N., 170
Shinozaki, A., 199, 233
Shirts, M. R., 75, 396
Shoichet, B. K., 112
Short, S., 293
Shrivastava, I. H., 160, 163, 168, 169, 170, 173
Shuker, S. B., 112, 114
Shurki, A., 286, 287, 289
Siani, M. A., 12
Siegbahn, P. E. M., 264
Sikorski, R. S., 296, 299
Silva, A. M., 141, 143
Silverstein, K. A. T., 3, 13
Simmerling, C., 35, 75, 99, 165
Simonson, T., 3, 4, 6, 47, 65, 67, 72, 75, 89, 105, 110, 319
Singh, U. C., 31, 88, 125
Sitkoff, D., 65, 68, 148
Skach, W. R., 202
Skeel, R., 199, 233
Skiff, W. M., 40, 50
Skillman, A. G., 8, 18
Sklenar, H., 112
Sligar, S. G., 197
Smals, A. O., 143, 144, 145, 146
Smith, B. L., 202
Smith, G. R., 160, 165, 166, 169, 170, 173, 176
Smith, J. C., 36, 319
Smith, L. J., 7
Smith, P. E., 135, 136, 378
Smith, R. H., Jr., 111
Smith, W., 50
Smondyrev, A. M., 165
Smopornpisut, P., 171
Snider, M. J., 274, 276, 313
Snow, C. D., 3, 4, 5, 9, 12, 75
Snyder, J. P., 285
Soares, C. M., 16, 176, 178
Sokalski, W. A., 46, 49
Soler, J. M., 185
Soman, K., 281
Sompornpisut, P., 171
Sorin, E. J., 75, 388, 396
Spassov, V. Z., 161, 165, 177
Spellmeyer, D. C., 12, 32, 134, 145

Spellmeyer, D. M., 14, 15
Spencer, R. H., 173
Splitt, H., 134
Sprik, M., 58
Sreerama, N., 378
Sridharan, S., 90, 91, 109
Srinivasan, J., 10, 73, 147, 148, 149
Stahlberg, H., 163
Stahle, L., 15
Stanton, R. V., 264, 283, 284
Starovasik, 384
States, D. J., 35, 164, 199, 316, 321, 339
Steckler, R., 321
Stefanovich, E. V., 67
Stehle, T., 237
Steigemann, W., 376
Steinbach, P. J., 37
Steitz, T. A., 198
Stepaniants, S., 197
Stern, H. A., 10, 14, 16, 40, 58, 60
Stevens, W. J., 263
Stewart, J. J. P., 338, 339
Stewart, J. P., 321
Still, W. C., 16, 69, 70, 72, 92, 93, 109, 147, 152, 260
Stillinger, F. H., 58, 62
Stock, A. M., 314
Stock, D., 214
Stockfisch, T. P., 39, 40, 45
Stone, A. J., 49
Stone, R., 291
Storm, D. R., 300
Stote, R., 36, 319
Stouch, T., 165
Stouch, T. R., 46
Stout, J. M., 52
Straatsma, T. P., 3, 63, 125, 131, 184
Štrajbl, M., 268, 284
Stratton, J. R., 299
Straub, J., 36, 319
Strockbine, B., 35, 75, 165
Stroud, R. M., 10, 11, 112, 160, 174, 197, 198, 202, 205, 208, 209
Stuart, D. I., 233
Stuart, S. J., 50
Suarez, D., 4, 5
Subramaniam, S., 103
Suenaga, A., 176, 179
Sugio, S., 314
Sugita, Y., 35, 75, 115, 116, 117
Suhai, S., 319
Sui, H., 202, 205
Sullivan, D. C., 5, 392
Sun, A.-Q., 325
Sun, D. P., 281
Sun, L., 378
Sun, S., 231
Sun, Y., 8, 9, 14, 18
Sunner, J., 47
Susnow, R., 103
Sussman, F., 262, 281, 289, 294
Sussman, J. L., 11
Sutcliffe, M. J., 289, 290
Svensson, B., 92, 103
Svensson, P., 139, 154
Swaminathan, S., 13, 35, 124, 164, 199, 316, 321, 339
Swart, M., 52, 54, 55
Sweet, R. M., 379
Swendsen, R. H., 43, 152, 153
Swenson, C. J., 38
Szabo, G., 181, 182
Szalma, S., 161, 165

T

Tajkhorshid, E., 160, 174, 197, 198, 202, 205, 206, 207, 208, 209, 211, 238
Takeyama, M., 210, 214
Takeyasu, K., 212
Tamm, L. K., 174, 181, 182, 183
Tan, Y. J., 381, 382
Tanford, C., 73, 89, 258
Tang, N., 18
Tantillo, D. J., 4
Tapia, O., 176, 178, 285
Tarakeshwar, P., 280, 299
Tarek, M., 164
Tashiro, M., 18
Tasumi, M., 170
Tate, J., 160, 163, 168, 169
Taylor, E. W., 18
Taylor, H. C., 11
Taylor, J. A., 90, 106
Taylor, S. S., 90
Teeter, M. M., 44
Tembe, B. L., 88, 96, 124, 129
Tempczyk, A., 16, 69, 70, 72, 92, 147, 152

Terstappen, G. C., 162
Thacher, T., 103, 104
Thacher, T. S., 103
Théry, V., 266, 267
Thiel, W., 266, 267, 268
Thole, B. T., 52, 54
Thomas, W. E., 232, 233, 236
Thompson, W. J., 18
Thornton, J. M., 378
Tickle, I. J., 11
Tidor, B., 151, 152
Tieleman, D. P., 164, 165, 169, 176, 179, 180
Tijian, R. J., 285
Tildesley, D. J., 251, 252, 253
Tilton, R. F., 314, 337, 338, 340
Tiraboschi, G., 61
Tirado-Rives, J., 4, 38, 43, 110, 111, 140, 141, 147, 164, 378, 379, 381
Tironi, I. G., 39, 40
Tobias, D. J., 164
Tobin, J. B., 293, 294, 297
Toda, M., 256
Toepert, F., 393, 394
Tokumasu, F., 212
Tomasi, J., 64, 68, 251, 320
Tomimoto, M., 31
Tomovick, R., 103
Tong, 318, 325, 333, 334, 362
Tong, L., 318, 324, 325
Topiol, S., 338
Torda, A. E., 39, 40
Torre, V., 169, 185
Torrie, G. M., 124, 152, 256, 257, 344
Totrov, M., 90
Toukmaji, A., 52
Tramontano, A., 5
Trevathan, M. W., 73
Trucks, G. W., 321, 338
Truhlar, D. G., 64, 67, 68, 72, 271, 320, 321, 348
Truong, T. H., 67
Tsai, J., 388, 390, 393, 394
Tsai, P., 316
Tsui, V., 9, 72, 75
Tu, K. C., 164
Tuckerman, M., 273
Tully, J. C., 343
Turner, M. A., 386

U

Uebayasi, M., 176, 179
Ulmschneider, 160
Ulyanov, N., 17
Unge, T., 141

V

Vacca, J. P., 18
Vagedes, P., 281
Vaidehi, N., 267
Vale, R. D., 8
Valiyaveetil, F. I., 165, 212
Valleau, J. P., 124, 152, 256, 257, 344
Vandeputte-Rutten, L., 175
van der Spoel, D., 199
van der Vaart, A., 4, 5, 47, 271
van Duijnen, P. T., 52, 54, 55
Van Gelder, P., 174, 175, 177
van Gunsteren, W. E., 4, 7, 8, 35, 62
van Gunsteren, W. F., 5, 6, 7, 28, 30, 39, 40, 58, 106, 127, 130, 131, 134, 135, 136, 137, 142, 145, 163, 164, 377, 382
van Lipzig, M. M. H., 137
Van Opdenbosch, N., 40
van Os, C. H., 202
van Schaik, R. C., 131
van Vlijmen, H. W. T., 72, 75, 94
Varadarajan, K., 199, 233
Varani, G., 300
Vargas, R., 59, 60
Vargha, A., 339
Várnai, P., 281
Vasu, T., 13
Vedani, A., 40
Veenstra, D., 34
Venable, R. M., 164
Venyaminov, S. Y., 378
Vergoten, G., 40
Verkhivker, G., 98, 99
Verkman, A. S., 202
Vesely, F. J., 51
Vijayakumar, M., 91
Vila, J. A., 69
Villa, J., 268, 270, 277, 279, 280, 281, 283, 284, 289, 290, 291, 300, 321, 343, 348
Vinter, J. G., 48

Viragh, C., 298
Vogel, V., 197, 232, 233, 234, 236
Vojtechovsky, J., 379
Volkmer-Engert, R., 393, 394
von Heijne, G., 162
Vorobjev, Y. N., 69, 75, 107, 108, 114
Voth, G. A., 209, 271, 291
Vrang, L., 141
Vreven, T., 267
Vuilleumier, R., 271
Vukobratovic, M., 103

W

Wacker, T., 175
Wada, Y., 213, 224
Wade, R. C., 90, 91, 109, 324
Wagener, M., 8
Wagner, F., 75
Wagner, R., 134
Wai, J. M., 18
Walian, P., 202, 205
Walker, J. E., 213, 214, 224, 225, 229, 230
Wall, I. D., 141
Wallace, C. J., 20
Wallin, E., 162
Wallqvist, A., 73, 75
Walz, T., 202, 205
Wang, H., 213, 214, 216, 224, 231
Wang, H. Y., 231
Wang, J. M., 5, 10, 34, 138, 140, 145, 146, 148, 149, 150
Wang, L., 130, 284
Wang, W., 5, 138, 140, 147, 148, 149
Wang, W. H., 16
Wang, Y. F., 175
Wang, Y. H., 16
Warne, N., 18
Warren, G. L., 3, 4, 6
Warshel, A., 30, 89, 124, 125, 128, 135, 251, 252, 253, 254, 255, 256, 257, 258, 259, 260, 261, 262, 264, 265, 266, 267, 268, 269, 270, 272, 273, 274, 276, 277, 279, 280, 281, 282, 283, 284, 285, 286, 289, 290, 291, 292, 293, 294, 295, 296, 297, 298, 299, 300, 318, 343, 344, 349
Warwicker, J., 148, 258
Watanabe, M., 28, 36, 176, 179, 319
Watson, H. C., 148, 258
Watts, C. R., 98
Webb, S. P., 348
Weber, J., 213, 225
Weckesser, J., 175, 177
Weiner, P. K., 31, 199
Weiner, S. J., 32
Weiss, G. H., 12, 15
Weiss, M. S., 175, 177
Weiss, R. M., 295
Weissig, H., 170
Weitzman, C., 202, 205, 208
Wells, I. A., 112
Wells, J. A., 294
Welte, W., 175, 177
Wendoloski, J. J., 164
Werbovetz, K. A., 141
Werten, P. J. L., 163
Wesolowski, T. A., 267, 272, 273
Westbrook, J., 170
Whatley, H., 18
White, J. M., 8
White, S. H., 161
Whitesides, R. A., 338
Whitt, S. A., 293, 294, 297
Wiberg, K. B., 46
Widmer, C., 178
Wieczorek, A., 18
Wieczorek, M., 18
Wiener, M. C., 161, 175
Wierenga, R. K., 318, 325, 337, 352, 389
Wight, C. A., 352
Williams, D. E., 31, 46, 47, 49
Williams, G., 170
Williams, J. C., 324, 325
Williams, M. A., 378
Williams, R. J. P., 300
Wilson, C., 18
Wilson, K. P., 377
Wilson, K. R., 40, 346
Wilson, R. J., 19
Wilson, T. H., 202
Wilusz, T., 18
Wimberly, B. T., 198
Winget, P., 67
Winn, P. J., 46
Winter, G., 274, 294
Wiórkiewicz-Kuczera, J., 36, 319
Wistow, G. J., 202

Wladkowski, B. D., 263
Wold, S., 15
Wolfenden, R., 274, 293, 313
Wolff, J., 145
Wolfson, H., 155
Wollmer, A., 223
Wolynes, P. G., 3, 373
Won, Y., 215
Wong, A. K., 12
Wong, C. F., 88, 90, 92, 102, 103, 104, 105, 106, 109, 125, 255, 256
Wong, K., 380
Wong, K.-B., 381, 382
Wong, L.-L., 16
Wong, M. W., 321
Woo, T. K., 273
Woodward, C. E., 92, 103
Woody, R. W., 4, 378
Woolf, T. B., 165
Word, J. M., 11
Wriggers, W., 197
Wright, P. E., 385
Wu, J., 16
Wu, N., 268, 281, 291, 292
Wyatt, R. E., 349
Wynn, R., 18

X

Xantheas, S. S., 55, 58
Xia, Z. X., 16
Xiang, Z., 7
Xie, X. S., 347
Xie, Y., 16
Xiong, G., 165
Xiong, J. P., 237
Xu, H., 14
Xu, W., 389
Xue, L. L., 16
Xuong, N. H., 90, 106, 314

Y

Yadav, A., 281, 283
Yamada, K., 232
Yamaguchi, A., 174
Yamashita, E., 174
Yamato, I., 176, 179
Yan, L., 161, 165
Yang, A.-S., 94
Yang, W., 266, 267, 268, 271, 281
Yao, P., 16
Yasuda, R., 213, 225, 227
Yasuda, T., 213
Yeates, T. O., 17
Yellen, G., 167
Yin, D., 36, 319
York, D., 3, 67, 199
Yoshida, M., 213, 225
Yoshimine, M., 58
You, T. J., 262
Young, M. M., 18
Young, W. S., 34, 68, 69
Yu, H., 58
Yu, W., 92
Yuh, Y. H., 30, 45
Yüksel, K. Ü., 325

Z

Zachariae, U., 177
Zacharias, M., 112, 131
Zacharias, N., 31
Zagrovic, B., 75, 388, 396
Zakrzewski, V. G., 321
Zalis, M. E., 11
Zeidel, M. L., 202
Zhang, H., 103
Zhang, J., 92
Zhang, L. Y., 65, 73, 110, 225
Zhang, Q. J., 175
Zhang, R., 237
Zhang, W., 18
Zhang, W. L., 18
Zhang, X., 318, 334
Zhang, Y., 212, 266, 267, 268, 281
Zhang, Z., 314
Zhong, Q. F., 164
Zhou, F., 197
Zhou, H.-X., 91
Zhou, R., 10, 14, 16, 40, 60, 147, 388
Zhou, Y., 163, 165, 168, 169
Zhu, F., 197, 205
Zhu, F. Q., 174
Zhu, S.-B., 103, 104
Zijlstra, R. W. J., 55

Zoebisch, E. G., 338, 339
Zonta, C., 48
Zou, X., 9, 14
Zuccarello, G., 138
Zucker, M., 5
Zwanzig, R. W., 96, 100, 105, 124, 125, 256

SUBJECT INDEX

A

Ab initio models, 5, 13, 15, 264, 268, 343–344
Ab initio surfaces, 268, 271, 281
ACE, 75
Adenosine triphosphate. *See* ATP synthase
ADP, 225
Aliphatic rings, 31
Alkanes, 65
Alkylammonium, 133–134
All-atom simulations, protein funding/unfolding
  barnase, 379–380, 380f
  bovine pancreatic trypsin inhibitor, 143–145, 375–377, 376f, 395
  chymotrypsin inhibitor 2, 381–383, 382f, 393
  Engrailed Homeodomain, 392, 392f
  fast, 391–393
  first systems of, 375–379
  folding problem, minimal models and, 371–373
  force field and dynamic protocols, 373–375
  future of unfolding simulations, 396–397
  heat capacity, 396
  hen egg white lysozyme, 377–378, 377f, 395
  infancy of direct simulation of folding, 395–396
  microscopic reversibility, 394
  modern, 379–391
  myoglobin, 378–379, 379f
  Protein A, 383–385, 384f, 395
  Protein G, 386–388, 387f
  protein unfolding by pathways, 393–395
  SH3 domain, 388–391, 389f
  ubiquitin, 385–386, 386f
  villin, 391–392, 391f
  WW domain, 393, 393f
AM1 methods, 319, 320, 321, 334, 339–340, 340f
Amber force fields, 30, 147, 164
  all-atom approach of, 31–32
  antechamber program for, 32
  bonded v. nonbonded interactions and, 34
  charge development problem of, 32–33
  CHARMM force field v., 38, 42t, 45t
  ff02-EP, 33, 33f, 47–48, 47t, 53, 60, 62f
  *ff*84, 41, 42t
  *ff*94 (Cornell *et al*), 32–35, 33f, 41, 42t, 43, 45t, 47t, 62f
  large scale simulation, 199
  Lennard-Jones parameters, 33–34
  NAMD, large scale simulations and, 199
  OPLS/, 38
  OPLS potentials, 32
  overpolarization, 32
  polar hydrogen only approach of, 31
  RESP, 33, 33f
  torsional potentials, 31, 34
Amide nitrogen, 59
Amino acids, minimal models of, 371–373
Ammonium ions, quaternary, 133, 134f
Anisotropic environment, 161, 162f
APBS package, 90
Apo-myoglobin, 379f
Applequist interactive polarization, 53, 54, 55
Aquaporins (AQP), 174
  aquaglyceroporins of, 202
  architecture/model of, 202–203, 203f, 204f
  crystallography, 208
  GlpF of, 202, 205, 207, 208–211, 210f
  large scale simulation and membrane water channel, 201–203, 203f, 204f, 205–211, 206f, 207f, 208f, 210f
  MD stimulation, 203, 209–211
  mercurial compounds, 201–202

Aquaporins (AQP) (*cont.*)
  NPA motifs, 205, 206, 206f, 207f
  physiological conditions, periodic boundary conditions and, 205
  PMF, 208–209, 208f
  water balance, 201–202
  water dynamics, 209–211, 210f
Arg6, 234
Arg23, 234
Arg78, 236f
Arg210, 216, 221, 221f
Arg-Glu salt bridge, 182–183
Arg-Gly-Asp (RGD), 236–237, 236f
Arginines, 206
Arrhenius-like expression (k_{cat}), 314, 325
Asp61, 216, 219, 221, 342
Asp71, 356
Asp96, 293
Asp102, 297
Asp654, 280
Asparagines, 206
Aspartic protease plasmepsin II, 141–143, 142f
Atomic force microscopy (AFM), 196–197, 232–234
Atomic partial charge, 103
Atomic point-charge, fixed
  electrostatics and limitations on, 46–48, 47t
Atomic simulations, 27–28
Atoms. *See also* All-atom simulations, protein funding/unfolding
  end-point catastrophe, 130–132, 132f
  propagation of, 252–253
ATP (adenosine triphosphate) synthase, 196, 198
  α/β stalk interactions, 212–213, 212f, 223, 227f, 228
  β_{DP}, 225, 229, 229f
  β_{E}, 229, 229f, 231
  β_{TP}, 225, 229–231, 229f
  energy conversion and, 211–232
  MD simulations, torque generation of F_o and, 214–221, 223
  mechanical model of F_o, 221–223, 221f
  PMF, 222, 222f
  POPE, 215
  proton translocation, 214, 215
  steered molecular dynamics (SMD), torque-drive catalysis in F_1 and, 224–227, 224f, 226f, 227f, 230, 231–232
  stochastic simulations of F_o, 222, 222f
  structural model of F_o, 218–219, 218f
  synthesis system of, 211–213, 212f
  targeted TMD simulations, torque-driven catalysis in F_1 and, 223–224, 224f
  top-down v. bottom-up modeling for, 231–232
  torque generation of F_o, 213–223, 217f, 218f, 220f, 221f, 222f
  torque-drive catalysis in F_1, 223–232, 224f, 226f, 227f, 228f, 229f, 230f
  transmembrane helices (TMH), 214, 216, 218–219, 221
Avidin, 107

B

B3LYP, 319
  CHARMM, 321, 327f, 330, 332f, 333, 336, 359f
Bacterial outer membrane protein, 160, 161
Balanol, 103
β-barrel, 163, 163f, 175, 181
Benzamidine, 100, 105
Benzene, 47t
Binding
  drug, 77–78
  free energy, 88–89, 90, 96–97, 103–111, 113, 117, 138, 261–263
  kinesin, 8
  ligand, 18, 27, 77–78, 89, 102, 110–111, 123–125
  nucleic acid, 5
Biological membranes, 161, 162*f*
Biology, structural
  aqueous solution, heavily hydrated crystals and, 3, 4
  biomolecular simulation in, 160
  error estimation, error propagation, experiment/theory and, 10–13
  gas-phase properties and, 2–3
  high throughput screening (HTS) for, 8
  models' role in protein simulations with, 13–15
  molecular recognition, docking and, 8–10, 112–113, 155
  parameterization of, 15–16, 18

Biology, structural (*cont.*)
  small molecules v. macromolecules computations in, 2–5
  testing theoretical models for, 16–18
  theory influencing experimentation in, 5–8
Biopolymer conformation forces, 238–239
Boltzmann distribution, 290
Born model, 128f
  ACE model and, 75
  alternative "pairwise" approaches in, 72
  ansatz of, 69
  CHARMM force field and Generalized, 109
  Coulomb field approximation (CFA), 71, 74, 261
  drug design, 92–94
  effective/perfect Born radius, 70–71, 74–75, 93–94, 260
  Generalized (GB), 67, 69–77, 89, 92–94, 108, 109, 152, 161, 165, 260–261
  goal of, 70
  integrals over spheres approach in, 72, 74
  molecular dynamics simulations with, 75–77, 76f
  native folds distinguished from misfolded decoys, 75, 77
  salt effects with, 73
  second generation generalized, 74–75
  S-GB (surface-GB) model, 72–73, 75
  solvation (free) energy, 70–71, 92–93
Born-Oppenheimer potential surface, 250
Born-Haber cycle, 255
Born-Oppenheimer ground-state energy, 27–28
BOSS program, 114
Bovine calbindin, 92
Bovine pancreatic trypsin inhibitor (BPTI), 143–145, 375–377, 376f, 395
Brownian dynamics, 92, 109
  membrane protein, ions, transport and, 177–178
  TIM, 324–325
BUFF force field, 40t, 41, 41t, 50

C

$c2_L$-helices, 221f
$c2_R$-helices, 217f, 220, 220f, 221f
Calcium ions, 92
Canonical ensemble (CTT), 320
α-carbons, 322–323, 323f, 380, 385
Carbonyl groups, 206–207, 292, 342
Carbonyl oxygen, 168, 171
Cartesian representation, 252, 284
CEDAR force fields, 39, 40t
C-F groups, 110
CFA. *See* Coulomb field approximation
C-H groups, 110
CHARMM force fields, 35–38, 164
  AM1-SRP/, 327–328, 328f, 334
  Amber force field v., 38, 42t, 45t
  B3LYP, 321, 327f, 330, 332f, 333, 336, 359f
  charge models for, 36–37, 37f
  CHARMM19, 35–36, 37f, 41, 42t
  CHARMM22, 36, 37–38, 41, 42t, 43, 44f, 319, 359
  CHARMM27, 37, 41, 42t, 43, 44f, 45t, 47t, 62f, 215
  gas-phase simulations, 36
  Generalized Born model, 109
  large scale simulation, 199, 200, 215
  Lennard-Jones parameters, 36
  neutral groups enforcement in, 37, 39
  parameterization of, 35, 37
  RESP, 36
  SCC-DFTB, 321, 356, 359f, 361
  TIM, enzymes and, 319, 320, 321, 327–328, 327f, 328f, 330, 332f, 333, 336, 339, 359f
  torsional potentials, 36–37
  TRAVEL, 321
Chemical space
  λ dynamics, CMC/MD and, 152–154
Chymotrypsin inhibitor 2 (CI2), 381–383, 382f, 393
Cl ions, 167, 172–173, 180
Computations
  computer limitations of, 5
  difficulty in experiments' similarity in, 12
  for drug design, 87–88
  drug design and force field energy, 88–89
  electrostatic potentials, 55–57
  enzymes, 279–280, 281, 289–299, 318–321
  error estimation, error propagation, experiment/theory and, 10–13
  finite-difference or boundary element numerical methods for, 64
  free energy, 96–97, 99, 99f, 100, 102–111, 123–155, 258–261

Computations (*cont.*)
geometric constraints of, 6–7, 11
of induced multipole moments model, 51–57
LBHB, poor use of 'high level,'' 298–299
MM/PBSA, 107–108, 109, 113, 114, 118
models' role in protein simulations with, 13–15
molecular recognition, docking and, 8–10
multipoles, electrostatics and, 49–50
parameterization and, 15–16, 18
small molecules v. macromolecules, 2–5, 49
in structural biology, 1–19, 41
testing theoretical models for, 16–18
theory influencing experimentation in, 5–8
thermodynamic properties v. molecular simulation, 6–7, 8, 9, 17–18
windowing, 97, 101
Computers
clusters of super, 226
drug design aided by, 87–88
large scale simulation and (NAMD) clustering of, 196, 199
limitations of, 5
power, 253
Conductor-like Screening Model. *See* COSMO model
Conjugate peak refinement (CPR), 180, 319
Continuum solvent models, 3, 9–10, 63–75, 66f, 78, 88, 89, 95, 108
COSMO model, 281
electrostatic energy of charges in, 65–66, 66f
solute-solvent interaction energy in, 66–67
solvent-solvent repulsion energy in, 66–67
Coulomb field approximation (CFA), 71
Coulomb's law, 29, 48, 70, 73, 93, 261
Covalent bonds, 29f
CPMD. *See* Enzymes Car-Parrinello Molecular Dynamics
Crystallography, 17, 76, 76f, 167, 169, 170, 213
AQP, 208
ATP synthase, torque generation, 213–214
lower-resolution macromolecular, 11
ultra-high-resolution, of small molecules, 11, 18
X-ray, 167, 184
Cs^+ ions, 169, 171
C-terminal helice, 228–231, 228f, 234, 234f, 377, 378, 386–397, 389f, 390, 393, 397f
CVFF force field, 40t, 145
CVT, 348, 350t, 353t
CVT-SCT, 348, 350t, 353t
CVT-ZCT, 348, 350t, 353t
cytoplasm, 221–222, 221f

D

Dang-Chang polarization model, 53
Debye-Hückel theory, 68, 73, 92
Dehyration energy, 167
DelPhi package, 90
Density functional theory (DFT), 170, 185, 267
constraint (CDFT), 272–273
embedding potential, 272
EVB, 273
frozen (FDFT), 267, 271–273
semi-empirical (SCC), 321
Deprotonation energy, 327
Desolvation
NAC, 288
RSD, 291–293, 292f
Detergent micelles, 182f, 183, 184
Detergent, zwitterionic, 164
DFT. *See* Density functional theory
DHAP. *See* Dihydroxyacetone phosphate
Dihedral angle, 323–324, 323f
DIHR inhibitor binding, 139
Dihydrofolate reductase. *See* DIHR inhibitor binding
Dihydroxyacetone phosphate (DHAP)
MGS, 316, 317f, 334, 358f
TIM, 314, 316, 317f, 318, 320, 329–330, 332f, 333, 334, 338, 340, 341, 342, 343, 347–348, 349, 352, 353t, 359, 361, 365–366
Dimer energetics, solute-water, 36
Dimer surface, *ab initio*, 36
Dimethylformamide (DMF), 59f, 60
Dipeptide, 34, 36, 43
Dipole moments, 103, 293
DISCOVER force field, 39, 40t
Discretized continuum (DC) model. *See* Poisson-Boltzmann (PB) model

Divide and conquer (D&C)
approach, 271–272
DMA values, 49
DMF. *See* Dimethylformamide (DMF)
DMPC, 165, 166f
DNA, 238, 262
Docking, 8–10, 112–113, 155
β-domain, 377
Domino effect, 224, 224f
Drude oscillator (DO) model, 51, 58
Drug binding, 77–78
Drug design
ligand binding, free energy calculations and, 123–125
Drug design, protein simulation and
chemical-scanning computational experiments, 109–110
computer aided, 87–88
dynamic pharmacophore method, 114–115
λ dynamics method, 99–100
explicit and implicit solvent models combined, 107–109
explicit-solvent models, parametrize implicit-sovlent models and, 111
fast approximate simulation for calculating binding free energies, 103–111, 117
fixed-conformation models for, 88–95, 117
force-field energy calculation for, 88–89, 117–118
free energy calculations, 96–97, 99, 99f, 100, 102
Generalized Born model, 92–94
Kirkwood integration formula, 97–98
locally enhanced sampling (LES), 98–99, 99f
mean-field sense, 92
parallel fast-growth method, 100–101
Poisson model, 90–91
Poisson-Boltzmann model, salt effects and, 91–92
protein flexibility in virtual screening, 111–115
protein/ligand flexibility, modeling molecular recognition and, 95–103
protonation states of proteins and ligands, 94–95
relaxed complex methods, 112–114, 118
replica-exchange method, enhanced sampling and, 115–117
semi-empirical linear response theory, 110–111, 118
single-step (window) Zwanzig's perturbation method, 105–106, 118
systematic sensitivity analysis, 103–105
time scale extended for, 88
translational and rotational entropy, 101–103
λ Dynamics method, 99–100
chemical MC/MD, 151–154

E

E. coli, 206, 212
E165, 358
E165D, 338, 340, 340t, 341
ECM. *See* Extracellular matrix
EDT2, 356, 358, 359, 359f, 360f, 361f
EDTI (Enediolate intermediate), 315, 317f, 326, 327f, 328, 328f, 329, 330, 330f, 331f, 332f, 333, 334, 335, 337, 352, 359, 359f, 365
EDTI. *See* Enediolate intermediate
Electrostatics, 16, 40, 91, 164
atomic multipole moments, 49
binding free energy calculations, 261–263
computations, multipoles and, 49–50
consistent treatment of, 52–57
continuum solvent models of, 3
COSMO model, 65–66, 66f, 281
via Coulomb's law, 29, 48
environmental effects via polarization, 50–52
enzyme catalysis, 282–283
FEP, 257
flexible models for static charge distribution, 48–50
free energy computational models, 258–261
of hydration, 28
ion transport, 177
limitations of fixed atomic point-charge, 46–48, 47t
lone pair sites for, 48–49
long range nature, 254
next generation, 59–60
pH and, 94
Poisson model, 90, 109, 177

Electrostatics (*cont.*)
polarization balance of inter- and intramolecular energetics, 57
solvation (free) energy, 70–71, 92–93, 296
solvent v. vacuum, 65, 70
Electrostatic potential charges (ESP), 31
Embden-Myerhoff glycolytic pathway, 316
Empirical valence bond. *See* EVB
ENCAD force field, 40, 40t
End-point catastrophe, 130–132, 132f
Enediolate intermediate. *See* EDT1; EDT2
Energy gap, between adjacent replicas, 116
Engrailed Homeodomain, 392, 392f
Entatic state proposal, 300
Entropy, 108
enzyme catalysis, 283–285
MM/PBSA method, 149
molecular configuration and, 8–9, 101–103
translational and rotational, 101–103
Enzymes
activation-free energy v., 274–275, 275f
Arrhenius-like expression (k_{cat}), 314, 325
chemical processes simulation in, 263–273
classical force fields, 250–258
computational methods of, 318–321
considerations of, catalysis, 273–277
dynamical proposals, 289–290
electrostatic contributions to, catalysis, 282–283
electrostatic free energies of macromolecules, 258–262
energy based analysis, LBHB and, 297–298, 300
entatic state proposal, 300
entropic proposal, catalysis and, 283–285
Enzymes Car-Parrinello Molecular Dynamics (CPMD), 273, 280
generic enzymatic reaction, 274, 276f
(ionic) HB, LBHB and, 293–299, 295f
high level calculations of incomplete models with low level results for, 279–280
induced fit, 300
introduction to, 249–250
LBHB, poor use of high level calculations, 289–299
Marcus' reorganization energy, 283
MDQT, 291
non-dynamical Monte Carlo methods, 289
orbital steering, 300
PMF, 292–293, 319
QM treatments of entire protein, 271–273
QM/MM approach, TIM and, 318–319, 321, 332f, 343–344, 347–348, 359, 362
QM/MM molecular orbital methods, 264–271, 343–344
quantum mechanical studies of reacting fragments in gas phase, 263–264
reactant state (RS), transition state (TS) and, 275–278, 281, 282–283, 284, 285, 286f, 298f
as reactor, 316, 318
reasonable calculations for, 281
reference state, 274, 276f, 277f
restrain-release (RRA) approach, 284
RS, RSD and, 275–278, 281, 285–289, 286f, 291–293, 292f, 300
RSD by desolvation, other electrostatic effects and, 291–293, 292f
solvation effect, 282
specificity, 316, 356–365
steric strain, confinement effect, NAC proposal and, 285–289, 286f, 287f
thermodynamic cycle, 275–276, 277f, 282
TIM, alternative catalytic pathways and, 326–329, 327f, 328f
TIM, classical transmission coefficient, and, 343–347, 346f, 347f
TIM comparison with model reactions for, 329, –330, 330f, 331f, 332f, 333–334
TIM "lid" transition, 322–325, 322f, 323f, 364
TIM mechanics, 314–316, 315f
TIM mechanics, activation barrier and, 325–342
TIM, MGS and, 316, 317f, 318, 356, 357f, 358–359, 359f, 360f, 361–362, 363f, 364, 365t, 366
TIM, perfect enzyme and, 313–314
TIM, perturbation analysis of important residues and, 334, 335, 336–337, 336f
TIM, protein motion effects, and, 349–350, 350t, 351f, 352, 353t354t–355t, 355
TIM, proton transfer, and, 315–316, 327–329, 349–350, 351f, 352, 354t–355t, 355, 357f, 358–359, 365

Enzymes (*cont.*)
TIM, proton tunneling and, 315–316, 320–321, 347–349, 350t, 353t, 355
TIM, QM/MM approach and, 318–319, 321, 326–327, 327f, 332f, 343–344, 347–348, 359, 362
TIM reaction, dynamic effects, and, 342–355
TIM, substrate polarization analysis by mutagenesis and, 337–342, 340t, 342f
TS, TSS and, 275–278, 282–283, 284, 285–289, 286f, 287f, 291–293, 292f, 294, 295, 298, 300
TST, 290, 344–345, 348–349, 350t, 353t
vibrationally enhanced tunneling not for large, catalysis, 290–291
water v., 274, 275f, 276f, 277, 277f, 278, 280
workings of, 4, 281–299, 313, 318, 364
Enzymes Car-Parrinello Molecular Dynamics (CPMD), 273, 280
Ergodicity, 134
Error
experiment/theory and estimation/ propagation of, 10–13
solvent models, 77
ESP. *See* Electrostatic potential charges
Estrogen, 137
EVB (empirical valence bond), 264, 268–271, 273, 280, 281, 294, 298, 301, 348
Ewald sums, 3
Ewald treatments, 255
Experiment
error estimation, error propagation, theory and, 10–13
model's role in, 13–15
theory influencing, 5–8
Extracellular matrix (ECM) with, 232

F

Fibrillogenesis, 232
Fibronectin (FN)
atomic force microscopy, 232–234
extracellular matrix (ECM) with, 232
FNIII modules, 232–237
integrin $\alpha_v\beta_3$, 237
mechanical signaling for, 232–234, 234f, 235f, 236–237, 236f
SMD, 233–234, 234f, 235f, 236
solvent models, 233
FKBP-12, 113–114
Flexibility, protein, 109
dynamic pharmacophore method, 114–115
relaxed complex methods, 112–114
in virtual screening, 111–115
Fluctuating charge (FQ) model, 50
FN. *See* Fibronectin
FN-III_{10}. *See* Fibronectin (FN-III_{10})
Folding problem, 371–373. *See also* Protein folding
Force fields, for protein simulation
Amber, 30–35, 41, 42t, 43, 44f, 47–48, 47t, 53, 145, 164, 199
atom-based, 31–32, 38, 38f, 39
BUFF, 40t, 41, 41t, 50
CEDAR, 39, 40t
CHARMM, 35–38, 41, 42t, 47t, 62f, 164, 200, 215m199, 319, 320, 321, 327–328, 327f, 328f, 330, 332f, 333, 334, 336, 339, 359f
Class I potential energy function of, 28–29, 41
classical, 250–258
comparisons among, 41, 42t, 43–45
CVFF, 40t, 145
definition of, 27
DISCOVER, 39, 40t
drug design, energy calculation and, 88–89, 117–118
electrostatics' consistent treatment in, 52–57
ENCAD, 40, 40t
environmental effects via polarization, 50–52
fixed atomic point-charge limitations with, 46–48, 47t
fixed partial charge model comparisons among, 41, 42t
flexible models for static charge distributions of, 48–50
free energy plot comparisons, 44t
"gold standards" not for, 41
GROMACS, 39
GROMOS, 39, 40t, 145–147, 164, 199
interactions of, 29, 29f
lone pair sites for, 48–49

Force fields, for protein simulation (*cont.*)
MD, direct simulations and, 252–254
MM3, 39–40, 40t, 45
MM4, 39–40, 40t, 45
MMFF, 39, 40t
NAMD and, 199–201
NEMO, 40t, 61
from 1980 to present, 30–40
nonpolarized, 61, 62f
OPLS, 38–39, 41, 42t, 43, 44f, 45t, 59–60, 59f, 60, 62t, 147, 164
in organic chemistry v. *sui generis,* 30
other, 39–40, 40t
parameterization of, 28
polarizable models for small molecules in, 58–60, 59f
polarization and, 28, 29, 32, 33, 33f, 41, 46, 47, 50–62, 78, 136
polarized, 60–61, 62f
potential functions, molecular mechanics, 250–252
proper/improper treatments of long-range effects of, 254–255
SIBFA, 40t, 60–61, 62f
solvent environment, molecular dynamics simulations with generalized Born model, and, 75–77, 76f
solvent models continuum for, 63–75, 78, 117
solvent models explicit for, 62–63
TINKER, 47–48, 47t, 58, 61, 62f, 72
torsional potentials, 29, 31, 34, 36, 37, 38, 43, 164
(explicit) water models for, 62–63
XED, 48
Fourier-transform infrared (FTIR) spectroscopy, 337–338
FQ. *See* Fluctuating charge (FQ) model
Free energy, 148, 327f
absolute, 138, 140, 141, 147
average, 148
binding, 88–89, 90, 96–97, 103–111, 113, 117, 138, 261–263
calculations, 96–97, 99, 99f, 100, 102–111, 258–263
chemical-scanning computational experiments, 109–110
computational models, 258–261
conformational, 372
difference of substrate to protein in, 255
discretized continuum (DC)/PB models, 258–259
enzymes and activation of, 285–288, 286f
explicit and implicit solvent models combined, 107–109
explicit-solvent models, parametrize implicit-solvent models and, 110
fast approximate simulation for calculating binding, 103–111
funnel-shaped, 393–394
Helmholtz, 96, 104–105, 108
Langevin dipoles (PDLP) model, 258–259, 260, 262
multiple-state, 151–153
plot comparisons, 44t
semi-empirical linear response theory, 110–111
single-step Zwanzig's perturbation method, 105–106
solvation, 70–71, 89, 92–93, 126, 152
steric forces, 285–287, 286f
stochastic boundary conditions, 254–255
systematic sensitivity analysis, 103–105
Tanford-Kirkwood (TK) model, 258
of transferring protein, 64–65
Free energy calculations, and ligand binding, 123–125
applications of, 129–135, 130f, 132f, 133f, 134f
deficient configurational sampling, 130–131
double-side sampling, 128–129
λ dynamics, chemical MC/MD and, 151–154
empirical/knowledge-based scoring approaches, 124
end-point catastrophe, 130–132, 132f
extrapolation of, 135–137
free energy differences for, 125–126
linear interaction energy (LIE) approaches, 137–147
MM/PBSA method, 147–149, 259
nonuniform λ-spacing, 131–132, 132f
P45cam, 145–147, 146f
plasmepsin II, 141–143, 142f
PROFEC method, 149–151
quadratic behavior, 128f
quantitative structure activity relationship (QSAR), 123–124

Free energy calculations, and
ligand binding (*cont.*)
thermodynamic cycle, 124–125, 134, 134f, 255–256, 287
thermodynamic integration, FEP and, 125–135, 153, 154–155
trypsin-BPT1, 143–145, 144f, 145f
Free energy perturbation (FEP), 3, 9–10, 105–106, 136, 150
binding large ligands, 256–257
drug design, ligand flexibility and, 96, 97, 99
LRA, all-atom models and, 255–258, 259
Monte Carlo (MC) methods, 127–128, 128f
PMF, 257–258
QM/MM, EVB and, 268
RSD, 293
single-step Zwanzig's perturbation method, 105–106
TI and, 125–135, 152, 153, 154–155
TI v., 128–129
umbrella sampling (US) method, 256, 257
Frozen bath assumption, 345–346
Funnel-shaped free energy, 393–394

G

GAP. *See* Glyceralde-hyde 3-phosphate
Gas phase
quantum mechanical studies of reacting fragments in enzymes, 263–264
solvated (SGP), 264
Gated channels, 167, 173f, 174
OmpA, 181f, 183
open v. closed, 171–172, 172f
voltage, 179
Gaussian94, 321
GB. *See* Born model, Generalized (GB)
Genes, 159, 162
Genome sequencing, 195, 239
Gln61, 280
Gln95, 340
GlpF (glycerol uptake facilitator), 202, 205, 208–211, 210f
MD stimulation, AQP and, 209–211, 210f
mutant of, 209
Glu, 183
Glu 165 (Glutamic acid), 314, 318, 320, 325, 326, 327f, 330, 332f, 333, 341, 342, 342f, 345, 349, 352, 353t
Glutamic acid. *See* Glu 165
Gly 210, 334
Glyceralde-hyde 3-phosphate (GAP), 314, 315f, 316, 317f, 334, 336–337, 356, 358, 361, 363, 364
Gramicidin A, 182
GROMACS force field, 39, 199
GROMOS force field, 39, 40t, 145–147, 164
Grotthuss mechanism, 202
GTP molecule, 279

H

H95N, 338
H95Q, 338–341, 342f
β-hairpin, 387–388
Haloalkane dehalogenase, 287–288, 287f
Hamiltonian, 349
EVB, QM/MM and, 269
formula, 101, 151
H(λ) classical, 96, 100, 108
QM and QM/MM, 265–266, 267
Hartree-Fock 6-31G, 32, 33f, 35, 36, 339
Hartree-Fock STO-3G, 31
HB. *See* Hydrogen Bond, 293
Helices
A, 377, 378–379
α, 163, 163f, 223, 376, 377, 385, 394
α4, 222
B, 377
C, 377
$c2_L$, 221f
$c2_R$ helices, 217f, 220, 220f, 221f
C-terminal, 228–231, 228f, 234, 234f, 377, 378, 386–387, 387f, 389f, 390, 393
D, 377
E, 378–379
F, 378
G, 378–379
H, 378–379
I, 283–285, 392
II, 283–285, 392
III, 283, 285, 392
native, 391

Helices (*cont.*)
N-terminal, 228–229, 229f, 231, 234, 234f, 378, 380, 381, 386–387, 389f, 390
peptide, 35, 45
transmembrane (TMH), 214, 216, 218–219, 221
Helmholtz free energy, 96, 100, 104–105, 108
Hen egg white lysozyme (HEWL), 377–378, 377f, 395
Heteropolymers, 393
HEWL. *See* Hen egg white lysozyme
High throughput screening (HTS), 8
His57, 297
His95, 318, 320, 326, 327f, 329, 330, 331, 332f, 334, 337, 340f, 341, 342f
Histidine, 284, 337
HIV proteases, 279, 280
HIV-1 RT, 153
Homology, 170, 172
Human Genome, 1
Hybrid potential energy function, 151
Hydrogen
bonding, 31, 34, 35, 38, 61, 180, 182, 202, 206, 206f, 208, 234, 235f, 324, 377
electronegativity of atoms, 34
Hydrogen Bond (HB), 293. *See also* Low Barrier Hydrogen Bond (LBHB)
LBHB and ionic, 295–296, 297, 299
oxyanion hole of subtilisin, 298
short, strong (SSHB), 296
Valence Bond, 294–295, 296
Hydrophobic collapse, 149
Hydrophobicity, 89, 140, 143, 147, 172, 216, 373, 375, 388, 396

I

Immumoglobulin-like (Ig), 232, 234
Induced multipole moments model, 51–57
Inhibitor aminoethyl phosphonate (IPP), 337
Integrals over spheres, 72, 74
Integrin $\alpha_v\beta_3$, 237
Interleukin-8, 76f
Inward rectifier (Kir) channels, 172–173
Ion binding energies, 254
Ion channels, 160, 161, 162, 167–174
brute force simulation, 167
ion permeability, selectivity in KcsA and, 167–171, 168f, 178
(mammalian) K channels, Kir and, 172–173
KcsA, MthK, gating models and, 171–172, 172f, 173f
Kv, 167
MscL, 173–174
OmpF porin, 181–182, 181f, 182f
related transporters, other channels and, 173–174
TWIK, 167
Ion-exclusion boundaries, 68
Ions
calcium, 92
Cl, 167, 172–173, 180
Cs^+, 169, 171
Na^+, 167, 168
Rb^+, 169
IPP. *See* Inhibitor aminoethyl phosphonate (IPP), 337

K

K channels, 167, 172–173, 180
KcsA, 163f
gating models, MthK and, 171–172, 172f, 173f
ion permeability and selectivity in, 167–171, 168f
Kir and, 172–173
potassium, 133, 133f, 167, 173f
Kinesin binding, 8
Kir channel, 172–173
Kirkwood integration formula, 97–98
Kv channels, 167

L

Lactate deydrogenase, 323
Lagrangian treatment, 51, 52
LamB porin, 175
Langevin
NAMD, 218
Langevin dipoles (PDLD) model, 258–259, 260, 262, 298
Langevin Dynamics, 205
Langevin thermostat damping, 217f

Large scale simulation
aquaporins–membrane water channels (AQP), 201–203, 203f, 204f, 205–211, 206f, 207f, 208f, 210f
ATP synthase, energy conversion and, 196, 198, 211–232
biopolymer conformation forces, 238–239
CHARMM, AMBER and GROMACS force field, 199, 200, 215
computing power used in, 196, 197t
DNA, 238
fibronectin's (FN) mechanical signaling for, 232–234, 234f, 235f, 236–237, 236f
future of, 237–239
genome sequencing, 195, 239
GlpF, 202, 205, 208–211, 210f
integrin $\alpha_v\beta_3$, 237
introduction of, 195–198, 197t
MD simulations, ATP synthase, and, 214–221, 223
mechanical model of F_o, ATP synthase and, 221–223, 221f
NAMD program, 196, 199–201, 200f
NPA motifs, 205, 206, 206f, 207f
RNA polymerase II, 198
steered molecular dynamics (SMD), 199, 207–208, 208f, 224–227, 224f, 226f, 227f, 230, 231–232, 233–234, 234f, 235f, 236, 238
structural model of F_o, ATP synthase and, 218–219, 218f
technology, 198–201, 198f, 200f
time scales of, 238
torque generation of F_o, ATP synthase and, 213–223, 217f, 218f, 220f, 221f, 222f
torque-drive catalysis in F_1 ATP synthase, 223–232, 224f, 226f, 227f, 228f, 229f, 230f
transmembrane helices (TMH), 214, 216, 218–219, 221
VMD program, 196, 199, 201, 215
water dynamics, 201–202, 209–211, 210f
LBHB. *See* Low Barrier Hydrogen Bond
Lennard-Jones parameters, 14, 16–17, 29, 111, 138, 150
for Amber force fields, 33–34
for CHARMM force fields, 36
of very small atom, 131
perturbation and, 106, 130
LES. *See* locally enhanced sampling (LES)
Levinthal's paradox, 371–372, 374
Ligand(s)
binding, 18, 27, 77–78, 89, 102, 110–111
binding large, 256–257
drug design, modeling molecular recognition, and flexibility of, 95–103
drug design, protonation states and, 94–95
floppy, 109
functional groups, 98–99, 99f
ghost, 152
MM/PBSA method, 107–108, 109, 114, 1198
ranking of, 99
receptor, binding energy and, 88–89
receptor complexes, 8–9, 10, 11, 18
soft, 137
Ligand binding, free energy calculations and, 123–125, 256–257
deficient configurational sampling, 131
double-side sampling, 128–129
λ dynamics, chemical MC/MD and, 151–154
empirical/knowledge-based scoring approaches, 124
end-point catastrophe, 130–132, 132f
extrapolation of, 135–137
free energy differences for, 125–126
ligand binding's applications and, 129–135, 130f, 132f, 133f, 134f
linear interaction energy approaches, 137–147
MM/PBSA method, 147–149
nonuniform λ-spacing, 131–132, 132f
P45cam, 145–147, 146f
plasmepsin II, 141–143, 142f
PROFEC method, 149–151
quadratic behavior, 128f
quantitative structure activity relationship (QSAR), 123–124
thermodynamic cycle, 124–125, 134, 134f
thermodynamic integration, free energy perturbation and, 125–135
trypsin-BPT1, 143–145, 144f, 145f
Linear interaction energy (LIE), 137–147, 148, 250, 262

Linear interaction energy (LIE) (*cont.*)
absolute binding free energy, 138, 140, 141, 147
β coefficient, 138–139
P45cam, 145–147, 146f
plasmepsin II, 141–143, 142f
PROFEC, 150
solvent accessible surface and solute-solvent van der Waals energy, 140–141, 140f
SRRRS isomers, 142
stereoselectivity, thrombin inhibitors and, 139–140
trypsin-BPT1, 143–145, 144f, 145f
Linear response approximation (LRA), 256–258, 259–262
Lipid bilayer, 161, 164–165, 166, 169, 179, 180, 182f, 204f
Local reaction field (LRF) model, 255, 257
Locally enhanced sampling (LES), 98–99
Loop L3, 178–179
Low Barrier Hydrogen Bond (LBHB), 293
deprotonated, 297
energy-based analysis of, 297–298
HB, VB and, 297
new element in, 294–296
NMR shift, 296
poor use of ''high level'' calculations, 298–299
single rather than double minimum, 294–295
TSS and, 294–295, 298
Valence Bond, 294–295, 295f, 296
Lys12, 330, 334, 341
Lys72, 293
Lysozyme, 124

M

Macromolecules
gas-phase properties of, 2–3
small molecules computations v., 2–5
Maltohexose, 180
Maltoporin, 180
Marcus' reorganization energy, 283
Maxwell distribution, 344
MC. *See* Monte Carlo (MC) methods
MD. *See* Molecular dynamics (MD)
MDQT. *See* Molecular dynamic with quantum transition
Membrane protein
biological membrane, 161, 162f
inner v. outer membrane, gram-negative cells and, 174
ion channels, 162
membrane protein structure, 159, 160, 162–163, 163f
solute transporters, 162
two classes of, 163f
Membrane protein simulations, 159–160, 163
ab initio molecular dynamics (AIMD), 184–185
bacterial outer, 160, 161
(lipid) bilayer, 165, 166f
classical molecular dynamics, 164–165, 166f
density functional theory (DFT), 170, 185
future prospects of, 183–185
gating models, KcsA, MthK and, 167, 171–172, 172f, 173f, 174
in vacuo, 164–165
ion channels, 160, 161, 162, 167–174
ion permeability, selectivity in KcsA and, 167–171, 168f
mammalian K channels, Kir, models and, 172–173
OmpA, OmpA^{NT,} elusive ion channel and, 181–183, 181f, 182f
outer (OMP), 174–183
porins as model of transport, selectivity, gating for, 175, 177–180
quantum-mechanical (QM/MM) methods, 185
transporters, MscL other channels and, 173–174
virtual outer, 184
X-ray diffraction, 160
Methyl glyoxal synthase (MGS), 316, 317f, 318
DHAP, 316, 317f, 334, 358f
EDT2, 358–359, 359f, 360f
substrate phosphate group, 362, 364f
TIM, 316, 317f, 334, 356, 357f, 358–359, 359f, 360f, 361–362, 363f, 364, 365t, 366
MGS. *See* Methyl glyoxal synthase
Michaelis-Mention constant, 314, 334
Microtubules, 8

Minimum energy path (MEP), 320–321
MM3 force field, 39–40, 40t, 45
MM4 force field, 39–40, 40t, 45
MMFF (Molecular Mechanics) force field, 39, 40t
MM/PBSA (Molecular Mechanics/Poisson-Boltzman/Surface Area) method
  free energy calculations, ligand binding and, 147–149, 259
  hydrophobic collapse, 149
  ligands, 107–108, 109, 113, 114, 118
  PROFEC, 150
Model(s)
  ab initio (first principle), 5, 13, 15, 264, 268, 271
  folding, 371–373
  semi-empirical, 13–14, 15
  testing theoretical, 16–18
  theory and role of, 13–15
Molecular dynamic with quantum transition (MDQT)
  surface-hopping method, 291
Molecular dynamics (MD), 124, 125, 138, 142, 148, 289, 344
  ab initio (AIMD), 184–185
  AQP, 203, 209–211
  ATP synthase, 214–221, 223–224, 224–227, 224f, 226f, 227f, 231–232
  Cartesian representation, 252
  direct simulations, enzymes and, 252–254
  λ dynamics, chemical MC/MD and, 151–154
  FEP, 127–128, 128f, 131, 132f
  fibronectin and steered (SMD), 233–234, 234f, 235f, 236
  KcsA potassium, 133, 133f
  LRA and, 259–262
  membrane protein and classical, 159–160, 164–165, 166f
  MscL channel, 173
  NAMD, 196, 199–201, 200f
  Ompa^{NT}, 182–183
  outer membrane protein and classical, 180
  PDLP and, 259–260, 262
  photoisomerization process, 253
  PROFEC, 150
  propagation of atoms, fixed total energy and, 252–253
  replica-exchange method, 115–116
  RS, TS and, 287
  steered (SMD), 199, 207–208, 208f, 224–227, 224f, 226f, 227f, 230, 231–232, 233–234, 234f, 235f, 236, 238
  targeted (TMD), 223–224, 224f
  torque generation F_0 ATP synthase, 214–221, 223
  torque-driven catalysis in F_1 ATP synthase and steered (SMD), 224–227, 224f, 226f, 227f, 230, 231–232
  torque-driven catalysis in F_1 ATP synthase and targeted (TMD), 223–224, 224f
  uncritical use of first principle of, 280
  VMD (visual), 196, 199, 201, 215
Molecular mechanics (MM). *See also* Quantum-mechanical molecular mechanics (QM/MM) method
  Cartesian representation, 252
  force fields, potential functions and, 250–252
  molecular crystals, 252
  valence bond-based (VB) surfaces an, 269
Molecular Mechanics force field. *See* MMFF force field
Molecular Mechanics/Poisson-Boltzman/Surface Area. *See* MM/PBSA method
Molecular potential functions, 250–252
Molecular recognition
  drug design and modeling of, 95–103
  λ dynamics method, 99–100
  free energy binding, 96–97
  free energy calculations, 96–97, 99, 99f, 100, 102
  Kirkwood integration formula, 97–98
  parallel fast-growth method, 100–101
  translational and rotational entropy, 101–103
Molecular simulation software, 16
Molecules
  condensed phase properties of, 2
  gas-phase properties of, 2
Molecules, small, 72, 74
  finite-difference or boundary element numerical methods for, 64
  macromolecules computations v., 2–5, 48
  polarizable models for, 58–60, 59f

Momentum relaxation time, 11
Monte Carlo (MC) methods, 9, 12, 17, 27, 30, 38, 48, 92, 115, 125, 252
  λ dynamics, chemical MC/MD and, 151–154
  enzymes, 289
  FEP, 127–128, 128f
  ion concentration, Grand Canonical, 178
MscL channel, 173–174
MthK
  gating models, KcsA, and, 171–172, 172f, 173f
MUSIC procedure, 114
Mutagenesis, 322
  substrate polarization analysis by, 337–342, 340t, 342f
Myoglobin, 378–379, 379f

N

Na^+ ions, 167, 168
NAC (near attack conformation), 285
  desolvation-induced, 288
  RSD and TSS, 285–289, 286f, 287f
NAMD program, 196, 233
  FFT-based (PME) mesh calculation, 199–200, 200f
  force fields, 199, 200, 201
  Langevin, 218
  NAMD2, 217f
  steered molecular dynamics (SMD), 199
  timestep limitation, 199–200, 200f
NCSA supercomputer clusters, 225
Near attack conformation. *See* NAC
NEMO force fields, 40t, 61
Neural nets, 14
N-methylacetamide, 33, 35, 37, 37f, 39, 62f
NMR, 35, 75, 114, 290, 382, 391, 395
  ATP synthase, 214, 219
  crystallographic experiment v., 7–8
  errors and, 11
  LBHB, 296
  membrane protein stimulations, 160, 165, 183, 184
  TIM, 325
  typical, 6, 7
NPA motifs, 205, 206, 206f, 207f
N-terminal helice, 228–229, 229f, 231, 234, 234f, 378, 380, 381, 386–387, 389f, 390
Nucleic acid
  binding, 5
  protein native structure and, 7–8
NVE, 233

O

ODCase, 291–293, 292f
Oligomer, 214, 222
OmpA, 166f, 176t
OmpF porin, 176t, 177–178, 179, 180
  as elusive ion channel, 181–182, 181f, 182f
  N-terminal domain ($Ompa^{NT}$), 181–183
OmpK36, 177
1-M KCI, 179
ONIOM model, 267
OpcA, 176t
OPLS (Optimized Potentials for Liquid Simulations) force fields, 38–39, 164
  all-atom (AA), 38, 39, 39f, 41, 42t, 43, 44f, 45t, 47t, 59–60, 59f, 60, 62f, 147
  AMBER/, 38
  charge models of, 39, 39f
  neutral groups enforcement in, 39
  polar hydrogen only approach of, 38
  potentials, 32
  torsional potentials, 38
  UA, 38, 39, 39f, 41, 42t, 43
Orbital steering, 300
Osmotic permeability (p_f), 210–211
Outer membrane proteins (OMP)
  inner v. outer membrane, gram-negative cells and, 174
  OmpA, $OmpA^{NT}$, elusive ion channel and, 181–183, 181f, 182f
  porins as model of transport, selectivity, gating for, 175, 176t, 177–180
  structure of, 175, 176t
Overpolarization
  of Amber force fields, 32

P

P1-ASP, 144
P21ras, 279

P45cam (camphor analogue complexes), 145–147, 146f
Parallel fast-growth method, 100–101
Parameterization
 of CHARMM force fields, 35, 37
 of force fields, 28, 43–44
 of structural biology, 15–16, 18
Particle mesh Ewald. *See* PME
PB. *See* Poisson-Boltzmann (PB) model
PCM model, 320
PDLD. *See* Langevin dipoles (PDLD) model
PDM programs, 49
Peptides, 41, 112, 116. *See also* Dipeptide; Tetrapeptide
 helical, 35
 helix-coil transition, 45
 poly, 4
 short/small, 34–35, 37
Periodic boundary conditions (PBC), 205
Perturbation. *See also* Free energy perturbation (FEP)
 double-side sampling, 128–129
 large atoms, 106
 Lennard-Jones parameters, 106, 130
 MD and FEP, 127–128, 128f, 131, 132f
 PROFEC, 106
 simulation, 9–10
 single-step, 137
 single-window Zwanzig's, method, 105–106, 118
 small, 153
 TIM, important residues and analysis of, 334, 335, 336–337, 336f
PGA, 356
PGH inhibitor, 320, 333, 337, 338, 342f, 364f
pH, 177, 381, 395
 dependent properties of proteins, 94
 establishing, 6
 low, 376, 378
Pharmacophore method, dynamic, 114–115
PhoE, 177
Photoisomerization, 253
Pictorial representation of free energy changes method. *See* PROFEC method
 π-cations interactions, 31
pK_a (protein affinity), 95, 177, 250, 254, 255, 262, 279, 280, 297, 298, 329
Plasmepsin II, 141–143, 142f
PM3 methods, 321
PME (particle mesh Ewald), 199, 200, 200f, 225
PMF. *See* Potential of mean force
Poisson equation, 319
Poisson model, 90–91
Poisson-Boltzmann (PB) model, 68–69, 71, 73, 75, 89, 107, 108, 152, 319
 as discretized continuum (DC) models, 258–259
 electrostatic forces, 90, 109, 177
 free energy, 258–259, 262
 protonationstates, 94–95
 salt effects included in, 91–92
Poisson-Nernst-Planck theory, 180
POL3 water model, 53
POL5/QZ model, 58
POL5/TZ model, 58
Polarizable electropole model, 58
Polarization
 additive model of, 52–53
 Applequist model of, 53, 54, 55
 balance of inter- and intramolecular energetics, 57
 catastrophe, 53–55
 current status of, force fields, 57
 damped induction interactive model of, 54–55, 58
 Dang-Chang model of, 53
 Drude oscillator (DO) model of, 51, 58
 environmental effects via, 50–52
 fluctuating charge (FQ) model of, 50
 force fields, 28, 29, 33f, 41, 47–48, 47t, 50–62, 78, 136
 induced multipole moments model of, 51–57
 interactive model of, 52–53, 54
 iteractive solution, 51–52
 models for small molecules, 58–60, 59f
 nonpolarizable protein force fields and, 61, 62f
 nonpolarization, prepolarization, water models and, 63
 permanent moments, 56
 polarizable protein force fields and, 60–61, 62f
 polarization analysis by mutagenesis, TIM and substrate, 337–342, 340t, 342f
 RESP, 33, 33f
 short range damping, 55
 solvent electrostatic, 107

Polarization (*cont.*)
  Thole water model, 54–55, 58
Polynomial path, 97–98
POPC (palmitoyl-oleoyl-phosphatidyl-choline), 165, 205
POPE (palmitoyl-oleoyl-phosphatidyl-ethanolamine), 205, 215
Pore proteins, 174
Porins, 174. *See also* Aquaporins
  conjugate peak refinement methods, 180
  LamB, 175
  Loop L3 and, 178–179
  maltoporin, 180
  OmpF, 176t, 177–178, 179, 180
  Rhodobacter capsulatus, 176t, 177, 178
  ScrY, 175
  transport selectivity, gating and, 175, 177–180
Potassium ion, 167, 173f. *See also* KcsA
Potentials of mean force (PMF)
  AQP, 208–209, 208f
  ATP synthase, 222, 222f
  barrier, 345, 345f
  enzyme, 292–293, 319
  FEP, 257–258
  RSD, 292–293
  solvent environment modeling, 63, 64–65, 68, 257–258
  TIM, 344–345, 345f
Pro64, 219
Pro82, 236f
PROFEC (pictorial representation of free energy changes) method, 149–151
Prokaryotic membrane protein, 160
Protein
  ab initio folding trajectories for small, 5
  amino acid substitution on, stability, 4
  crystal structures, 17
  electrostatic properties of, 16
  folded structure of arbitrary, 4
  free energy of transferring, 64–65
  mass spectrometry of, 18
  models' role in simulations of, 13–15
  nucleic acid native structure and, 7–8
  TIM, enzymes, and motion of, 349–350, 350t, 351f, 352, 353t, 354t–355t, 355
  typical calculation of, 6
Protein A, 383–385, 384f, 395
Protein Databank, 1, 11
Protein degree of freedom, 64
Protein folding
  barnase, 379–380, 380f
  bovine pancreatic trypsin inhibitor, 143–145, 375–377, 376f, 395
  chymotrypsin inhibitor 2, 381–383, 382f, 393
  Engrailed Homeodomain, 392, 392f
  fast, 391–393
  first systems of, 375–379
  folding problem, minimal models and, 371–373
  force field and dynamic protocols, 373–375
  future of unfolding simulations, 396–397
  heat capacity, 3
  hen egg white lysozyme, 377–378, 377f, 395
  infancy of direct simulation of folding, 395–396
  kinetics, 3
  microscopic reversibility, 394
  minimal models, 373
  modern, 379–391
  myoglobin, 378–379, 379f
  pathway and funnel views of, 372
  Protein A, 383–385, 384f, 395
  Protein G, 386–388, 387f
  protein unfolding by pathways, 393–395
  SH3 domain, 388–391, 389f
  timescales of, 374
  ubiquitin, 385–386, 386f
  unfolding, 371, 374, 379–380, 383–384, 385–386, 386, 386f, 387, 387f
  villin, 391–392, 391f
  WW domain, 393, 393f
Protein G, 386–388, 387f
Protein kinase A, 103
Proton sponge, 297
Proton translocation, 214, 215
Proton transfer (PT), 280, 315–316, 327–329, 358–359
  coupling process of, 349–350, 351f, 352, 354t–355t, 355, 365
  essential and path modes, 352, 353t, 357f
  lower effective, 355
  saddle point modes, 352, 354t–355t, 355, 357f, 365
  uncoupled, 352

Proton tunneling
minimum energy path (MEP), 320–321
small-curvature tunneling (SCT), 320–321, 348, 350, 350t, 353t
TIM, 315–316, 320–321, 347–349, 350t, 353t, 355
Protonation, 213
artificial, 279
calculation of, 262
states, of proteins and ligands, 94–95
PSC supercomputer clusters, 225
Pseudoration, 31

Q

Quantitative structure activity relationship (QSAR), 14, 123–124
Quantum-mechanical molecular mechanics (QM/MM) method, 185, 332f. *See also* Molecular dynamic with quantum transition
ab initio, 268, 273, 281, 301, 343–344
COSMO model and, 281
DFT, FDFT, CDFT and, 267, 271–273
divide and conquer (D&C) approach, 271–272
enzyme and substrate divided in, 264–265
Enzymes Car-Parrinello Molecular Dynamics (CPMD), 273, 280
enzymes, reacting fragments in gas phase and, 263–264
EVB as reliable, 268–271, 280, 281, 301, 348
EVB potential, FEP and, 268
hybrid orbital, linked atom problem and, 266–267
inner region I, surrounding region II and, 265, 267
linear relationship between free energies and reaction energies (LFER), 270–271
molecular orbital, 264–271
ONIOM model, 267
QM-FE approach, 264
quantum and classical connection in, 266
region I, region II connectivity, 266–267, 272
solute-solvent coupling, 265–266, 267, 269–270
TIM and, 318–319, 321, 326–327, 327f, 332f, 343–344, 347–348, 359, 362
total potential, 264–265

R

Rb^+ ions, 169
Reactant state (RS), 275–277, 284, 300
destabilization (RSD), 275–278, 281, 285–289, 286f, 291–293, 292f
NAC, RSD and, 285–289, 286f
RSD, desolvation and, 291–293, 292f
Relaxed complex methods, 112–114
automated docking small libraries, 112–113
rapid docking of minilibraries, 112–113
single v. double ligand method, 113
slower docking algorithm, 113
Replica-exchange method, 115–117, 118
RESP. *See* Restrained electrostatic potential fit
Restrained electrostatic potential fit (RESP), 15, 33, 33f, 36, 46
Restrain-release (RRA) approach, 284
RGD loop, 236–237, 236f
Rhodobacter capsulatus porin, 176t, 177, 178
RMS values, 46, 59f
Root mean square deviation (RMSD), 171, 201
RS. *See* Reactant state (RS)
RSD. *See* Reactant state, destabilization

S

Salt
Born model, 73
bridge, 182–183, 220–221, 220f
Poisson-Boltzmann model, 91–92
β-sandwich motif, 232
SCC-DFT. *See* Semi-empirical density functional theory (SCC-DFT)
ScrY porin, 175
Semi-empirical density functional theory (SCC-DFT), 321, 356, 359f, 361
Semi-empirical linear response theory, 110–111, 118

Serine proteases, 144f.143, 297–298, 299
S-GB (surface-GB) model, 72–73, 75
SH3 domain, 388–391, 389f
SHAKE-like algorithm, 344
β–sheet, 376, 380, 381, 385, 387, 390, 394
Shell model. *See* Drude oscillator (DO) model
SIBFA Force fields, 40t, 60–61, 62f
Single-window Zwanzig's perturbation method, 105–106, 118
Small-curvature tunneling (SCT), 320–321, 348, 350, 353t
  CVT-SCT, 348, 350t, 353t
Solute transporters, 162
Solute-solvent coupling
  QM/MM, 265–266, 267, 269–271
Solute-solvent interaction energy, 66–67
Solute-solvent van der Waals energy, 140–141, 140f
Solvated gas phase (SGP), 264
Solvation
  absolute energy, 147
  effect on enzymes, 282
  (free) energy, 70–71, 89, 92–93, 126, 152
  offset, 152
Solvent dielectric, 64
Solvent environment modeling, 3, 62–77
  all-atom, 251, 254
  (generalized) Born (GB) model in, 67, 69–77, 89, 92–94
  continuum solvent models explicit for, 63–75, 66f, 78, 88, 89, 95, 108
  COSMO model in, 65–68, 66f, 281
  electrostatic energy of charges, 65–68, 66f, 89
  explicit and implicit solvent models combined, free energy and, 107–109
  explicit-solvent models, parametrize implicit-solvent models and, 110
  hydrophobic contributions, 89
  molecular dynamics simulations with generalized Born model in, 75–77, 76f
  native folds distinguished form misfolded decoys, 75, 77
  PMF and, 63, 64–65, 68, 257
  Poisson model in, 90–91
  Poisson-Boltzmann (PB) model in, 68–69, 71, 73, 91–92
  potential energy function, 69
  salt effects, Born model and, 73
  second generation generalized Born model, 74–75
  solute-solvent interaction energy, 65–66
  stochastic boundary conditions, 254–255
  surface constraint all-atom solvent (SCASS), 254–256
  water models explicit for, 62–63
  water simulations v., 75
Solvent-solvent repulsion energy, 66–67
SOR. *See* Successive overrealization (SOR) technique
SPC model, 62–63
SPC/E model, 63
α-spectrin SH# domain, 390
SRRRS isomer, 142
ST2 model, 62
Standard deviation
  among runs v. within stimulation, 12
Staphylococcal protein A, 384f
Staphylococcal protein G, 386
Steered molecular dynamics (SMD)
  AQP, 207–208, 208f
  ATP synthase, torque-drive catalysis in F_1 and, 224–227, 224f, 226f, 227f, 230, 231–232
  fibronectin and, 233–234, 234f, 235f, 236
  NAMD program, 199
  TMD and, 224, 224f
Stereochemistry, 142f
Stereoselectivity, 139, 207
Steric forces, 285–288, 286f
Stochastic boundary conditions, 254–255, 318
A-strands, 234, 235f
B-strands, 234, 235f
F-strands, 236, 236f
G-strands, 234, 235f, 236, 236f
β-strands, 234, 235f, 236, 390
Streptavidin, 107
Successive overrealization (SOR) technique, 52
Sugar molecules, 175, 180, 202
Surface constraint all-atom solvent (SCASS), 254–256

T

T4 lysozyme, 91
T7 DNA polymerase, 280
Tanford-Kirkwood (TK) model, 258
Targeted molecular dynamics (TMD), 223–224, 224f
Taylor's series of expansion, 104–105, 127, 137
Temperate gap, between adjacent replicas, 116–117
Tetrapeptide, 34
Thermostat, 253, 254
Theory
  error estimation, error propagation, experiment and, 10–13
  experimentation influenced by, 5–8
  model's role in, 13–15
  testing theoretical models with, 16–18
Thermodynamic cycle
  enzymes, 275–276, 277f, 282
  free energy calculations, FEP, and ligand binding, 124–125, 129–130, 130f, 134f, 255–256
  free energy simulations with, 124–125, 287
  LES, 99
Thermodynamic integration (TI)
  FEP, 128–131, 153, 154–155
  formalism, 101
  formula, 127
  free energy calculations, ligand binding and, 125–135
  as slow growth, 127–128
Thermodynamic properties
  computations, 6–7, 8, 9, 17–18, 104
  OPLS, 38
Thole-type (TTM) water model, 54–55, 58–59
Thrombin inhibitor, 139
TI. *See* Thermodynamic integration (TI)
TIBO, 153
TIM. *See* Triosephosphate isomerase
Time-dependent transmission coefficient (κ(t)), 344–346, 346f
Timestep limitation, 199–200, 200f, 203
TINKER force fields, 47–48, 47t, 58–59, 59f, 72
TIP3P model, 35, 36, 53, 58, 62–63, 215
TIP4F model, 63
TIP4P model, 63
TIP5P model, 63
TK. *See* Tanford-Kirkwood (TK) model
TMD. *See* Targeted molecular dynamics
Torque generation, of F_o,
  ATP synthase, 213–223, 217f, 218f, 220f, 221f, 222f
Torque-drive catalysis, in F_1
  ATP synthase, 223–232, 224f, 226f, 227f, 228f, 229f, 230f
Torsion, 29, 43, 164
  Amber force field, 31, 34
  CHARMM force field, 36–37
  OPLS, 38
Transition state (TS), 275–277, 291–293, 292f
  charging, 282–283
  LBHB, TSS and, 294–295, 298
  NAC, TSS and, 285–289, 286f, 287f
  stabilization (TSS), 278, 281, 285–289, 286f, 294–295, 300
Transmembrane helices (TMH), 214, 216, 218–219, 221
Transport proteins, 174
Triosephosphate isomerase (TIM)
  activation barrier and mechanics of, 325–342
  alternative catalytic pathways, 326–329, 327f, 328f
  Brownian dynamics, 324–325
  classical transmission coefficient, 343–347, 346f, 347f
  comparison with model reactions for, 329, 330, 330f, 331f, 332f, 333–334
  DHAP, 314, 316, 317f, 318, 320, 329–330, 332f, 333, 334, 338, 340, 341, 342, 343, 347–348, 349, 352, 353t, 359, 361, 364–366
  dihedral angle, 323–324, 323f
  dimeric enzyme, 314
  dynamic effects in reaction of, 342–355
  EDT2, 356, 358, 359, 359f, 360f, 361f
  EDTI, 315, 317f, 326, 327f, 328, 328f, 329, 330, 330f, 331f, 332f, 333, 334, 335, 337, 352, 359, 359f, 365
  frozen bath assumption, 345–346
  GAP, 314, 315f, 316, 317f, 334, 336–337, 356, 358, 361, 363, 364
  hydrolytic phosphate elimination reaction, 322
  instantaneous barrier, 346–347

Triosephosphate isomerase (TIM) (*cont.*)
isomerization, 316, 317f, 358
"lid" transition, 322–325, 322f, 323f, 364
mechanics, 314–316, 315f
MGS, 316, 317f, 318, 356, 357f, 358–359, 359f, 360f, 361–362, 363f, 364, 365t, 366
NMR techniques, 325
as perfect enzyme, 313–314
perturbation analysis of important residues, 334, 335, 336–337, 336f
PGH, 320, 333, 337, 338, 342f
phosphate elimination, 362
PMF, 344–345, 345f
protein motion effects, 349–350, 350t, 351f, 352, 353t354t–355t, 355
proton transfer (PT), 315–316, 327–329, 349–350, 351f, 352, 354t–355t, 355, 357f, 358–359, 365
proton tunneling, 315–316, 320–321, 347–349, 350t, 353t, 355
QM/MM approach, 318–319, 321, 326–327, 327f, 332f, 343–344, 347–348, 359
substrate phosphate group, 362, 364f
substrate polarization analysis by mutagenesis, 337–342, 340t, 342f
time-dependent transmission coefficient (κ(t)), 344–346, 346f
TST, 290, 344–345, 348–349, 350t, 353t
Trp 168, 325
Trypsin, 100, 106, 146, 198f
Trypsin-benzamide, 145
Trypsin-BPT1, 143–145, 144f, 145f
TS. *See* Transition state
TSS. *See* Transition state, stabilization
TST, 290, 344–345, 348–349, 350t, 353t
2-D WHAM, 43, 44f
Tyr16, 298
Tyrosine, 133

U

Ubiquitin, 385–386, 386f
UHBD package, 90
Umbrella sampling (US), 256, 257

V

Valence Bond (VB)
Low Barrier Hydrogen Bond, HB and, 294–295, 295f, 296
Valence bond-based (VB) surfaces, 269
Van der Waals forces, 3, 11, 14, 16–17, 29, 31, 90–91, 131, 138, 140f, 150, 164, 266, 324
Variational transition state theory (VTST), 320, 321
VET. *See* Vibrationally enhanced tunneling
Vibrationally enhanced tunneling (VET), 290–291
Villin, 391–392, 391f
Viral fusion, 8
Virtual outer membrane, 184
Virtual screening, 8
protein flexibility, 111–115
Virtual spring, 180
VMD (visual molecular dynamics) program, 196, 199, 201, 215

W

Water, 3
AQP, MD simulation and, 209–211, 210f
balance, 201
enzymes v., 274, 275f, 276f, 277, 277f, 278, 280
explicit models of, 62–63
generalized Born model v. simulations in, 75
model, 14, 185
nonpolarization, prepolarization and, 63
osmotic permeability (p_f), 210–211
PMF, 63, 64
POL3 model for, 53
POL5/QZ model for, 58
POL5/TZ model for, 58
SPC model for, 62–63
SPC/E model for, 63
ST2 model for, 62
Thole-type (TTM) model for, 54–55, 58–59

Water (*cont.*)
TINKER model for, 58, 72
TIP3P model for, 35, 36, 53, 58, 62–63, 215
TIP4F model for, 63
TIP4P model for, 63
TIP5P model for, 63
Windowing computations, 97, 101
WW domain, 393, 393f

X

X-PLOR, 199
X-ray crystallographic methods, 167, 184
X-ray diffraction, 160

Z

Zwanzig's perturbation method, single-window, 105–106, 118

ISBN: 0-12-034266-9

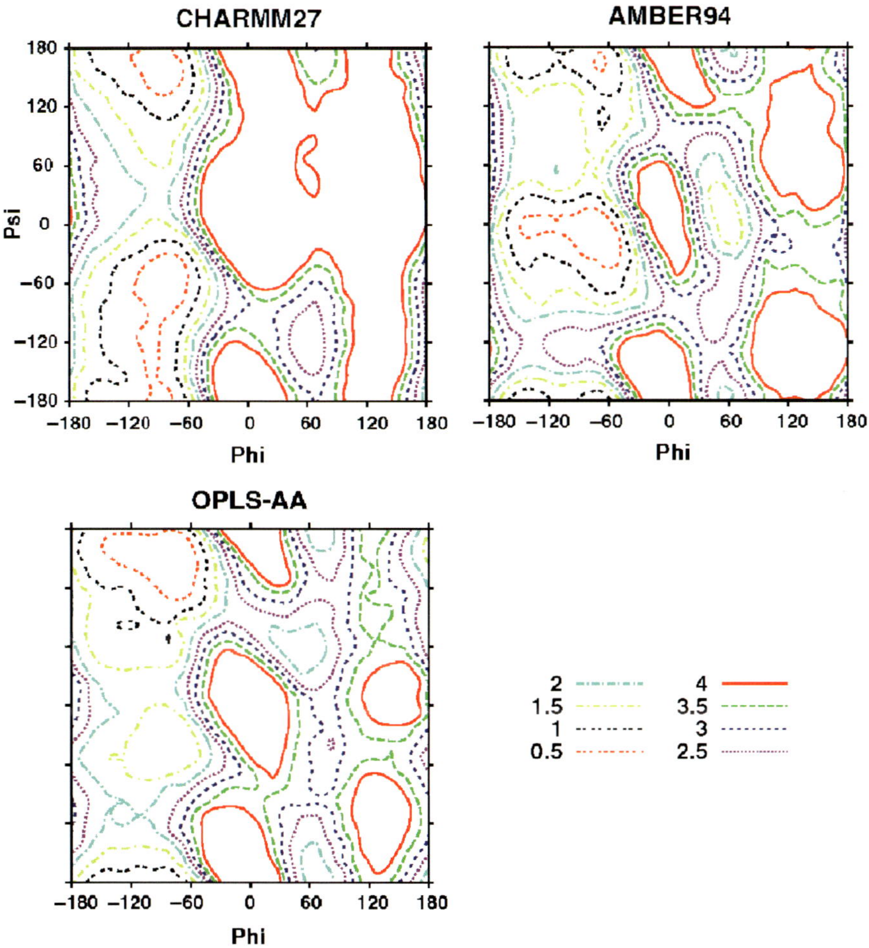

PONDER AND CASE, FIG. 5. Comparison of ϕ-ψ free energy plots from 2-D WHAM analysis of umbrella sampling trajectories for alanine dipeptide as computed with the AMBER *ff94*, CHARMM27 and OPLS-AA pairwise force fields. The contours shown are at 0.5 kcal/mol intervals from the global minimum of each Ramachandran map.

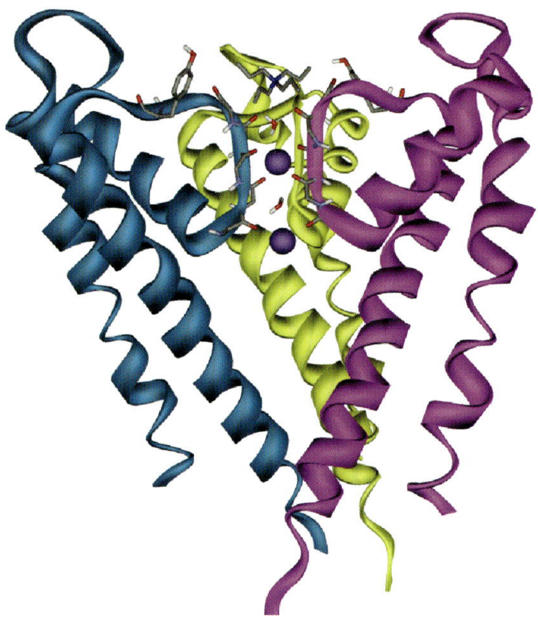

BRANDSDAL *ET AL.*, FIG. 4. Snapshot from an MD simulation of the KcsA potassium channel in complex with tetrapropylammonium ion at the extracellular binding site (one channel subunit has been removed for clarity). Two K^+ ions in the selectivity filter are also shown.

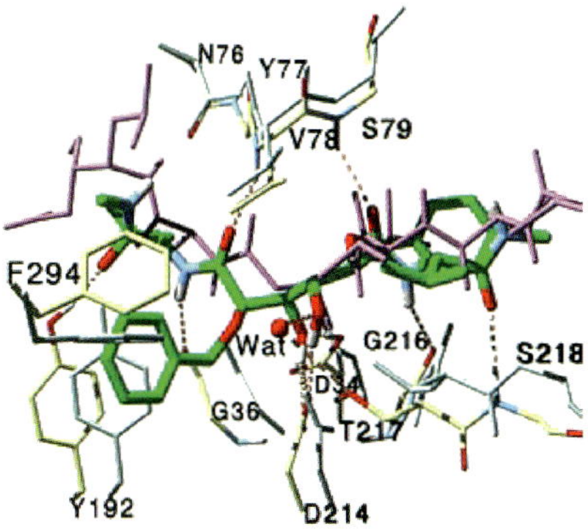

BRANDSDAL *ET AL.*, FIG. 8. Stereo view of the predicted (MD average) binding conformation of an active inhibitor (*top, rightmost in Fig. 7;* green) in complex with Plm II (yellow), superimposed on the x-ray structure of pepstatin A (purple) complexed with Plm II (light blue) (Silva *et al.*, 1996).

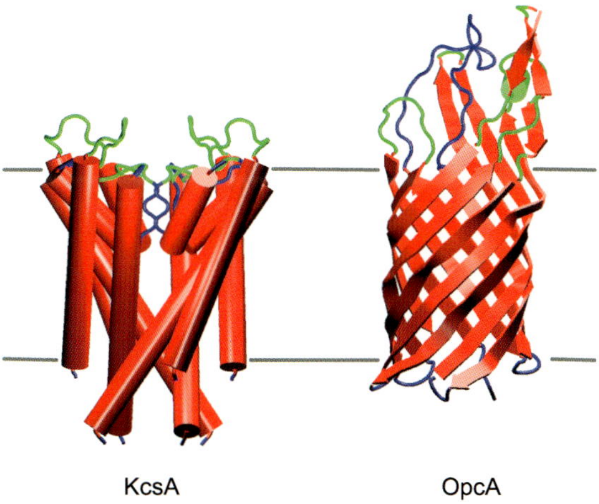

DOMENE *ET AL.*, FIG. 2. The two classes of membrane protein: an α-helix bundle illustrated by KcsA (Zhou *et al.*, 2001); and a β-barrel, illustrated by OpcA (Prince *et al.*, 2002). The grey horizontal lines indicate the presumed location of the membrane.

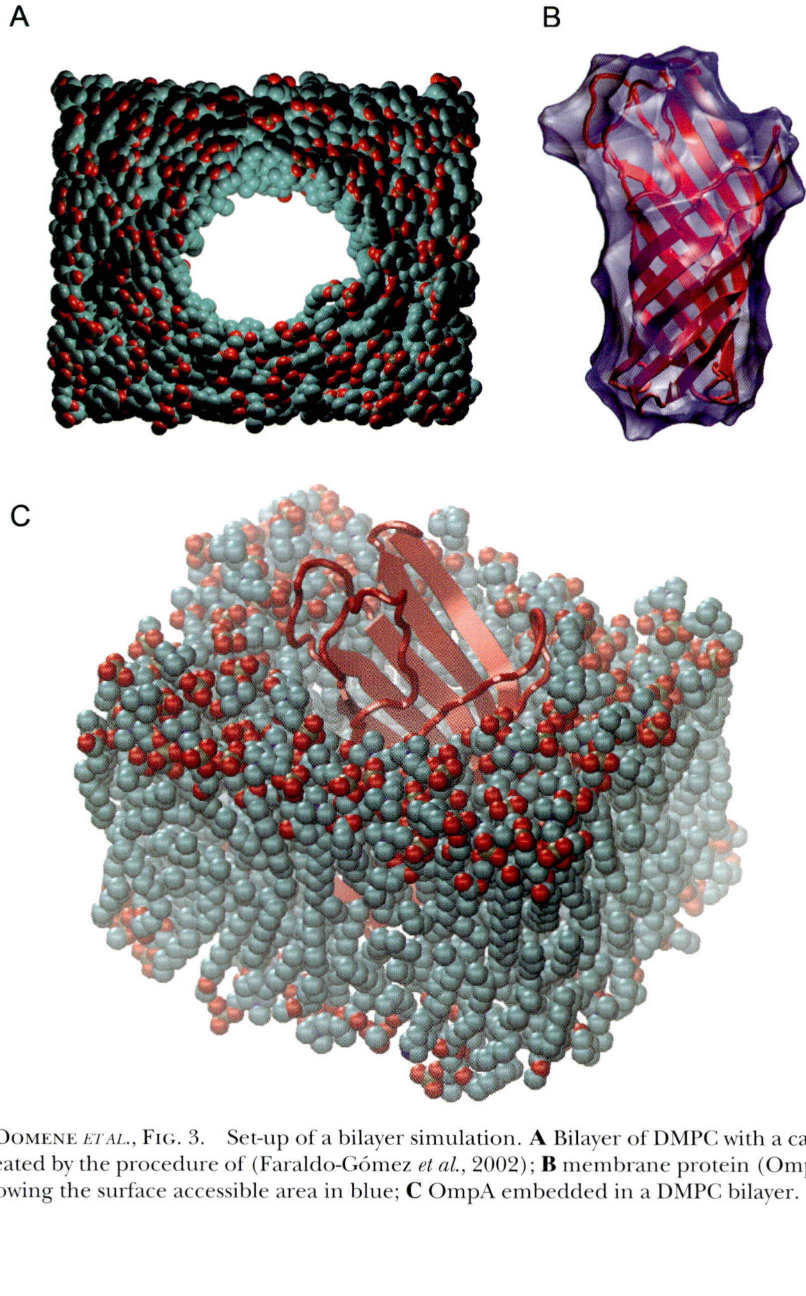

DOMENE *ET AL.*, FIG. 3. Set-up of a bilayer simulation. **A** Bilayer of DMPC with a cavity created by the procedure of (Faraldo-Gómez *et al.*, 2002); **B** membrane protein (OmpA) showing the surface accessible area in blue; **C** OmpA embedded in a DMPC bilayer.

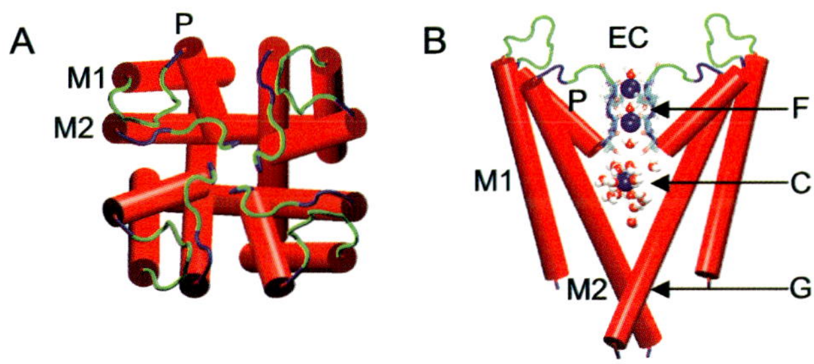

DOMENE *ET AL.*, FIG. 4. KcsA: **A** is a view down the pore axis from the extracellular mouth, showing the M1, P, and M2 helices; **B** is a view (of just two subunits) down a perpendicular to the pore axis, showing the extracellular mouth (*EC*), the filter (*F*), the central cavity (*C*), and the gate (*G*) regions. Two K^+ ions (plus waters) are shown in the filter (at sites S1 and S3) along with a third K^+ ion in the central cavity.

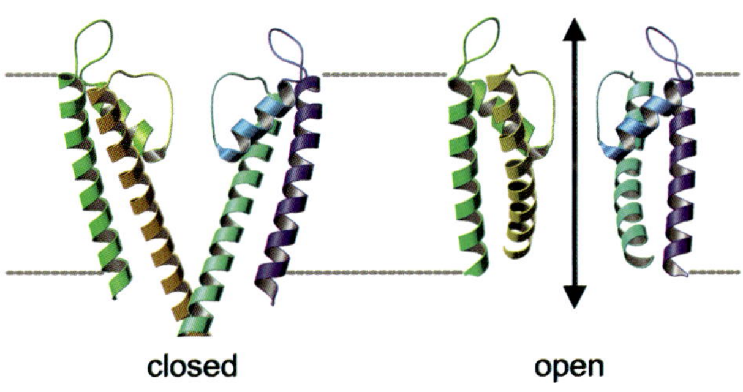

DOMENE *ET AL.*, FIG. 5. Comparison of the closed (x-ray) vs. open (modelled on MthK) structures of KcsA. Two of the four subunits are shown.

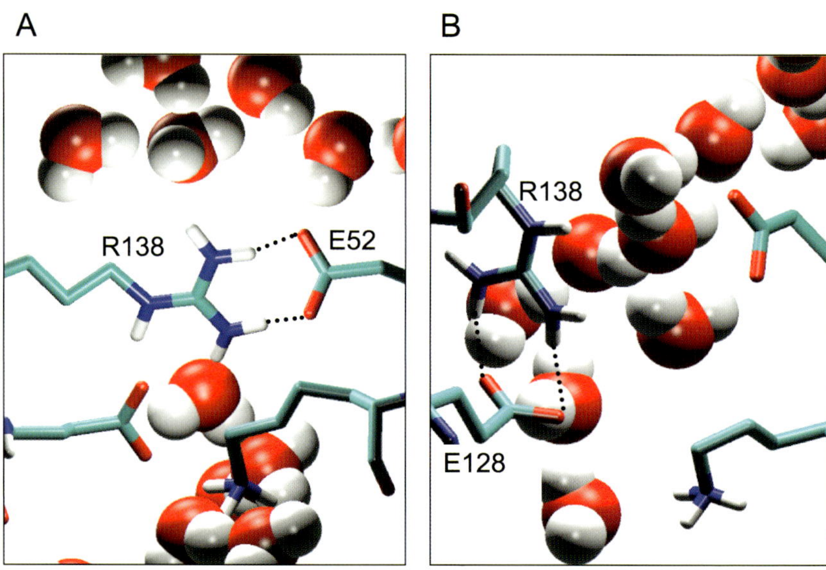

DOMENE *ET AL.*, FIG. 7. The gate region in the outer membrane protein OmpA (Bond *et al.*, 2002). Structures are shown from (**A**) a closed state simulation in which water is unable to pass the R138-E52 salt bridge; and (**B**) an open state simulation, in which R138 interacts with E128 (broken black lines); note that an uninterrupted column of water molecules is present.

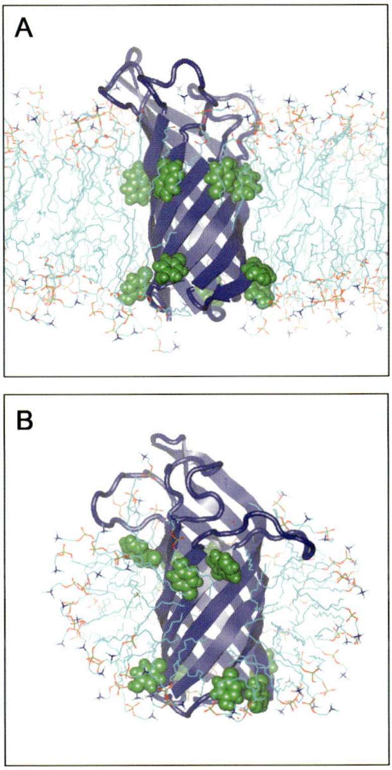

DOMENE *ET AL.*, FIG. 8. Simulations of OmpA in (**A**) a lipid bilayer and (**B**) a detergent micelle (Bond and Sansom, 2003). Both structures are taken at 10 ns of simulation. Note how the external loops of OmpA have altered conformation in **B** to interact with the surface of the micelle.

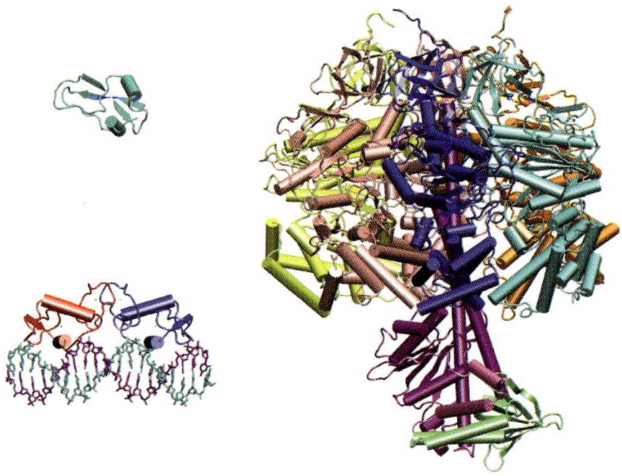

TAJKHORSHID *ET AL.*, FIG. 1. Simulations have increased in size, from bovine pancreatic trypsin inhibitor (upper left, about 3K atoms), through the estrogen receptor (lower left, 36K atoms), to P_1-ATPase (right, 327K atoms). (Atom counts include solvent.)

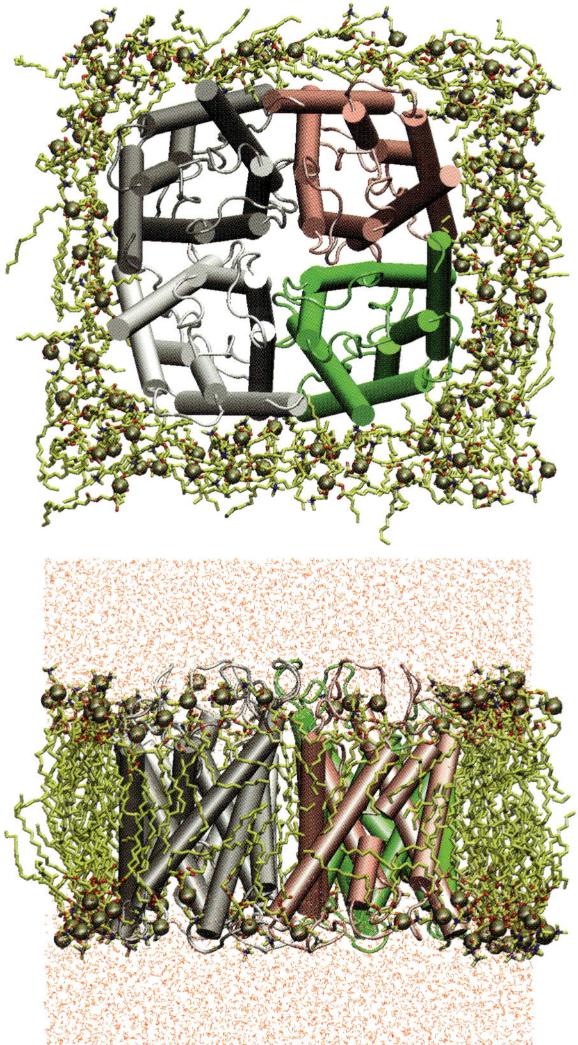

TAJKHORSHID *ET AL.*, FIG. 4. Top and side views of the tetrameric model of an AQP modeled in a fully hydrated patch of lipid bilayer (water molecules are not shown in the top view) [152]. The monomers are shown using cartoon representations and in different colors. In the side view (bottom), the positions of phosphate groups of lipids are specified using vdW representation.

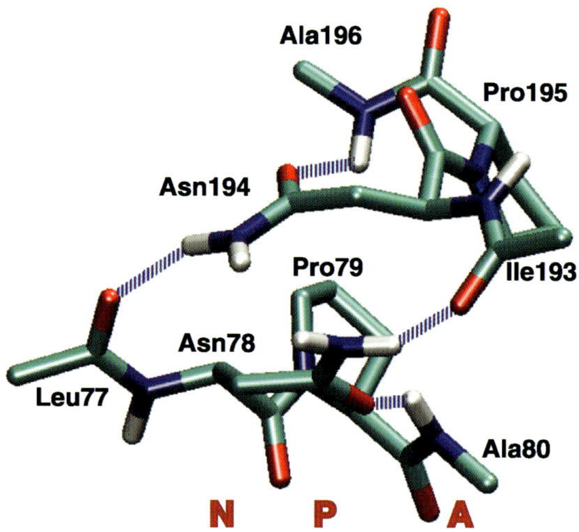

TAJKHORSHID *ET AL.*, FIG. 5. Hydrogen bonds between the two NPA motifs are essential for the stability of the structure and the function of AQPs. Due to two stable hydrogen bonds between the amido group of the asparagine side chain of each NPA motif with neighboring side chains, one of the amido hydrogens is restrained to be fully exposed toward the interior of the channel, where it forms hydrogen bonds with the permeating substrate. Hydrogen bonds between the two NPA motifs are also important for the stability of the two reentrant loops.

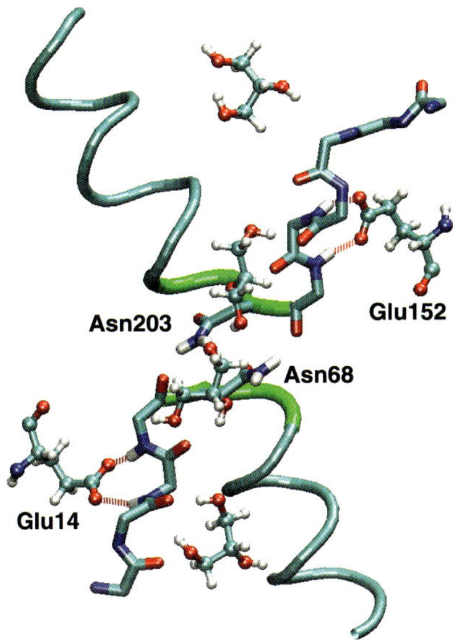

TAJKHORSHID *ET AL.*, FIG. 6. The curve-linear pathway of substrate conduction in AQPs formed by backbone oxygen atoms of reentrant loops. Carbonyl oxygens are shown in red. This peculiar secondary structure of the non-helical halves of the loops is stabilized by hydrogen bonds with Glu14:O_ε and Glu152:O_ε atoms. Also shown are the location of the NPA motifs (green), the α-helical parts of the reentrant loops (blue), and the side-chains of Asn68 and Asn203 at the center of the channel. Snapshots of the four glycerol positions (shown in ball and stick representation) from the simulations are included to illustrate the conduction pathway.

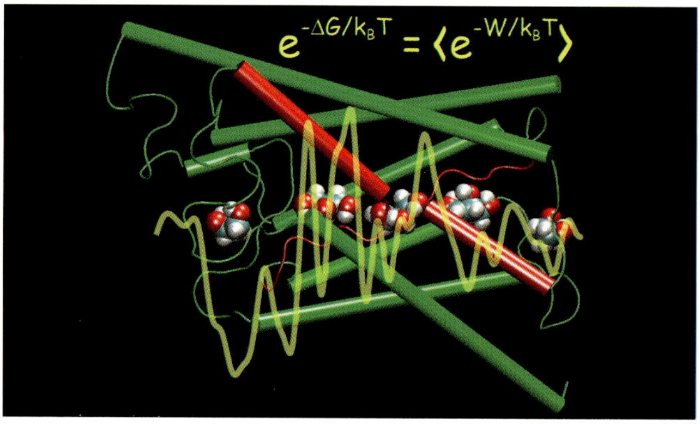

TAJKHORSHID *ET AL.*, FIG. 7. Potential of mean force (PMF) along the pathway of glycerol conduction in GlpF, constructed from SMD simulations [64]. The PMF is superimposed on a GlpF channel, in which several positions adopted by glycerol along its transport are highlighted.

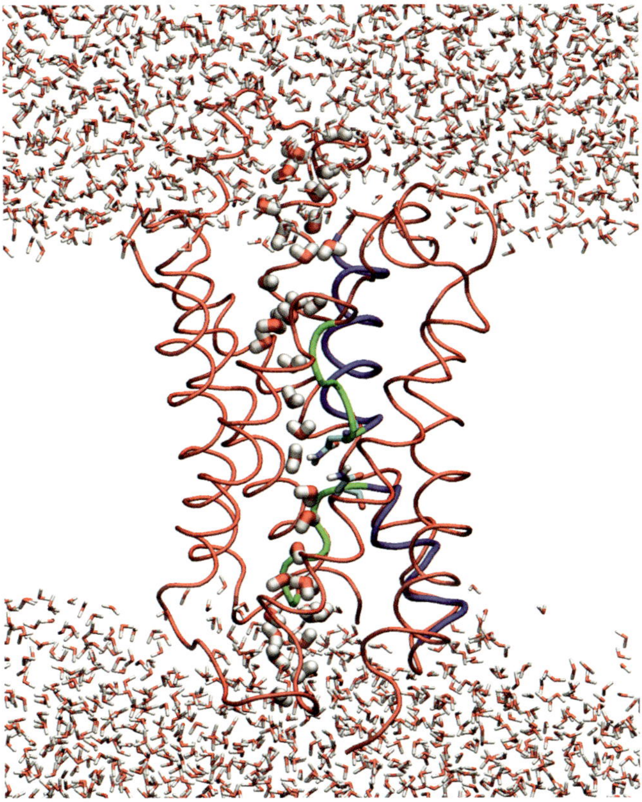

TAJKHORSHID *ET AL.*, FIG. 8. A snapshot from MD simulation of GlpF. Only one monomer is shown. A single file of water forms in the channel during the simulation. The orientation of water molecules in the single file is reversed in the two halves of the channel due to the electric field of the protein. This bipolar configuration of water (wing configuration) prevents proton conduction [129].

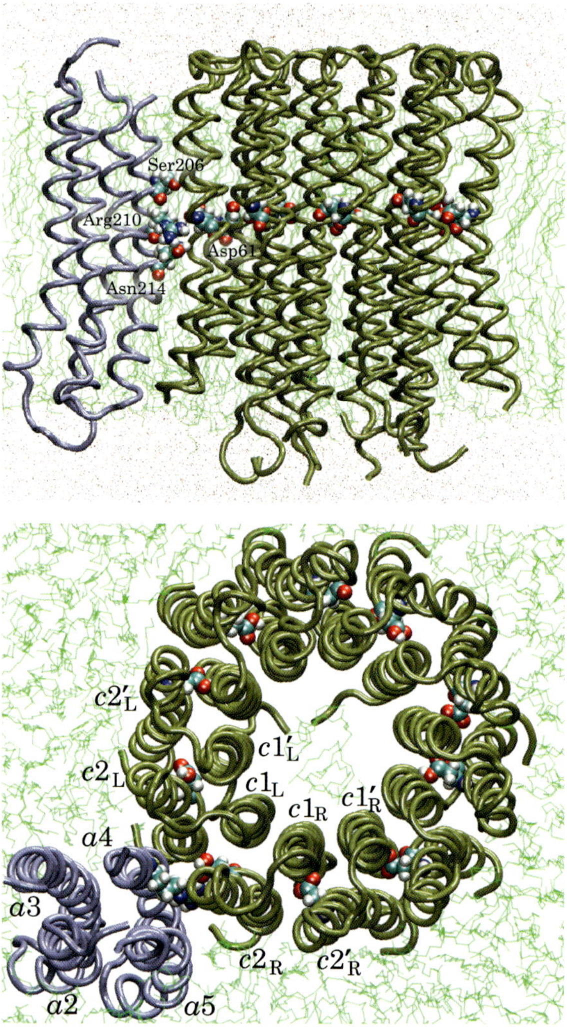

TAJKHORSHID *ET AL.*, FIG. 10. Structural model of F_0 ATP synthase. Top: side view (cytoplasmic side up, periplasmic side down); bottom: front view from the cytoplasmic side. The F_1 unit is not shown for clarity.

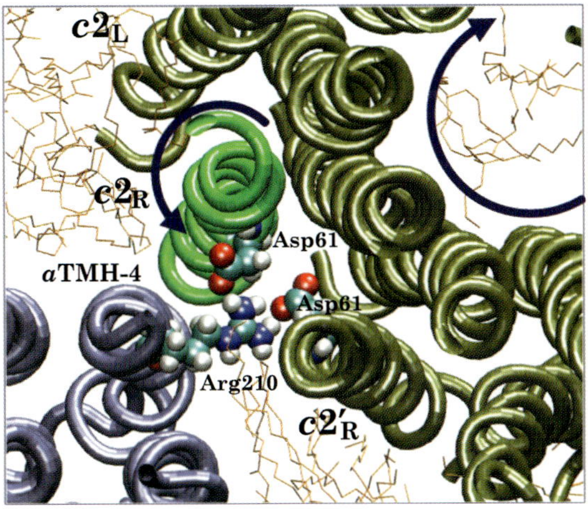

TAJKHORSHID *ET AL.*, FIG. 12. Concerted rotation of the *c* subunit outer helix and the c_{10} complex in a lipid bilayer. The highlighted helix has been forced to rotate counterclockwise by 220°. The moment of the salt bridge transfer between the two neighboring *c* subunits is shown. The initial structure is shown in Fig. 10.

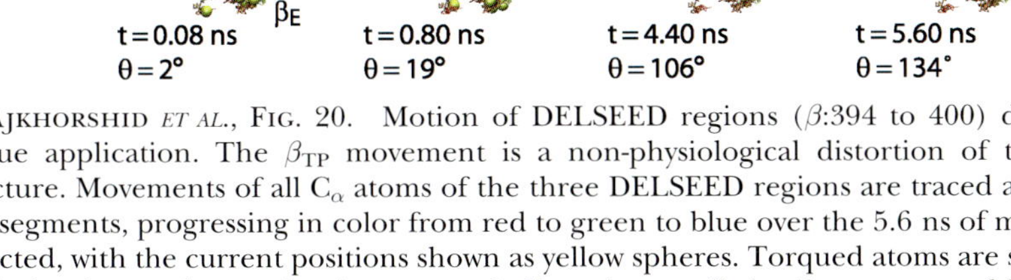

TAJKHORSHID *ET AL.*, FIG. 20. Motion of DELSEED regions (β:394 to 400) during torque application. The β_{TP} movement is a non-physiological distortion of the F_1 structure. Movements of all C_α atoms of the three DELSEED regions are traced as thin line segments, progressing in color from red to green to blue over the 5.6 ns of motion depicted, with the current positions shown as yellow spheres. Torqued atoms are shown as black spheres, the γ subunit as a purple line, the γ coiled-coil as green and brown tubes. The distortion of the β_{TP} DELSEED region is seen in the movement from the green trace area at 4.40 ns to the blue trace area at 5.60 ns.

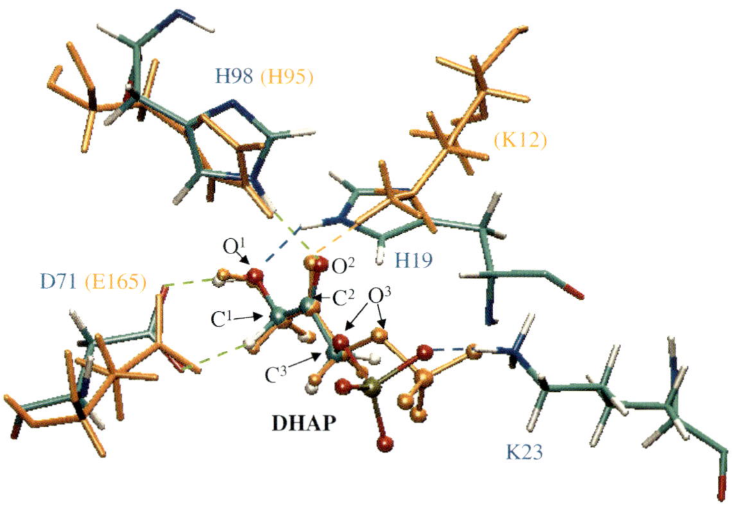

CUI AND KARPLUS, FIG. 14. A comparison of active site arrangements in methylglyoxal synthase (MGS) and triosephosphate isomerase (TIM) based on the x-ray structures with PGH as the substrate analogue (replaced by DHAP in the figure). The substrate is shown in ball-and-stick, and protein residues are shown in line-forms; the MGS residues are colored according to atom types, and those in TIM are shown in orange. The hydrogen bonding interactions found in both enzymes are shown in green, these found specifically in MGS and TIM are shown in blue and orange, respectively. Note the similarity in the orientations of the catalytic base (Asp 71 in MGS and Glu 165 in TIM) and the distal histidine (H98 in MGS and H95 in TIM) relative to the substrate. The phosphate torsional angle (O^2-C^2-C^3-O^3) is rather different in the two enzymes.

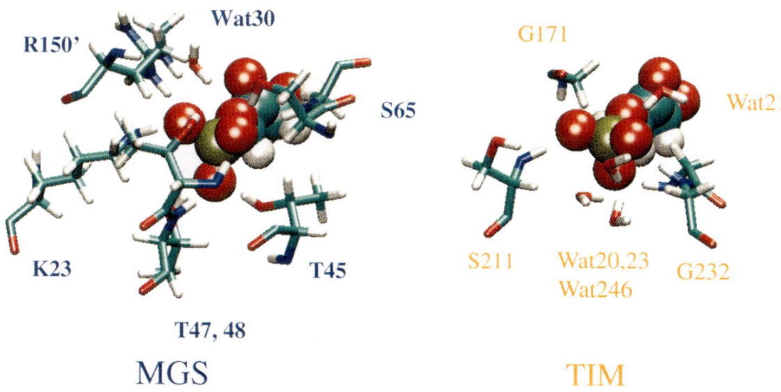

CUI AND KARPLUS, FIG. 19. The environment of the substrate phosphate group in (a) MGS and (b) TIM, based on the relevant x-ray structures with PGH as the substrate inhibitor. Residues and water molecules within 3.5 Å of the substrate P atom are shown. Note that there are many more hydrogen bonding interactions between protein residues and the phosphate group in MGS than in TIM.